A-LEVEL PHYSICS

COURSE COMPANION

Jim Breithaupt MSc
Head of Physics, Wigan College of Technology

Ken Dunn BSc
Head of Physics, Richard Taunton College, Southampton

BPP Letts Educational Ltd

First published 1983

Revised 1985, 1988, 1991
Reprinted 1983, 1985, 1986, 1992

Illustrations: Tek-Art

Aldine House
Aldine Place
142-144 Uxbridge Road
London W12 8AW

British Library Cataloguing in Publication Data
Breithaupt, J.
A-level physics : course companion.
1. Physics
I. Title II. Dunn, Ken
530 QC32

ISBN 85758 027 3

Preface

The key aim of this book is to help students in their 'approach' to A-level Physics. It should serve as a study guide, workbook and revision aid throughout *any* A-level/AS-level Physics Course (or its equivalent) regardless of the syllabus followed. It is not intended to provide a complete guide to the subject and should be used as a companion to a textbook, which it is designed to complement rather than to duplicate.

In A-level Physics the recent trend in examinations has been to ask questions that require more 'understanding' and less 'recall' than was previously expected. It is no longer usually possible to 'look up' answers to such questions in a textbook. An important objective of this book is to teach students to tackle this style of examination more successfully.

At AS-level all Physics syllabuses involve a reduced range of topics compared with A-level Physics. However, the depth of treatment of topics at AS level is the same as at A level. Hence the philosophy, upon which this book is based, is equally suitable for both courses.

The largest section of this study aid (Part II – The core of A-level Physics) has been divided into Units that reflect the latest syllabus revisions. Each Unit contains a section of essential information followed by extensive question practice relating to that unit. The essential information omits full details of experiments together with many proofs of equations; such material is well presented in most textbooks. Instead, great emphasis has been placed upon instilling understanding of key concepts, formulae and diagrams. Guidance has been provided for the correct use of equations and graphs, care being taken to identify traps and pitfalls often encountered by underprepared candidates. Where unusually detailed information is supplied it has either been prompted by Examiners' reports or by the authors' own experience of topics frequently misunderstood (especially by the weaker students).

Recent examinations from all the Boards (including Scottish Higher and SYS) provide the question practice together with A-level style questions written by the authors. Each question is accompanied by a commentary designed to help with the approach to a topic in general, to a style of question in particular, and to a good answer instructively. Selective use of the questions (aided by the index) is the most effective way of working with the book.

Apart from the Units and question practice (Part II) there are four other sections. Part I is entitled 'Working for A level' and includes a full 'Guide on how to use this book' together with information on 'Study Techniques' and complete syllabus analysis for all Boards. This analysis relates syllabus content to the topics within the book and includes all current AS syllabuses.

Part III ('Examination Technique') provides advice that encompasses general examination techniques and all the particular types of examination paper and question that may be encountered.

Questions covering the whole A-level Physics spectrum appear in Part IV ('Test yourself'). As well as multiple-choice, long- and short-answer questions, there are also Comprehension Papers and Practical Problems. There are 388 questions in the book, all with commentaries. The origin of each question is identified together with all the other Boards for which the subject material of the question may be suitable.

The final section, Part V ('For reference'), contains the answers to all multiple-choice questions and calculations, an unusually detailed index, and the table of constants and values. As well as giving conventional references, the index identifies key equations, questions and diagrams. The table of constants and values provides the data needed for the questions together with all the physical constants and many other important values. It also serves as a useful and unusual revision aid and as a means of checking whether a sensible answer has been obtained to a calculation.

Some other features of the book are worthy of note. Each Unit is accompanied by an introduction that outlines the importance of its subject material to Physics in general and A-level study in particular; it also clarifies what background information is needed before the Unit can be tackled – in terms of either other units in the book or concepts and material from GCSE. There is an entire Unit devoted to data analysis (with question practice), to assist with practical and project work as well as written exam questions. There is considerable emphasis on interpreting diagrams and graphs, all of which (like the questions and equations) have been coded for easy reference. The use of a second colour has been exploited throughout.

Most study aids/revision guides attempt a complete coverage of the A-level syllabuses. Such books take the form of 'extended notes' and offer few advantages over a standard textbook. In this Letts Study Aid the authors have made a genuine attempt to help those students who feel that conventional texts do not provide adequate assistance in their approach to modern A-level Physics and its examination.

Preface to the fourth edition

Most Examination Boards have revised their A-level Physics syllabuses at least twice since the first publication of this book in 1983. Previous revisions of the *A-level Physics course companion* have catered for these changes and this fourth edition takes account of the latest syllabus alterations. The emphasis in this book has always been on helping students to understand, explain and apply the central concepts of A-level Physics. Since these concepts remain major features of the latest A-level syllabuses, little change in the text has been necessary.

Question styles *have* changed since 1983. The introduction of examination data sheets by most Boards means that more demands are made on the understanding and explaining of concepts and theories. All the questions have been reviewed to ensure that students will gain maximum benefit from this book; thus, new questions have been introduced to replace old questions on topics which no longer have central importance at A-level.

Acknowledgements

This book was inspired by the efforts and needs of the authors' A-level students. It is dedicated to them.

Many people other than the authors have been involved in the preparation of this study aid. Much of the material on examination and study techniques was written by Graham George, and the syllabus analysis was supplied by Derek Pickersgill; both of them provided valuable criticism of the text and contributed to the question practice. Criticism and question practice were also carried out by Wesley Starr and Jim Wilson who advised on Irish and Scottish Examinations, respectively. In addition to the general acknowledgement to our students, a special mention should be made of two of them, Richard Mayell and Tim Morten, who were exceptionally helpful and willing 'guinea pigs'. We thank the staff of BPP (Letts Educational) Ltd for their excellent advice, unfailing courtesy and determination to try and 'get things right', regardless of the time and effort involved. Assistance with checking the manuscript was given by June Dunn and Jonathan Haughton.

We are most grateful to the following Examination Boards for permission to use questions from their recent examinations; Associated Examining Board, University of Cambridge Local Examinations Syndicate, Joint Matriculation Board, University Entrance and Schools Examination Council – University of London, Northern Ireland Schools Examination Council, Oxford and Cambridge Schools Examination Board (including Nuffield questions), Oxford Delegacy of Local Examinations, Scottish Examinations Board (Higher and Sixth Year Studies questions), Southern Universities Joint Board for School Examinations, and the Welsh Joint Education Committee.

Jim Breithaupt and Ken Dunn, 1983

Contents

Throughout the text Greek alphas are represented by the symbol α. This is not an italic 'a' but an alpha in the particular typeface used.

Part I Working for A Level

Guide to using this book

This book can be used at *all* stages of your A-level Physics course (or its equivalent). Although students may develop their own methods and ideas, the authors envisage three main methods of use. First it may be used as a workbook; ie, when you are studying a subject you can use the questions to assist you to improve and check your understanding. Secondly you can use it as a study aid; ie, when you are trying to teach yourself a topic (or to research information that has proved confusing during your normal teaching), the text and questions will provide ideal resource material. Finally you can use it as a revision aid; when examinations approach (and even well before), the text, questions, syllabus analysis and index will become a useful basis for your revision programme. Fuller details about these three methods are given in this section of the book. However the section starts with a guide to the organization of the contents; those who are prepared fully to understand the coding and numbering system used will get far more out of this book. To Scottish users an apology is offered; although this book is also designed to deal with the Scottish 'Higher' and 'Sixth Year Studies' courses, for simplicity throughout the text the term 'A level' is used.

Study units, topics and your syllabus

Unfortunately all the examination Boards have different methods of classifying the subject material on their syllabuses; hence it is unlikely that the core information in this book (Part II) will be laid out in identical fashion to your own syllabus. The authors have chosen nine main headings, the Units 1 to 9; these Units are divided into Topics. Hence 6.2 is a reference to Topic 2 in Unit 6; from the contents list this can be seen to be 'Heat Capacities'.

Each Topic may contain several items, and this greater detail about the specific content of each Topic can be found in the syllabus analysis which starts on p. 7. For example, Topic 8.2 is entitled 'Free electrons in metals', but the syllabus analysis reveals that items within the topic include 'thermionic emission' and the 'photoelectric effect'. Each Examination Board syllabus has been analysed against the Topics and items from Part II of this book. This analysis also includes material that does not appear in the book; information of this nature appears in colour, and some of it is covered in the question practice. Although the analysis will provide a good basic guide as to the requirements of your Board, it is strongly recommended that you obtain an official copy of your syllabus from your teacher or direct from the Board. There is a list of addresses of all the Examination Boards on p. 16, together with information about the papers that they set.

Equations, figures and questions

With only one exception, all equations, figures (diagrams and graphs), and questions are numbered according to the Unit in which they appear. All figures have the letter 'F' in front of their Unit and Topic code, equations use the letter 'E'. Questions have a letter after their Unit and Topic code according to the following scheme:

M Multiple-choice question (coded answer)
S Short-answer question
L Long-answer question
P Practical-problem question
C Comprehension paper
D Data-analysis paper

Hence the following codes:

6.2 refers to Topic 2 in Unit 6
F6.2 refers to the second figure in Unit 6
E6.2 refers to the second equation in Unit 6
6.2M refers to the second question in Unit 6 that happens to be of the multiple-choice type:

At the end of each unit are the questions (in colour) and their commentaries (in black). There is a set of multiple-choice questions, followed by a set of short answers and a set of long answers. The questions in each set reflect the particular unit.

The one exception to the coding scheme occurs in Part IV (Test yourself) where the questions cover the whole range of A level, and include practical problems and comprehension. As this section has no Unit number, neither do the questions. Hence 86P is the eighty-sixth question in Part IV, and is a practical problem.

At the end of each question (in bold type) its source is given according to the code for the Boards on p. 16, those written by the authors are indicated with a blank (–). When questions were not available from certain Boards the authors have written questions in the style of those Boards. Together with the source of each question is identification of all other Boards for which the subject material of the question may also be suitable; however this does not imply that questions of that style are set by the Boards so mentioned. For example the practical problems questions are taken from the Nuffield examination, yet, despite the fact that this style of question is not used by most of the other Boards, they will be useful to virtually all candidates. An asterisked (*) Board indicates that only candidates for the Option relevant to the question should attempt it.

Although the commentaries to the questions have mostly been written by Examiners, it must be pointed out that they are not model answers and are not 'official' opinions from the Boards whose questions they describe. The commentaries vary in both depth and nature; some concentrate on developing the approach to a particular style of question, some focus upon vital physical aspects of the questions, some identify traps and pitfalls that the question provokes, and some simply supply references to key equations, diagrams and Topics within the Units. Occasionally more than one approach to a question is discussed, that adopted by the student will depend normally upon the way in which he/she has been taught.

All necessary constants and values required, in addition to those given in the question, can be found in the Table at the back of the book. The answers to all calculations and multiple choice questions are also supplied.

Using the table of constants and numerical values and the index

This table (pp. 186–89) has other uses apart from supplying the values and constants needed for the questions. In particular, if you use it regularly you will get to know 'where' on the table a value appears, and this knowledge coupled with the method with which the table is constructed will aid you to memorize roughly how large the values are. Important physical constants are printed in colour, and often the letters with which constants and values are commonly represented in equations are given. The units of each quantity are supplied.

The index is very detailed, and using it fully will enable you to get the most out of the book. References may be given to Topics, figures, equations or questions using the codes already supplied. References to questions are printed in colour. Sometimes, if a reference involves only part of a Topic, page numbers are given; eg 2.3 is a reference to Topic 3 in Unit 2, but 23 refers to page 23. Important references are highlighted by bold type. As the index reveals, it is likely that a particular subject may be mentioned in many Topics other than those in which it mainly features. However do not forget that the list of contents can be used to help you locate more general information.

Using it as a workbook

The obvious method is to locate questions that reflect the subject material with which you wish to work. Such questions can be found at the end of the appropriate Unit, or they may be located in Part IV (Test yourself). Several subjects may be encountered in more than one Unit so full use of the index to identify questions is recommended. Although it is important to cover all the questions from your own Board, also include others if they do not venture outside your syllabus (as indicated by the information at the end of the question). Even questions of a style different to that used by your Board can provide valuable practice.

At first, try to attempt questions without using the commentary, then examine the commentary if you are 'stuck', or compare it with your own efforts if you have made some progress. A textbook may prove essential if you really get 'bogged down'. Answers to numerical and multiple choice questions can be checked. However, for a numerical question you should first ask yourself whether your answer seems reasonable; often you will need to use the table of values and constants to help you ascertain this, as well as (very important) checking that you have the correct units for your answer. If the answer is clearly wrong, try and locate your error; only venture to the answers in the back when 'totally stuck' or when you think your answer may be correct. Avoid the temptation to write down the eight decimal places your calculator will probably supply; it annoys examiners and costs marks. Try to identify the errors you make, if you keep repeating errors, then keep a checklist of your 'favourite mistakes' that you can refer to before tackling questions (failing to convert centimetres to metres will probably appear on it!).

You can also read through a Unit/Topic. Often in the text there are exercises recommended for the student to carry out, based on equations and diagrams. It is a good idea to make up lists or cards reflecting a theme (see pp. 148–9); the following are possible suggestions: equations, key words and terms, numerical values with names and units, diagrams and graphs. The equation list (for example) may just give equations with their reference numbers (eg, E2.6 $g = -GM/R^2$); then you may look at the list and test yourself by asking what each term stands for, what the units are, how big might each number be (table of constants and values), and how the equation is derived (you may need your textbook). List the key words with references as to where to find them (in this book or your textbook) and also use these lists for testing yourself. Perhaps you may draw diagrams and graphs with almost no labelling apart from their reference numbers (eg, F7.5), then you can ask yourself how the graph/diagram is produced, exactly what is plotted on each axis, what are the rough numerical values on the axes (with units), and finally what is the significance of the slope, area, intercepts, maxima and minima, or any other features of interest. These lists become of great value at revision time: it is so much more valuable and interesting to revise by 'doing' something rather than simply to try and absorb information by continual reading.

Using it as a study aid

If you are trying to learn a new topic or to understand one that makes little sense to you, then it is wisest to start by using a textbook. However if some understanding has been already achieved then this book can be very useful. Some students who were 'stuck' on subjects have claimed that ideas in this book have enabled them to return to their textbooks/lessons and make progress. This is because an attempt has been made in the Units to avoid duplicating too much of the information readily available in a textbook; instead many of the topics that regularly 'stump' students have been highlighted and sympathetically explained. Further explanation is provided in some of the question commentaries that can hence be used as learning aids. The coverage of equations is linked to the need of having an understanding of each of the terms involved, their units, and the problems that may be encountered when handling a particular formula. Diagrams and graphs get a full treatment, often with numbered axes and full information on slopes, areas (etc.) that can lead to useful further exercises for the student.

Using it for revision

You will need a full understanding of those topics and questions that are relevant to your syllabus; the subject requirements can be obtained from the syllabus analysis, preferably supplemented with an official copy of the Board's syllabus. You will often need to refer to your teacher for interpretation of the syllabus information or fuller details of the examination papers you are due to sit. Read the section on examination technique (Part III) and make sure that you have carried out plenty of question practice under exam conditions; Part IV (Test yourself) is designed for this. The question practice with the Units always features the same subject material, but you also need practice at tackling sets of questions covering the whole syllabus; for this reason the questions in Part IV have been suitably 'mixed'. Further, try and limit the time allowance to that you would receive in the examination when practising in this way; time allocations are given at the start of each set of questions in Part IV.

If you have not made up 'lists/cards' as suggested on p. 6 and pp. 148–9, then your revision period is the time to do so. The table of constants and numerical values can be a great aid in list-making, or in identifying terms and concepts that may have slipped your memory. You will find your notes and/or a textbook, and your laboratory notebook, essential to your revision; this book alone cannot provide a complete revision programme.

Make sure that you have read the relevant parts of Unit 9 (Data analysis), much of it will feature on your syllabus, but may not have been formally taught (in that most of the information would be acquired as part of other teaching and during practical work). The Comprehension papers (Part IV 94C and 95C) are useful to all, but if your examination includes such a paper make sure that you have obtained and tried some of the papers set by your own Board. It is wise when at the final revision stage to concentrate on questions from your Board; obtain more past papers from your teacher or by applying to the Board (addresses p. 16). When the exam is imminent your revision programme should be complete; however if you have made revision lists then a final check through to ensure familiarity is not a bad idea.

Other ideas

The authors have included their own ideas on how to use this book. However you (or your teacher) may come up with other schemes that the book may (or may not!) suit. If you have such ideas please write to the authors (care of the Publishers) with such ideas or information about vital omissions or errors that you may encounter. They will be taken into account in future revisions. It is hoped that this study aid will serve you well and that you will be successful in your examinations; good luck!

STUDY TECHNIQUES

Too many students sit down to a session of work having given little thought as to 'how' they will carry out their study. This section is designed to guide you in this area. If you read the information supplied carefully and develop your own study skills you will learn to work effectively; not only will you learn better, but you will also learn faster. Students who have taken courses in 'How to study' unanimously point out that acquiring a good technique has not only improved their understanding, but has also helped to increase their leisure time! There are many good books on 'studying' that are recommended to those who find this necessarily concise section too brief for their needs.

Studying beyond GCSE

'In GCSE you mainly need to know the facts and remember them, but in A level you need to know much more.'

'In A level you are encouraged to think much more deeply.'

If you imagine that our quest for knowledge in the physical world

is like climbing a mountain, then success at GCSE Physics means that you have got as far as the foothills, with a mountain range towering above you – challenging, inspiring and exciting. The next step, A level, will need a carefully planned approach, new techniques to be learned, determination and, not least, a lot of hard work. Success will be rewarded by the satisfaction of achievement, looking back at the panorama of knowledge below and an occasional glimpse upward through the clouds at even more exciting peaks above. This book will act as your guide, helping you along the route, and this chapter in particular will show you some of the skills you will need, the pitfalls to avoid and how best to plan your work for a successful approach to the A-level examination.

The climb to GCSE is mainly a smooth one, a brisk stroll up gentle slopes. The way is long but not too arduous. Knowledge is gradually accumulated, basic principles developed and apparently difficult problems usually have a fairly straightforward solution. The journey to A level is much longer, and the climb more steep. You need to be properly equipped with conceptual ability, an enquiring mind, mathematical techniques and practical skills. Only then will you reach your objective without stumbling along the way.

What then are the requirements to tackle A-level Physics successfully? The volume of knowledge required is very much greater than at GCSE and the depth of treatment much more profound – you must be prepared to do plenty of background reading and make your own notes (more of that later). You must get to grips with the basic concepts and have a thorough understanding of the fundamental principles and laws, so that you can apply them to familiar and unfamiliar situations. In order to explain the behaviour of matter you will need to acquire the ability to develop conceptual and mathematical 'models', modifying and shaping them as you find out more about the physical nature of the world around you. You will discover, and have to learn to accept, that there is not always a 'correct' answer to every question and you must be prepared to amend, or even reject, previously held theories in the light of further evidence. Only in such a way does science move forward.

GCSE requires little more than a basic grasp of mathematics and a certain degree of numeracy. You begin to see at A level that the universe appears to be governed by physical laws that have a mathematical basis. Consequently, you need to develop your mathematical abilities in order to understand more fully the simplistic beauty of these laws. You must be able to carry out complex computations quickly and accurately with the help of an electronic calculator and have some idea of the order of magnitude of quantities (the Table of values on pp. 186–9 is designed to help you with this). The ability to make sensible approximations should be developed and the significance of experimental errors must be appreciated. A basic working knowledge of calculus is advantageous, although not essential.

Throughout the A-level course you should be involved in practical work. It is much easier to understand ideas that relate to your own experience, and so experiments relating closely to the concepts that you are learning will help reinforce the learning process. Some experiments will lead you to find out and question discoveries for yourself, and others will help you develop relevant skills in manipulating apparatus, making careful observations and interpreting data.

These, then, are the skills that you must acquire in order to climb the A-level path. The rest of this chapter will help you to achieve them by showing you:

how to STUDY along the way, and

how to REVISE for the final ascent.

Part III of the book explains the necessary examination technique.

When and where to study

Whether you are studying for your A-level Physics at college, school, evening class or at home, you will need to spend a considerable amount of time working on your own, possibly without a great deal of guidance. Quite often students find this transition from the more formal pattern of GCSE teaching a difficult step to make:

> 'A level requires you to work more by yourself and also to think for yourself.'
>
> 'The emphasis on pupils learning by themselves is greater.'

Do you recognize the student who leaves for home with a bag full of books intent on an evening of hard work? He gets home, has a cup of tea and decides he ought just to play a couple of tracks of the LP his friend has lent him. By the time he has enjoyed both sides, tea is ready. After a satisfying repast he sits down and has a look at the paper, just to keep abreast with current affairs. He remembers that he has got to make a 'phone call – better do that before buckling down to work. Of course there is lots of news to chat about on the 'phone, so it is quite a long session. At last he is ready to do some work. Better have a cup of coffee to keep going. So, with a cup of coffee by his side, he settles down in a comfortable armchair, props up his feet, gets out his Physics book and starts to read. The first couple of pages are heavy going and not much seems to sink in, so he starts to re-read them. Nothing seems to stick. 'Oh well! There's a good film on TV, so the Physics can wait until tomorrow night; after all it's not due in for a couple of days . . .'

How do you avoid getting into such a situation? It is essential that you plan ahead, perhaps as much as a week in advance. At GCSE, it is often sensible to schedule your work so that it is done the day before it is due to be handed in. At A level such an approach is disastrous. Two problems arise; firstly, the nature of the work means that you often encounter difficulties that will need advice from your teacher or the aid of books unobtainable until the following day; secondly, the demands of A level are unpredictable and you may find that you have also been given some other tasks to complete overnight. In all sciences, the need frequently arises to analyse results and plot graphs from one day's practical/demonstration in time for the next lesson; the student who is always unable to carry out such work (because he has too much other work to complete by the following morning) will struggle with his A-level Course and incur the wrath of his teachers!

For successful planning you must decide how much work you need to do, and how much time you intend to spend on it. You will not study effectively if you work for long periods without a break. After about 40 minutes concentration begins to wane. A break of a few minutes, in which you do something completely different, can soon re-charge your batteries and enable you to continue your studying to greater effect. So, try to plan your work so that you fit in these breaks at convenient intervals. **When** you work is very much a personal matter. Some people find that they work best early in the morning, others seem to work better late at night – find out when suits you best and plan accordingly.

Where you work is also important. It is helpful if you can have somewhere that you can call 'your own', even if it is only a table in the corner of your bedroom, where you can work without being disturbed. Make the area attractive, perhaps with posters of your favourite pop group or football team. You may find music helpful – some people do actually work better accompanied by background music, despite popular opinion to the contrary. Your surroundings should be comfortable, warm and, most important, well lit. Make sure that you have essential materials readily at hand: writing paper, pens (several colours are helpful), drawing instruments, calculator and books should be kept conveniently nearby. Now you are ready to start!

Reading for A level

'... more reading is necessary.'

'At GCSE you were given more information. At A level you have to find out more for yourself.'

Since you have been able to read for as long as you can remember, it probably does not occur to you to question your ability to read **effectively**. You can almost certainly improve your reading skills by developing some fairly simple techniques that can be found in specialist books on the subject.

You will read more purposefully if you are:
motivated: without a target, reading lacks drive.
interested: concentration relates to degree of interest.
active: think for yourself – don't expect the author to think for you. Read critically and question what is said.

During your A-level course you will encounter many references to books and magazines and other resource material. Some of it will only be 'background reading' and the temptation is great to ignore such texts and only concentrate on the 'necessary'. However, modern examinations strongly favour the student who has developed his reading skills by doing such work. The Comprehension papers set by some Boards are handled far better by students who are accustomed to appraising scientific writing critically. Further, modern exam questions in Physics tend to be wide ranging and often take a considerable amount of time and skill to comprehend fully. Students who have 'read little' frequently run into 'time trouble' in written examinations, and have an unfortunate habit of misinterpreting questions.

Note taking

'Teaching methods are different ... making notes for your own use is essential.'

'More responsibility on pupil to make own notes.'

The amount of note taking that you will be required to do will very much depend on the particular methods adopted by your teacher. However, you should remember that your notes are for **your** benefit, and should eventually form the basis of your revision plan. Bearing this in mind, the way in which you present your notes is going to have a considerable effect on their efficiency in helping you to understand and learn the work.
The example on p.5 illustrates the difference between the right and wrong way to write notes. Notes should be visually outstanding, achieved in the example by the use of CAPITALS, Underlining, symbols (the √ and), by using a box and by the use of colour.

Your notes should be succinct; remember that you should already have a textbook, so there is no point trying to write another one. Wherever possible try to illustrate your notes with diagrams (a good diagram can save volumes of words), charts and tables. You should always write up your notes, with the help of books if necessary, as soon after the lesson as possible while the subject matter is still fresh in your mind. Make sure you **understand** the work at this stage, or else further development and, later on, revision will be difficult. Seek help from other books or your teacher if necessary.

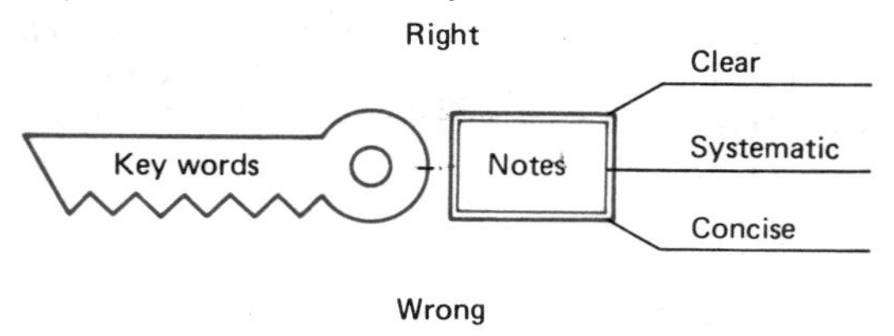

When you are making your notes do not write long sentences containing a large number of unnecessary words, so that it is difficult to pick out the really important ideas, but rather you should set out your notes in a concise and orderly fashion so that they can be clearly followed and understood.

Learning

'There is a greater emphasis on learning rather than being told in A level.'

'. . . more to memorize.'

'In GCSE you mainly need to know the facts and remember them, but in A level you need to know much more . . . you need to know most of the basic principles.'

Physics is a demanding discipline. There are certain basic laws and principles that must be memorized. Key equations and formulae should be known by rote, even if a data sheet or booklet is supplied to you during your examination. A data booklet is helpful if you happen to forget a particular formula which you need to be able to start answering a question. Searching frantically through a data booklet to find a likely formula is a waste of valuable time in an examination. The data booklet should not really be needed for the key questions.

Keep to a minimum the basic laws and equations that you need to memorize. Highlight them in your notes: by boxing-in colouring; or even by making a separate list of basic facts that must be known.

You must thoroughly understand fundamental **principles** and **concepts** before you attempt to commit them to memory. Certain mathematical and analytical **skills** are needed. These can only be acquired through **practice**, particularly by answering questions, which is why a large part of this book is devoted to just this.

The important thing is to have the right attitude to learning. If you **want** to learn you **can** learn. To this end you should set yourself goals at which to aim so that the satisfaction of attaining them acts as a spur for the next task ahead of you.

Revision

'Dividing the course up into Units makes revision easier.'

'Unit tests (however irksome!) are a good idea because they let you know what you do or don't understand'.

'I like the idea of frequent tests . . . makes revision easier'.

Revision is not something that is to be left to the end of the course, a few weeks before you sit your A-level examination. Rather, it should be an on-going process that actually begins from your very first Physics lesson. The amount of knowledge to be accumulated is vast and must therefore be assimilated and gradually consolidated as you go along. Indeed, revision is really part of the learning process, although it will of necessity become more concentrated as the examination approaches.

Recall improves immediately after learning, when you have had time to process the information properly, then declines sharply. By use of reviews, memory can be maintained at the original high level as shown by the graphs.

Of course, these times are not absolute, and some reviewing will take place automatically, for example when doing homework questions or class tests. The important principle is **regular review**, which means a continuous, planned revision programme.

There are several important points concerning memory that can be illustrated by performing the following experiment for yourself. Please read the instructions first.

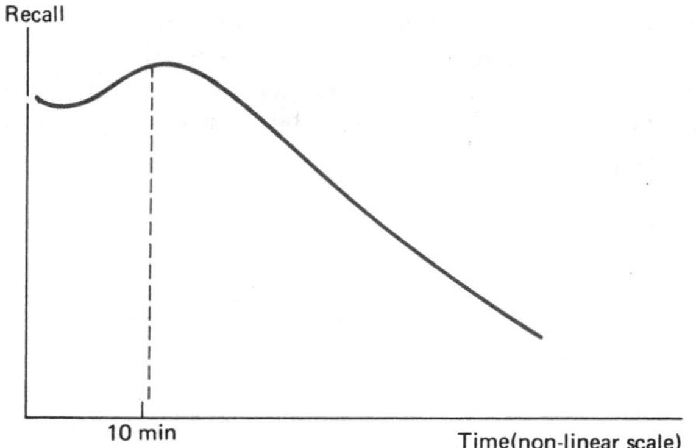

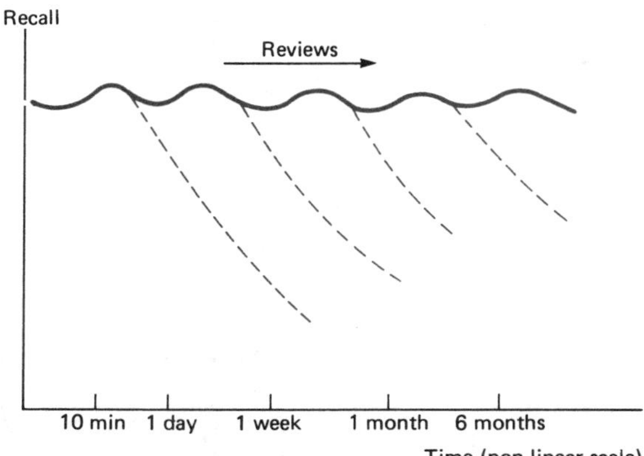

Cover the list of words opposite with a piece of paper. Read each word on the list once, quickly, then cover the list

On another sheet of paper write down as many of the words as you can remember.

Check your answers against the original list and answer the following questions:

1 How many words did you remember from the beginning and end of the list, compared with the middle?
2 What about words that appeared more than once?
3 Outstanding items?
4 Phrases?

of
move
and
crayon
work
bad
the
of
end
blue
own
Albert Einstein
green
paper
and
pencil
ride
BED
and
the
out
how to study
friend
the
book

This simple task shows that you remember more if:

1 Items are at the beginning or end of a list.
2 Repetitions are made.
3 Something is outstanding.
4 Things are linked, as in phrases.

These principles should be used as far as possible during the whole of your A-level course, but in particular when revising.

First of all, get **organized**. Devise some form of revision time table, bearing in mind principle (1) above and that learning periods of about 30–40 min produce the best recall. Plan your time carefully, giving yourself definite goals to aim at (for example completion of a specific topic), the attainment of which provide you with satisfaction and encouragement.

Next, be **active**. Simply reading a text book in a comfortable armchair with your feet up is not conducive to successful learning. You should write out important definitions, formulae and equations; draw diagrams and graphs checking them against your text book or notes; describe experiments in note form as though you were performing them; above all practise lots of questions (repetition!).

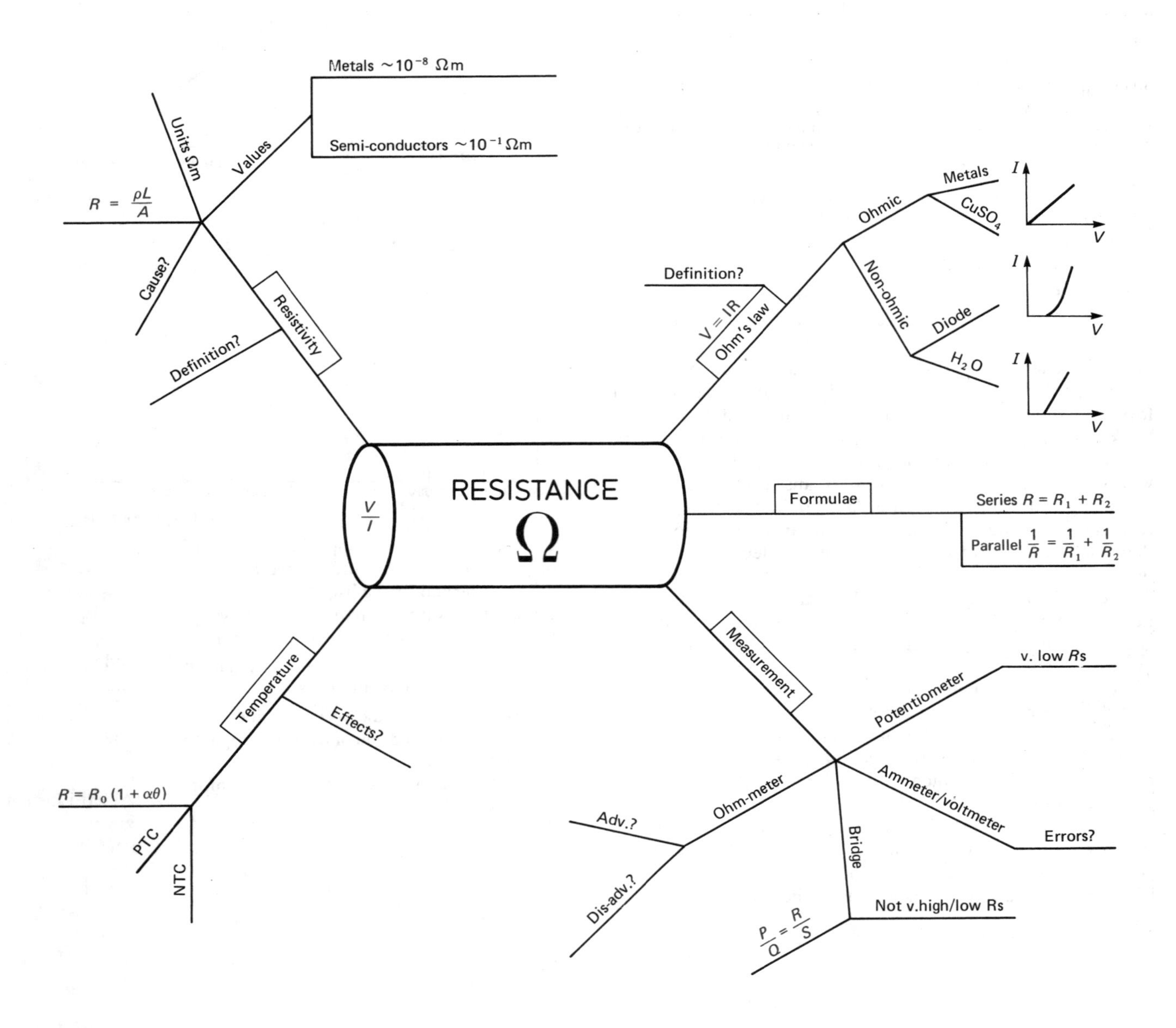

Hopefully by now you can begin to see the importance of the section on note taking, and how well-organized notes can help during your revision. You can, at this stage, make special revision notes in a form suitable for quick reference; perhaps colourful charts to hang on the wall of the room where you work, or postcards with essential information that you can glance at on the bus or walking to school. For example, a card for resistance might look something like the diagram.

The above diagram is by no means exhaustive. You can probably think of many other points that could be usefully added. This is an advantage of such a card – you can add new ideas as you think of them or come across them in your revision. Charts like this involve the principles of highlighting the important areas, ie, making them **outstanding**, and **linking** together main ideas, which as we have seen is important in the process of recall, which is the key to unlocking your memory.

Finally mention should be made of a revision method/learning aid that can be both valuable and entertaining. Whenever you find it particularly difficult to remember equations or information it is suggested that you invent codes to assist you; such codes are called 'mnemonics'. Some examples that have been used by A-level Physics students are ROY G BIV (Rainbows), CIVIL (Reactance), Talking Of Animals Some Ostriches Have Curiously Adjusted Heads (Trigonometry), non-listeners go to the sign theatre ($n\lambda = 2d \sin\theta$), Blue Bends Best (Dispersion), I am a knave (I = nAve).

ANALYSIS OF THE SYLLABUSES OF THE EXAMINATION BOARDS

The analysis is based upon the topic titles of Part II of this book. Although considerable detail has been included, to print the entire syllabus of each Board would be impracticable. Hence it is strongly recommended that students obtain an official copy of the syllabus for their examination and consult it when the need arises. Information has been included about the subject material required for S-level and AS-level courses.

INTRODUCTION AND KEY TO THE TABLE OF ANALYSIS

Information covered in the units of this book is printed in black, so colour print refers to items that are not featured in the main text of Part II. The table shows whether subject material is compulsory or optional. Sometimes only part of an item is needed, or the approach of the particular Examination Board to the item requires special attention; a symbol has been devised to indicate this by suggesting that you consult the official syllabus. The code used is as follows:

	Covered in this book	Not covered in this book
Compulsory	●	●
Compulsory, but consult official syllabus	■	■
Optional	○	○
Optional, but consult official syllabus	□	□

The analysis is based upon syllabuses for the 1991 A-level and AS-level examinations.

A-Level syllabus comments

JMB There are now two syllabuses, A and B. The former 'Engineering Science' syllabus has been retitled Physics Syllabus B. The code covers both syllabuses, except where symbols ⓐ and ⓑ are used to indicate those items on one syllabus only. For Syllabus A, five options have been analysed, indicated as follows: Ⓐ Waves & Fluids, Ⓑ Nuclear Physics, Ⓒ Optical Instruments, Ⓓ Rotational Dynamics, Ⓔ Electronics. The other options are too wide-ranging to cover here. Each candidate must attempt one option only. The rest of the indicated JMB material is the compulsory core.

London In addition to the basic syllabus, there are five optional topics which are:
Ⓐ Energy and its Uses, Ⓑ Solid Materials, Ⓒ Telecommunications, Ⓓ Medical Physics, Ⓔ Amplifiers and Analogue Electronics. Each candidate must attempt two topics. Only topics A, B and E have been analysed here.

Cambridge In addition to the compulsory core syllabus, each candidate is expected to study one of four options:
Ⓢ Sound and Music; Ⓒ Communications; Ⓜ Medical Physics; Ⓣ Physics of Transport. Options C and T only have been analysed here.

Oxford and Cambridge Each candidate is expected to study the core syllabus and two of the four options as follows:
Ⓐ Rotational Dynamics; Ⓑ Properties of Solids; Ⓒ a. c. and Electronics; Ⓓ Physics in Medical Diagnosis. Options A, B and C have been analysed here.

SEB The Higher Grade syllabus (Scottish H) is analysed using the normal symbols. The Sixth Year Studies (Scottish SYS) syllabus consists of a compulsory core syllabus, indicated by S in the syllabus analysis, and four optional topics shown in the syllabus analysis as follows:
Ⓐ Analogue Electronics; Ⓓ Digital Electronics; Ⓡ Radio Communications; Ⓞ Optical Instruments. Candidates must study one optional topic.

S level The Special paper of each Board covers the same items as the Advanced papers of the Board except where the syllabus includes optional topics. These are **not** included on S-Level papers.

AS-Level syllabus comments

The following syllabuses have been included in the syllabus analysis:

AEB All topics on the syllabus are compulsory. Teacher-assessed practical coursework carries 30 per cent of the total marks.

COSSEC (University of Cambridge Local Examinations Syndicate; Oxford and Cambridge School Examinations Board; Southern Universities Joint Board.) Candidates must study the core syllabus and option A or B and option C or D. The option topics are as follows:
Ⓐ Energy; Ⓑ Physics of Transport; Ⓒ Communications, Ⓓ Medical Physics. Options A, B and C have been analysed here. Internal assessment of practical work comprises 20 per cent of the total mark.

JMB All topics on the syllabus are compulsory. Assessment of practical skills, either by a practical examination at the end of the course or by internal assessment during the course, carries 15 per cent of the total marks.

London The course is composed of a basic syllabus and two options A and B. Each candidate should attempt one option only. Both options are analysed here. Teacher-assessed practical coursework carries 20 per cent of the total marks.

Oxford The syllabus consists of a core and four options. Each candidate must attempt one option. All four options have been analysed, indicated by the following codes.
Ⓐ Physics of Movement, Ⓑ Wave Motion, Ⓒ Electronics, Ⓓ Radioactivity and Nuclear Physics. Teacher-assessed practical work carries 20 per cent of the total marks.

Analysis of examination syllabuses	AS LEVELS					A LEVELS									
	JMB	COSSEC	AEB	London	Oxford	London	JMB	O and C Nuffield	AEB	Oxford	Cambridge	O and C	WJEC	Scottish	N. Ireland
UNIT 1 FORCE AND MOTION															
1.1 Motion in a straight line															
Dynamics, equations & graphs	●	■🄱	●	■	●	●	●	●	●	●	●🅃	●	●	●Ⓢ	●
Measurement of g	●				■	●			●	■	●	●	■	●	■
1.2 Projectile motion															
Flight-path equations					●	■		■	●	●	●	●		●	●
Proof of equations					■				●	●	●	●		●	●
1.3 Force and momentum															
Newton's laws & momentum	●	●🄱	●	●🄱	●	●	●	●	●	●	●🅃	●	●	●	●
Impulse	●	■🄱	●	🄱	■	●	●	●	●	●	●🅃	●	●	●	●
Conservation of momentum	●	●🄱	●	🄱	●	■	●	●	●	●	●🅃	●	●	●	■
Weight and weightlessness	●	●	●	●	●	●	●	●	●	●	●	●	●	●	●
1.4 Work, energy and power															
Work, energy and power	●	●🄱	●	●	●	●	●	●	●	●	●	●	●	●	●
Energy resources	●	●🄰	■	🄰		●🄰		●			●				
Conservation of energy	●	●	●	●	●	●	●	●	●	●	●	●	●	●	●
ke $= \frac{1}{2}mv^2$	●	●	●	●	●	■	■	●	■	■	■	■	■	●	■
Collisions	●	●			●	●	●	●	●	●	●	●	●	●	●
1.5 Statics															
Resolving and combining forces	●	●🄱	●		●	●	●	●	●	●	●🅃	●	●	●	●
Couples	●	■🄱	●		Ⓐ	●	●🄳	●	●	●	●🅃	●	■	■	●
Principle of moments	●	●	●		Ⓐ	●	●	●	●	●	●	●	●	■	●
Closed polygon of forces as a condition for equilibrium	●	■	●		🄰	●	●	●	●	■	●	●	●		■
1.6 Simple harmonic motion															
shm equation and definition					Ⓐ	●	●	●	●	●	●	●	●	Ⓢ	●
General solution of the shm equation					Ⓐ	●	●	●	●	●	●	●	●	🅂	●
Theory of the simple pendulum					🄰	●	●	●	●		●	●	●	🅂	●
Theory of the oscillations of a loaded spring					🄰	●	●	●	●	●	●	●	●	🅂	●
1.7 Uniform circular motion															
Angular speed (ω)	●	●		●	Ⓐ	●	●	●	●	●	●	●	●	Ⓢ	●
Centripetal acceleration $a = \omega^2 r$	●	●			Ⓐ	●	●	●	●	●	●	●	●	Ⓢ	●
Proof of $a = \omega^2 r$	■						●						■	🅂	
Centripetal force: simple examples	●	●			Ⓐ	●	●	●	●	●	●	●	●	Ⓢ	●
1.8 & 1.9 Rotation of a rigid body															
Moment of inertia					Ⓐ		🄳ⓑ		●		●	🄰	●	🅂	
Angular momentum $= I\omega$					Ⓐ		🄳ⓑ		●		●	🄰	●	Ⓢ	
Rotational ke $= \frac{1}{2}I\omega^2$				■	Ⓐ		🄳ⓑ		■		■	🄰	■	Ⓢ	
Formulae for I of various simple objects							🄳							🅂	
Theorems of parallel and perpendicular axes							🄳								
Conservation of angular momentum					Ⓐ		🄳ⓑ		■		■	🄰	■	Ⓢ	
Rotation oscillations							🄳								
Determination of I by experiment							🄳					🄰	■	🅂	
1.10 Beyond Newton's laws															
Quantum theory				●🄱		●	●	●	●	●	●	●	●	●	●
Photon energy $E = hf$				●🄱		●	●	●	●	●	●	●	●	●	●
Relativistic effects; $E = mc^2$				🄱		■	🄱	●		●	●	●	●	●🅂	■
De Broglie's equation				🄱			ⓐ	●	■	●	●		■	●🅂	
UNIT 2 FIELDS AND THEIR EFFECTS															
2.1 Fields: their nature, strength and forces															
Gravitational, electrical & magnetic				■		●	●	●	●	●	●	●🄰	●	■🅂	●
2.2 Potential & pe in fields															
Zero of pe	●					■	●	●	●	■	●	■	●	Ⓢ	●
Potential at a point	●					●	●	●	●	■	●	■	●	Ⓢ	●
Relationship between field strength and potential	●			■		●	●	●	●	●	●	■	●	Ⓢ	●
2.3 Graphs & diagrams of field strength & potential															
Field lines	●			●		●	●	●	●	●	●	■	●	●Ⓢ	●
Equipotentials	●			●		●	●	●	●	■	●	■	●	🅂	●
2.4 Intermolecular fields															
Force and energy curves						🄱	ⓐ	■		●	●	🄱	●		
Hooke's law behaviour						●🄱	ⓐ	■			●	🄱	●		
Expansion of solids, liquids & gases						■🄱	[a]				●	🄱	●		
2.5 Electric & gravitational fields from spheres & points															
Inverse-square law for force	●			●🄱		●	●	●	●	●	●	●	●	🅂	●
$1/r$ variation of potential	●						●	●	■	●	●	●	●	Ⓢ	●
Experimental verification of Coulomb's law								■						Ⓢ	
2.6 Earth's 'uniform' gravitational field															
$g = GM/R^2$	●			●		●	●	●	●	●	●	■	●	Ⓢ	●
Experimental determination of G						■						🄰		🅂	
2.7 Launching satellites from the Earth															
Escape speed	●						ⓐ	●		●				Ⓢ	■
Speed of orbit	●						●	●	■	●	●	🄰		Ⓢ	●
Geostationary orbits	●		■			■	●	●	■	●	■🄲	🄰		Ⓢ	●
Kepler's laws	■			🄱			■					🄰		Ⓢ	■
2.8 Uniform electric fields															
Relationship between field strength & pd				●		●	●	●	●	●	●	●	●	●Ⓢ	●
Field strength between parallel plates				■		■	●	●	●	●	●	●	●	■Ⓢ	●
Permittivity of free space ε_0						●	●	●	●	●		●	●	Ⓢ	●
Relative permittivity ε_r						●	●	●	●	●		■	●		
Experimental determination of ε_0 or ε_r						■	■	●					■		
2.9 Capacitance of parallel plates and isolated spheres															
Definition of capacitance					●	●	●	●	●	●	●	●	●	●	●
$C = \varepsilon_r \varepsilon_0 A/d$ (parallel plate)						●	●	●	●	●	●	●	●		

Analysis of examination syllabuses/ continued	AS LEVELS					A LEVELS									
	JMB	COSSEC	AEB	London	Oxford	London	JMB	O and C Nuffield	AEB	Oxford	Cambridge	O and C	WJEC	Scottish	N. Ireland
$C = 4\pi\varepsilon_0 r$ (conducting sphere)							ⓐ	●		■					
Experimental investigation of $C = \varepsilon_r\varepsilon_0 A/d$						●	●	●	●	■		■	■		
Measurement of C					●	●	■	●	●	●	●	Ⓢ	■		
Energy stored in a capacitor					●	●	●	●	●	●	●	●	●	●	●
Derivation of $C = \varepsilon_r\varepsilon_0 A/d$								■							
2.10 Measuring magnetic fields and forces															
Magnetic field strength	●	●	●		●	●	●	●	●	●	●	●	●	Ⓢ	●
Force exerted on a current-carrying conductor	●	●	●	■	●	●	●	●	●	●	●	●	●	Ⓢ	●
Force exerted on a moving charge	●				■	●	●	●	●	●	●	●	●	Ⓢ	●
Field due to: a long straight wire	●	●			■	●	■	●	●	●	■	●	●	Ⓢ	■
a solenoid	●				■	●	■	●	●	●	■	●	●	[S]	■
a flat circular coil	●				■	■	■	■	●	●	■	●	●	[S]	■
Measurement of flux density (magnetic field strength)	■					■	■	■	■	■	■	■	●	[S]	■
Experimental comparision of magnetic fields	■					■	■	●	■		●	●	■	[S]	●
The current balance			■			■	■	●	●	■	●	■	■	Ⓢ	■
Moving-coil galvanometer	●						ⓐ		■						
dc motor		■	■				[a][b]	■			■	■		Ⓢ	
Loudspeaker	■	■													
Ferromagnetism	■				■	■	[b]			■					■
Hysteresis							[b]								
Torque on a plane coil in a uniform field	●				●		ⓐ		●	●	●				
2.11 Using magnetic field formulae															
Force between two current-carrying conductors	●					●	●	●	●	●	●	●	●	[S]	●
Definition of the ampere						●	■	●	●	●	●	■	●	[S]	●
Permeability of free space μ_0	■					●	●	●	●	●	●	●	●	Ⓢ	●
Relative permeability μ_r					■		[b]	●	●				●		
Hall effect						■	●	●	●		■	●[B]	■	Ⓢ	
The voltage generator		■	■	■		■	●	●	■	■	■	■	●	Ⓢ	■
2.12 Principles of electromagnetic induction															
Magnetic flux & flux linkage		●	■	■	■	●	●	●	●	●	●	●	●	[S]	●
Faraday's law		●	■	■	■	●	●	●	●	●	●	●	●	[S]	●
Lenz's law		●	●	●	●	●	●	●	●	●	●	●	●	[S]	●
emf induced by rotating coil		■	■	●[A]	■		●	●	●	●	■	●	●	[S]	●
2.13 Calculating flux: magnetic/electric analogy															
Flux					●		ⓑ	●						Ⓢ	
Reluctance							ⓑ	■							
Magnetic conductivity							[b]	■							
2.14 Inductance and transformers															
Self inductance							●	●	●	●	■	●	■	Ⓢ	●
Energy stored in an inductor							ⓐ	●	●			●			
Mutual inductance				[A]			●	●	●	●	●	■	■	Ⓢ	●
Efficiency of transformers			●	[A]		[A]	ⓐⓑ	●	●	■	●				■
Energy losses			●	[A]		●[A]	ⓐⓑ	●	●	■	●				■
$V_p/V_S = N_p/N_S$			●	[A]		●[A]	ⓐⓑ	●	●	■	●	■			●
Eddy currents							●ⓐ	●	■		■	■	●	Ⓢ	
Induction motor								■							
Growth and decay of current in LR circuits							ⓑ	■	■		■	■		[S]	
2.15 Electromagnetic fields															
Electromagnetic waves	●			●		●	ⓐ	●	●	●	●	●			●
Velocity of electromagnetic waves				[B]				■			■	■		Ⓢ	■
UNIT 3 OSCILLATIONS AND WAVES															
3.1 General information about oscillations															
Phase, amplitude, period & frequency		●	■		●	●	●	●	●	●	●	●	●	●	●
3.2 Energy stored in oscillators															
ke & pe in simple harmonic oscillator					[A]	●	●	●	●	●	●	●	●	Ⓢ	●
Total energy in simple harmonic oscillator					[A]	●	●	●	●	●	●	●	●	ⓈⓇ	●
Free & damped oscillations			●		●	●	●	●	●	●	●	●	●	Ⓡ	●
Critical & over-damping			●		●		●		●	●	●	●	●	Ⓡ	●
Resonance			●		■	●	●	●	●	●	●	●	●	Ⓡ	●
3.3 Frequences of different simple harmonic oscillator															
(i) Simple pendulum					[A]■	●	●	●	●	■	●	●	●		●
(ii) Water column in U-tube															
(iii) Mass on spring					[A]	●	●	●	●	■	●	●	●		●
(iv) Electric current							■	●	■			●	●		■
(v) Floating cylinder					[A]					■					
Derivations							■		●	■	●	■			●
3.4 General information about waves															
Progressive waves	●	●			●	●	●	●	●	●	●	●	●	[S]	●
Standing waves	●				●	●	●	●	●	●	●	●	●	[S]	●
Relationship between velocity, frequency & wavelength	●	●			●	●	●	●	●	●	●	●	●	●	●
Intensity								●			●	■		●	■
Equation for a progressive wave										●			●	●Ⓢ	
Relationship between phase angle & path difference						●	■	●	●	●	■	●	■	[S]	■
3.5 Properties of waves															
Reflection					●	●	●	●	●	●	●	●	●	●	●
Refraction					●	●	●		●	●	●	●		●	●
Principles of superposition		●[C]			Ⓑ	●	●	●	●	●	●Ⓒ	●	●	●	●
Interference	●				Ⓑ	●	●	●	●	●	●	●	●	●	●
Diffraction	●				Ⓑ	●	●	●	●	●	●	●	●	●	●
Polarization		■			■	●	●	●	●	●	●	●	●		■
Experimental demonstration of interference, diffraction & polarization	■				[B]	●	●	●	●	●	●	●	■	[S]	■
Applications of polarization						●	[b]	■							

Analysis of examination syllabuses/ continued

	AS LEVELS					A LEVELS									
	JMB	COSSEC	AEB	London	Oxford	London	JMB	O and C Nuffield	AEB	Oxford	Cambridge	O and C	WJEC	Scottish	N. Ireland
3.6 Sound															
Experimental determination of velocity					●	●	[a]		●	●	■		■	[S]	■
Beats		●			Ⓑ		ⓐ			●		●	●	Ⓢ	●
Doppler effct					Ⓑ		Ⓐ			●					
Standing waves on strings and in pipes					●	■	●	●	●	●	●	●	●	Ⓢ	●
Formulae for fundamental frequency					■	■	ⓐ	■	●	●	●	■	●	Ⓢ	■
Experimental verification of $f = (1/2l)\sqrt{(T/m)}$					■	■				■	■		■	[S]	●
Decibels						■									
3.7 Electromagnetic waves															
Photons	●			●		●	●	●	●	●	●	●	●	●	●
$E = hf$	●			●		●	●	●	●	●	●	●	●	●	●
Measurement of velocity								●					●		
Electromagnetic spectrum: properties	●	●		●	●	●	●	●	●	●	●	●	●		●
: detection				■	■	●		■							■
Radio waves: modulation and bandwidth		[c]				Ⓒ					[c]			[A][R]	
: aerials and propagation						Ⓒ								[R]	
3.8 Wave-speed formulae															
$v = f\lambda$	●	●				●	●	●	●	●	●	●	●	●	●
Water waves								■							
Mechanical waves						■	ⓐ	■	■		■	■			
Sound waves						■		■	■				■		
Electromagnetic waves				[B]				■				■		[S]	●
UNIT 4 OPTICS															
4.1 Wave nature of light															
Huygen's wave theory						●	ⓐ	●	●	●	●	●	●	■	●
Interference: Young's fringes	●	●			[B]	●	●	●	●	●	●	●	●	●	●
Coherence	●	●			[B]	●	●	●	●	●	●	●	●	●	●
Lloyd's mirror					[B]					●					
Holography								●						■	
Polarization by reflection – the Brewster angle												■			
4.2 Effects due to gaps and films															
Diffraction grating	●				Ⓑ	●	●	●	●	■	●	●	●	●Ⓢ	●
Thin-film interference: parallel sides							Ⓐ				●			■	
: non-parallel sides–wedge					[B]		Ⓐ								
: n–p sides–Newton's rings					[B]		Ⓐ								
Multiple slits							Ⓐ	●							
Single-slit diffraction	●				[B]	■	■Ⓐ	●	■	■	●	●	■		■
Circular aperture diffraction (qualitative)							ⓒ	●			■				
4.3 Ray optics															
Waves and rays					■	●	ⓐ	●	●	●	●	●	●	●	●
Huygen's explanation of reflection					■	●	ⓐ			●					
Formula relating u, v & f for a spherical mirror					●					●					
Huygen's explanation of refraction					●	●	ⓐ			●					
Snell's law					■	●	ⓐ		■	●				●	●
Refractive index		Ⓒ			■	●	ⓐ		■	●				●	■
Total internal reflection					●	●	ⓐ		■	●	[C]			●	●
Optical fibres		[c]					●[c]				[c]			●	■
$n_1 \sin i_2 \sin i_2$					●	●	ⓐ			■	●			■	
Measurement of reflective index							[c]		■					■	■
Deviation by triangular prism						■	[c]		■	■				■	■
Lenses					●	■	ⓐ		■	●		■	■	●ⓞ	●
Formula relating u, v, & f for a simple lens					●		[a]		■	●		■		ⓞ	■
Experimental determination of focal length					■		ⓐ		■	■				■[o]	■
Image defects							ⓒ							■[o]	
Power of a lens							ⓐ								●
Dispersion							ⓐ[c]							■	
Intensity of light								●				■		●	
Application of total internal reflection		ⓒ			■		ⓐ			■				●	■
4.4 Optical instruments															
Image formation & magnifying power for:															
(i) the simple microscope							[a][c]			■				[o]	
(ii) the compound microscope							[c]							[o]	
(iii) the refracting telescope in normal adjustment							[a][c]			■				[o]	
The eye-ring							[c]			●				[o]	
Resolving power							[A][c]	●			■				
Reflecting telescopes							[c]							[o]	
The spectrometer					[B]	■	[a][c]			■	■	■	■	[S]	
The camera, f-number, depth of field							[a][c]							[o]	
Projector, binoculars							[c]							[o]	
Interferometer														●Ⓢ	
Spectacle lenses							[a]							[o]	■
UNIT 5 MATERIALS															
5.1 Strength of solids															
Stress	●		●		●	●[B]	●	●	●	●	●	●	●		●
Strain	●		●		●	●[B]	●	●	●	●	●	●	●		■
Stress – strain curves for metals & non-metals	●		●		■	●[B]	■	●	●	●	●	●	■		■
Hooke's law	●		●		■	●	●	●	●	●	●	●	●		●
Limit of proportionality	●		●		■	●	●	●	●	●	●	●	●		●
Elastic limit	●		●		■	●	●	●	●	●	●	●	●		●
Elastic deformation			●			●[B]	●	●	●	●	●	●	●		●

Analysis of examination syllabuses/ continued	AS LEVELS					A LEVELS									
	JMB	COSSEC	AEB	London	Oxford	London	JMB	O and C Nuffield	AEB	Oxford	Cambridge	O and C	WJEC	Scottish	N. Ireland
Plastic deformation			●			●[B]	■	●	●	●	●	[B]	●		■
5.2 Elastic behaviour															
Young's modulus	●		●		●	●	●	●	●	●	●	●	●		●
Spring constant	●		●		■	●	●	●	●	●	●	●	●		●
Work done in stretching a wire	●				●	●	●	●	●	■	●	●	●		●
Experimental determination of Young's modulus	●				●	●	●	●	●	●	●	●	●		●
Stored energy	●				●	●	●	●	●	●	●	●	●		●
5.3 Structure of solids															
Atoms in solids						■[B]	■	●	●	●	●	●[B]	●		■
Ionic, covalent, metallic & Van der Waal's forces							[b]	●					●		
Atoms and elasticity			■			●[B]	●	●		●	●	●[B]	●		■
Explanation of Hooke's law						●[B]	(a)	●			●	●[B]	●		■
Atomic spacing						■[B]	●	●			●	●[B]	■		
X-ray crystallography						[B]		■				[B]			■
Bragg's law						[B]						[B]			
Microwave analogue of Bragg reflection						[B]						[B]			
Amorphous solids & polymers			■			■	[a]	■	■	■	■	[B]	■		■
Defects in solids								■	■			[B]	■		
5.4 Fluids															
Static pressure		[B]			●	●	●		●	●	●[T]			●	
Archimedes' principle		[B]				■			●	■	[T]			■	
Laminar & turbulent flow		[B]					(A)		■		[T]				
Continuity							(A)[b]		●						
5.5 Non-viscous flow															
Bernoulli's equation		[B]					(A)		●		[T]				
Pitot tubes & Venturi meters							[A]		●						
5.6 Viscous flow															
Fluid friction		[B]				■	(a)		●	●	●[T]				
Cofficient of viscosity							[A]		■						
Velocity gradient							[A]		■						
Poiseuille's law									■						
Stoke's law							[A]								
Terminal velocity						●	(a)	●	●	●	■				
5.7 Surfaces															
Surface tension and surface energy													●		
Molecular explanation of surface tension													●		
UNIT 6 HEAT AND GASES															
6.1 Temperature															
Temperature and heat		●	●	●	●	●	●	●	●	●	●	●	●	■	●
Heat		●	●	●	●	●	●	●	●	●	●	●	●	■	●
Thermometric property		●	●	●	●	●	●		●	●	●	●	●		■
Fixed points		●	●	●	●	●	●		●	●	●	●	●	■	●
Centigrade scales		●	●	●	●	■	●		●	●	●	●	●		●
Ideal gas scale		■	■	■	●	■	●		●	●	■	●	●	■	●
Celsius scale		●	●	●	●	●	●		●	●	■		●	■	●
Zeroth law							[b]	●	●						
Liquid in glass thermometer		●			●	■	■		■	●	●		■		■
Resistance thermometer		●	●		●		■		■	●	●		■		■
Thermoelectric thermometer		●			●	●	■		■	■	●		■		■
Constant volume gas thermometer					■	■	■		■	●	■		■		
Advantages & disadvantages of different thermometers					■					●	●				
6.2 Heat capacities															
Specific heat capacity (c)		●	●	●	●	●	●	●	●	●	●	●	■		
Specific latent heat (l)		●	●		●	●	●		●	●	●	●	■		
Molecular interpretation of latent heat						■				●	■	■	■		
c of a liquid by electrical heating			■		●	●	[a]		■	■	●				
c of a solid			■		●	●	[a]		■	■	●				
c of a liquid by continuous flow					■					■					
l of a liquid			■		●	●	[a]		■	■	●				
Cooling correction									■						
6.3 Thermal coefficients															
Linear expansion							[b]		●	●					
Resistance									■	●		■[B]			
6.4 Thermal conduction															
Temperature gradient			●	●	●	●	●	●	●	●	●	●	●		●
Coefficient of thermal conductivity (k)			●	●	●	●	●	●	●	●	●	●	●		●
Thermal conductors in series			■	■		■	●	■	●	■	●	■	●		■
Thermal resistance							(a)	●	■		■				
Energy conservation and U-values			●	[A]		■(A)	(a)(b)	●							
Nature of conduction			■		●	●	●		■	●	●	●	●		●
Experiment to determine k: for good solid conductor									■		■	■			
: for a poor solid conductor									■		■	■			
6.5 Thermal radiation															
General properties				[B]		■	■	■	■	■	■	■	■		■
Black-body radiation				[B]											
Distribution of energy in black-body spectrum				[B]											

Analysis of examination syllabuses/ continued

	AS LEVELS					A LEVELS									
	JMB	COSSEC	AEB	London	Oxford	London	JMB	O and C Nuffield	AEB	Oxford	Cambridge	O and C	WJEC	Scottish	N. Ireland
Laws of radiation				[B]											
6.6 Ideal gases															
Ideal gas equation ($PV = nRT$)					■	●	●	●	●	●	●	●	●	■	●
Boyle's law					■	■	●	●	●	■	■	●	●	■	●
Charles' law					■	■	●	■	●	■	■	●	●	■	●
Combined gas law					■	■	●	●	●	●	●	●	●	■	●
Brownian motion			●			●	●	●	●						●
Assumptions of the kinetic theory of gases					■	●	●	●	●	●	●	●	●	■	●
$PV = \frac{1}{3} Nmc^2$						●	●	●	●	●	●	●	●		●
Derivation of $PV = \frac{1}{3} Nmc^2$						●	●	●	●	●	■	●	●		●
Kinetic energy and temperature					■	●	●	●	●	●	●	●	●	■	●
Molar heat capacity for ideal monatomic gas							ⓑ	●	●		●	■			
Boltzmann constant						●	●	●	●	●	●	●	●		●
Diffusion (qualitative)							ⓑ	●	●					■	
Dalton's law of partial pressures									●						
Mean free path								■					●		
Distribution of molecular speeds						●	[a]	●	■						
6.7 Thermodynamics of ideal gases															
First law		●	■	●		●	●	●	●	●	●	●	●		●
Changes at constant volume $\Delta U = nc_v\Delta T$							[b]		●	■	●	●	●		
Changes at constant pressure $c_p - c_v = R$							[b]		●	■	●	■	●		
Adiabatic changes		●		■		■	[b]		●	●	■	■	■		
Isothermal changes		●				●	[b]		●	●	●	■	●		
$\gamma = c_p/c_v$							ⓑ		●				●		
Derivation of $c_p - c_v = R$							[b]				■		●		
6.8 Real gases															
Andrews' experiments							[A]								
Critical temperature							[A]								
Explanation of real gas behaviour							[A]		■			■		■	
Van der Waals' equation							[A]								
6.9 Vapours															
Definition of saturated vapour									●						
Saturation vapour pressure									●						
Independence of volume									●						
Dependence on temperature									●						
Boiling									●						
Molecular explanation of vapour phenomena			■						●						
Measurement of svp									■						
Experimental investigation of dependence of svp on temperature									■						
6.10 Entropy & equilibrium															
Energy transformations		Ⓐ				●	●	●	■						
Efficiency		Ⓐ	■	■[A]		Ⓐ	[b]	●	■						
The second law of thermodynamics		Ⓐ	■	■[A]		Ⓐ	[b]	●	■						
Heat engines		[A]	■	■[A]		Ⓐ	[b]	●	■						
Chance							[b]	●	■						
Thermal equilibrium				■			[b]	●	■						
Attaining equilibrium							[b]	●	■						
Entropy							[b]	●	■						
UNIT 7 ELECTRICITY AND ELECTRONICS															
7.1 Current and charge															
Unit of charge	●	●	●	●	●	●	●	●	●	●	●	●	●	●	●
$I = dQ/dt$	●	●	●	●	●	●	●	●	●	●	●	●	●	●	●
Demonstration of equivalence of static & current electricity					■			●		■	●				
Conduction in metals		●	●		●	●	■	■	●	●	■	●[B]	●	■	●
$I = nuAe$						●	●	■	●	■	●	●[B]	●		●
Effect of temperature on metallic condition	●		●			●	■	●	●	●	●	●[B]	●		●
Electrolysis						●					■	■			■
Electrostatic phenomena						●			●	●					
Van de Graaff generator									●	●	●				
7.2 Potential difference															
The volt	●	●	●	●	●	●	●	●	●	●	●	●	●	●	●
Power	●	●	●	●	●	●	●	●	●	●	●	●	●	●	●
EMF	●	●	●	■	●	●	●	●	●	●	●	●	●	●	●
Internal resistance	●	●	●	■	●	●	●	●	●	●	●	●	■	●	●
Maximum power theorem							■	●			■			■[A]	■
Kirchoff's laws	●					■	●	■	●	●		●	●	■	■
7.3 Limits & uses of Ohm's law															
Resistance	●	●	●	●	●	●	●	●	●	●	●	●	●	●	●
Resistivity and conductivity	●		●		●	●	●	●	●	●	●	●	●		●
I–V curves	●	●	●	■	●	●	●	●	●	●	●	●	●		●
Ohm's Law	●	●	●	●	●	●	●	●	●	●	●	●	●	●	●
Resistor combination rules	●	■	●	■	●	●	●	●	●	●	●	●	●	●	●
Meter conversion	●	■			■	●	●		●	■	■	■	●	■	●

Analysis of examination syllabuses/ continued

	AS LEVELS					A LEVELS									
	JMB	COSSEC	AEB	London	Oxford	London	JMB	O and C Nuffield	AEB	Oxford	Cambridge	O and C	WJEC	Scottish	N. Ireland
7.4 Potential dividers and potentiometers															
Potential dividers	●		●			●	●	●	●	●	●	●	●	■	●
The principle of the potentiometer	■					●	■	●	●	●	●	●	●		●
Comparison of cell emfs	■					■	●		●	●	●	●	●		●
Measurement of a small emf							■		●	■	■	■			■
Comparison of two resistances															■
7.5 The Wheatstone bridge															
Condition for balance	●						ⓐ			●		●		●	
Metre bridge arrangement	●						ⓐ			●		■		●	
Uses	●						ⓐ			■		●		■	
7.6 Capacitors in dc circuits															
Capacitance defined		●			●	●	ⓐ	●	●	●	●	●	●	●	●
Combination rules					●	●	ⓐ	●	●	●	●	●	●		●
Discharging through a resistor					[c] ■	■	■	■	■	■	■	■	■	■	■
Charging through a resistor					[c] ■	■	■	■	■	■	■	■	■	■	■
Time constant					[c] ■	■	■	■	■	■	■	■	■	■	●
Practical capacitors					■	●		■	●	●				●	
Action of dielectric (qualitative)							ⓐ		■	●	■	●	●	●	●
Blocking capacitors							ⓐ	■						●	
7.7 Measuring ac															
Frequency		ⓒ	●	Ⓐ	●	●	●	●	●	●	●	●	●	●	●
Amplitude		ⓒ	●	Ⓐ	●	●	●	●	●	●	●	●	●	●	●
rms current and pd		ⓒ	●	Ⓐ	●	●	●	■	●	●	●	●	●	●	●
Representation of sine wave signals		ⓒ	●	Ⓐ	●	●	●	■	●	●	●	●	●	●	●
Rectifiers		ⓒ	●		ⓒ	[E]	●		●	●	●	■	●		■
Smoothing		ⓒ	●		ⓒ	[E]	●	■	●	●	●	●	●		■
Description of the cro	●		■		■		●	■	■	●	■	■	■	Ⓢ	■
Uses of the cro	●	●	●		■	● [E]	●	■	●	●	●	●	●	Ⓢ	●
Lissajous' figures							[a]		■	■					
Bridge rectifier meters		ⓒ			ⓒ	[E]	■		●	●	●	●	■		■
7.8 ac circuits															
Pure resistance		ⓒ				●	●	■	●	■	ⓒ		●	● Ⓡ	●
Capacitative reactance		ⓒ					●	■	●	●	ⓒ	ⓒ	●	ⒶⓇ	●
Inductive reactance		ⓒ					●	■	●	●	ⓒ	ⓒ	●	Ⓡ	●
Phase relationships		ⓒ					●	●	●	●	ⓒ	ⓒ	●	Ⓡ	●
Filter circuits												[C]		ⒶⓇ	
Oscillators								■						ⒶⓇ	
LCR circuits		ⓒ					●	■	●		ⓒ	ⓒ	●	Ⓡ	●
Impedance		ⓒ				[E]	●	■	●		ⓒ	ⓒ	●	Ⓡ	●
Phase angle		ⓒ					●	■	●		ⓒ	ⓒ	●	Ⓡ	●
Power dissipation				[A]		■	■	■	●	■		ⓒ	■	Ⓡ	■
Series resonance		ⓒ					●		●		ⓒ	ⓒ	●	Ⓡ	●
7.9 Transistors															
Semiconductors			■			●	●	●	●	●	●	●	●	●	●
Effect of temperature on resistance			●			●	■	●	●	■	●	●	●	●	●
Intrinsic conduction							●		●		●	Ⓑ	●	●	■
Extrinsic conduction							●					Ⓑ	●	●	
p and n materials							●	■				Ⓑ	●	●	
Transistor action							ⓐ								
Audio amplifier														[A]	
p-n junction diode							ⓑ							■	
Zener diode														[A]	
7.10 Operational amplifiers															
Open-loop voltage comparator					ⓒ	[E]	●	●	●	●	●	ⓒ	■	■ [S]	■
Analysis of feedback					[c]	[E]	[E] [b]	■	■	■	■	ⓒ	■	■ [S]	■
Inverting amplifier					ⓒ	[E]	■ [E]	●		■	●	ⓒ	■	●	■
Non-Inverting amplifier					ⓒ	[E]	[E] [b]	●	●	■	●	ⓒ	■	●	■
Voltage follower							[E]	●							
Astable multivibrator					ⓒ		■	■	●	●		ⓒ	■		■
Integrators & analogue computing						[E]	[E]	●						[A]	
Frequency response, band width					[c]	[E]	[E] [b]	■	●	■	■	ⓒ	■	■ [A]	
Transducers					■	[E]	■ [E]	■	■	■				■ [S]	■
7.11 Logic circuits															
Transistor as a logic switch							ⓐ	■			■				■
Logic gates and truth tables					ⓒ		■ [E]	■	●	●	●		■	[D]	■
Simple combinations of logic gates					ⓒ		■ [E]	■	●	●	●		■	[D]	■
Binary adders					[c]		[E]			■	■			[D]	■
Counters							[E]	■		■	■		■	[D]	■
Digital displays							[E]	■			●		■	[D]	■
Bistable multivibrator (flip-flop)							[E]	■						[D]	■
Monostable multivibrator								■							
Astable multivibrator								■							
Digital transmission		ⓒ									ⓒ			[D][R]	

Analysis of examination syllabuses/ continued

	AS LEVELS					A LEVELS									
	JMB	COSSEC	AEB	London	Oxford	London	JMB	O and C Nuffield	AEB	Oxford	Cambridge	O and C	WJEC	Scottish	N. Ireland
UNIT 8 STRUCTURE OF THE ATOM															
8.1 Properties of the electron															
Charge: Millikan's experiment				[B]			ⓑ	●	●	■	●		●	Ⓢ	●
Charge/mass ratio: methods of measurement				[B]			ⓑ	■	●	●	●	●	●	Ⓢ	●
Mass, speed and energy						■	■	●	●	●	●	●	●	Ⓢ	●
Velocity selector						■	■	■	●	■	●	●	■	Ⓢ	●
Electron diffraction				[B]			[a]	●	●	●	●		■	■	■
8.2 Free electrons in metals															
Thermionic emission	●					●	ⓐ	●	●	■	●		●		■
Work function					●	●	●	●	●	●	●	●	●	●	●
Photoelectric effect				●[B]	●	●	●	●	●	●	●	●	●	●	●
Threshold frequency				●[B]	●	●	●	●	●	●	●	●	●	●	●
Einstein's equation				●[B]	●	●	●	●	●	●	●	●	●	●	●
Experimental determination of *h*/*e* or *h*				■[B]		●	■	●	●		■	■	●		●
Experimental determination of work function							■	●			■	■	●		
Photocells & their applications													●		
8.3 Electrons within atoms															
Energy levels, ionization & excitation	●			●	■	●	●	●	●	●	●	●	●	●	●
Experimental evidence for energy levels	●				■	■	■	●	●	■	■	■	■	■	■
Energy levels in hydrogen	●			[B]	■	■	[a]	●		■	■	■	■	Ⓢ	■
Pauli exclusion principle and electron shells								■							
Bohr theory				[B]				■					●	■[S]	
8.4 Energy emitted by electrons															
Change of energy levels				[B]	●	●	●	●	●	●	●	●	●	●	●
Emission spectra				[B]	●	●	●	●	●	●	●	●	●	●	●
Line spectra				[B]	●	●	●	●	●	●	●	●	●	●Ⓢ	●
Absorption spectra				[B]	●		●	●	●		●	●	●	■	■
Continuous optical spectra				[B]	●		ⓐ	●	●	●	■	●	●	●	●
Continuous X-ray spectrum	■						ⓐ						●		●
X-ray line spectrum	■						ⓐ						●		●
Minimum X-ray wavelength	■						ⓐ						●		●
Production of X-rays	■					■	[a]						●		●
Principles of lasers														■	
8.5 Properties of the nucleus															
The nuclear model: alpha scattering evidence	●	●	■	●	■	●	ⓐ	●	●	●	●	●	●	●Ⓢ	●
Neutrons, protons & isotopes	●	●	●	●	●	●	●Ⓑ	●	●	●	●	●	●	●	●
Proton number ~~Z~~, nucleon number A, nuclear equations	●	●	■	●	●	●	●	●	●	●	●	●	●	●	●
Nuclear size & density					■		Ⓑ								
Properties of the nuclear force							Ⓑ								
The 'liquid drop' model							Ⓑ								
Mass spectrometry								●			●			[S]	
Accelerators and particle physics				[B]				■							
8.6 Spontaneous emission of nuclear enegy: radioactivity															
Stable & radioactive isotopes	●	●	●	●	●	●	●Ⓑ	●	●	●	●	●	●	●	●
α-particle emission	●	●	●	●	●	●	●Ⓑ	●	●	●	●	●	●	●	●
β-particle emission	●	●	●	●	●Ⓓ	●	●Ⓑ	●	●	●	●	■	●	●	●
Radioactive series	●					●	●Ⓑ	●	■		●				■
γ-ray emission	●	●	●		●Ⓓ	●	●Ⓑ	●	●	●	●	●	●	●	●
$N-Z$ curve for stable nuclei		■		●	[D]		Ⓑ			■					
Energy balance in nuclear decay & Q values					[D]		Ⓑ			■	■	●			
Electron capture							Ⓑ								
Detection of ionizing radiations	●	■	■	■		■	●	●	●	●	●	●	●	■	■
Experiments using:															
(i) Cloud chambers	■	■	■	■	●	■	ⓐ	●	■	●	■		■		
(ii) Geiger–Muller tubes	■	■	■	■	●	●	●	●	■	●	■	●	■		
(iii) Ionization chambers and dc amplifiers	■	■	■	■	●	■	ⓐ	●		■	■		■		
Distinction between α, β, γ by relative ionizing power and absorption	●	●	■		■	■	●	●	●	●	●	●	●		■
Mathematical model of radioactivity: $dN/dt=-\lambda N$	●	■		●	■[D]	●	●	●	●	●	●	●	●		●
Decay constant	●	■		●	●	●	●	●	●	●	●	●	●		●
Half-life	●	●	●	●	●	●	●	●	●	●	●	●	●		●
$N_t=N_0\exp(-\lambda t)$	●	■		■	●	●	●	●	●	●	●	●	●		●
Experimental determination of half-life	●		■		[D]	●	●	●	●	●	●	●	■		■
Waste disposal	●	●					■[B]				●				■
Uses & hazards of radiation	●	●	■				■[B]		■		■		■	■	●
8.7 Induced emission of nuclear energy: fission and fusion															
Binding energy of nucleon and nucleus		Ⓐ		●	[D]	[A]●	Ⓑ	●		■	●	●	●	■	●
Equivalence of mass and energy		Ⓐ		●[B]	[D]	[A]●	Ⓑ	●		●	●	●	●	■	●
Atomic mass unit		Ⓐ		●[B]	[D]	●	●Ⓑ	●		●	●	●	●	■	●
Mass defect		Ⓐ		●	[D]	[A]●	[B]	●		●	●	●	●	■	●
Fission		Ⓐ		●	[D]	[A]■	[B]	●		●	■	■	●	■	●
Nuclear reactor		Ⓐ		[A]		[A]●	[B]	●		■				■	■
Production of isotopes							[B]							[N]	
Fusion		Ⓐ		●[B]	[D]		[B]	●		●	■		■	■	■

Analysis of examination syllabuses/ continued	AS LEVELS					A LEVELS									
	JMB	COSSEC	AEB	London	Oxford	London	JMB	O and C Nuffield	AEB	Oxford	Cambridge	O and C	WJEC	Scottish	N. Ireland
UNIT 9 DATA ANALYSIS															
9.1 Units and dimensions															
Base units (not definitions)	●	●	●	■	●	●	●	●	●	●	●	●	●	●ⓢ	●
Definition of the ampere						●	●	●	■	●	●	●	●	ⓢ	●
Linking derived units to base units		●	●		●	●	●	●	●	●	●	●	●	●ⓢ	●
Dimensions			●			■	●	■	●	●	●	●	●	●ⓢ	●
Dimensional analysis			■			■	●	■	●	●	●	●	●	🅂	●
Deriving formulae using dimensions								■					●		
9.2 Graphs															
Graph plotting	●	●	●	●	●	●	●	●	●	●	●	●	●	●ⓢ	●
Slopes, areas and intercepts	●	●	●	●	●	●	●	●	●	●	●	●	●	●ⓢ	●
Common graph shapes	●	●	●	●	●	●	●	●	●	●	●	●	●	●ⓢ	●
Formulae for common graph shapes	●	●	●	●	●	●	●	●	●	●	●	●	●	■ⓢ	●
Log scales	●				●	●	●	●	●	●	●	●	●	🅂	●
9.3 Errors															
General appreciation	■	■	●	■	●	●	●	●	●	●	●	●	●	●ⓢ	●
Errors and measurements			■		●	●	ⓐ	●	●	●	●	●	●	●ⓢ	■
Systematic and random errors			●			■	ⓐ	■	■	●	■	■	■	●🅂	■
Combining errors for formulae						■	ⓐ	■	■	●	●	■	■	🅂	■

ADDRESSES AND CODES FOR THE EXAMINATION BOARDS

AEB	The Associated Examining Board Stag Hill House, Guildford, Surrey GU2 5XJ
Cambridge	University of Cambridge Local Examinations Syndicate Syndicate Buildings, 1 Hills Road, Cambridge CB1 2EU
COSSEC (AS only)	As for Cambridge, Oxford and Cambridge or SUJB
JMB	Joint Matriculation Board Devas Street, Manchester M15 6EU
London	University of London Schools Examinations Board Stewart House, 32 Russell Square, London WC1B 5DN
NISEC	Northern Ireland Schools Examinations Council Beechill House, 42 Beechill Road, Belfast BT8 4RS
Oxford	Oxford Delegacy of Local Examinations Ewert Place, Summertown, Oxford OX2 7BZ
O and C	Oxford and Cambridge Schools Examinations Board 10 Trumpington Street, Cambridge and Elsfield Way, Oxford
SEB	Scottish Examinations Board Ironmills Road, Dalkeith, Midlothian EH22 1BR
SUJB	Southern Universities' Joint Board for School Examinations Cotham Road, Bristol BS6 6DD
WJEC	Welsh Joint Education Committee 245 Western Avenue, Cardiff CF5 2YX

Questions from the Boards are indicated by the appropriate codes for the Boards. However the following should be noted:

O and C Nuffield – for Nuffield A-level questions
SEB H – for Scottish Higher Grade
SEB SYS – for Scottish Sixth Year Studies

SUMMARY OF QUESTION STYLES OF EACH EXAMINATION BOARD

AEB – M, S, L, D, PE
Cambridge – M, S, L, PE or PI
JMB – M, S, L, C, PE or PI
London – M, S, L, C, D, PE
NISEC – M, S, L, PE, C, D
O and C – M, S, L, PE or PI
O and C Nuffield – M, S, L, C, PE, PI
Oxford – M, S, L, PE
SEB H – M, S, L
SEB SYS – S, L, PI
WJEC – S, L, PE

KEY M = Multiple choice, S = Short answer questions, L = Longer questions, C = Comprehension questions, D = Data analysis paper, PE = Practical examination, PI = Internal assessment of practical work

NOTE: Consult your teacher *or* write to your Board for exact details of the form of your examination.

Part II The core of A-Level Physics

Unit 1 Force and motion

The topics in this Unit are very important in any A-level course because other branches of physics almost always refer back to aspects of it. Mechanics is a well developed, yet changing subject; basic concepts are continually being re-examined in attempts to understand the extremes of nature from atoms to galaxies. Detailed study of this Unit should enable key concepts of mechanics to be understood, both for their own sake and for use in other Units.

An introduction to basic ideas of mechanics is a feature of most GCSE courses, and the earlier Topics (1.1 to 1.5) of this Unit build on such coursework. Remember to look back at your GCSE course notes if you find difficulty with any points in these topics. A thorough understanding of these early Topics is essential before you tackle the Topics (1.6 to 1.8) of shm and rotation. You will find that the study of shm reappears prominently in Unit 3 (Oscillations and waves); likewise, the topic of Satellite motion in Unit 2 refers back to Unit 1 (Circular motion). The final part of the Unit deals with those recent developments in mechanics that are included in A-level courses nowadays.

1.1 MOTION IN A STRAIGHT LINE

Displacement is a distance moved in a given direction.
Speed is rate of change of distance.
Velocity is rate of change of displacement.
Acceleration is rate of change of velocity.

If the above definitions are applied to an object with initial velocity u that is **uniformly accelerated** to a final velocity v in time t, the following equations can be derived:

E1.1 $a = \frac{(v-u)}{t}$

E1.2 $s = \frac{(u+v)t}{2}$

E1.3 $s = ut + \frac{1}{2}at^2$

E1.4 $v^2 = u^2 + 2as$

where a = acceleration (ms^{-2}), u = initial velocity (ms^{-1}), v = final velocity (ms^{-1}), s = displacement (m), t = time (s)

Velocity/time graphs and Displacement/time graphs are valuable ways of depicting motion in a straight line. For example, the motion of a ball thrown directly upwards is shown by the graphs of F1.1. Because velocity and displacement are **vectors**, care must be taken with the sign convention for direction; in the diagram, the $+y$-axis is used for upwards direction (and $-y$ for downwards).

As the ball rises its velocity becomes less and less + ve, becoming zero at its maximum height; then, its velocity becomes more and more − ve because it is now falling at increasing speed until it returns to 'ground level'. Note that for an object acted on by gravity only (ie, negligible air resistance), its acceleration, g, is equal to the slope of the graph and this is constant. Also, since

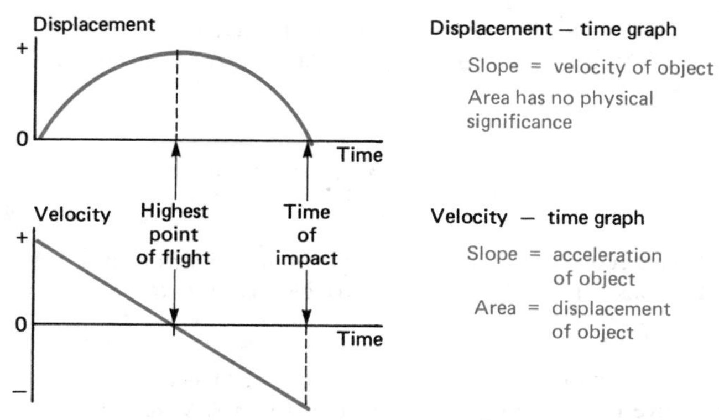

F1.1 Graphs of displacement and velocity for a ball thrown in the air

the area under the line gives the displacement, it is important to remember that area above the time axis gives + ve displacement and area below the time axis gives − ve displacement. In this example, there is as much area above the time axis as there is below, hence the total displacement from 'launch' to 'impact' is zero.

1.2 PROJECTILE MOTION

The key point in dealing with projectile motion is to remember that gravity always acts downwards only. There is **no** horizontal acceleration, assuming negligible air resistance.

Consider a projectile launched from 0 with initial velocity of magnitude u in a direction at angle A above the horizontal. Let x and y denote horizontal and vertical components of displacement respectively. The path is shown in F1.2.

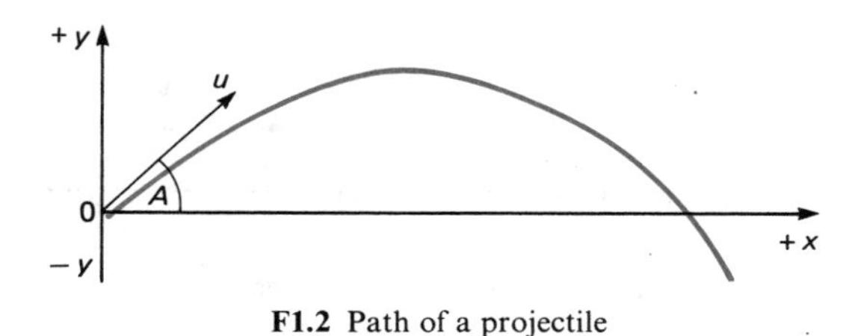

F1.2 Path of a projectile

Horizontal motion is analysed separately from vertical motion, as follows:

Horizontal motion	**Vertical motion**
Initial velocity $= u\cos A$	Initial velocity $= u\sin A$
Acceleration $= 0$	Acceleration $= -g$ (− for downwards)
Time taken $= t$	Time taken $= t$
Distance moved $= x$	Distance moved $= y$
From E1.3, $x = ut\cos A$	From E1.3, $y = (ut\sin A) - \frac{1}{2}gt^2$
From E1.1, final speed $v_x = u\cos A$	From E1.1, final speed $v_y = (u\sin A) - gt$

The horizontal velocity remains constant, but the vertical velocity decreases steadily from its initial value at a rate of g. Note that the magnitudes of the displacement and the velocity after time t, are given respectively by:

displacement magnitude (ie, distance) $= \sqrt{(x^2 + y^2)}$

velocity magnitude (ie, speed) $= \sqrt{(v_x^2 + v_y^2)}$

1.3 FORCE AND MOMENTUM

Momentum is defined as mass × velocity. Since velocity is a vector, momentum is also a vector. Its unit is kg m s^{-1}.

Newton's laws of motion are:

(i) An object stays at rest or at uniform velocity unless acted on by a resultant force.

(ii) The change of momentum per unit time is proportional to the resultant force, and in the same direction.

(iii) Action and reaction are equal and opposite.

The first law tells us what a force is (ie, that which changes an object's state of motion). The second law gives a useful equation:

E1.5 $F = \frac{mv - mu}{t} = ma$

where F = Force (N), m = mass (kg), u = initial velocity (ms^{-1}), v = final velocity (ms^{-1}), a = acceleration (ms^{-2})
t = time (s).

This equation applies when the mass m remains constant. In situations where the mass is variable, E1.5 takes the form $F = \frac{d}{dt}(mv)$.

The unit of force is the **newton** (N), defined as that force that will give a mass of 1 kg an acceleration of $1 ms^{-2}$. Always use these units.

The **impulse** of a force is defined as force × time. From E1.5 it can be shown that the change of momentum ($mv - mu$) is equal to the impulse.

The **Principle of conservation of momentum** states that when no resultant force acts upon a system of bodies, the total momentum of the system remains unchanged. This is a very important principle of physics, and it is closely linked with Newton's third law. This can be seen by considering a collision between mass m, initially moving at velocity u, and mass M initially at rest. Suppose, after the collision, that the two masses move off in the same direction as that in which mass m was initially moving, as represented in F1.3. Let the final velocity of mass m be v, and the final velocity of mass M be V. Let t denote the time of duration of the impact.

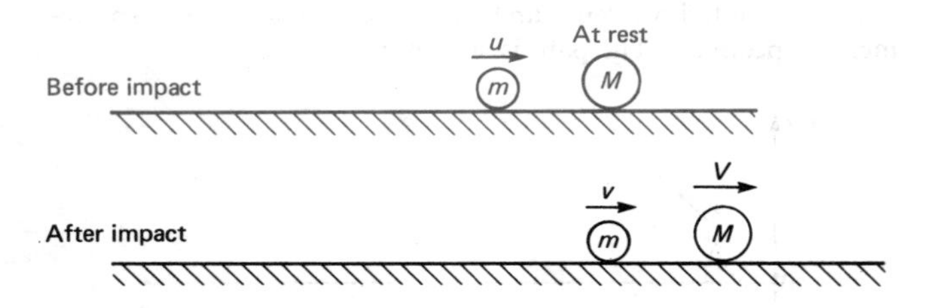

F1.3 Collision between two masses

For mass m, its change of momentum per unit time is $\frac{(mv - mu)}{t}$; for mass M, its change of momentum per unit time is $\frac{MV}{t}$. Since change of momentum per unit time is equal to force, then the force acting upon m is $\frac{(mv - mu)}{t}$, and the force acting upon M is $\frac{MV}{t}$. From Newton's third law, the force acting upon m is equal and **opposite** to the force acting upon M; thus $\frac{(mv - mu)}{t} = -\frac{MV}{t}$ where the $-$ sign represents the opposite directions of the two forces. From the last equation, it can be shown that:

Total initial momentum (mu) = Total final momentum ($mv + MV$)

In other words, the total momentum is unchanged; the principle of conservation of momentum applies to **all** collision and explosion situations, but remember that momentum is a vector so can either be + or − (according to which direction is defined as +) in straight line (ie, one dimensional) problems.

Weight is the force of gravity acting upon an object. For a mass m, its weight is mg, since if the object were allowed to fall freely, it would accelerate at g due to the force of gravity only. The correct unit for weight is the newton. Since g can change from one place to another, then the weight of a mass can alter from one place to another; however, mass does not vary in this way because it is a measure of the **inertia** (ie, resistance to change of velocity) of the object.

A person in 'free-fall' flight experiences the sensation of **weightlessness** because that person is without support. The person is still acted upon by the force of gravity, causing an acceleration g. To appreciate this situation, consider a person in a lift that accelerates downwards at a. F1.4 shows that the person is acted upon by weight mg and by the **reaction** R from the floor of the lift. Thus the force that accelerates that person downwards is $mg - R$; hence, $ma = mg - R$, so giving reaction $R = mg - ma$. Assuming a is less than g, the reaction R will be less than the weight mg so that the person will feel less support from the floor than usual; in other words, the person feels lighter than usual.

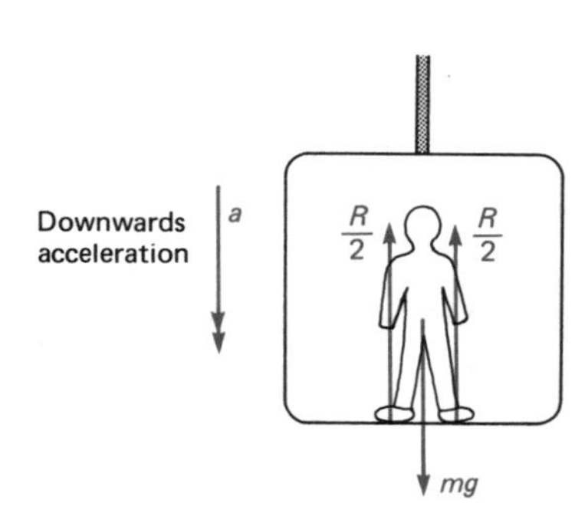

F1.4 Weightlessness: a person in a lift

1.4 WORK, ENERGY AND POWER

Work done by a force is defined as force × distance moved by the force in the direction of the force. **Energy** is the capacity of a body to do work. When work is done, energy is transformed from one form to another, so energy can be thought of as 'stored work', to be used (or stored) according to whether work is done by (or on) the object in possession of the energy. The **joule** (J) is the unit of work and of energy; 1 J of work is done when 1 N of force moves its point of application by 1 m in the direction of the force.

Power is the rate at which work is done. The unit of power is the **watt** (W) which is equal to a rate of 'doing work' of $1\,J\,s^{-1}$. For a constant force F moving at constant velocity v in the same direction as the line of action of the force, the power is equal to the force × velocity; ie, Power = $F \times v$.

The **Principle of conservation of energy** states that energy can never be created or destroyed. It is a very important principle of physics, and there has never been any experimental evidence to suggest otherwise. Energy can be changed from one form to other forms, but the total energy after the change must always be the same as before.

Kinetic energy (ke) is the energy possessed by a mass because of its motion. For a mass m, moving at speed v its ke = $\frac{1}{2}mv^2$; a simple proof will be found in any A-level textbook, and it would be useful to look it up at this stage.

Potential energy (pe) is energy due to position. See p. 31.

An **elastic** collision is one in which the total ke before the collision is the same as the total ke after the collision. An **inelastic** collision is one in which some or all of the initial ke is transformed into other forms of energy, usually heat or sound or both. Remember that conservation of momentum applies to **both** elastic **and** inelastic collisions.

1.5 STATICS

When an object is in static equilibrium, the forces that act upon it balance one another out, so that there is no resultant force upon the object.

Types of forces The more common forces met in equilibrium situations include:

1 weight, always considered to act at **the centre of gravity** of an object.

2 tension (or compression).

3 friction, which acts between surfaces (parallel to the surfaces at the point of contact) when the surfaces move (or try to move) relative to one another.

4 normal forces, which act on an object when it is pushing into (or trying to push into) another object. 'Normal' forces act 'perpendicular to' the surface.

The situation shown in F1.5 shows the forces acting on a beam resting against a smooth wall on a rough floor. The combination of F_1 and N_1 is the **reaction** of the floor upon the beam.

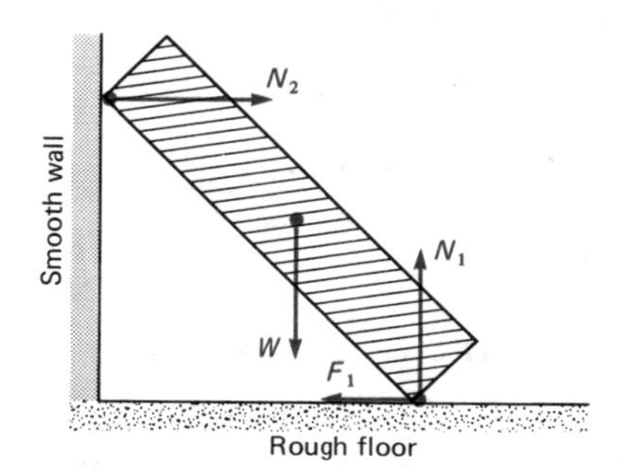

F1.5 Beam resting against a wall

Vector nature of force Two important techniques in 'handling' forces are:

1 Resolving a force into two perpendicular components, as shown in F1.6.

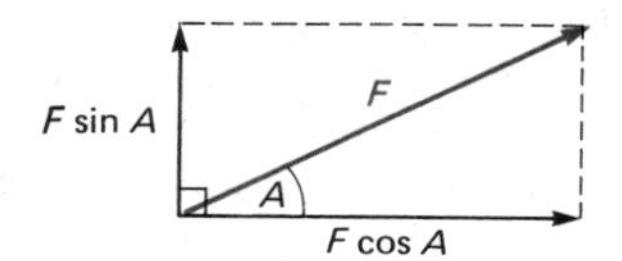

F1.6 Resolving a force F

2 Combining, as vectors, several forces; to do this for two forces, either draw a vector diagram and complete the parallelogram, as in F1.7 (i) or resolve one of the forces into two perpendicular components, one component being along the line of action of the other force as in F1.7 (ii); the resultant force R is then given by $R^2 = (F_1 + F_2 \cos A)^2 + (F_2 \sin A)^2$.

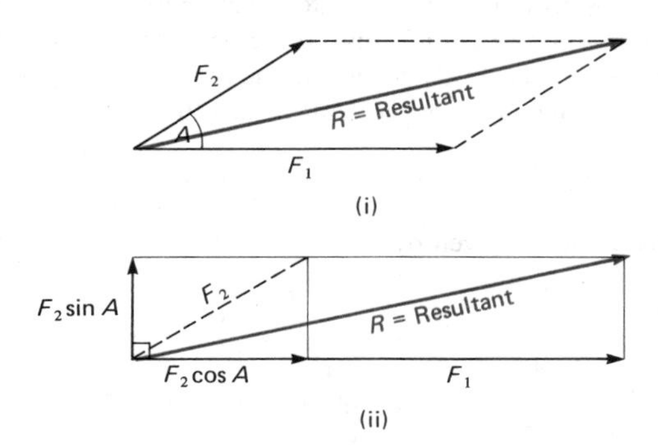

F1.7 Adding forces: (i) parallelogram method; (ii) combining componei

Turning forces Any force that tends to turn an object about a given point may be called a turning force. The **moment** of a turning force about a fixed point is defined as force × perpendicular distance from the line of the force to the point. Note that the unit of moment is Nm; the joule is not used because in a static situation, no work is done.

A **couple** is a pair of equal and opposite forces that do not act along the same line. The moment of a couple is always the same about any fixed point, and is always equal to the product of one of the forces and the perpendicular distance between the lines of action. The moment of the couple shown in F1.8 is Fd.

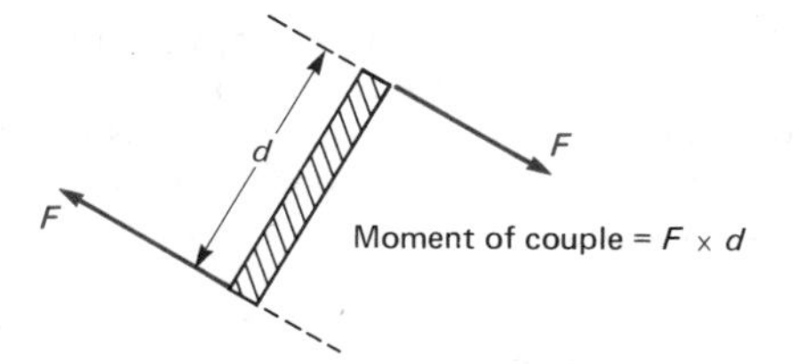

F1.8 Turning effects

Note that the term **torque** is usually used for a moment about a fixed axis.

The **Principle of moments** states that for a body in equilibrium, the total clockwise moment about a given point equals the total anticlockwise moment about that same point.

Analysing equilibrium forces When analysing the forces that act upon a body in equilibrium, always start by making a sketch diagram showing all the forces acting on the object. Then, use of the following three rules ought to enable the analysis to be completed:

1 Resolve all forces into components in two mutually perpendicular directions. Choose the two perpendicular directions according to the situation. For the beam shown in F1.5, the best directions are horizontal and vertical; for a mass on a slope, the best directions might be parallel and perpendicular to the slope.

2 Balance the force components for each of these two directions. In other words, for each direction equate the force components one way to those acting the opposite way.

3 Apply the principle of moments about the most convenient fixed point. Choose the fixed point as that point through which one (or more if possible) of the 'unknown' forces has its line of action.

1.6 SIMPLE HARMONIC MOTION

Simple harmonic motion (shm) is defined as oscillating motion of an object about a fixed point, such that the acceleration (a) of the object is:

1 proportional to the displacement (x) from the fixed point,

2 always directed towards the fixed point.

The definition can be summarized by the following equation:

E1.6 $a = -\omega^2 x$

where a = acceleration (ms^{-2}), x = displacement (m), ω = angular frequency (rad s^{-1}).

The **angular frequency** (ω) is a constant of the motion. (Note that the − ve sign in E1.6 indicates that the acceleration is always directed towards the fixed point.)

According to Newton's second law, the force producing the motion is proportional to the acceleration (ie, $F = ma$). Therefore, the restoring force of an shm system can be written as $F = -m\omega^2 x = -\textit{constant}\ x$. In other words, the **restoring force** must be proportional to the displacement from the fixed point, and directed towards the fixed point. Thus, any oscillating system for which the restoring force meets these conditions will move in shm.

The **solution of the shm equation** can be evaluated with the aid of knowledge of differentiation. The general solution is

$$x = A \sin(\omega t) + B \cos(\omega t),$$

where A and B are constants determined by the initial values of displacement and velocity (v). Note that v is obtained by differentiating x with respect to time (ie, $v = \frac{dx}{dt}$), and a is given by differentiating v with respect to time (ie, $a = \frac{dv}{dt} = \frac{d^2x}{dt^2}$). If the initial conditions are $x = x_0$, $v = 0$ and $t = 0$, then the general solution becomes $x = x_0 \cos(\omega t)$.

For a mass on a vertical spring, as in F1.9, if the mass is displaced from equilibrium and then released from rest, then the displacement is given by the equation $x = x_0 \cos(\omega t)$, and its variation with time is represented by the displacement/time graph of F1.10. Remember that the slope of the displacement/time curve gives the velocity, and the slope of the velocity/time curve gives the acceleration.

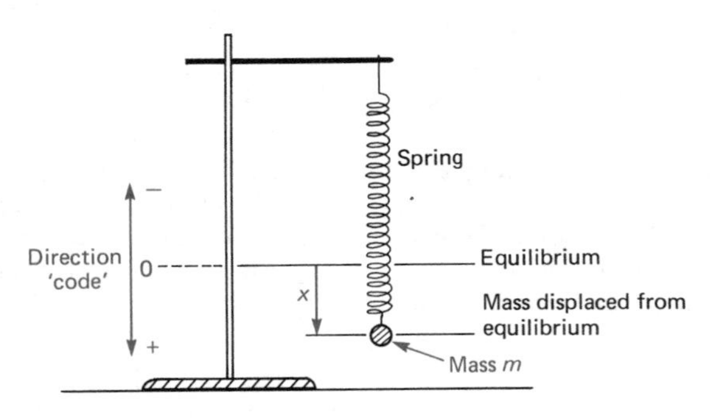

F1.9 Mass on a spring (shm)

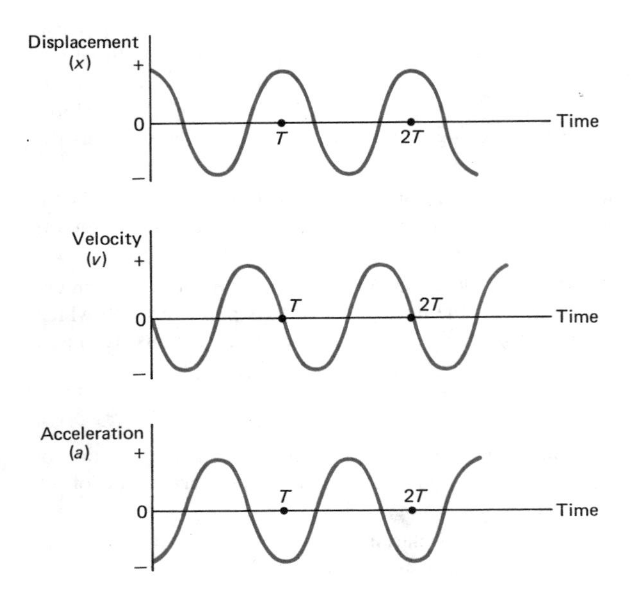

F1.10 Graphs of Simple Harmonic Motion

The **Time period** (T) is the time taken to move through one complete cycle of oscillation. From the above example, one complete cycle corresponds to $\omega T = 2\pi$ so giving $T = 2\pi/\omega$. Note that the **frequency** (f), the number of complete cycles per second, is given by $f = 1/T = \omega/2\pi$.

Oscillations of a simple pendulum When a pendulum bob is displaced from equilibrium and then released with the string taut, it oscillates in a vertical plane with its motion along the arc of a circle, as in F1.11.

The diagram shows the pendulum, with angular displacement A from its equilibrium position, as it moves along its path. The restoring force is provided by the tangential component of the weight (ie, $mg \sin A$) so that the acceleration a is given by $a = -g \sin A$, which can be written as $a = -gA$, provided angle A is small enough for $\sin A$ to be approximated to angle A in **radians**.

To show that the motion is shm, angle A must now be linked to the arc length x; in fact $x = LA$ (remember that when $A = 2\pi$ radians, then $x = 2\pi L$ = circumference; a useful check!).

The equation for acceleration can now be written as $a = -\frac{g}{L}x$ or as $a = -\omega^2 x$ where $\omega^2 = \frac{g}{L}$. Thus, the acceleration (a) meets the conditions set out in the definition (E1.6), so the simple pendulum moves with shm, provided its angular displacement (A) is small. In practice, this means never greater than

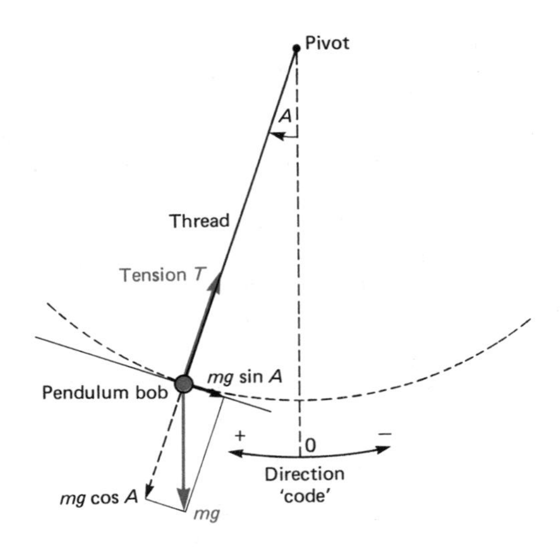

F1.11 Simple pendulum (shm)

about 10°. Finally, the time period, T, can be expressed in terms of L and g:

E1.7 $T = 2\pi\sqrt{L/g}$

where L = length of thread from pivot to 'bob' centre (m), g = acceleration due to gravity (ms^{-2}), T = time period (s).

Always remember that the time period T is the time for the bob to pass from one extreme to the opposite extreme and **back again**.

Oscillations of a loaded vertical spring fixed at its upper end The spring obeys **Hooke's law**; consequently, when the length of the spring changes, there is a change in the tension in the spring **in proportion** to the change of length. When the mass is displaced downwards from its equilibrium position and then released from rest, the mass oscillates about the equilibrium position, as in F1.10.

To show that the motion is shm, consider the forces on the mass:

1 In equilibrium, the weight (mg) = equilibrium tension (kl), (l = extension, k = tension per unit extension).

2 At displacement x from equilibrium:

resultant force = weight (mg) − tension ($kl + kx$)

so resultant force = $-kx$ since $mg = kl$

The resultant force is in the **opposite** direction to the displacement (ie, if mass is above equilibrium, force on it is downwards, and vice versa) and acts to try to **restore** the mass to equilibrium. By Newton's second law, the acceleration a is then given by

$$a = -\frac{k}{m}x \text{ or as } a = -\omega^2 x, \text{ where } \omega^2 = \frac{k}{m}.$$

The time period is given by:

E1.8 $T = 2\pi\sqrt{m/k}$

where T = time period, m = mass (kg), k = spring constant (Nm^{-1}).

For both the loaded spring and the simple pendulum, the displacement x varies with time as given by $x = x_0 \cos\omega t$, assuming an initial displacement x_0 and zero initial velocity. The amplitude of oscillations is x_0 (ie, the maximum displacement).

Energy in shm For a mass moving with shm, there is a continual interchange between its ke and the pe associated with the restoring force. If the mass is initially released at rest from a non-zero displacement (measured from equilibrium), then the initial pe stored in the system due to displacement of the mass from equilibrium changes to ke and back to pe after one half-cycle. At any instant during the motion, **ke + pe = initial pe**, as indicated by F3.2 (See p. 51).

Since the restoring force F is given by $F = -$ constant x, then **the pe is given by** $\frac{1}{2}kx^2$ where k is the force constant; this can be understood by considering the area under the force/displacement graph in F1.12.

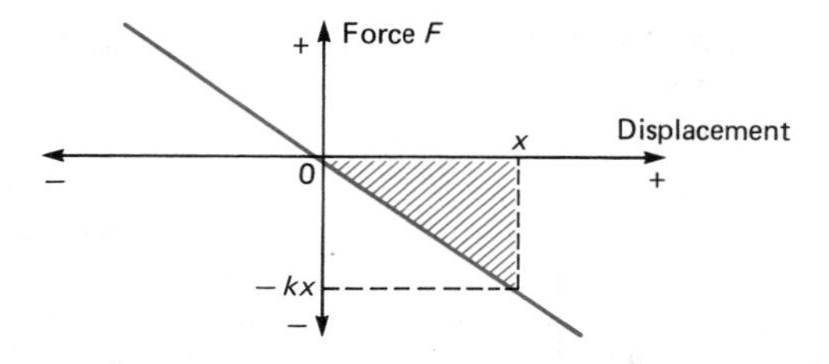

F1.12 Force-displacement graph (shm)

For displacement x, the area of the shaded triangle $= \frac{1}{2}$ base$(x) \times$ height(kx); area under a force/displacement curve is equal to the work done, which in this case gives the pe stored.

Since pe + ke = initial pe, it then follows that when the displacement is x and the velocity is v, $\frac{1}{2}kx^2 + \frac{1}{2}mv^2 = \frac{1}{2}kx_0^2$, and because $k = m\omega^2$, then the following useful equation is obtained:

E1.9 $v^2 = \omega^2(x_0^2 - x^2)$

where v = velocity (ms^{-1}), ω = angular frequency (rad s^{-1}), x_0 = amplitude (m), x = displacement (m).

When **frictional forces** are present, the total mechanical energy (ie, pe + ke) of the system is gradually changed to heat energy by the frictional forces. As a result, the amplitude of the oscillations gradually becomes less and less, as shown in F3.3.

1.7 UNIFORM CIRCULAR MOTION

When a single particle is moving at steady speed (v) on a circular path (radius, r), its velocity is constantly changing as its direction changes. Its time period (T), the time for one complete rotation, is equal to its circumference ($2\pi r$)/speed (v):

E1.10 Angular speed $\omega = \frac{2\pi}{T}$ **for steady rotation**

It follows from the above definition that the speed (ie, tangential speed), v is linked to the angular speed (ω) by:

E1.11 $v = \omega r$

where v = speed (ms^{-1}), ω = angular speed (rad s^{-1}), r = radius (m).

To cause the continual change of direction of a particle moving in uniform circular motion, a resultant force must act upon the particle; if the force suddenly ceases to act, then the particle will move off at a tangent to the circle. Whatever the reason for the resultant force (eg, tension in a string, magnetic field force on a moving charge), the resultant force is referred to as the **centripetal** force, and its direction is always **towards** the centre of the circle.

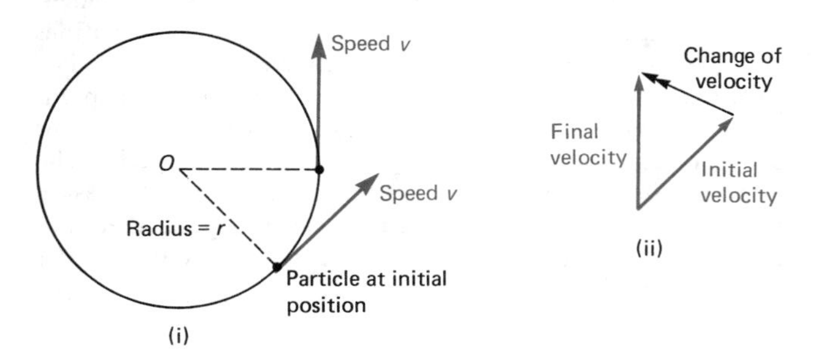

F1.13 Circular Motion: (i) speed at two points in motion; (ii) velocity change between the two points

In F1.13(i), the particle is shown at two positions on its path. The two positions are separated by a short interval of time, so the change of velocity (from one position to the other position) is directed to the centre of the circle, as in vector diagram F1.13(ii). Remember that the centripetal acceleration is equal to the change of velocity per unit time. The centripetal acceleration is given by:

E1.12 $a = -\frac{v^2}{r} = -\omega^2 r$

where a = centripetal acceleration (ms^{-2}), v = (tangential) speed (ms^{-1}), r = radius of rotation (m), ω = angular speed (rad s^{-1}).

Note that the $-$ sign in the equation indicates that the centripetal acceleration is directed **inwards** (ie, to the centre of rotation). Also, the angular acceleration is zero because the angular speed is constant. Take care not to confuse angular acceleration with centripetal acceleration.

1.8 ROTATION OF A RIGID BODY

Consider a rigid body that can rotate about a fixed axis. To change the angular speed, a **torque** must be applied to the body; the link between torque applied and angular acceleration produced is one that can be established by treating the rigid body as a structure of point masses $m_1, m_2, m_3, \ldots$ at distances $r_1, r_2, r_3 \ldots$, respectively, from the axis of rotation, as shown in F1.14.

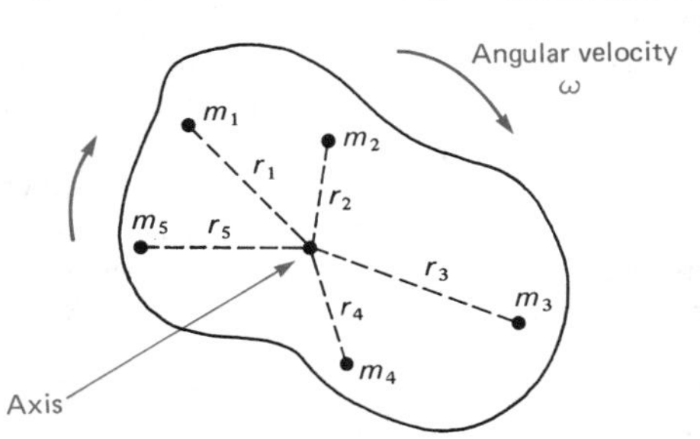

F1.14 Moment of inertia

Let the angular speed of the body be ω, and since the body is rigid, then the angular speed of each point mass will also be ω. The (tangential) speed for each point mass will be ωr_i (using E1.11 where $i = 1$ or 2 or 3 or...), so that the (tangential) acceleration of each point mass when ω changes will be ar_i where a is the angular acceleration ($= \frac{d\omega}{dt}$). Hence, the force acting upon each point mass must be $(m_i a r_i)$, requiring an individual torque (ie, couple) of moment $(m_i a r_i)r_i$ for each point mass. This 'individual moment' can be written as $(m_i r_i^2)a$, so that the total torque required to produce angular acceleration of **all** the particles of the body is $\sum_i (m_i r_i^2)a$. The symbol $\sum_i$ is the summation symbol, so that $\sum_i m_i r_i^2$ is a shorthand expression for $m_1 r_1^2 + m_2 r_2^2 + m_3 r_3^2 + \ldots$

The **moment of inertia** (I) about a given axis of a rigid body is defined as $\sum_i m_i r_i^2$. The unit of I is kg m^2.

Therefore, the moment of inertia is equal to the torque required to produce unit angular acceleration, so giving the following equation:

E1.13 $T = Ia$

where T = torque (N m), I = moment of inertia (kg m^2), a = angular acceleration (rad s^{-2}).

This is a useful equation when a constant torque is applied to a rigid body of known moment of inertia. The body will then experience a constant angular acceleration, which will either increase or decrease its angular velocity at a steady rate.

Since I is defined as $\sum_i m_i r_i^2$ then the further away from the axis the mass is, the greater will be the moment of inertia. This is illustrated in F1.15, which shows a hoop of mass M and a uniform disc of the **same mass**. Because the distribution of mass of the disc is closer to the axis than the mass of the hoop, then the moment of inertia of the disc must be smaller than that of the hoop. In fact, the hoop's moment of inertia is simply given by MR^2 because each part of the hoop's mass (M) is at a distance R from the axis.

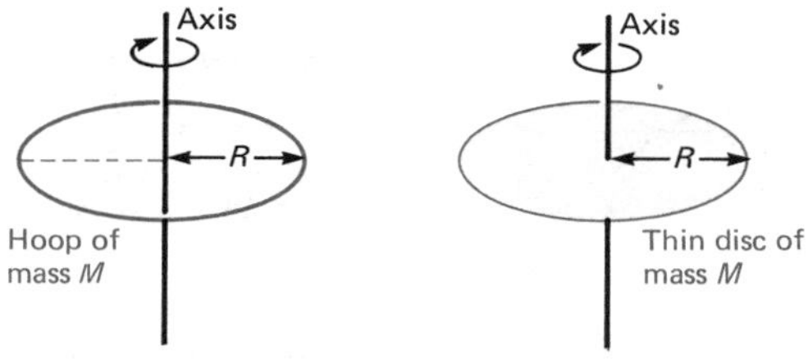

F1.15 Moments of inertia: hoops and discs

Angular momentum of a single particle that rotates about a given point is defined as momentum (mv) × radius of rotation (r). Since $v = \omega r$, then the angular momentum is equal to $m\omega r^2$. From this expression, the angular momentum of a rigid rotating body can be written as $\sum_i m_i \omega r_i^2$ or $\sum_i (m_i r_i^2)\omega$ which then gives:

E1.14 angular momentum $= I\omega$

where I = moment of inertia (kg m^2), ω = angular speed (rad s^{-1}).

The equation $T = I\alpha$ can be written $T = \frac{d}{dt}(I\omega)$ since $\alpha = \frac{d\omega}{dt}$, so that the link between torque and angular momentum can be established as:

torque = rate of change of angular momentum.

If there is no torque acting upon a rigid rotating body, then its angular momentum (and its angular velocity) does not change.

The same rule (ie, constant angular momentum if no resultant torque acts) applies to a system of rotating bodies in a more subtle form, for even though the total angular momentum must remain the same, one body can lose angular momentum to another body of the system. Thus one part might speed up at the expense of other parts, or a reduction of I for the whole system will cause an increase of ω (but $I\omega$ must remain the same for the whole system). This generalization of the rule from a single rigid body to a system of rotating bodies is known as the **principle of conservation of angular momentum**, and is usually stated in the form: if no resultant couple acts upon a system, then its total angular momentum remains constant.

When an object is in rotational motion, it possesses ke on account of that motion. For a rigid body, since the speed of each point mass of the body is given by ωr_i, then the ke of each point mass is $\frac{1}{2}m_i\omega^2 r_i^2$, so that the total rotational ke is given by $\frac{1}{2}\sum_i (m_i r_i^2)\omega^2$ or $\frac{1}{2}I\omega^2$ since $I = \sum_i m_i r_i^2$.

E1.15 Rotational ke $= \frac{1}{2}I\omega^2$

where I = moment of inertia (kg m^2), ω = angular speed (rad s^{-1}).

An example of an energy transformation involving rotational ke is provided by a cylinder rolling down a slope. Suppose the cylinder is released from rest at the top of the slope; let v and ω represent, respectively, its speed and angular speed at the base of the slope.

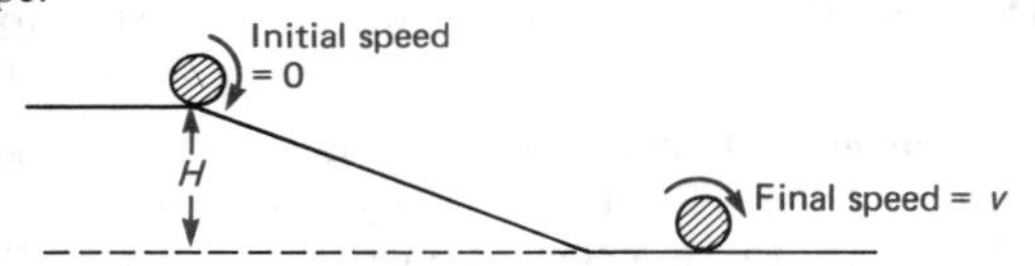

F1.16 Cylinder rolling down a slope

From F1.16, the loss of pe (mgh) = gain of ke, so that $mgh = \frac{1}{2}mv^2 + \frac{1}{2}I\omega^2$, and using $v = \omega R$, then the speed (v) at the base of the slope can be calculated (given all other values). The final ke is shared between ke of rotation ($= \frac{1}{2}I\omega^2$) and ke of translation ($= \frac{1}{2}mv^2$). For the cylinder in the above example, $I = \frac{1}{2}mr^2$ (assuming solid rather than hollow) so the ke of rotation can be shown to be equal to $\frac{1}{4}mv^2$. Hence the final ke is shared between rotation and translation in the ratio 1:2.

In order to change the rotational ke of a rigid body, a couple must be applied; it can be shown that the **work done by a couple** = **moment of couple** × **angular displacement**, as shown by the simple arrangement of F1.17. Thus, the work done by a couple = change of rotational ke.

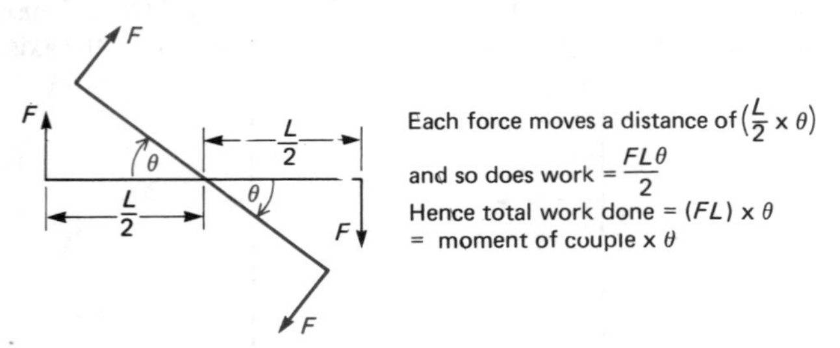

F1.17 Work done by a couple

1.9 COMPARISON OF LINEAR AND ANGULAR MOTION

A useful comparison can be made by considering:

A mass m moving along a straight line		A rigid body rotating about a fixed axis
1 Displacement (s)	is equivalent to	Angular displacement (θ)
2 Velocity (v)		Angular velocity (ω)
3 Acceleration (a)		Angular acceleration (α)
4 Mass (m)		Moment of Inertia (I)
5 Momentum (mv)		Angular momentum ($I\omega$)
6 Force (F)		Torque (T)
7 $F = ma$		$T = I\alpha$
8 ke $= \frac{1}{2}mv^2$		ke $= \frac{1}{2}I\omega^2$
9 Work $= Fs$		Work $= T\theta$

The above comparisons can be extended to the equations of dynamics; for example the equation $v = u + at$ (E1.1) can be 'translated' into angular motion as $\omega = \omega_0 + \alpha t$, where ω and ω_0 are the initial and final values of angular velocity.

1.10 BEYOND NEWTON'S LAWS

Classical mechanics, sometimes called Newtonian mechanics, is based upon Newton's laws as applied to particles. All the preceding topics are based upon classical mechanics. However, there are many situations in which Newton's laws are inadequate, and new theories are needed. This section outlines some of these situations met in your studies.

Quantum theory Quantities in classical mechanics can take any value (eg, any value of ke may be obtained by putting suitable values of m and v into $\frac{1}{2}mv^2$). However, it was first suggested by Planck (in attempting to explain black body radiation curves; see p. 91), that in some systems, quantities could only take certain specific values. Such systems are said to be **quantized**, an individual value being called one quantum (plural, quanta). Planck proposed that an oscillator can only gain or lose energy by discrete amounts (quanta) given by:

E1.16 $E = hf$

where E = one quantum of oscillator energy (J), f = oscillator frequency (Hz), h = Planck's constant (J s).

This formula started a whole new branch of physics called **quantum mechanics** and touches your studies through radiation, photoelectricity (see p. 124), energy levels and spectra (see p. 125) and heat distribution in solids (see p. 94). Other quantum effects include Millikan's discovery that electric charge is quantized (in units of e, the charge of an electron). Beyond your present studies, quantum mechanics opens up a range of theories and equations far more complex and wide than classical mechanics.

Relativistic effects Newton's laws have to be modified for speeds approaching the speed of light (at speeds greater than 0.9 c is the usual criterion). Einstein developed the theory of special relativity and showed, amongst other things, that the two classical conservation laws for mass and energy need to be handled with care. This is because energy has mass according to this conversion formula:

E1.17 $E = mc^2$

where E = energy (J) equivalent to mass m, m = mass (kg) equivalent to energy E, c = speed of light in a vacuum (ms^{-1}).

In small scale events involving comparatively large energies the 'mass' of the energy involved becomes significant. In your studies, the main use of E1.17 is in explaining mass defects in nuclear physics (see p. 130) and the extraordinary quantities of energy that can be released by nuclear fission and fusion.

The starting point of special relativity is the experimental result that the speed of light (c) is the same for all observers (with constant relative velocities). Many important classical equations can be adapted to deal with relativistic effects (eg, the Doppler effect, see p. 54). The full consequences and theory of the principles of relativity lie outside the scope of your present studies.

Wave mechanics Classical mechanics is based upon the study of particles, each particle having non-zero mass. De Broglie showed that under the right experimental conditions all particles could be shown to behave like waves (and conversely all waves could be shown to behave like particles). He expressed the link by the following equation:

E1.18 $\lambda = h/mv$ or $\lambda = h/p$

where λ = wavelength (m), h = Planck's constant (J s), m = mass of particle (kg), v = velocity of particle (m s^{-1}), p = momentum associated with the wave (kg m s^{-1}).

This is a fundamental equation of wave mechanics, and is used to explain the particle properties of waves (eg, photon nature of light), the wave properties of electrons, and the impossibility of finding electrons in nuclei. This wave-particle duality, as expressed by **de Broglie's equation** is an important concept because we tend to explain effects involving sub-microscopic particles by large scale models (eg, billiard ball models for atomic collisions, wave model for electron diffraction). The concept of particles having wave properties is perhaps easier to accept than the reverse. F1.18 can help although it must be emphasized that it is intended as a 'thinking model' and **not** as an explanation of wave/particle duality.

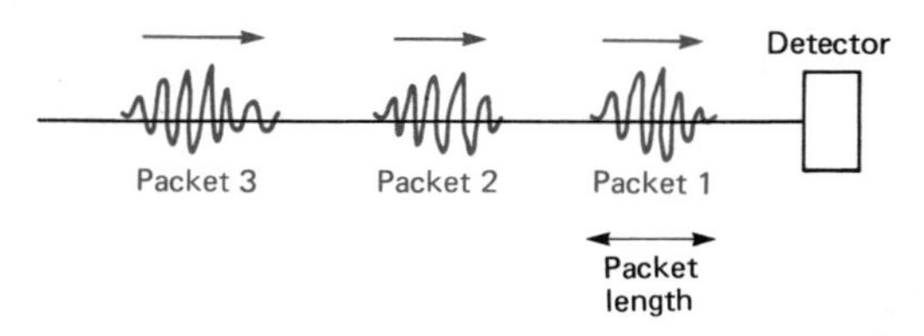

F1.18 Packets of wave – photons

In the diagram, the wave detector will receive three separate 'packets' of waves in turn; with very little energy arriving between the 'packets' the detector receives three separate impulses of wave energy, as if 'hit' by three particles; if the waves were 'quantized' into packets like this, then a wave/particle combined model makes sense. In electromagnetic radiation, the packets of waves are called **photons**, each with energy given by E1.16, and with packet lengths up to 0.4 m for light (see p. 64).

Statistical mechanics Classical physics studies individual particles or bodies. The study of systems containing many bodies often requires the mathematical laws of statistics to predict macroscopic (ie, large scale) behaviour. Where probability and chance are involved, the type of treatment is referred to as statistical mechanics. Examples in your studies may include kinetic theory of gases (see p. 91), heat flow and distribution of quanta (see p. 94) and radioactive decay (see p. 129). Normally a model of the system is set up that involves random behaviour; this results in many of the equations of statistical mechanics involving exponential functions.

Into the future Already theoretical physicists have found it helpful to combine these theories to probe beyond classical mechanics. For example, we find books on relativistic quantum mechanics. New concepts are being proposed all the time.

UNIT 1 QUESTIONS

1.1M The magnitude of the acceleration of a moving object is equal to the:

A gradient of a displacement – time graph.
B gradient of a velocity – time graph.
C area below a force – time graph.
D area below a displacement – time graph.
E area below a velocity – time graph.

(**London**: *all other Boards*)

See p. 17 if necessary.

1.2M A stone is thrown from P and follows a parabolic path. The highest point reached is T. The vertical component of acceleration of the stone:

A is zero at T.
B is greatest at T.
C is greatest at P.
D is the same at P as at T.
E decreases at a constant rate.

(**Cambridge**: *all other Boards*)

Assume air resistance is negligible and that the earth's gravitational field is uniform over the range of the flight path. See p. 17 if necessary.

1.3M Which of the graphs below correctly shows how the acceleration (a) and velocity of a perfectly elastic ball bouncing on a horizontal surface varies with time?

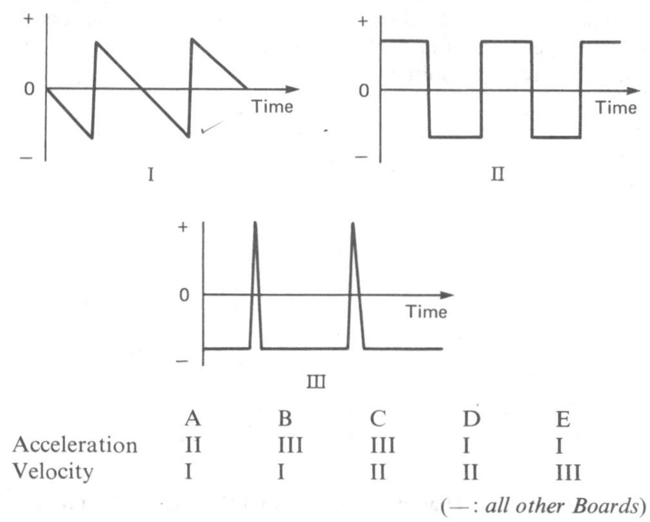

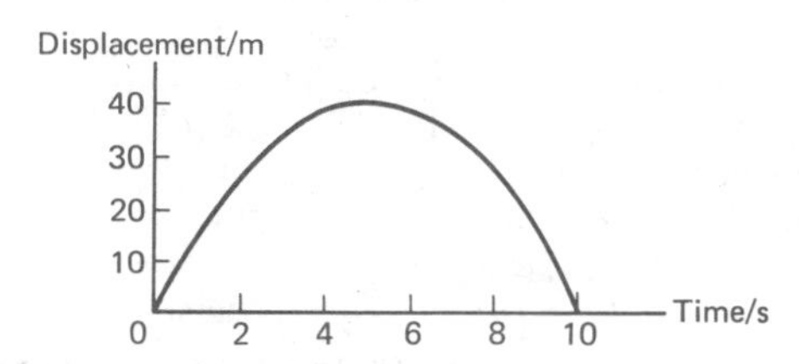

	A	B	C	D	E
Acceleration	II	III	III	I	I
Velocity	I	I	II	II	III

(—: *all other Boards*)

Remember that between impacts, the acceleration is due to gravity only. Also, the slope of the velocity graph gives the acceleration.

1.4M The displacement – time graph for a moving body is shown below.

Displacement/m
40
30
20
10
0 2 4 6 8 10
Time/s

Which of the following velocity – time graphs could represent the motion of the same body?

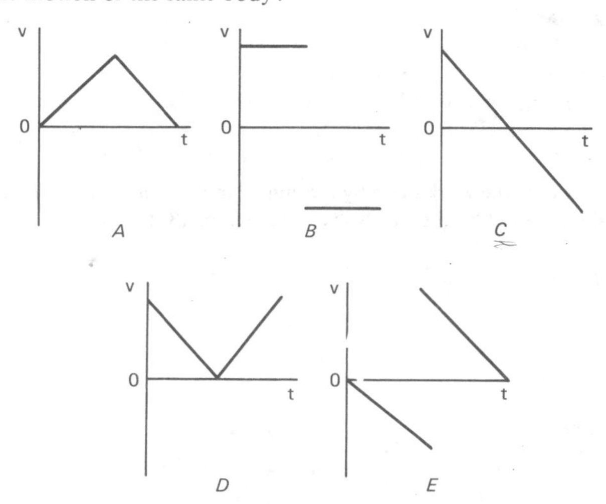

(**SEB H**: *all other Boards*)

Remember that the slope of a displacement–time graph gives the velocity. Choose your answer by considering the slope of the displacement–time graph at time = 0, 5 and 10 s thus giving approximate values of velocity at the beginning, the middle, and the last part of the motion.

1.5M A cricket ball and a tennis ball are simultaneously thrown vertically upwards with equal initial speeds. Ignoring air resistance which **one** of the following statements is correct?
A The two balls travel the same distance in the same time.
B The tennis ball rises the highest.
C The tennis ball is in flight the longest.
D The tennis ball has a smaller impact speed.
E The two balls have equal pe at maximum height.
(—: *all Boards*)

Remember that acceleration due to gravity (g) does not depend upon mass. Neglect air resistance, and if you consider E, don't forget that initial ke becomes pe at maximum height – do the balls have equal initial ke?

1.6M Force is applied to an object of mass 2 kg at rest on a friction-free horizontal surface as indicated on the graph.

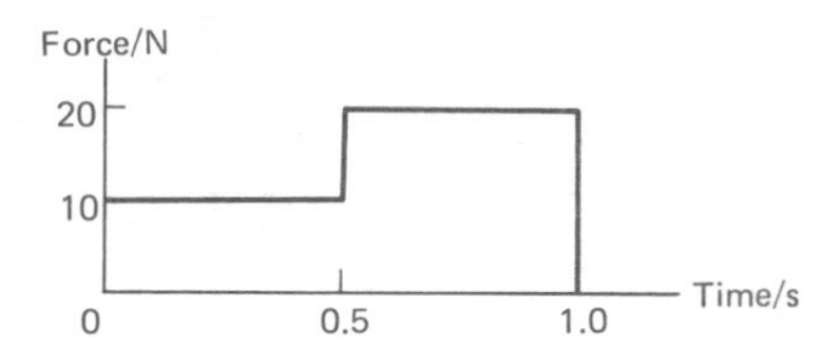

After 1 second, the speed of the object in m s^{-1} is:
A 7.5 B 12.5 C 15 D 25 E 30
(**SEB H**: *all other Boards*)

This question is based on the idea that the area under a force–time graph is equal to the change in momentum. The area under the graph is readily calculated, the mass of the object is known, and hence the speed (ie, change in speed from rest) can be calculated.

1.7M Four identical railway trucks, each of mass m, are coupled together and rest on a smooth horizontal track. A fifth truck of mass $2m$ and moving at 5 m s^{-1} collides and couples with the stationary trucks. After impact the speed of the trucks is v, where v equals:
A 5/6 m s^{-1}. B 1 m s^{-1}. C 5/4 m s^{-1}.
D 5/3 m s^{-1}. E 5/2 m s^{-1}.
(**London**: *all other Boards*)

Draw a diagram similar to F1.3 and then equate the initial momentum to the momentum after impact and solve to find v.

1.8M A car of mass m has an engine which can deliver power P. The minimum time in which the car can be accelerated from rest to a speed v is:
A mv/P. B P/mv. C $2P/mv^2$
D $mv^2/2P$. E $Pmv^2/2$.
(**NISEC**: *all other Boards*)

Equate the work done by the engine in t s to the kinetic energy gain acquired by the car. Solve for t. See p. 18 if necessary.

1.9M The diagram below shows a cord supporting a picture.

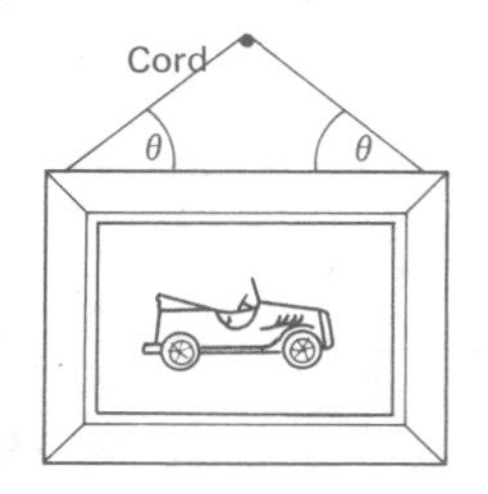

Which of the following graphs correctly represents the relationship between the tension in the cord, T, and the angle θ?

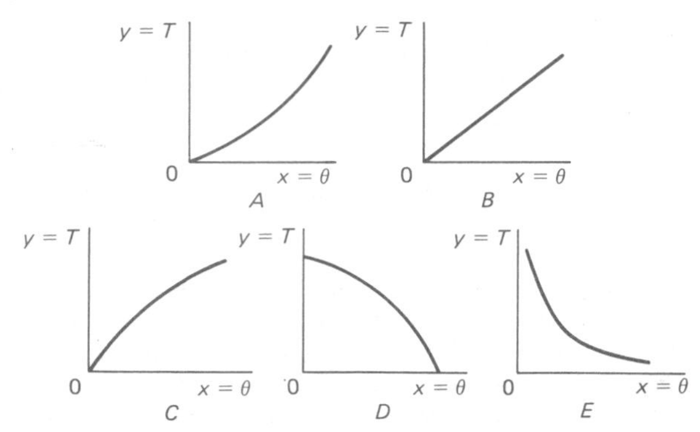

(**London**: *all other Boards*)

Draw the forces acting (T in each string, and mg through the centre of the picture). Resolve T into vertical and horizontal components. Obtain an equation for T by equating the vertical forces. Then consider the value of T when $\theta = 0$; only one graph satisfies this condition.

1.10M A 0.1 kg mass hanging from a light helical spring produces an equilibrium extension of 0.1 m. The mass is pulled vertically downwards by a distance of 0.02 m and then released. Taking g as 10 m s^{-2}, the equation relating displacement x of the mass from its equilibrium position and the time t after release is:
A $x/\text{m} = 0.1 \sin[10(t/\text{s})]$.
B $x/\text{m} = 0.1 \cos[0.2\pi(t/\text{s})]$.
C $x/\text{m} = 0.02 \sin[0.2\pi(t/\text{s})]$.
D $x/\text{m} = 0.02 \cos[0.1(t/\text{s})]$.
E $x/\text{m} = 0.02 \cos[10(t/\text{s})]$.
(**Cambridge**: *all other Boards except SEB H*)

The shm solution for a mass displaced from equilibrium and then released is discussed on p. 20. Remember that the amplitude is the maximum displacement from equilibrium. Also, ω must be calculated from the spring constant and the mass; see the comments before E1.8. To determine the spring constant, make use of the information in the first sentence of the question.

1.11M The following are quantities associated with a body performing shm:
1 The velocity of the body.
2 The accelerating force acting on the body.
3 The acceleration of the body.
Which of these quantities are exactly in phase with each other?
A None of these. B 1 and 2 only.
C 1 and 3 only. D 2 and 3 only.
E 1, 2 and 3.
(**NISEC**: *all other Boards except SEB H*)

See p. 20. The force producing motion is proportional to the acceleration, so acceleration and force must change together. When the acceleration is at its maximum what is the value of the velocity? When the acceleration is zero what is the velocity? Can a and v be in phase?

1.12M A bead, X, resting on a smooth horizontal surface, is connected to two identical springs and is made to oscillate to and fro along the line of the springs.

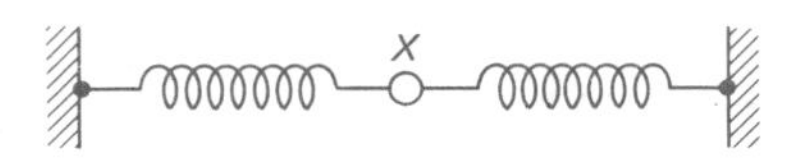

When the bead passes through the central position, its energy is:
A zero.
B mostly potential energy.
C all potential energy.
D half potential and half kinetic energy.
E all kinetic energy.
(**London**: *all other Boards*)

See p. 51 and F3.2.

1.13M For a simple pendulum undergoing shm with small oscillations, which of the following correctly describes the ke of the bob and the Tension of the thread supporting the bob at zero displacement (ie, passing through the equilibrium position)?

	A	B	C	D	E
ke =	Max	0	Max	0	Min
Tension =	Min	Max	Max	Min	0

(–: *all other Boards except SEB H*)

See F1.11 if necessary. Start by considering the ke at zero displacement – is the object moving fastest or slowest at this point? Thus choose from A – E on the basis of ke so eliminating some alternatives. Then consider the tension, and think of the 'strain' on the thread as the bob moves through a half cycle. Hence make your final choice.

1.14M A particle rotates clockwise in a horizontal circle of radius r with constant angular velocity ω as shown below. The particle is at S at time zero and at P at time t. Q represents the projection of point P on to the diameter through S. Measured with respect to the origin O, the displacement, linear velocity and linear acceleration of Q in the direction OS are y, v, and a, respectively.

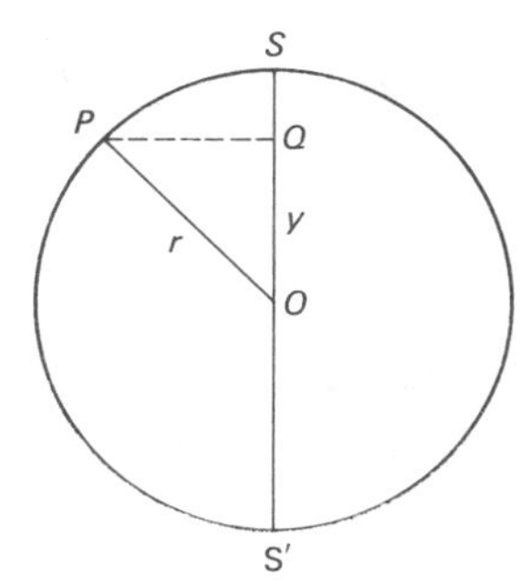

Which one of the following sets of expressions is correct?

A $y = r\cos\omega t$; $v = -r\omega\sin\omega t$; $a = r\omega^2\cos\omega t$.
B $y = r\cos\omega t$; $v = -r\omega\sin\omega t$; $a = -r\omega^2\cos\omega t$.
C $y = r\cos\omega t$; $v = -r\omega\cos\omega t$; $a = -r\omega^2\sin\omega t$.
D $y = r\sin\omega t$; $v = -r\omega\cos\omega t$; $a = -r\omega^2\sin\omega t$.
E $y = r\sin\omega t$; $v = r\omega\cos\omega t$; $a = -r\omega^2\sin\omega t$.

(**Cambridge**: *all other Boards except SEB H*)

Start by considering the y-value: at $t = 0$, the y-value is equal to r so you can narrow your choice down to **either** one of A, B, C **or** one of D, E. Then, remember that circular motion involves an equation linking centripetal acceleration and angular speed (E1.12) so that centripetal acceleration is always directed inwards. Thus the component of centripetal acceleration, a, is given by $a = -\omega^2 y$. Hence, make your final choice.

1.15M A flywheel of moment of inertia 0.5 kg m² is brought to rest in 10 s by a constant couple of 10 N m acting alone on the flywheel. What was the initial angular velocity of the flywheel, in rad s^{-1}?

A 0.5. B 2. C 50. D 100. E 200.

(—: *O and C*, JMB*, AEB, Cambridge, WJEC SEB, SYS*)

One possible approach if you forget the 'angular' equation is to think of the corresponding 'linear' situation; a mass of 0.5 kg brought to rest in 10 s by a constant force of 10 N – to find the initial speed, you would doubtless calculate the acceleration and then use the appropriate dynamics equation to find the initial speed. Solve this problem by using the same approach with the appropriate angular quantities and equations. See p. 22 if necessary.

1.16M In 1923, de Broglie suggested that an electron of momentum p has properties corresponding to a wave of wavelength λ. Which one of the following graphs correctly shows the relationship between λ and p?

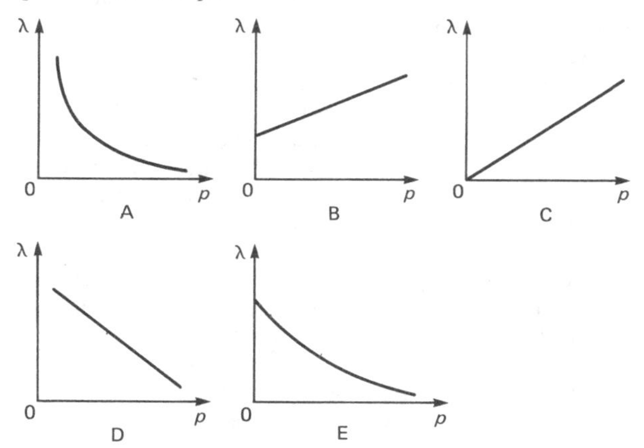

(**Cambridge:** *all other Boards except London and NISEC*)

See E1.18 if necessary.

1.17S A stone of mass 80 g is released (from rest) at the top of a vertical cliff. After falling for 3 s, it reaches the foot of the cliff, and penetrates 9 cm into the ground. What is **(a)** the height of the cliff, **(b)** the average force resisting penetration of the ground by the stone?

(**SUJB**: *all other Boards*)

Draw a simple sketch, and show the point where free-fall ends and penetration into the ground starts. For part **(a)**, use E1.3 with initial velocity of zero and constant acceleration of $g = 10$ m s^{-2}. The distance fallen from rest in 3 s is then equal to the cliff height (H).

For part **(b)**, you can **either** (i) calculate the impact velocity (v) using E1.1, then consider penetration with initial velocity v, final velocity = 0, and distance moved = 0.09 m. Calculate the acceleration (which will be – ve) using E1.4, then determine the force from $F = ma$; remember that m must be in kg **or** (ii) calculate the pe of the mass, taking ground level as zero pe, when the mass is at the cliff top. Then assume all the initial pe is converted into ke just before impact: the stone then loses all this ke in doing work burrowing into the ground; ie, work done = ke above, and since average force × penetration distance = work done, then the average force can be calculated.

1.18S A trolley of mass 0.80 kg is held in equilibrium between two fixed supports by identical springs (S_1 and S_2) as shown in the diagram; each spring has an extension of 0.10 m. In the second diagram, the trolley is shown moving to the right a distance of 0.05 m. The relation between the force (F) in newtons and the extension (x) in metres for each spring is given by $F = 20x$.

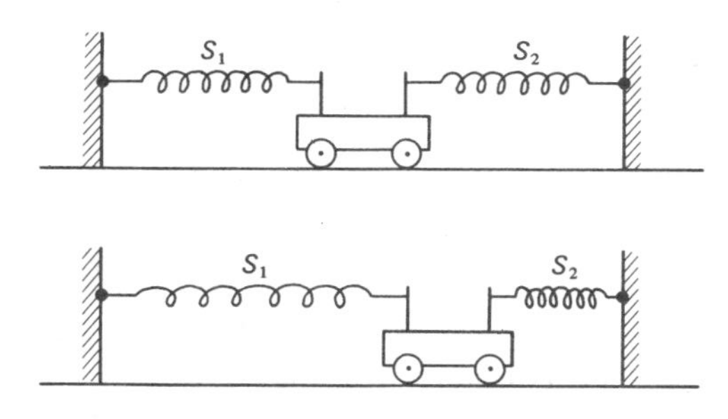

(a) What is the **change** in the force exerted by spring S_1 caused by moving the trolley to the right as in the second diagram?
(b) If the trolley is now released what will be the magnitudes of (i) the resultant force acting upon the trolley at the moment of release, and (ii) the initial acceleration of the trolley?
(c) Showing the steps in your calculation, determine the total energy stored in S_1 and S_2 when the springs are stretched (i) as in diagram 1, (ii) as in diagram 2.

(d) What is the kinetic energy of the trolley as it passes through the equilibrium position?

(**O and C Nuffield**: *all other Boards*)

(a) The **change** in the force F exerted by S_1 is wanted, corresponding to a **change** of extension of S_1 from 0.10 m to 0.15 m. Use the given equation $F = 20x$.

(b) When the trolley was in the centre, as in diagram 1, the force in S_1 was equal and opposite to the force in S_2. With the trolley moved to the side, S_1 exerts an extra force, as calculated above; S_2 exerts a reduced force, and the reduction in S_2 is equal to the increase of force in S_1. For (i), calculate the force in S_1 and then in S_2, then determine their resultant. For (ii), use E1.5.

(c) (i) Use energy stored = average force × extension; remember that the average force is $\frac{1}{2}$ × force value which gives that extension. (ii) The same method as above can be used, but remember that both force and extension for S_1 is greater than for S_2, so you will have to calculate energy stored in S_1 first, then repeat the method with different figures for S_2.

(d) At any point after release, ke + pe = initial pe; remember that the pe at equilibrium is non-zero, given by (c) (i) above.

1.19S A child's toy boat is built with two floats and is propelled by water draining out from a high tank through a hole, as indicated in the perspective and sectional drawings.

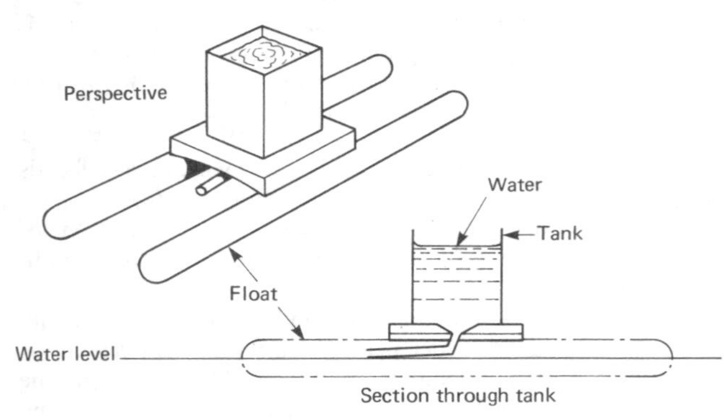

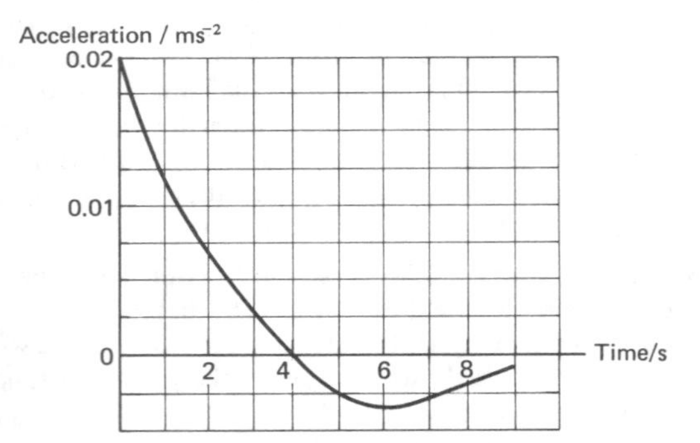

(a) The boat is initially at rest. Why does the water draining out of the tank cause it to accelerate?

(b) The graph shows how the boat's acceleration varies with time. (i) Find from the graph an approximate value for the maximum speed reached by the boat. Give your reasoning and show clearly how you arrive at your answer. (ii) It can be deduced from the graph that the tank takes at least six seconds to empty. How is this deduction made?

(**O and C Nuffield**: *all other Boards*)

(a) Acceleration requires force; force involves change of momentum. Relate the mass flow at the outlet to the change of momentum, and thus explain why a force acts to accelerate the boat.

(b) (i) Just as displacement is given by the area under a velocity–time graph, so velocity is given by the area under an acceleration–time graph; using the graph, you can for example check that the velocity after 1.0 s is given by 6 × 0.0025 m s^{-1} since there are 6 squares, each of area corresponding to velocity 0.0025 m s^{-1}, under the curve between time = 0 and time = 1.0 s. Remember that area under the time axis counts as − ve, so the maximum velocity is reached at 4.0 s. (ii) When the tank becomes empty, the 'thrust' will become zero, so that the boat will gradually be brought to rest by the viscous drag of the water. Consider how the velocity will change with time during this stage; make a simple sketch graph. Its gradient will then tell you how the acceleration will change. By comparing with the boat's acceleration graph, you should be able to answer the question.

1.20S A bullet of mass 0.020 kg is fired from a rifle. The barrel of the rifle is 0.50 m long with an internal diameter of 8.0 mm and the average excess pressure of the gas in the barrel is 5.0 × 10^3 atmospheres. Assuming that the recoil velocity of the rifle is negligible, and neglecting friction and any rotational energy acquired by the bullet,

(i) show that the average force on the bullet is 3.5 × 10^4 N.
(ii) calculate the acceleration of the bullet.
(iii) calculate the muzzle velocity of the bullet.
(Assume 1 atmosphere = 1.4 × 10^5 Pa.)

(**AEB** Nov 80: *all other Boards*)

(i) The excess pressure acts over the whole of the area of cross-section of the barrel. Calculate the area in m^2 and the excess pressure in Pa, then calculate the force.

(ii) Use Newton's second law. See p. 18 if necessary.

(iii) The muzzle velocity is the velocity with which the bullet emerges from the barrel. It can be calculated by using your value of acceleration and a distance equal to the length of the barrel. See E1.4 if necessary.

1.21S A cable-operated lift of total mass 500 kg moves upwards from rest in a vertical shaft. The graph shows how its velocity varies with time.

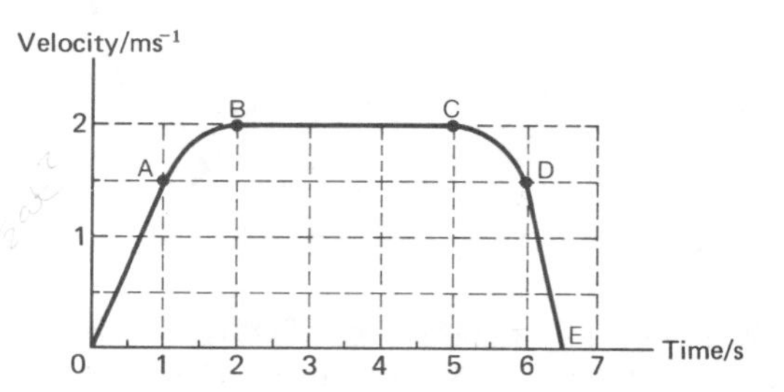

(a) For the period of time indicated by DE, determine (i) the distance travelled, (ii) the acceleration of the lift.

(b) Calculate the tension in the cable during the interval (i) $0A$, (ii) BC. Assume that the cable has negligible mass compared with that of the lift, and that friction between the lift and the shaft can be ignored.

(**JMB**: *all other Boards*)

(a) Use either graphical methods, as described on p. 17, or use the dynamics equations E1.2 and E1.1.

(b) (i) Express the resultant force (assume it acts upwards) in terms of the tension and the weight. Since the resultant force = mass × acceleration, then you ought to be able to calculate the tension, after calculating the acceleration from 0 to A from the graph. (ii) Start by calculating the acceleration from B to C. Then use the same method as in (i).

1.22S A particle rests on a horizontal platform which is moving vertically in simple harmonic motion with an amplitude of 50 mm. Above a certain frequency, the particle ceases to remain in contact with the platform throughout the motion.

(a) Find the lowest frequency at which this occurs.

(b) At this minimum frequency, at what point in the motion does contact cease? (Take the acceleration of free fall, g, as 10 m s^{-2}).

(**Cambridge**: *all other Boards except SEB H*)

Remember that for shm, the resultant force is always in proportion to the displacement from equilibrium; using the defining equation for shm, E1.6, the resultant force F can be written as

$F = -m\omega^2 x$ where ω is the angular frequency, m is the particle mass, and x is the displacement from equilibrium. When the mass first leaves the platform during shm, only the weight mg provides the restoring force since the support of the platform is non-existent at that point; with the sign convention of '+ is downwards', the restoring force is thus $+mg$ at that point.

By considering part **(b)** first, you should be able to write down the displacement in terms of the amplitude (x_0) when contact ceases. Remember the sign convention; $-$ is up. Then you should be able to deal with **(a)**.

1.23S The diagram shows a thread of liquid of length l and density ρ contained in a U-tube. The limbs of the U-tube, each of cross-sectional area A, are held vertically with the open ends upwards. The level of the liquid in one limb of the tube is depressed a small distance x and is then released. Show that the thread of liquid executes simple harmonic motion and derive an expression for its period of oscillation (neglect surface tension effects).

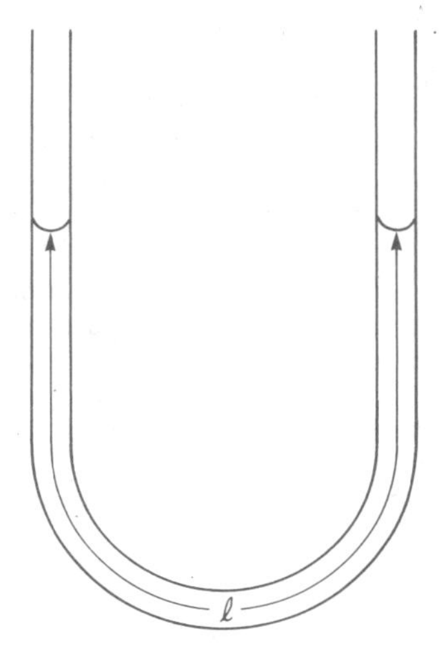

(**AEB** June 81: *all other Boards except SEB H*)

Remember that shm is defined as oscillating motion in which (i) the acceleration (a) is always proportional to the displacement (x) from equilibrium, (ii) the acceleration is always directed towards the equilibrium point. In other words, you must prove that the equation $a = -\omega^2 x$ (E1.6) applies in order to prove that the motion is shm (ω = angular frequency).

Derive expressions for the total mass being moved (in terms of A, ρ, l) and for the restoring force. In the latter case, remember that the restoring force is provided by the 'head' of liquid between the levels on each side; the expression for the restoring force is therefore in terms of A, ρ, x and g.

Then use Newton's second law in the form of $F = ma$, to give an expression for the acceleration (= restoring force/mass being moved) in terms of g, x and l; then, since $a = -\omega^2 x$ is the defining equation for shm, you should be able to prove that the motion is indeed shm and also be able to give an expression for ω in terms of x and l. Finally, using the equation on p. 21 that links time period T and ω, you should be able to prove that the time period is given by $T = 2\pi\sqrt{l/2g}$.

1.24S A man stands at the earth's equator. Find **(a)** his angular velocity, **(b)** his linear speed, **(c)** his acceleration, due to the rotation of the earth about its axis. (1 day = 8.6×10^4 s; radius of earth = 6.4×10^6 m).

(**Cambridge**: *all other Boards except SEB H*)

(a) Use E1.10. Remember to give your answer in rad s^{-1}.

(b) Use E1.11.

(c) Use E1.12 for centripetal acceleration, with the radius of rotation equal to the earth's radius and with the value of angular velocity as in (a).

1.25S A long but light string is attached to, and wrapped many times round, a stationary flywheel mounted with its axis horizontal. A mass of 20 kg is attached to the free end of the taut string, and is released from rest to fall 50 m down a pit shaft. When the mass arrives at the bottom of the pit with velocity 30 m s^{-1}, the flywheel is found to have acquired an angular velocity of 4π rad s^{-1}. Assuming no energy losses, find the moment of inertia of the flywheel about its axis.

(**WJEC**: *and O and C*, JMB*, AEB, Cambridge, SEB SYS*)

Given that there are no energy losses from the system, then you can best answer the question by considering the energy changes from pe of the mass to ke of the mass and flywheel. In this way, you should be able to make up an equation containing terms for, (i) pe loss of the 20 kg mass, (ii) ke gain of the 20 kg mass, (iii) ke gain of the flywheel, in terms of its moment of inertia I and its angular velocity ω; see E1.15 if necessary.

1.26S A man travelling in a lift notes that a mass suspended vertically from a light spiral spring within the lift is producing a steady extension of the spring which is 4/5 of that produced when the lift is stationary. What can you deduce about the motion of the lift? If the man's mass is 60 kg, what force is exerted by the lift upon the man? (**WJEC**: *all other Boards*)

Remember that the extension of the spring is in proportion to the tension; start by making a simple sketch showing the spring and mass when (i) the lift is stationary, (ii) the lift is moving such that the extension is as above in the question. Before considering the motion of the lift in detail, remember that only accelerated motion (either up or down) will give rise to a change of the spring's length from its length when the lift is stationary.

For (ii) above, indicate the forces on the mass on your sketch; in terms of these forces (tension and weight), obtain an expression for the resultant force assuming downwards acceleration. Use $F = ma$ then to link acceleration (a) with the tension and weight. If the lift accelerates downwards, you should see from your expression for acceleration that the tension is less than the weight.

For the force exerted by the lift on the man, you must first calculate the acceleration a in terms of g; since the extension is only 4/5 of its 'stationary' value, then the tension must only be 4/5 of its stationary value. Hence show that the acceleration $a = g/5$. The final step can be aided by a simple sketch showing the man in the lift, with the forces acting upon the man indicated. There are two forces, which are his weight Mg and the reaction R from the floor of the lift. Using the same idea as in paragraph 2 above, you ought to be able to calculate R, in terms of Mg and then determine its value in newtons.

1.27L This question is about measuring the acceleration, and so the velocity and displacement, of a moving vehicle, by making observations on masses carried within the vehicle.

The diagram shows the principle of one sort of accelerometer (device for measuring acceleration). A mass m is free to move horizontally within a case, but is restrained by springs fixed to the case. A pointer on the mass can move over a scale fixed to the case. When the case and mass are at rest, the pointer is opposite the zero mark on the scale. When the pointer shows a displacement, x, from zero, the net force exerted by the springs is kx.

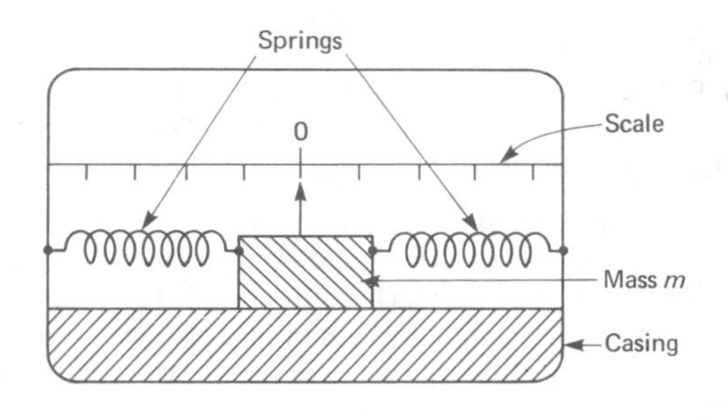

(a) Explain why, when the mass and case are moving at constant velocity in the horizontal direction, the pointer still reads zero. Assume that the velocity has been constant for a long time.

(b) If the casing is in a state of steady acceleration, a, to the left, explain carefully in words why the pointer now has a fixed displacement, saying whether the displacement is to the left or to the right and explaining why. Assume that the acceleration has been constant for a long time.

(c) Give an expression for the magnitude of the displacement in (b).

(d) If the casing were to be suddenly displaced from rest by a sharp blow from a hammer, for example, and then held at rest, describe the subsequent motion of the mass if there is a small amount of friction between it and the casing.

(e) It is suggested that, in use to measure varying accelerations, it would be a good idea to have zero friction between the mass and the casing. Argue briefly for or against this idea.

(f) Suppose that, in use, appreciable changes of acceleration are expected to occur over times not exceeding time t. Give an argument to help decide whether the period T of natural oscillations of the mass and springs should be large, or should be small, compared with t.

(g) In designing an accelerometer for use in a car, a period T of $\pi/5$ seconds was chosen and it was assumed that accelerations up to $2\,m\,s^{-2}$ should be measured. What would be the displacement at an acceleration of $2\,m\,s^{-2}$? (The values of m and k are not needed.)

(h) Suppose that it is decided that an accelerometer for use in a car accelerating at up to $2\,m\,s^{-2}$ should have a period of 2π seconds. What problems would arise in designing this accelerometer?

(**O and C Nuffield**: *all other Boards except SEB H*)

(a) Zero reading of the pointer on the scale means no resultant force on the mass. Consideration of Newton's first law (see p. 17) should enable you to explain.

(b) Steady acceleration of the mass to the left means that there must be a steady force on the mass to the left. Since the force is provided by the springs there must be a steady displacement from zero; since displacement one way involves a resultant force due to the springs in the other direction, you should be able to explain (i) the steady displacement, (ii) its direction.

(c) Since the net force is given as kx, you should obtain an expression for the magnitude of x in terms of m, k and a.

(d) The friction would cause the subsequent oscillations to be 'damped'. See p. 51 if necessary. With a small amount of friction, the damping is 'light'.

(e) With zero friction and steady acceleration, the mass would oscillate with shm freely, assuming acceleration from rest or constant velocity. The mean position of the pointer would be in proportion to the acceleration. However, here you must consider the further complication of **varying** accelerations, involving a varying 'mean' (ie, the midpoint between the extreme positions at either end varies).

(f) If the period T is large compared with t, then the pointer will not have reached its correct position by the time the next change of acceleration takes place. The speed of response to applied changes should be discussed in terms of first T large then T small compared with t.

(g) Use E1.6. The value of angular frequency ω is calculated from the given value of T.

(h) Repeat the procedure in (g) to obtain the displacement corresponding to a period of 2π seconds. The problems arise from the displacement value obtained.

1.28L State the principle of conservation of momentum. A particle of mass m moving with speed v makes a head-on collision with an identical particle which is initially at rest. How would you tell from the subsequent motion of the particles whether they had made (a) an elastic, (b) a completely inelastic, collision? In each case, work out how (if at all) the kinetic energy of the first particle, and the kinetic energy of the system as a whole, is affected by the collision.

The neutrons in a beam from a reactor have an average energy of 6.0×10^{-13}J. This is reduced to 6.0×10^{-21}J by causing the neutrons to make a series of collisions with carbon nuclei in a moderator. On average, the fractional loss of kinetic energy of a neutron at each collision in the moderator is 0.14. About how many collisions must a neutron make in this process?

(**Cambridge**: *all other Boards*)

See p. 18 for the principle of conservation of momentum.

A head-on collision means that both particles move along the same line after the collision as before. Remember that in an elastic collision between two particles, they 'rebound' off one another without loss of **total** kinetic energy; for a completely inelastic collision, there is no 'rebound' at all, and the two particles stick together. For (a), you need to state how you would tell no loss of total ke had taken place; state what you would measure in order to find out if total ke remains constant. To work out ke changes in each case, you should apply the principle of conservation of momentum to each in turn. In this way, you ought to be able to show for (a) that the particles 'exchange' velocities, ie, the target moves off at speed v and the initial particle halts; hence, after your working, comment on ke changes as required. For (b), application of the conservation of momentum gives the final speed of the 'joint' particles; then, give ke changes.

Consider a series of 'average' collisions for one neutron with initial energy as given. Each collision reduces its energy to 0.86 of its energy before the collision; hence, after two 'average' collisions, the energy becomes 0.86×0.86 of its energy before the first collision. Deal with the problem neatly by considering a neutron making n 'average' collisions; given the initial energy and the final energy after n collisions, you ought to be able to make up an equation for n.

1.29L

(a) Define 'simple harmonic motion'. Give three examples of systems which vibrate with approximately simple harmonic motion. How does the displacement of a simple harmonic oscillator vary with time? What is meant by 'the phase difference' between two simple harmonic motions of the same frequency? Illustrate your answer graphically, by considering the variation of the displacement with time of two motions vibrating with simple harmonic motion of the same frequency but which have phase differences of (i) 90°, and (ii) 180°.

(b) At what points in a simple harmonic motion are (i) the acceleration, (ii) the kinetic energy, and (iii) the potential energy of the system each at (1) a maximum, and (2) a minimum? Sketch graphs showing how (iv) the kinetic energy, (v) the potential energy, and (vi) the sum of the kinetic and potential energies for a simple harmonic oscillator each vary with displacement.

(c) Calculate the period of oscillation of a pendulum of length 1.8 m, with a bob of mass 2.2 kg. What assumption is made in this calculation? If the bob of the pendulum is pulled aside a horizontal distance of 20 cm and then released, what will be the values of (i) the kinetic energy, and (ii) the velocity of the bob at the lowest point of its swing?

(**London**: *all other Boards except SEB H*)

(a) See p. 19 and F1.10. Approximate shm means a system in which there are frictional forces so giving lightly damped oscillations. For the meaning of phase difference, give a brief written explanation followed by the two graphs. See p. 50 if necessary.

(b) (i), (ii) and (iii) See F3.2. Give your answers in terms of the displacement, eg, at the point of extreme displacement. (iv), (v), (vi) see p. 51. Remember that the total energy remains constant, and that the pe is proportional to (displacement)^2.

(c) See p. 20. Use $g = 10\,m\,s^{-2}$. The assumption is to do with the approximation for sin A. For (i) and (ii), draw a simple sketch showing the pendulum at its lowest point and at the moment of release. Then calculate the vertical displacement of the bob between the two positions using Pythagoras' theorem. You can then calculate the pe difference of the bob between the two positions. The energy changes as outlined in **(b)** will then enable you to determine the ke and hence the velocity at its lowest point.

1.30L

(a) (i) Write an expression for the centripetal acceleration of a particle of mass m moving in a circle of radius r with a uniform speed v. (ii) Show that the centripetal force F can be written in the form $F = \frac{4\pi^2 mr}{T^2}$ where T is the period of rotation.

(b) Describe how you would verify experimentally the relationship between the centripetal force F and the period T.

(c) A car of mass 800 kg moves with a uniform speed of 15 m s^{-1} in a circular path of radius 100 m on a horizontal surface. (i) What is the magnitude and direction of the centripetal force acting on the car? (ii) How is this centripetal force produced?

(d) By considering the forces acting on a car when it moves round a curve, explain, with the aid of a diagram, why roads are banked at corners.

(**SEB SYS**: *all other Boards except SEB H*)

(a) (i) See E1.12. (ii) Use your expression for (i) and the fact that v = circumference/time period.

(b) See your textbook for a suitable experiment. Your description will be aided by a sketch, and you should indicate how the centripetal force is supplied and measured, how T is measured (time over 20 cycles, etc) and how r is measured. In each case, state what instruments are used. You must indicate how you would ensure r is the same for each force value. With several pairs of readings of F and T, at constant r, a graph of $y = F$ against $x = 1/T^2$ will give a straight line passing through the origin.

(c) Use E1.12 to obtain the centripetal acceleration. Remember that the centripetal force acts towards the centre of the circle; this force must be supplied by frictional forces between the tyres and the road. To help understand the nature of these forces it should be noted that if they did not exist the car would travel in a straight line.

(d) Without banking all the centripetal force has to be supplied by the friction force between tyres and road and if the car travels too quickly... However, on a banked road the centripetal force can be supplied entirely by the horizontal component of the normal reaction, provided the speed is at the 'right' value. At this speed, there will not be any sideways friction; give a diagram to aid your explanation, showing how the weight and normal reaction combine to give the centripetal force.

1.31L

(a) Describe an experiment to determine the moment of inertia of a flywheel about its axis. The account should include details of the measurements taken, explain how the effects of friction are allowed for, and show how the result is calculated.

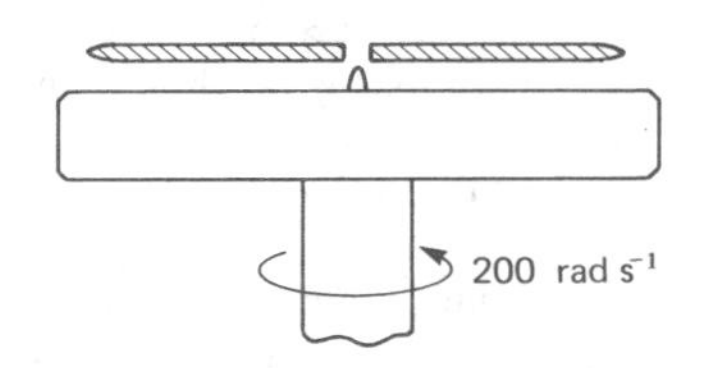

(b) A flywheel is driven about a vertical axis at a constant angular speed of 200 rad s^{-1}. A stationary metal disc with a hole through its centre is dropped onto the rotating flywheel, as shown in the diagram. The friction between the disc and the flywheel causes the disc to accelerate uniformly to an angular speed of 200 rad s^{-1} in a time of 0.5 s. The moment of inertia of the disc about an axis through its centre and perpendicular to its plane is 6×10^{-3} kg m^2. Find (i) The couple acting on the disc, (ii) the ke gained by the disc, (iii) the total energy transferred by the couple. Explain why the answers to (ii) and (iii) are not equal.

(**Oxford:** *O and C*, JMB*, AEB, Cambridge, WJEC, SEB SYS*)

(a) See your textbook.

(b) (i) Calculate the angular acceleration of the disc, then use E1.13 to calculate the couple which causes that acceleration. (ii) Use E1.15. (iii) Calculate the total angle, in radians, which the **turntable** moves through in 0.5 s. Then use the equation; work done = couple × angle moved through. See p. 22 if necessary.

There is an equal and opposite couple acting on the turntable due to the disc; work done by this couple will produce heat energy. Give your answer by discussing these points.

1.32L The motion of a body is said to be 'simple harmonic' if the acceleration of the body is directly proportional to its distance from a fixed point and is always directed towards that point.

(a) Show that the vertical motion of a mass on the end of a helical spring is a case of simple harmonic motion.

(b) A body of mass 0.20 kg hangs from the end of a helical spring, the axis of which is vertical, as shown in the sketch. When the body is raised 0.10 m above its equilibrium position and is then released, it executes shm with a period of 1.0 s. Calculate (i) the force constant of the spring (ie, the force required to produce unit extension); (ii) the speed of the body as it passes through the equilibrium position; (iii) the maximum kinetic energy of the body; (iv) the magnitude and direction of the acceleration of the body at the lower extremity of its first oscillation; (v) the maximum value of the upward force which the spring exerts on the mass. Find also the distance by which the equilibrium position is lowered if the mass is increased to 0.30 kg. (Acceleration of free fall $g = 9.8$ m s^{-2}).

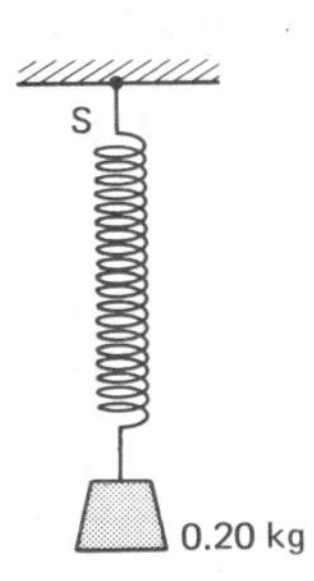

(c) Oscillations of a mass and spring system, as in (b), could be induced if the support S is caused to vibrate vertically through a very small constant distance. For a given frequency of the driving force the amplitude of the resulting oscillations will attain a steady value after some time. (i) Sketch a graph to show how this amplitude may be expected to change if the frequency of vibration of S is varied over a wide range. (ii) On the same axes sketch a second graph to indicate the results if the mass had been immersed in a liquid. (iii) Explain carefully the reasons for the shapes of the graphs.

(**NISEC**: *all other Boards except SEB H*)

(a) See p. 20.

(b) (i) Use E1.8. (ii) and (iii) Use E1.9 to calculate speed v, then calculate ke. (iv) Use E1.6 with ω calculated from T. (v) When the mass is at its lowest position, the restoring force (ie, tension − weight) is equal to the mass × acceleration. Hence calculate tension.

(c) (i) This is a resonance situation. See p. 51 and your textbook. (ii) Increased damping will give a less sharp resonance curve. (iii) See p. 51. Give your answer in descriptive terms by explaining why the amplitude builds up at the resonant frequency.

1.33L

(a) A steady stream of balls each of mass 0.2 kg hits a vertical wall at right angles. If the speed of the balls is 15 m s^{-1} and 600 hit the wall in 12 s and rebound at the same speed, what is the average force acting on the wall? Sketch a graph to show how the actual force on the wall varies with time over a period of 0.10 s.

Explain how the average force on the wall could be obtained from this graph. Explain briefly how the above problem can lead to an understanding of how a gas exerts a pressure on the walls of its container.

(b) A car travelling along a level road at a speed of $10\,m\,s^{-1}$ crashes head-on into a wall. If the mass of the car is 1000 kg, calculate the kinetic energy and momentum of the car just before the collision.
If the impact time (the time taken for the car to come to rest) is 0.2 s, calculate the average force acting on the wall and explain why it is an average. Why is it advantageous for a passenger if the impact time is increased? Make a calculation to support your point. How in practice could car design be improved to achieve an increase of impact time?

(**London**: *all other Boards*)

(a) Use 'Force = change of momentum per second', remembering that the **velocity** will change by $2v$ as it is a **vector** quantity. As the stream is regular, you can work out how many balls hit the wall in 0.1 s, assuming the force acts for a very short time interval. You do not have sufficient information to find out the force of each impact, but the average force will be the sum of all the impacts in 0.1 s, divided by 0.1 s (what does 'sum of' mean for a graph?). Compare the above situation with the kinetic theory of gases; see p. 92.

(b) Simply substitute the data into the basic equations for kinetic energy and momentum, remembering to give the units of each. Think what you would **see** and **hear** when the car crashes!

You have already found the momentum, so the average force will be the change in momentum (ie, to zero) divided by the time taken. Strictly speaking, force is the rate of change of momentum, but in this case the momentum changes over a finite time so only an average can be found. What will happen to the magnitude of the force of impact if the momentum can somehow be caused to change over a longer time interval?

1.34L State Newton's laws of motion and outline an experimental method of verifying the second law.
A bead of mass 5 kg is free to slide along a perfectly smooth rigid wire which lies in a vertical plane. The wire is curved, and when the bead is released from rest under the influence of gravity, its speed (v) varies with time (t) as shown in the graph.

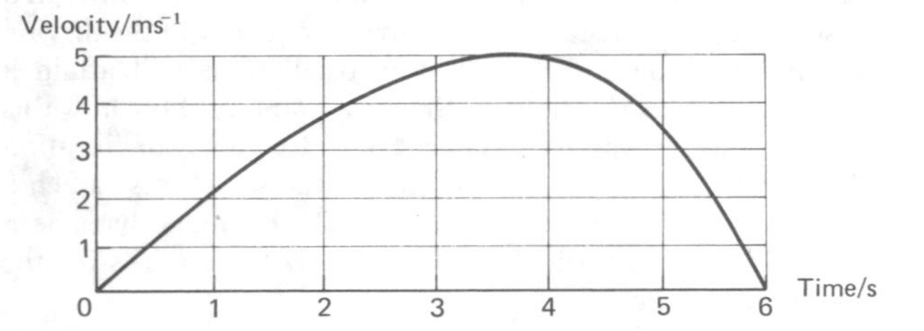

(a) What physical quantities are represented by (i) the total area under the graph, (ii) the slope of the graph at a given point?

(b) At what approximate angles to the vertical is the bead moving at times $t = 0$, $t = 3.75$ s and $t = 6$ s?

(c) What is the difference between the potential energy of the bead at its highest and lowest points?

(d) Give a rough sketch of the shape of the wire, indicating the point from which the bead is released, and its position at times $t = 3.75$ s and $t = 6$ s.

(**O and C**: *all other Boards*)

For Newton's laws of motion, see p. 17 if necessary. Refer to your textbook for experimental verification.

(a) The graph is a speed – time graph. See p. 17.

(b) Since the only forces on the bead are its weight and the normal reaction from the wire, then consideration of the sketch below will show you that the force which accelerates the bead is $mg \cos A$ where m is the bead mass and A its direction to the vertical. Thus, its acceleration is $g \cos A$; since acceleration can be found from the graph, then angle A can be found. Assume $g = 9.8\,m\,s^{-2}$.

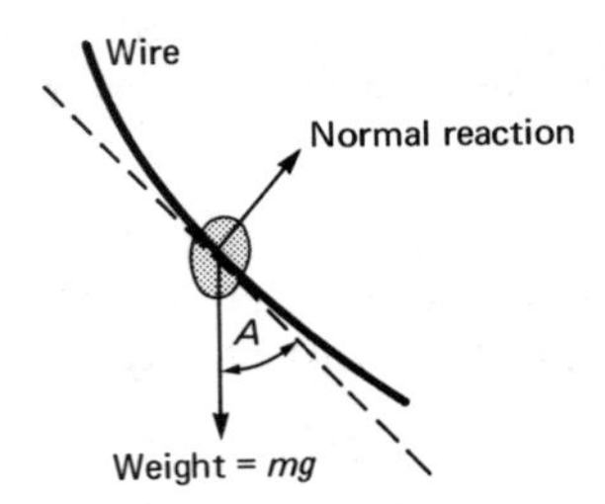

(c) At the highest point its speed is zero. At the lowest point, its speed is greatest. Calculate the pe difference from the ke difference since no friction acts.

(d) From **(b)**, you should appreciate that the wire is steepest at the final part. The lowest point is where the speed is greatest. At $t = 6$ s, the bead should be shown having regained its initial height since its speed is then zero (ie, all its initial pe at $t = 0$ has been recovered as pe at $t = 6$ s).

1.35L This question is about the ideas and the evidence in the wave theory of atoms. The passage below consists of numbered statements **(i)** to **(v)**. For each of these you are asked to give arguments that 'explain' and 'support' them. The arguments that you use may be of many kinds and you should indicate what kind they are – eg, theories, models, calculations, evidence.
Passage **(i)** It can be shown that electrons have wave properties, having a wavelength related to their momentum by $mv = h/\lambda$.
(ii) If a wavelike electron is confined in a 'box' of size 10^{-10} m, its momentum can't be less than a certain size, and so its kinetic energy has a lower limit, too. The smaller the box, the bigger the ke of the electrons.
(iii) If the atom is to be stable, the sum of the pe and ke of the electron must be negative. We know that the electrical potential energy of an electron is -10eV when it is about 10^{-10} m from a proton. All this leads to an explanation of why atoms cannot be much smaller than about 10^{-10} m.
(iv) Consideration of its spectrum shows that a hydrogen atom has a whole series of energy levels with energies given by c/n^2, where c is a constant and n has values 0, 1, 2, 3, . . . etc.
(v) The simplest idea is that electrons in hydrogen behave like standing waves on a string of length r as in the diagram. This would explain why there are discrete values of kinetic energy associated with integers n, but it would give the wrong rule for the way energy depends upon n.

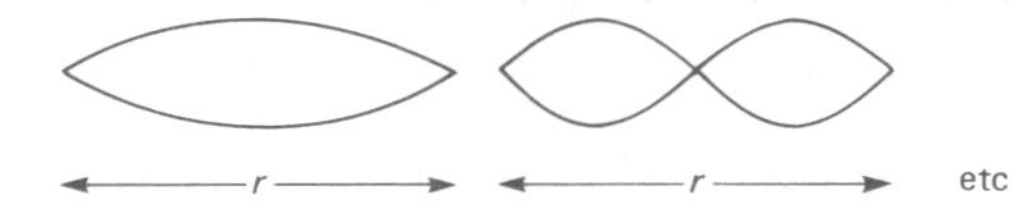

(**O and C Nuffield** *only*)

(i) To explain the statement, see p. 23. Support for the statement will be found in the account of electron diffraction experiments in your textbook; you should indicate how the momentum is calculated and how the electron wavelength may be determined from the 'diffraction angle' and the known value of crystal plane spacing.

(ii) The key point is that the smaller the box, then the smaller is the electron wavelength. The given equation $mv = h/\lambda$ then enables momentum (and then ke) to be brought in. Assume ke ($= \frac{1}{2}mv^2$) is proportional to (momentum)^2.

(iii) For a stable atom, the electron(s) cannot escape. Use E2.3 to support the statement that the pe $= -10$ eV at 10^{-10} m between an electron and a proton. Remember that

$$1\text{eV} = 1.6 \times 10^{-19}\text{J}.$$

To explain why atoms cannot be much smaller than 10^{-10} m approximately, use your arguments in **(ii)** to estimate the ke of an electron with wavelength 10^{-10} m. Then, repeat for a box of size 10^{-11} m; ke will be × 100 whereas pe will only be × 10, compared with values for 10^{-10} m. It should be clear that ke + pe will become + ve if the box becomes too small.

(iv) First of all, explain the link between energy levels and line (emission) spectra. Then, by reference to an energy 'ladder' for hydrogen, explain why spectrum frequencies are given by $f = \text{constant} \times (\frac{1}{n_1^2} - \frac{1}{n_2^2})$ where n_1 and n_2 are +ve integers.

(v) Use a general expression for possible wavelengths of standing waves on a string of length r; see p. 55. Then use the given equation $mv \doteq h/\lambda$ to obtain an expression for momentum, and hence ke. In this way, show that the ke depends upon integer n as given by ke = constant × n^2/r^2. The final point, that the energy (= ke + pe) has a different dependence upon integer n, is because pe is given by − constant/r. It can be shown that the minimum energy is given by r = constant × n^2. Hence, the energy level formula E = (ke + pe) = − constant/n^2 follows.

1.36L

(a) (i) Define the 'moment of inertia' of a rigid body about a fixed axis. (ii) A body has mass M and moment of inertia I about an axis through its centre of gravity. The centre of gravity of the body moves with linear velocity v and the body rotates with angular velocity ω about the axis through the centre of gravity. Give an expression for the total kinetic energy of the body.

(b) A solid cylinder of radius 0.12 m and mass 3.0 kg starts from rest and rolls without slipping through a distance of 0.70 m down a slope inclined at an angle of 30° to the horizontal. Neglecting rolling friction, calculate (i) the total kinetic energy of the cylinder, (ii) the linear velocity of the centre of gravity of the system, (iii) the angular velocity of the cylinder about its axis. The moment of inertia of a solid cylinder of mass M and radius a about its axis is $\frac{1}{2}Ma^2$.

(c) (i) Calculate the linear acceleration of the cylinder down the slope. (ii) If there were no friction and the cylinder slid without rolling what would then be the acceleration down the slope?

(**JMB***: *AEB, Cambridge, WJEC, O and C*, SEB SYS*)

(a) (i) See p. 21. (ii) See p. 22.

(b) Start by making a simple sketch to establish the situation in your mind. (i) Use ke gained = pe lost to calculate the total ke of the cylinder. (ii) Use the expression from (a)(ii), the given expression for I of a cylinder, and the fact that angular speed $\omega = v/r$ to obtain an expression for the total ke in terms of M and v. Then, use the value of ke deduced in (b)(i) to calculate the speed v. (iii) Use $\omega = v/r$.

(c) (i) Use E1.4 to calculate the acceleration, using the value of (b)(ii) for the final speed. Remember the distance travelled is 0.7 m. (ii) Without friction, the acceleration is given by (the weight component acting down the slope) / mass.

Unit 2 Fields and their effects

A 'field' is simply a 'thinking model' used by scientists to explain the behaviour of objects that can exert forces on one another when they are not in contact. The four fields are electric, magnetic, gravitational, and the field that exists between nucleons (neutrons and protons) when they are very close to one another. Theoretical physicists have attempted to connect all these Fields by producing 'unified field theories', but at the moment only the links between electric and magnetic fields are clearly defined. The term 'electromagnetic fields' is used to refer to the special case of changing electric and magnetic fields. The vital importance of a detailed study of fields can be illustrated by pointing out that all matter is 'held together' by field forces and that much of modern technology is based upon the applications of electric and magnetic fields.

Most GCSE courses introduce field concepts in a descriptive, non-mathematical fashion, and it is important that basic ideas and rules such as 'like charges repel' are well understood. At A level, the emphasis is upon a deeper understanding of fields, involving measurements, equations and calculations. A good understanding of the link between force and energy is very important for successful study of this Unit, so it may be helpful to look back at this link in Unit 1. Field concepts from this Unit are applied in almost all of the other units, most prominently in Unit 8 (Structure of the atom) in Unit 7 (Electricity and electronics) and in Unit 3 (Oscillations and waves). The topic of intermolecular fields is used in Unit 5 (Solids etc.) and in Unit 6 (Heat and gases).

2.1 FIELDS: THEIR NATURE, STRENGTH AND FORCES

Objects can exert forces upon each other without touching. Scientists call this 'field' behaviour, and talk about each of the objects producing a field; it is the interaction between the fields that creates the force. The three most important fields for A level are gravitational, electric, and magnetic fields; the objects that produce these fields are masses (gravitational fields), charges (electric fields), and objects carrying electric currents (magnetic fields).

A **field** is said to exist around one object if a second similar object placed into that field experiences a force due to the presence of the first object.

Forces between objects can be attractive or repulsive. In field physics, **attractive forces** are given negative values and repulsive forces positive values; eg, + 3N would represent a **repulsive force** and − 20N would represent fan attractive force.

The **strength of a field** is measured by determining the size of the force upon a 'test object' placed in the field.

The 'test object' is always 'one unit' of whatever produces that particular field; a **gravitational field** is tested with a unit mass (1 kg), an **electric field** is tested with a unit charge (1 C), and a **magnetic field** is tested with a wire 1 m long carrying an electric current of 1A. A further point only concerns magnetic fields when the direction of the field relative to the current carrying wire is important.

The **force exerted by a field** upon an object can be found from a knowledge of the strength of the field and the nature of the object placed in it. If a charge of + 3C were placed in an electric field of strength 4NC^{-1}, then the charge would experience a force of + 12N (a repulsive force indicated by the positive sign).

2.2 POTENTIAL AND POTENTIAL ENERGY IN FIELDS

It is a familiar experience that objects placed in fields can acquire energy; eg, two repelling magnets placed close together acquire ke as they spring apart. At first energy is used to place the objects into the fields; eg, energy has to be supplied to push the two repelling magnets so that they are close to one another. This energy is then stored in the field, the amount of energy stored depends upon the position of the object in the field (as well as upon the strength of the field and the nature of the object involved); eg, if the repelling magnets are pushed even closer together more energy has clearly been stored as work is done in pushing them closer together (and they 'spring' apart faster when released). Energy that is stored according to an object's position is referred to as 'potential energy'. There are other forms of pe apart from that stored in fields; eg, stored according to the length (extension or compression) of a spring.

pe is measured by comparing the potential energy of the object at its chosen position with its energy at a reference position. The reference position is usually chosen to be 'infinity' where the pe is defined to be zero.

There are cases when it is more convenient to choose a different reference position; eg, zero extension of a spring (rather than infinite extension!), and on the Earth, gravitational potential energy is often measured (using *mgh*) relative to the surface of the Earth.

The **potential** at a point in a field is the pe that a suitable 'test object' would have at that point.

The nature of the 'test objects' is explained in 2.1; eg, if a mass of 2 kg were found to have a gravitational pe of − 8J at a point in a field, then the gravitational potential at that point would be $-4\ \mathrm{J\,kg^{-1}}$ (which is the pe energy that the 'test object', 1 kg, would acquire at that point). In an electric field, the potential would be measured in joules per coulomb (or volts). The difference in potential between two points is referred to as the potential difference (pd) between them; this term can be applied to any type of field, not just to electric fields.

The **pe** of an object placed at a point in a field can be found from a knowledge of the potential at that point and the nature of the object placed there. Thus a charge of + 0.1C placed at a point of potential $+300\ \mathrm{V}(+300\ \mathrm{JC^{-1}})$ in an electric field will have a pe of +30 J. The sign is again important as it still indicates the nature of the force (repulsive in this case) and the direction in which an object will move if allowed to lose its pe converting it into ke.

It is important to note that as soon as a field exists it is possible to talk about the field strength or potential at any point; however, it is necessary to place a suitable object into the field before one can start talking about forces or potential energies.

There is one important general equation of field physics that links field strength and potential (or force and pe):

E2.1
$$E = -\frac{dV}{dr} \quad \text{or} \quad F = -\frac{d(pe)}{dr}$$

$$\text{So } V = -\int E\,dr \text{ or pe} = -\int F\,dr$$

where E = strength of field (various units), V = field potential (various units), F = force exerted by a field (N), pe = potential energy in a field (J), r = distance (m).

The distance r is normally measured to the centre of the object producing the field. The negative sign is needed to maintain the sign convention that negative forces are attractive. Beware of the trap of assuming that a field with a negative potential must always produce an attractive force on an object placed in that field. For example, a negative point charge creates a field with negative potential values, but another NEGATIVE charge placed in that field will experience a repulsive force.

At A-level the main use of E2.1 comes not in manipulating the calculus integrals and differentials but in being able to interpret graphs according to the equation. The negative value of the slope at any point of a graph of potential (V on y-axis) plotted against distance (r on x-axis) gives the strength of the field at that point (or the actual force is given by the slope of the pe − distance graph). If the area beneath a graph of field strength (E on y-axis) plotted against distance (r on x-axis) is measured between two points, then the pd between those points has been found (or the pe difference comes from a graph of force against distance).

2.3 GRAPHS AND DIAGRAMS OF FIELD STRENGTH AND POTENTIAL

F2.1 is a diagram of the electric field that exists between the parallel plates of a capacitor charged up to 400 V. The field is illustrated by the solid black lines which show the path of travel of a positive charge released from rest within the field. By definition, the arrows show the direction in which positive charge would move. The strength of the field is indicated by the distance between the **'field lines'**. As the lines are the same distance apart, this shows that the field strength is constant (a **uniform field**),

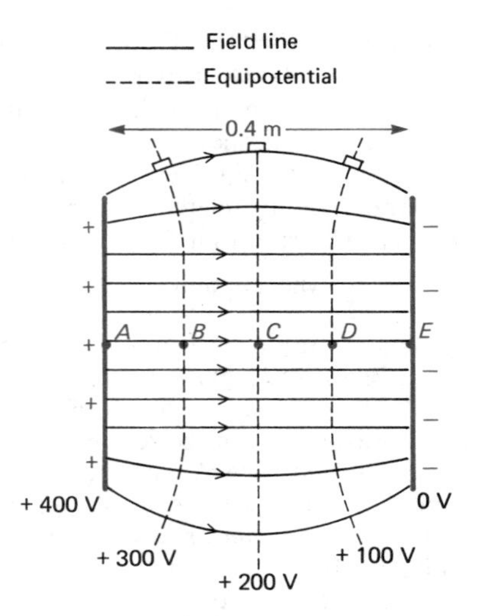

F2.1 Field lines and equipotentials for a parallel-plate capacitor

except at the edges where the lines are further apart and hence the field is weaker. A charge placed anywhere in the uniform field will experience a constant force, despite the fact that it may be put at points of different potential. It is possible to represent a field by lines (**equipotentials**) joining points in the field that are at the same potential; these are the dotted black lines. The equipotentials are always at right angles to the field lines; this makes it easy to draw an equipotential diagram if the field 'shape' is known. There are further ways of drawing field diagrams; eg, by using depth of shading to illustrate the strength of the field. In F2.1 path *ABCDE* (0.4 m long) has been marked and F2.2(a) and F2.2(b) show the variation of field strength (E_E) and potential (V_E) plotted along that path with the distance (r) measured from the positive capacitor plate.

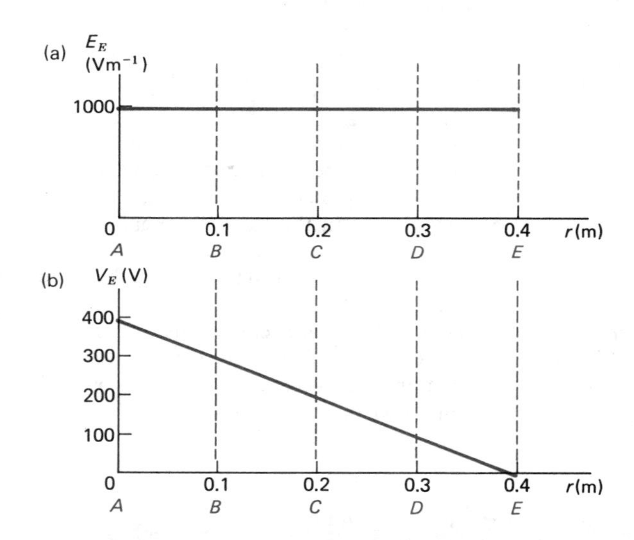

F2.2 (a) Field strength plotted against distance for F2.1.
(b) Potential plotted against distance for F2.1

In F2.2(a): slope, no great physical significance apart from the fact that zero slope means a uniform field; area, the negative value of the area under the graph between any two points measures the pd between them (E2.1); intercept, the field strength is $1000\ \mathrm{N\,C^{-1}}$ (or $1000\ \mathrm{V\,m^{-1}}$) E2.9.

In F2.2(b): slope, the negative value of the slope gives the field strength, which is a constant value ($1000\ \mathrm{V\,m^{-1}}$) as the graph is a straight line (E2.1); area, no great physical significance; intercepts, the right-hand plate has been chosen as the zero of potential (0 V), and the left-hand plate is at 400 V.

Note that measuring the slope of F2.2(b) and taking its negative value leads to F2.2(a), and measuring the area under the graph of F2.2(a) and taking its negative value leads to F2.2(b).

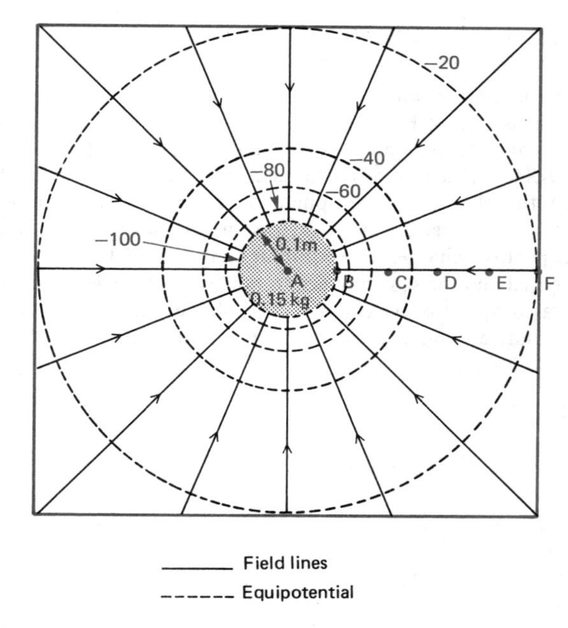

F2.3 Field lines and equipotentials around a sphere

F2.3 shows the field lines and equipotentials around a sphere of mass 0.15 kg due to its gravitational field. Note that the field lines get further apart the greater the distance from the centre of the sphere, indicating that the field strength is getting weaker. The equipotentials are marked in equal steps $(20, 40, 60, 80, 100 \times 10^{-12}\,\mathrm{J\,kg^{-1}})$ but they are not equal distances apart on the diagram (as they would be in a uniform field). They are of course still at right angles to the field lines. The arrows show the direction of the force on a mass placed into the field. The force is always attractive (unlike electric and magnetic fields when both attractive and repulsive forces are possible).

In F2.4(a): slope, no great physical significance; area, the area under the graph between any two points measures the gravitational pd between them (E2.1); intercepts, etc., the graph starts from the field strength at the surface of the sphere $(1000\,\mathrm{p\,N\,kg^{-1}})$ and asymptotically approaches zero field when $r \to \infty$.

In F2.4(b): slope, the negative value of the slope gives the field strength at any point (E2.1); area, no great physical significance; intercepts, etc., the graph starts from the surface of the sphere $(100\,\mathrm{p\,J\,kg^{-1}})$, asymptotically approaching zero potential as $r \to \infty$ (the usual definition of zero potential).

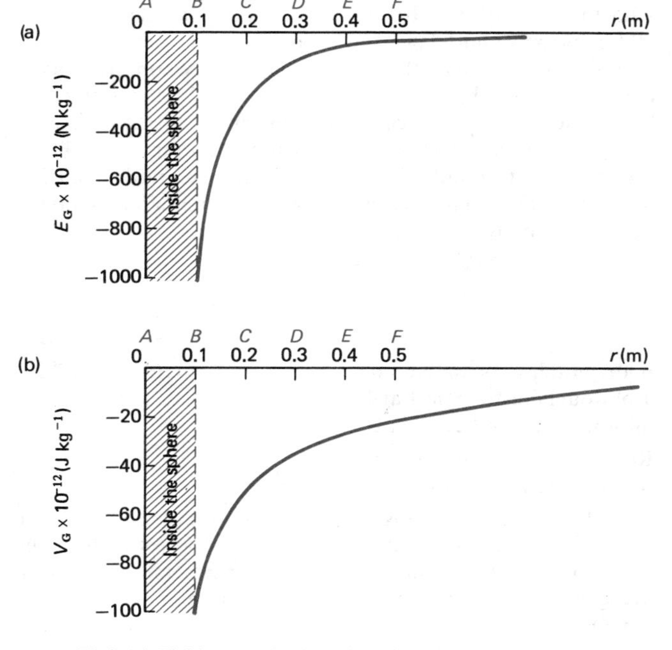

F2.4 (a) Field strength plotted against distance for F2.3; (b) potential plotted against distance for F2.3

In F2.3–4(b) the negative values of field and potential are important as they result in negative values for forces and potential energy (as mass values are always positive) correctly indicating an attractive force field. The largest value that the potential can have is zero (at $r = \infty$), as all negative values are less than zero! This is correct as to gain gravitational pe a mass has to be taken further away from the centre of the field; just think of a mass in the Earth's gravitational field gaining pe as it is taken further away from the surface of the Earth. Ultimately the mass will end up at infinity, when it will have its maximum pe; but if the value at infinity is only zero then all other values must be less than zero, ie, negative! Graphs F2.4(a) and F2.4(b) can be produced from E2.4 and E2.5, respectively.

2.4 INTERMOLECULAR FIELDS

In many systems an object may be influenced by more than one field; eg, a charged mass can experience forces due to both electric and gravitational fields. It is further found that a molecule can experience both short-range repulsive forces and long-range attractive forces when near to a single neighbouring molecule. Much can be learnt from studying the force–separation and pe–separation curves for two neighbouring molecules. These are shown as F2.5(a) and F2.5(b).

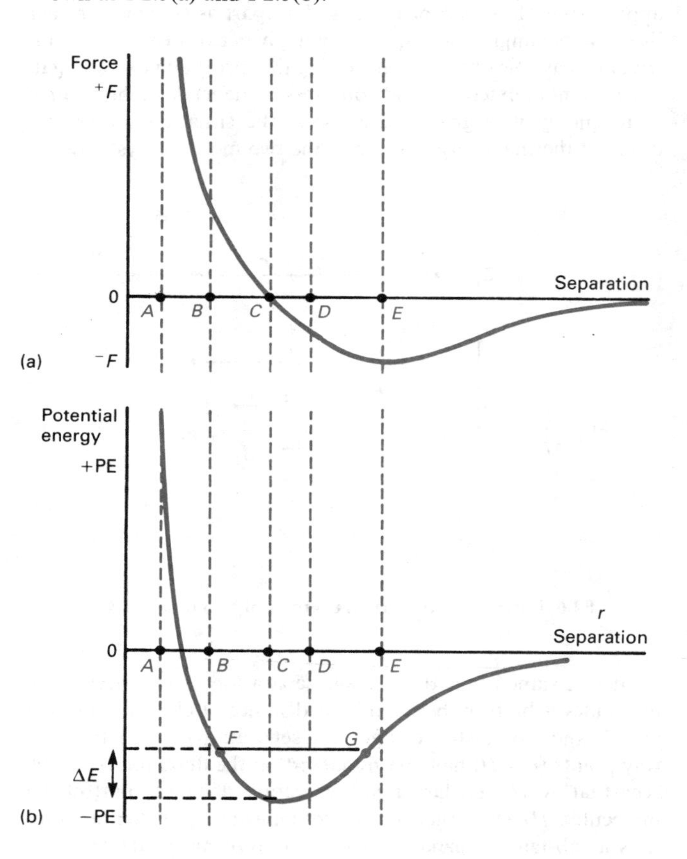

F2.5 (a) Force plotted against separation for two molecules; (b) Potential Energy plotted against separation for two molecules

In F2.5(a): slope, of no great physical significance except for the part between $r = B$ and $r = D$ where the graph is roughly a 'straight line' indicates Hooke's law behaviour; area, the area under the graph between any two points indicates the pe difference between them (E.1); intercepts, etc., there is zero force between the molecules at $r = \infty$, the molecules cannot get so close that $r = A$, equilibrium occurs when the force between the molecules is zero at $r = C$, the force is repulsive if $r < C$ and attractive if $r > C$, with maximum attractive force at $r = E$.

In F2.5(b): slope, the negative value of the slope at any separation is the force between the molecules at that separation (E2.1), if $r < C$ then the slope is always negative and the force positive (repulsive), if $r > C$ then the slope is always positive and the force negative (attractive). At $r = C$ the slope and force are zero, the pe is at a minimum and the two molecules are in equilibrium (no net force between them) and at $r = E$ the slope of the graph, although still positive, stops increasing and starts decreasing (point of inflexion) which means that the attractive force reaches its maximum value at this point; area, no great physical significance; intercepts, etc., pe is zero at $r = \infty$, molecules cannot get so close that $r = A$ (infinite energy required).

F2.5(a) and F2.5(b) only represent the forces between TWO neighbouring molecules and do not take account of the fact that a molecule may be influenced by many neighbours. However this simple microscopic model makes several useful predictions. The equilibrium separation occurs when $r = C$ (about 0.2 nm) and a pair of molecules with no thermal energy (ie, not vibrating and consequently at **absolute Zero**) should be this far apart. Beware of the mistake of confusing the two curves and saying that equilibrium occurs on the force curve at $r = E$ because it is a minimum here, or that it occurs between $r = A$ and $r = B$ on the pe curve because the pe is zero there! We can allow for **thermal energy** by giving the two molecules some extra energy ΔE as shown on F2.5(b). As in any vibrating system the energy is continually being converted from kinetic to potential and vice versa (see 3.2). The extra energy ΔE allows the molecules to approach as close as F or to get as far apart as G. This is rather like a mass hanging on a spring being given extra energy when it becomes capable of both compressing the spring and extending it, the amount of extension and compression depends on how much extra energy it is given. F2.6 shows the situation for several different thermal energies given to the two molecule system.

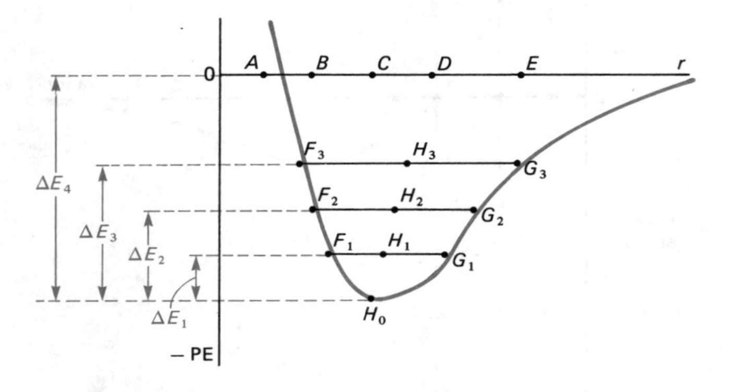

F2.6 Potential and thermal energies of two molecules

It is assumed that the average separation of two oscillating molecules is halfway between their distance of closest approach ($r = F$) and their distance of greatest separation ($r = G$), the half-way points ($r = H$) have been marked on the diagram. It can be seen that as ΔE gets larger so the average distance apart of the molecules (H) gets larger; or macroscopically as the temperature rises a substance **expands**. Another macroscopic prediction is of **Hooke's law** from F2.5(a) between $r = B$ and $r = D$, where the straight line indicates force is proportional to separation. Thermal energies of molecules are very small compared with their pes so a thermal energy corresponding to ΔE_3 in F2.6 corresponds to a very high temperature. At this sort of temperature the maximum separation allowed ($r = G_3$) is so great that it is possible in the bulk of a substance for the molecules to change places with other molecules; this type of exchange is thought to be typical of the behaviour that occurs in a **liquid**. Increase the thermal energy to ΔE_4 the molecules have enough energy for $r = \infty$ and they can break completely free of one another forming a **gas**. See 6.2 for the molecular interpretation of latent heat.

2.5 ELECTRIC AND GRAVITATIONAL FIELDS FROM SPHERES AND POINTS

The basic shape of an electric or gravitational field from a sphere or point is that of F2.3, although the diagram is 2-dimensional and it must be remembered that the fields are 3-dimensional. The 'field line' diagram would be correct for an electric field in which the charge on the sphere or at the centre point were negative: the direction of the arrows would have to be reversed for a positive charge. The equipotentials would have to be re-numbered and the sign changed to positive if the field were being generated by a positive charge. The following equations provide mathematical models for these fields and the field and potential graphs F2.4(a), (b).

E2.2 $E_E = \frac{1}{4\pi\varepsilon_0} \cdot \frac{Q}{r^2}$ (or $F_E = \frac{1}{4\pi\varepsilon_0} \cdot \frac{Q_1 Q_2}{r^2}$)

E2.3 $V_E = \frac{1}{4\pi\varepsilon_0} \cdot \frac{Q}{r}$ (or $pe_E = \frac{1}{4\pi\varepsilon_0} \cdot \frac{Q_1 Q_2}{r}$)

E2.4 $E_G = -G \cdot \frac{m}{r^2}$ (or $F_G = -G \cdot \frac{m_1 m_2}{r^2}$)

E2.5 $V_G = -G \cdot \frac{m}{r}$ (or $pe_G = -G \cdot \frac{m_1 m_2}{r}$)

where E_E = electric field strength ($N C^{-1}$) due to a charge Q(C), ε_0 = permittivity of free space ($CV^{-1} m^{-1}$), F_E = force (N) exerted by the interaction of the electric fields of two charges Q_1 and Q_2, V_E = electric field potential (V) or ($J C^{-1}$) due to a charge Q (C), pe_E = electric field pe (J), E_G = gravitational field strength ($N kg^{-1}$) due to a mass m (kg), G = gravitational constant ($N m^2 kg^{-2}$), F_G = force (N) exerted by the interaction of the gravitational fields of the two masses m_1 and m_2, V_G = gravitational field potential ($J kg^{-1}$) due to a mass m (kg), pe_G = gravitational field pe (J).

The distance r(m) in the case of field strengths and potentials is simply the distance from the centre of the field to the point at which the measurement is to be made. When two objects are involved, as when measuring pe and forces r (m) is the distance between their centres. The constant G is universal, but the value of ε_0 is valid only for fields in vacuums, and although there is virtually no change for its value in air, it must be modified by using the appropriate ε_r for experiments in any other medium (see E2.12). Be careful to watch the sign conventions which define whether forces are attractive or repulsive, and whether the pe increases or decreases as $r \to \infty$ (see 2.1 and 2.2). Note that fields and forces follow a $1/r^2$ (inverse square law) whereas potentials and potential energies obey a $1/r$ rule. You can use E2.4 and E2.5 to check F2.3–4(b). Remember when calculating forces that Newton's third law (action and reaction are equal and opposite) applies and although you may only be concerned with the force on an object placed in a field, the object creating the field must experience an equal and opposite force. Experiments to determine the values of G and ε_0 are important and will be found in your textbook. Note that E2.2 can be derived from E2.3 (using E2.1 and a knowledge of how to differentiate); similarly E2.4 can be derived from E2.5.

2.6 EARTH'S 'UNIFORM' GRAVITATIONAL FIELD

If the graph axes and equipotentials were re-numbered, F2.3–4(b) would illustrate the Earth's gravitational field. The following information (and E2.4 and E2.5) are needed:

Radius of the Earth (R) $= 6.4 \times 10^6$ m
Mass of the Earth (M) $= 6.0 \times 10^{24}$ kg

F2.3 shows that the field is clearly not uniform, but if a section is examined close to the surface of the Earth it looks as shown in F2.7. To a good approximation the field appears uniform, and its strength (g) is the force on 1 kg at the surface of the earth, so using E2.4:

E2.6 $g = -\frac{GM}{R^2}$

where G = gravitational constant ($N\,m^2\,kg^{-2}$), M = mass of the Earth (kg), R = radius of the Earth (m).

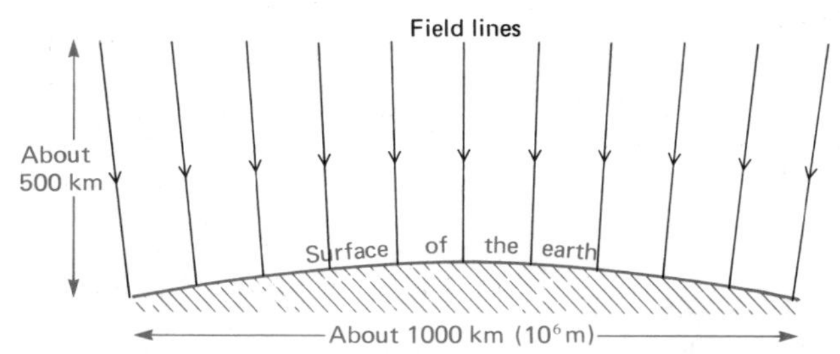

F2.7 Gravitational field close to the surface of the Earth

From $F = ma$ (E1.5) $F/m = a$, so any value of force per unit mass (newtons per kilogram, the units of gravitational field) is also a measure of acceleration. So g can be regarded as the gravitational field strength at (and just above) the surface of the Earth, or as the acceleration of a mass falling in the uniform gravitational field. The weight (mg) of an object (mass m) is the force of gravity upon it.

The value of g varies slightly owing to the Earth not being a perfect sphere and, more importantly, because of its rotation.

2.7 LAUNCHING SATELLITES FROM THE EARTH

What goes up will not come down if we throw it up fast enough! As something leaves the surface of the Earth it gains gravitational pe as it loses ke. F2.4(b) can be used to study this effect: if an object is given ke per kg of 80 pJ (p = pico = 10^{-12}) when it leaves the surface of the sphere, eventually its gravitational potential will increase by this amount. This will change its gravitational potential from − 100 to − 20 pJ kg^{-1} and the object will stop rising 0.5 m away from the CENTRE of the sphere and 0.4 m away from the SURFACE of the sphere. However if AT LEAST 100 pJ kg^{-1} of ke per kg are given to the object it will have enough energy to escape to $r = \infty$. For the Earth's gravitational field this produces E2.7.

E2.7 $v = \sqrt{\dfrac{2GM}{R}}$

where v = escape velocity ($m\,s^{-1}$), G = gravitational constant ($N\,m^2 kg^{-2}$), M = mass of the Earth (kg), R = radius of the Earth (m).

The equation is derived from taking the ke ($\frac{1}{2}mv^2$) of a mass m travelling at a speed v and equating it to the pe difference of the mass between being at the surface of the Earth ($-GmM/R$) and at infinity (zero). The escape velocity in the Earth's gravitational field is about 11 km s^{-1}. The escape velocity from the 0.1 m sphere of 0.15 kg mass featured in F2.3 – 4(b) can be checked to be about 45×10^{-6}m s^{-1}. The **escape velocity** is the minimum speed away from the centre of a gravitational field that an object must be given to escape completely the influence of the field. The escape velocity may be directed at any angle as long as it is 'away' from the field.

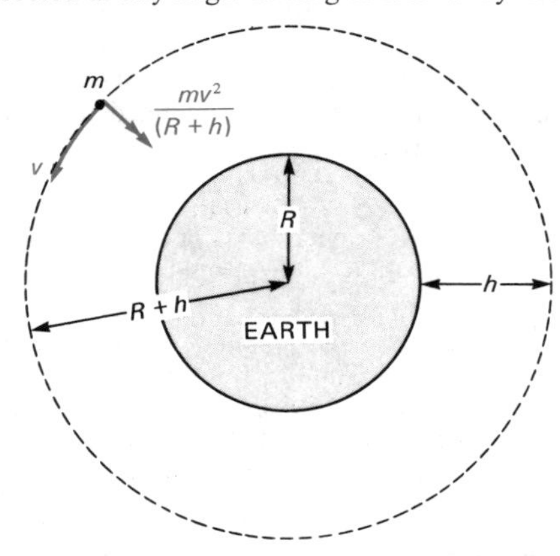

F2.8 Satellite orbiting the Earth

A satellite can be maintained in a stable circular orbit (F2.8) if a steady force of the right value($= mv^2/(R+h)$) is always directed towards the centre of the circle in which it is travelling. The centripetal force is provided by the gravitational attraction ($GmM/(R+h)^2$) which is always acting in this direction, so if a satellite can be made to travel at the correct speed then the gravitational force upon it can keep it in orbit, giving E2.8

E2.8 $v = \sqrt{\dfrac{GM}{R+h}}$

where v = speed of satellite in orbit ($m\,s^{-1}$), G = gravitational constant ($N\,m^2\,kg^{-2}$), M = mass of the Earth (kg), R = radius of the Earth (m), h = height of orbit above SURFACE of the Earth (m).

Note that neither the escape velocity nor the speed of orbit for a particular height depend upon the mass of the projectile involved. It is a popular mistake to confuse the distance of an object from the centre of the Earth ($R + h$) with its height above the surface of the Earth (h). It is possible to calculate the period (time for 1 revolution) of an orbit from its circumference ($2\pi(R+h)$) and the satellite speed (v). If the period is 24 hours then the Earth rotates with the same period as the satellite; this means that the satellite can stay stationary above a fixed point on the Earth; this is called a **parking orbit**. Calculation will show the height of such an orbit to be 36 000 km. Gravitational field equations have further applications in a study of Kepler's laws and planetary astronomy, etc.; textbooks will provide more information if needed.

2.8 UNIFORM ELECTRIC FIELDS

It can be shown that uniform electric fields exist between parallel charged plates or close to large planes with an even distribution of charge over their surface (see your textbook). Consider a charge of $+q$ being moved from R to S through the pd between the plates of V volts. The charge will gain electric pe of qV joules as work is done against the repelling constant force F exerted by the field. The situation is shown in F2.9. So work done in moving

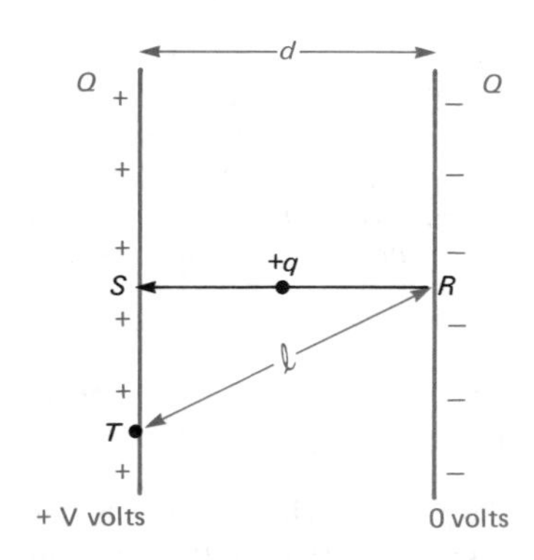

F2.9 Moving a charge in a uniform field

charge from R to S = the constant force × distance involved, or $qV = Fd$. But in any electric field of strength E, the force F upon a charge q is given by Eq, or $qV = Eqd$, hence

E2.9 $E = V/d$

where E = strength of uniform electric field ($N\,C^{-1}$ or $V\,m^{-1}$), V = pd between two points in the field ($J\,C^{-1}$ or V), d = distance between the two points (m).

It is important to remember that this formula applies only to uniform fields, and the distance d must be measured along a field line that joins two points at the correct pd V. The strength of the field between points R and T (a pd of V volts apart) is NOT V/l, for although l is the distance between these points it is not measured along a field line. Beware also of calculating the pd incorrectly; eg, if point R were at $-V$ volts and point S at $+V$ volts then the pd involved is $2V$ volts and **not** 0 volts.

If the capacitor plates have surface area A, it would seem sensible to assume that the field generated would depend on the charge per unit area (Q/A) upon the plates; this quantity is frequently referred to as the **charge density** (σ). So we have $E \propto Q/A$. Taking the proportionality we convert it into an equation with the aid of a suitable constant called ε_0, so

E2.10 $\varepsilon_0 E = Q/A$ (or $\varepsilon_r \varepsilon_0 E = Q/A$)

where ε_0 = permittivity of free space ($\mathrm{C\,V^{-1}m^{-1}}$), E = strength of uniform electric field ($\mathrm{NC^{-1}}$ or $\mathrm{Vm^{-1}}$), Q = total charge on each capacitor plate (C), A = area of each capacitor plate ($\mathrm{m^2}$), ε_r = relative permittivity of a medium (no units).

This equation defines ε_0, which applies to a vacuum (free space). The strength of the field is found to vary if the medium is changed, in this case the value of ε_r for the medium concerned must be used. The larger the value of ε_r then a bigger charge will be stored on a pair of capacitor plates for a given pd, the medium between the plates plays an important role in the storing of energy in the field and charge on the plates. This definition of ε_0 applies to parallel plates and rectangular symmetry, so it is no surprise to find 4π appearing in E2.2 and E2.3 which describe a spherically symmetrical situation (eg, area of a square of side $r = r^2$, but surface area of a sphere of radius $r = 4\pi r^2$). ε_0 is also known as the **electric constant**.

2.9 CAPACITANCE OF PARALLEL PLATES AND ISOLATED SPHERES

The **capacitance** of a capacitor is a measure of the charge (in C) stored in it per volt of potential difference that is used to charge it giving:

E2.11 $C = Q/V$ (or $Q = CV$)

where C = capacitance in farads (F or $\mathrm{C\,V^{-1}}$), Q = charge stored (C), V = pd used to charge capacitor (V).

It is possible to combine E2.9 and E2.10 so $\varepsilon_r \varepsilon_0 V/d = Q/A$, and rearranging this we get $Q/V = \varepsilon_r \varepsilon_0 A/d$ which defines the capacitance of a parallel plate capacitor:

E2.12 $C = \dfrac{\varepsilon_r \varepsilon_0 A}{d}$

where C = capacitance of parallel plate capacitor (F), ε_r = relative permittivity of medium between plates (no units), ε_0 = permittivity of free space ($\mathrm{C\,V^{-1}\,m^{-1}}$), A = surface area of each plate ($\mathrm{m^2}$), d = separation of the plates (m).

If a conducting sphere with a charge Q residing on its surface is considered, it must obey E2.3 which can be rewritten as $Q/V = 4\pi\varepsilon_0 r$, which defines the capacitance of an isolated conducting sphere:

E2.13 $C = 4\pi\varepsilon_0 r$

where C = capacitance of isolated conducting sphere (F), ε_0 = permittivity of free space ($\mathrm{C\,V^{-1}\,m^{-1}}$), r = radius of the sphere (m).

In this case the pd to which the sphere is charged is assumed to be the pd between the surface of the sphere and infinity (zero volts as usual).

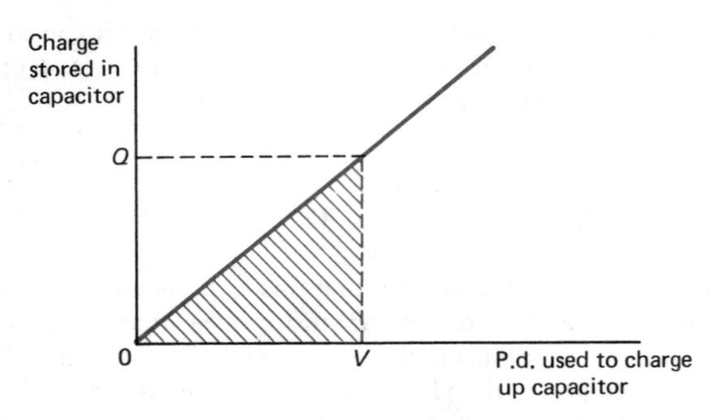

F2.10 Charge and energy stored in a capacitor

In F2.10: slope, constant, its value (Q/V) is the capacitance; area, electrical energy stored in capacitor; intercepts, etc., graph passes through origin as for zero charging voltage no charge is stored.

From F2.10 it is shown that as the area under the graph is the energy stored in the capacitor then as the shaded area is $\frac{1}{2}QV$ and as $Q = CV$ (E2.11):

E2.14 Energy stored $= \frac{1}{2}QV = \frac{1}{2}CV^2 = \frac{1}{2}Q^2/C$

where electrical energy stored in capacitor is in joules, C = capacitance of capacitor (F), Q = charge stored in capacitor (C), V = pd used to charge capacitor (V).

2.10 MEASURING MAGNETIC FIELDS AND THEIR FORCES

At A-level the concepts of magnetic potential and pe are not required; so an understanding only of magnetic fields and the forces they exert is needed (together with some applications). Magnetic fields are created by electric currents (moving charges); to understand how this can be applied to permanent magnets it is necessary to examine them microscopically and explain that the currents involved can be linked to the electrons moving (moving charge is electric current) within the atoms. The **strength of a magnetic field** is found by measuring the force on a wire 1 m long carrying a current of 1A at right angles to the field. The magnetic field strength is sometimes called the 'magnetic flux density' or the 'magnetic induction'. So magnetic field strengths (B) are measured in Newtons per Ampere per metre ($\mathrm{NA^{-1}m^{-1}}$); these units are abbreviated by being called **Teslas** (T). However a major complication arises when dealing with magnetic fields (as opposed to electric or gravitational fields), the direction of the current being used to test the field related to the direction of the field itself is vital. They must be at right angles to one another, otherwise the force involved will be reduced; the force becomes zero if the field and current are parallel to one another.

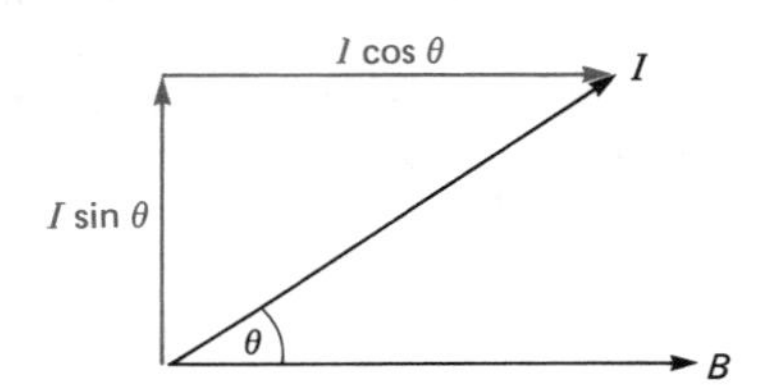

F2.11 Current (I) and magnetic field (B) at an angle (θ)

F2.11 shows that if the angle between the magnetic field (B) and current (I) is θ, then the vector components of the current at right angles and parallel to the field are $I\sin\theta$ and $I\cos\theta$, respectively. The force on the $I\cos\theta$ component is zero (as it is parallel to the field) and the force on the ('right angle') $I\sin\theta$ component is 1 Newton per Ampere of current per metre length of wire per Tesla of magnetic field giving:

E2.15 $F = BIL\sin\theta$

where F = force exerted by magnetic field on wire (N), B = strength of the magnetic field (T or $\mathrm{N\,A^{-1}m^{-1}}$), I = current in the wire (A), L = length of wire in the field (m), θ = angle between the current and field (see F2.11).

This formula is based upon the macroscopic concept of electric current, but it can easily be changed to a microscopic version which deals with moving charges:

E2.16 $F = Bqv\sin\theta$

where F = force exerted by magnetic field on charged particle (N), B = strength of the magnetic field (T or $\mathrm{N\,A^{-1}\,m^{-1}}$), q = size of charge upon particle (C), v = speed of the particle ($\mathrm{m\,s^{-1}}$), θ = angle between the magnetic field and the direction in which the particle is travelling.

Not only are the current and field at right angles to one another to enable the resulting magnetic force to be measured, but the resulting force is at right angles to both the current and the field. To get the correct direction of the force, the left-hand rule has to be used (check in your textbook if you need to remember how to use the rule).

The usual way of remembering the **left hand rule** is via the **M** in thu**M**b which represents the **M**otion (direction of the force), the **F** in **F**irst finger (**F**ield), and the **C** in se**C**ond finger (**C**urrent). Beware of the trap of thinking that the **F** stands for Force. Another popular error occurs when applying the rule to moving charged particles; remember that electric current has its direction defined from the direction in which positive charge moves, so electrons travelling to the 'left' represent a current 'travelling to the right'.

There are three fields created by currents in wires that are particularly important at A-level. They are:

1 A long straight wire – a field of concentric circles getting weaker as you get further away from the wire (direction comes from the corkscrew rule).

2 A solenoid – a solenoid is a long coil filled with air (or a vacuum). The field is like that of a bar magnet except that the inside of the solenoid is accessible where the field is found to be **uniform**. The direction of the field is found by treating it like a bar magnet and saying that the North-seeking end is where the current flows anticlockwise and vice versa.

3 A flat circular coil (ie, one of zero length) – the field is similar to that of a 'squashed solenoid'. Most importantly these coils are used in pairs (**Helmholtz Coils**) when they can produce a uniform field near their axis over a reasonably large volume.

Further information and diagrams of the three fields are available in your textbook. Remember that the field direction is defined so that it fits in with the left-hand rule. The following formulae give means of calculating the field strengths for the three cases:

E2.17 $B = \dfrac{\mu_0 I}{2\pi r}$

where B = field strength at distance r from a long uniform wire (T), μ_0 = permeability of free space (N A^{-2}), I = current in wire (A), r = radial distance from the wire (m).

E2.18 $B = \dfrac{\mu_0 NI}{L}$

where B = strength of uniform field inside a solenoid (T), μ_0 = permeability of free space (N A^{-2}), N = total number of turns of wire in solenoid (no units), L = length of solenoid (m), I = current passing through the solenoid (A).

E2.19 $B = \dfrac{\mu_0 NI}{2r}$

where B = strength of the field at the centre of a flat circular coil (T), μ_0 = permeability of free space (N A^{-2}), N = total number of turns of wire in coil (no units), I = current passing through the coil (A), r = radius of the circular coil (m).

2.11 USING MAGNETIC FIELD FORMULAE

Parallel wires If two wires are parallel a distance r apart, then the field at one wire (carrying a current I_1) due to the other wire (carrying a current I_2) is given by $\mu_0 I_2/2\pi r$ (E2.17). So:

E2.20 $\dfrac{F}{L} = \dfrac{\mu_0 I_1 I_2}{2\pi r}$

where F/L = force per unit length on each wire (N m^{-1}), μ_0 = permeability of free space (N A^{-2}), I_1 = current in one wire (A), I_2 = current in the other wire (A), r = distance between the wires (m).

Two points of importance arise when using this formula. First it must be noted that the formula gives the force per unit length (NOT the force) on the wires. Second both wires experience an equal and opposite force (Newton's third law), and the force between the wires is attractive if the parallel currents run in the same direction (repulsive if the currents run in opposite directions). Students are accustomed to 'like' charges 'repelling', hence they wrongly assume that currents in the same direction also repel.

Electric currents give rise to several effects; magnetic, heating, chemical, dynamic, light production, etc. It could be possible to define the Ampere using any of them; however the magnetic effect has been chosen to define the Ampere.

One **Ampere** is the constant current that if maintained in two parallel conductors of infinite length, of negligible circular cross-section, and placed 1 m apart in a vacuum, would produce between those conductors a force of 2×10^{-7}N per metre length.

It can be seen from E2.20 that the definition of the Ampere also fixes the value of μ_0 (**the permeability of free space**) to be $4\pi \times 10^{-7}$ N A^{-2}. μ_0, although defined for a vacuum has almost exactly the same value in air; however, the medium can affect the strength of magnetic field and a term μ_r (relative permeability) needs to be introduced to allow for the nature of the medium. Hence to allow for a change of medium $\mu_0\mu_r$ should be put into all the formulae in place of μ_0; note that relative permeability has no units. The similarity between $\mu_r\mu_0$ and $\varepsilon_r\varepsilon_0$ (E2.10) is obvious; both sets of constants help define the strength of field produced appropriate to the medium involved (the larger the value of the constants the larger the field strengths in any given situation). μ_0 is also called the **magnetic constant**.

Hall effect A piece of conducting material has a current I passing through it and a magnetic field of strength B is applied at right angles to the current. The current is passing to the right and the magnetic field is directed into the paper so the force on the current is 'upwards' (left-hand rule, E2.16). In F2.12 it has been assumed that the current to the right is being carried by negative charge carriers moving to the left (each carrying $-q$ of charge). Hence the size of the upward force on them is Bqv (E2.16) where v is the drift velocity of the charge carriers (E7.3) carrying the current.

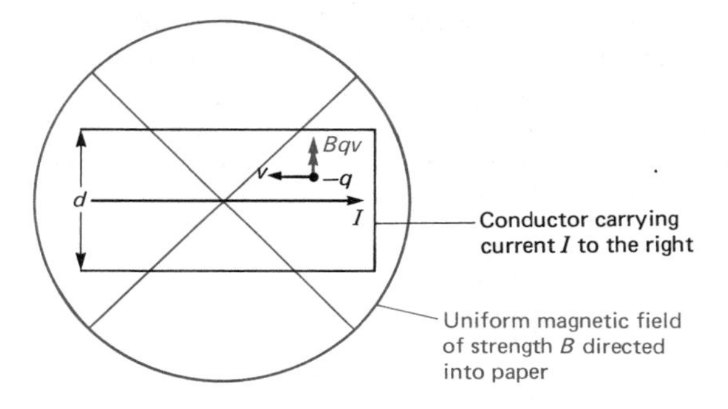

F2.12 Force on an electron carrying a current in a magnetic field

Hence the charge carriers will move under the influence of the magnetic field towards the top of the conductor, creating a negative region at the top of the conductor (surplus negative charge carriers) and a positive region at the bottom (deficiency of negative charge carriers). This means that an electric field is created that tends to attract the charges back to the bottom of the

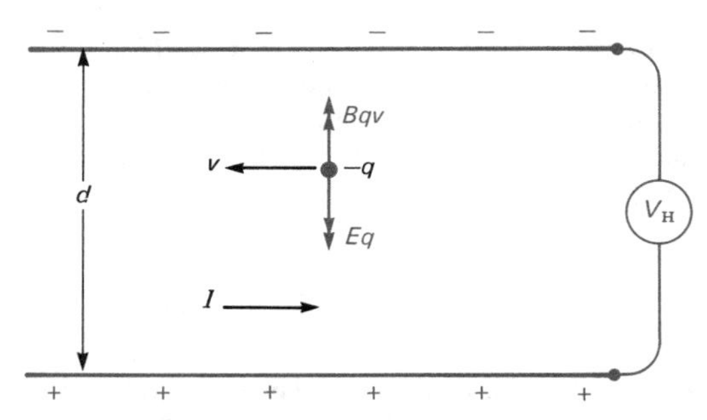

F2.13 The Hall Effect

conductor as in F2.13. With the top of the conductor negative and the bottom positive a pd V_H (called the **Hall voltage**) exists between the top and bottom of the slice. A uniform electric field of strength V_H/d (E2.9) will be created that will result in a downward force Eq on each of the negative charge carriers as shown in F2.13. As more and more charge carriers move to the top of the conductor so the electric field (and V_H) increase. Eventually equilibrium is established when the two forces on the charge carriers become equal and cancel each other out, when a steady Hall voltage will appear between the top and bottom of the slice. So eventually $Bqv = Eq = V_H q/d$ or rearranging:

E2.21 $V_H = Bvd$

where V_H = Hall voltage (V), B = strength of the magnetic field (T), v = drift velocity of charge carriers (m s^{-1}), d = thickness of the conductor (m).

Using the drift velocity formula (E7.3) the formula can be rewritten:

E2.22 $V_H = \dfrac{BId}{nAq}$

where I = current passing through material (A), n = number of charge carriers per cubic metre (m^{-3}), A = cross-sectional area of material (m^2), q = charge on each charge carrier (C).

The Hall effect can be observed in metals, but semiconductors provide much larger Hall voltages owing to the higher drift velocities that result from the smaller values of n in these materials. (see E2.21/22). The theory is correct for a n-type semiconductor in which most charge carriers are electrons. However to convert to positive charge carriers (eg, holes – as in p-type semiconductors) requires little modification; with the current in the same direction it is positive charge carriers that move upwards and so the only modification needed is to the direction of the electric field (and hence the direction of the Hall voltage) which will be reversed in this type of material. Several uses have been found for the Hall effect; they include determining the sign (positive or negative) of charge carriers and (most importantly) the measurement of magnetic field strengths.

Voltage generator Consider a metal rod of length d being moved to the left at a steady speed v in a uniform magnetic field of strength B directed into the paper as shown in F2.14. The only

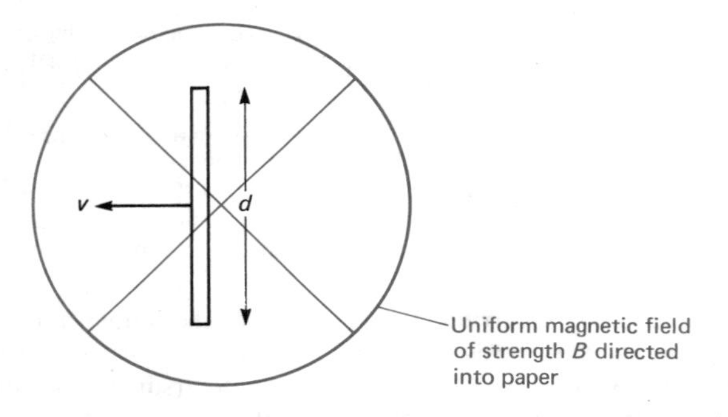

F2.14 Conductor moving in a uniform magnetic field

charge carriers free to move anywhere in the metal are negative electrons which are being carried to the left at a steady speed v. But this is exactly the same situation that has been discussed in the Hall effect theory (where the movement was provided by an electric current); so we know that a steady pd of V volts will be set up across the rod with the top of the rod negative and the bottom positive such that:

E2.23 $V = Bvd$

where V = induced voltage across rod (V), B = strength of uniform magnetic field (T), d = length of rod (m), v = speed of rod at right angles to the field (m s^{-1}).

Remember that the direction of the voltage can be influenced by the direction of either the magnetic field or the movement, and also by whether the charge carriers are positive or negative.

2.12 PRINCIPLES OF ELECTROMAGNETIC INDUCTION

The voltage generator (2.11) illustrates the basic theme of electromagnetic induction: the creation of an emf (voltage) using a magnetic field, charges free to move AND relative movement between the field and the charges. In 2.11 the relative movement is created by moving the charges; there is no reason why the magnetic field should not be moved instead (indeed this happens in many electromagnetic devices when the field is often 'moved' by changing its strength, eg, using ac to generate it). When emfs are induced by these methods, the process is referred to as **electromagnetic induction**. A study of the Hall effect theory (2.11) shows the role played by both electric and magnetic fields in creating the pd; hence the suitability of the term electromagnetic.

It is helpful when studying electromagnetism to introduce the concept of magnetic flux (φ) measured in **Webers** (Wb or Tm^2). **Magetic flux** is measured by multiplying an area by the strength of the magnetic field passing at right angles through that area. This gives:

E2.24 $\varphi = BA \sin\theta$

where φ = amount of magnetic flux (Wb), B = strength of the uniform magnetic field (T), A = area through which the field is passing (m^2), θ = angle between the plane of the area and the direction of the magnetic field.

When measuring flux it is important to remember the $\sin\theta$ term, which allows for the fact that the area and field must be at right angles to one another. Note that B (T) could also have its units written as Webers per square metre (Wb m^{-2}). To see how using magnetic flux can be useful it is valuable to reconsider the term vd in E2.23, which can be represented as in F2.15. The

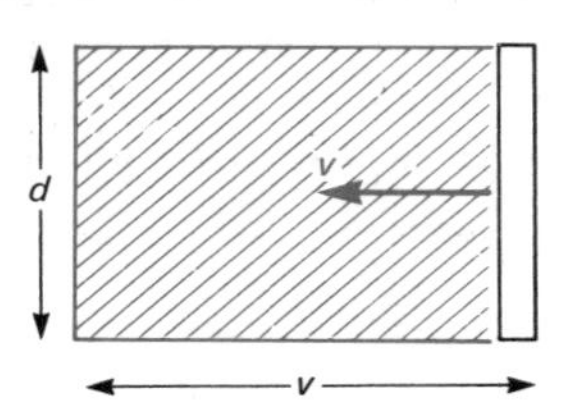

F2.15 Area swept out per second by a conductor

shaded area in F2.15 is vd, which is the area swept out by the moving conductor per second (as it moves a distance v in one second and has a length d). This means that the flux that the rod moves through in one second (Bvd) is the area that the rod moves through in a second (vd) multiplied by the strength of the field (B) at right angles to that area giving:

E2.25 Induced voltage = flux swept out per second,
or Induced voltage = rate of change of flux linkage,
or $V = -\mathrm{d}\varphi/\mathrm{d}t$

Care has to be taken when using E2.25 in calculating the flux 'linking' the circuit. In a coil of cross-sectional area A with a field B passing at right angles down the axis of the coil the flux passing through the coil is BA. However this flux passes through each turn of the coil, so if there are N turns to the coil then the flux 'linkage' is NBA. It is the change in flux linkage that gives rise to induced emfs.

The size of the **induced voltage** ($\mathrm{d}\varphi/\mathrm{d}t$) is often referred to as the **Faraday law** (occasionally the **Faraday – Neumann law**), and the direction of the induced voltage (as indicated by the minus sign in E.2.25) is a consequence of Lenz's law. **Lenz's law** states that the direction of any induced emf is always such that it opposes the change that created the emf; a result necessary for the conservation of energy. In the voltage generator/Hall effect (2.11) the direction of the magnetic force that redistributes the charges in the conductor (creating the emf) is upward (F2.13), however; the emf created generates an electric field that applies a force on the charge carriers in the opposite (downward) direction; hence Lenz's law is being obeyed. Thus the direction of the induced emf can be determined either by an application of the left-hand rule (2.10) or by use of Lenz's law.

2.13 CALCULATING FLUX: THE MAGNETIC/ELECTRIC CIRCUIT ANALOGY

The key to determining induced emfs is to obtain an expression for the magnetic flux (φ) in any situation. The modern approach

Magnetic terminology		Analagous electrical quantities
Flux (Wb) Magnetic flux is driven around a magnetic 'circuit'; all the flux entering a box must leave that box.	$\varphi \ldots I$	**Electric current** (A) Electric current is 'driven' around an electric circuit; all the current entering a junction in the circuit must leave that junction.
Current-turns (A) It is the current-turns of a coil that drive the flux around a magnetic circuit (N = number of turns in coil, I = current in coil). NI is sometimes called the magnetomotive force or mmf.	$NI \ldots V$	**Potential difference** (V) It is the pd in an electric circuit that drives the electric current around that circuit; V can be referred to as the electromotive force or emf.
Reluctance (H^{-1} or $A\,V^{-1}s^{-1}$) Increased reluctance causes a reduction in the amount of flux that can be created for a given value of current-turns,	$R_m \ldots R$	**Resistance** (Ω) Increased resistance causes a reduction in the amount of electric current that can be created for a given value of pd.

to calculating flux is to use the magnetic/electric circuit analogy as shown in the Table

By definition analogous quantities must obey the same mathematics. So once a good understanding of electrical circuit theory has been acquired, this can then be applied to magnetic circuits. The equations can be adapted simply by replacing the electrical quantities with their analogous magnetic quantities, eg, $I = V/R$ in an electrical circuit; so in a magnetic circuit:

E2.26 $\varphi = NI/R_m$

where φ = amount of flux created (Wb), N = number of turns in the coil (no units), I = current passing through the coil (A), R_m = reluctance of the circuit (H^{-1}).

A problem still exists in deciding how the reluctance is to be measured. The analogy can be used again as follows.

$R = \rho L/A$ in an electrical circuit (E7.8), so in a magnetic circuit:

E2.27 $R_m = \left(\frac{1}{\mu_r \mu_0}\right) L/A$

where R_m = reluctance of sample of material (H^{-1}), μ_r = relative permeability of material (no units), μ_0 = permeability of free space ($N\,A^{-2}$), L = total length of sample (m), A = uniform cross-sectional area of sample (m^2).

In electrical circuits $1/\rho$ is known as the **electrical conductivity** of a material: it can be seen from E2.27 to be analogous to $\mu_0\mu_r$ which hence can be regarded as the **'magnetic conductivity'** of a material (ie, a measure of how easy it is to generate a given flux or field in a material, see 2.11). It is possible to pursue the analogy much further using more basic electrical equations; however this is not usually done at A-level. An important practical example of the use of the analogy may help. All the magnetic flux in F2.16

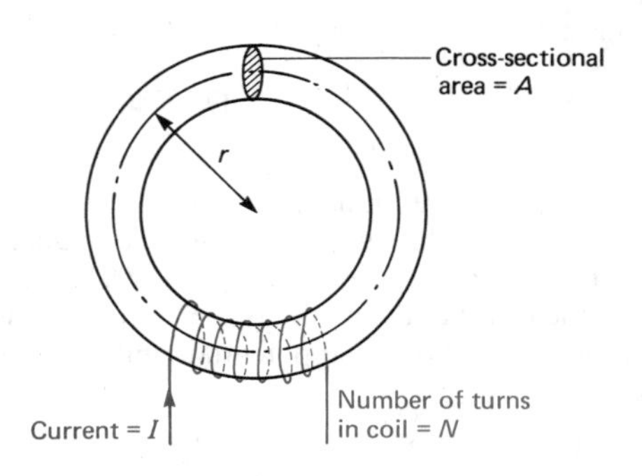

F2.16 Generating flux in an iron ring

will be contained in the iron ring, so the total length (L) of the magnetic circuit is the circumference ($2\pi r$) and the flux passes through a uniform cross-sectional area (A), so combining E2.26 and E2.27 we get:

E2.28 $\varphi = NI\mu_0\mu_r \frac{A}{2\pi r}$

where φ = flux generated in a closed ring (Wb), N = number of turns of wire around ring (no units), I = current passing through wire (A), μ_0 = permeability of free space ($N\,A^{-2}$), μ_r = relative permeability of ring material (no units), A = uniform cross-sectional area of ring (m^2), r = average radius of the ring (m).

This formula will hold true for any ring as long as all the flux generated by the coils stays within the ring. If we imagine the equivalent electric circuit as perhaps consisting of a wire (the ring) shorting out two terminals of a battery (the coil and current); then the reason the current passes through the wire and not through the surrounding air is because the resistance of the wire is so much lower than that of the air. Hence in the magnetic circuit for all the flux to remain within the coil, the reluctance of the coil must be far lower than that of the surrounding air. As iron for example has a μ_r of about 1000, meaning that it is about 1000 times better at 'conducting' magnetic flux than the air ($\mu_r \approx 1$) it is clear why virtually no flux 'travels outside' an iron ring.

2.14 INDUCTANCE AND TRANSFORMERS

A changing flux within a coil will generate an emf across the ends; but a changing flux ($d\varphi/dt$) is proportional to a changing magnetic field (dB/dt), and a changing magnetic field is created by a changing electric current (dI/dt). So if a steady dc current ($dI/dt = 0$) is passed through a coil of zero resistance there will be no pd across the coil, owing either to $V = IR$ or an emf being induced. However if an ac current ($dI/dt \neq 0$) is passed through the coil an emf will be induced across the ends of the coil. This voltage (the induced emf) must be proportional to rate of change of current giving:

E2.29 $V_L = -L dI/dt$

where V_L = voltage across an inductor (V), L = inductance of the coil in henrys (H or $V\,s\,A^{-1}$), I = current passing through the coil (A), t = time (s).

Because the coil is inducing a voltage in itself, the property L is often referred to as the **self-inductance** of the coil. The minus sign in the formula is an aid to remembering that the induced emf acts in such a way as to oppose the changing current that causes it (Lenz's law – 2.12). Often dI/dt is measured directly off a graph, or an oscilloscope trace is used to display a plot of current against time.

Proof of the following formula for the energy stored in an inductor can be found in your textbook:

E2.30 Energy stored $= \frac{1}{2} L I^2$

where electrical energy is stored by the magnetic field around the inductor (J), L = inductance of the inductor (H), I = current in inductor (A).

Often a changing current in one coil is used to induce a voltage in another coil. It can be arranged that the same flux can pass through both coils; this can be done by winding the coils directly on top of one another or by using a flux circuit to link the coils as in the **transformer** shown in F2.17. If a changing current is

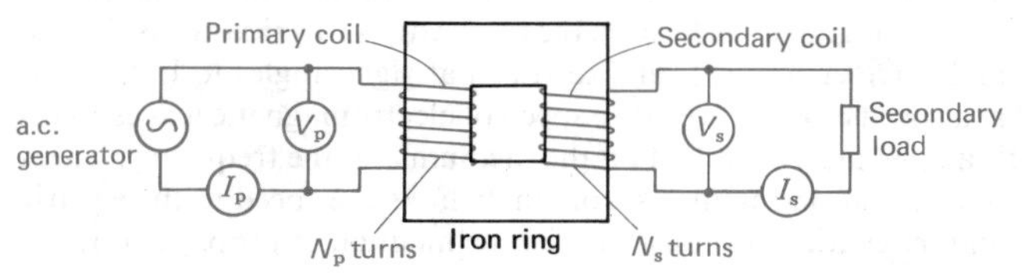

F2.17 A transformer

passed through the primary coil (dI_p/dt) all the flux generated will also link the secondary coil, so similarly to E2.29 the voltage induced in the secondary is proportional to the changing current in the primary:

E2.31 $V_s = -M dI_p/dt$

where V_s = voltage induced in secondary coil (V), M = mutual inductance (H), I_p = current in the primary coil (A), t = time (s).

Remember that this formula does not only apply to transformers but to any pair of coils that share the same flux and can mutually induce voltages across one another; hence the term **'mutual inductance'**.

Transformers are easily the most efficient machines built by man (98% or better) so it is reasonable to say that the power input is equal to the power output giving:

E2.32 $V_p I_p = V_s I_s$

where V_p = rms voltage across primary (V), I_p = rms current through primary (A), V_s = rms voltage across secondary (V), I_s = rms current through secondary (A).

It is important to realize that all electromagnetic effects involve changing voltages and currents (ie, ac), and this must be recognized when making measurements. The formula E2.32 would still be valid if all the rms values were replaced by peak values (see 7.7).

The other important transformer formula recognizes that with the same flux through each coil the voltage across each coil must be proportional to the number of turns:

E2.33 $V_p/V_s = N_p/N_s$

where V_p = rms voltage across primary (V), V_s = rms voltage across secondary (V), N_p = number of turns in primary (no units), N_s = number of turns in secondary (no units).

Again BOTH values could be replaced by their peak values. Always check that the bigger voltage appears across the coil with more turns!

2.15 ELECTROMAGNETIC FIELDS

It is unfortunate that permanent magnets were discovered long before the magnetic effect of an electric current; hence 'wrongly' creating two separate subjects of electricity and magnetism. Modern day scientists are left with many awkward legacies arising from the confusion of terminology. It is clear that magnetic fields are created by electric currents which are themselves maintained by electric fields (or pds). Hence associated with every changing magnetic field (B) must be a changing electric field (E). Such changing 'joint' fields are called **'electromagnetic fields'**.

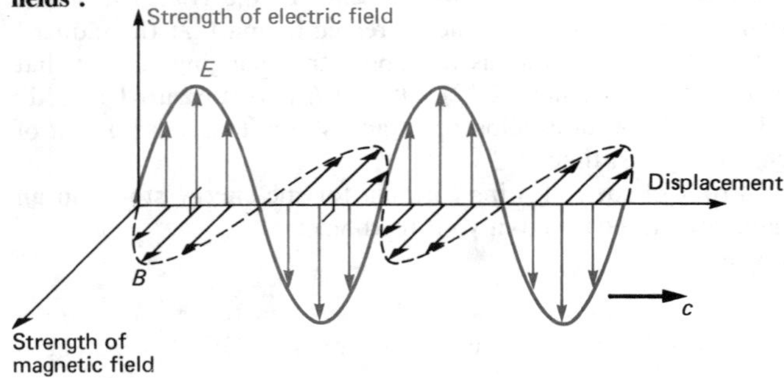

F2.18 Electric and magnetic fields of a plane polarized electromagnetic wave

F2.18 shows how the strength of the electric and magnetic fields vary (represented by the lengths of the arrows) with displacement in space. Note that the two fields are at right angles to one another, but their magnitudes are in phase. When an electromagnetic field exists, the fields vary sinusoidally (as shown) and wave energy is found to travel at right angles to both fields at a speed c. This speed (the speed of electromagnetic waves/light) is a constant dependent on the medium and the frequency. There is an equation that not surprisingly links the speed to the electric and magnetic field constants for the medium of propagation:

E2.34 $c = 1/\sqrt{\varepsilon_0 \mu_0}$

where c = speed of light in a vacuum ($m\,s^{-1}$), ε_0 = permittivity of free space ($CV^{-1}\,m^{-1}$), μ_0 = permeability of free space ($N\,A^{-2}$).

Note that 'vacuum' and 'free space' mean the same. All electromagnetic waves travel at the same speed in a vacuum, regardless of frequency.

UNIT 2 QUESTIONS

2.1M Suppose a particle P is projected in a uniform field, which can be magnetic, electric or gravitational, as shown in the diagram. For the particle to move in the plane of the paper in the **circular** path indicated, the conditions would have to be:

	Particle	Field
A	Positively charged	Electric
B	Positively charged	Magnetic
C	Negatively charged	Electric
D	Uncharged	Gravitational
E	There are no conditions which could produce such a motion.	

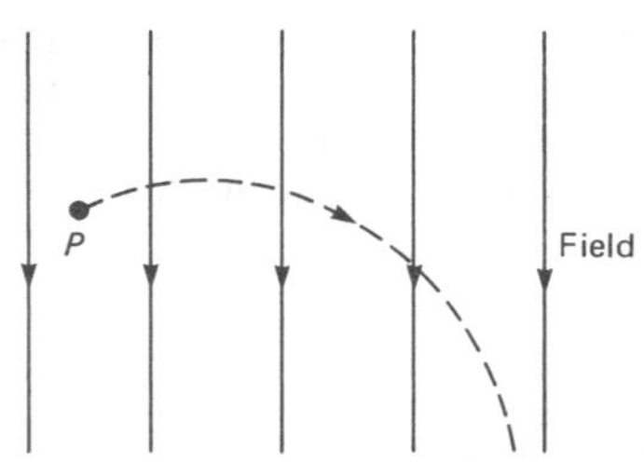

(**London**: *all other Boards except SEB H*)

A charge in a uniform electric field will behave in the same way as a mass in a uniform gravitational field; is a circular path possible in either of these situations? Also, in the question, the path, field and force are all in the **same plane**, which makes one answer impossible (see note after E2.16).

2.2M Mars has a diameter of approximately 0.5 that of Earth, and a mass of 0.1 that of Earth. The surface gravitational field strength on Mars compared with that on Earth is greater by a factor of:

A × 0.1 B × 0.2 C × 0.4 D × 2.0 E × 10.0

(—: *all other Boards except SEB H and SEB SYS*)

Use E2.6.

2.3M X and Y are two points at respective distances R and $2R$ from the centre of the Earth, where R is greater than the radius of the Earth. The gravitational potential at X is $-800\,kJ\,kg^{-1}$. When a 1 kg mass is taken from X to Y, the work done on the mass is:

A −400 kJ. B −200 kJ. C +200 kJ. D +400 kJ. E +800 kJ.

(**Cambridge**: *all other Boards except SEB H*)

Remember that gravitational potential V near a planet varies with the inverse of the distance from the planet's centre. Determine V at point Y, given that $V = -800\,kJ\,kg^{-1}$ at point X. Then use the fact that the work done per unit mass is equal to the change of potential (ie, final potential – initial potential).

2.4M The gravitational pd between a point at the surface of a certain planet and a point 10 m above the surface is $4.0\,J\,kg^{-1}$. Assuming the gravitational field is uniform, the work done, in J, in moving a mass of 2.0 kg from the surface to a point 5.0 m above the surface is:

A 0.16 B 1.0 C 2.0 D 4.0 E 16

(**NISEC**: *all other Boards except SEB H*)

Field strength in a gravitational field is the force in N acting on a mass of 1 kg. Since work done equals force × distance, field strength = pd ÷ distance. Applying this information in the first sentence gives the field strength (which is uniform). Use this result with the information in the second sentence to obtain the work done in J. It may be helpful to realize that, in a gravitational field, the field strength in $N\,kg^{-1}$ is equivalent to the acceleration due to gravity in $m\,s^{-2}$. See p. 44 if necessary.

2.5M Suppose that the acceleration of free fall at the surface of a distant planet were found to be equal to that at the surface of the Earth. If the diameter of the planet were twice the diameter of the Earth, then the ratio of the mean density of the planet to that of the Earth would be:
A 4:1. B 2:1. C 1:1. D 1:2. E 1:4.

(**London**: *all other Boards except SEB H and SEB SYS*)

Use E2.6 and the fact that M = density × volume (for a sphere = $4\pi R^3/3$). Equate: g for Earth = g for planet. Then solve to find the ratio of the densities.

2.6M A small positively charged ball suspended from an insulating thread is raised vertically from near the negatively charged 'dome' of a Van der Graaf machine. Which curve, A–E, correctly shows the **change** of pe (gravitational + electrostatic) of the ball with height, H, above its initial position?

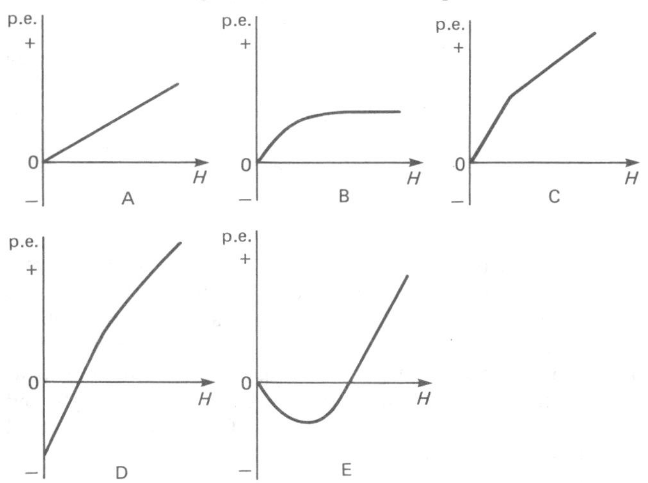

(—: *all Boards except SEB H*)

Assume height changes are small so that the Earth's gravitational field may be considered uniform over the range of H; thus, gravitational pe increases in proportion to H. Now consider how the electric pe changes with H; because + ve charge is being moved away from − ve charge, the electric pe increases with separation, but the increase becomes less and less as the separation increases.

Make your choice by considering how the electric pe changes modify the gravitational pe changes as H increases from its initial value.

2.7M An isolated hollow pear-shaped conductor is given a charge. Which of the following statements is true?
1 The potential is the same at all points on the conductor.
2 The charge is concentrated at the pointed end of the conductor.
3 No electric field exists inside the conductor.

Answer: A if **1**, **2**, **3** correct
B if **1**, **2** only.
C if **2**, **3** only.
D if **1** only.
E if **3** only.

(**London**: *all other Boards except SEB H*)

Details of the distribution of charge on a conductor can be found in most textbooks. Remember that a conductor must be an equipotential surface or else a current would flow until it is.

2.8M The diagram shows two equal and opposite charges which are at a fixed distance apart in a non-uniform electric field E, for which lines of force are shown. The pair of charges experiences:
A a resultant force in the plane of the paper but no couple.
B a resultant force in the plane of the paper and a couple.
C a resultant force normal to the plane of the paper but no couple.
D a resultant force normal to the plane of the paper and a couple.
E a couple but no resultant force.

(**Cambridge**: *all other Boards except SEB H*)

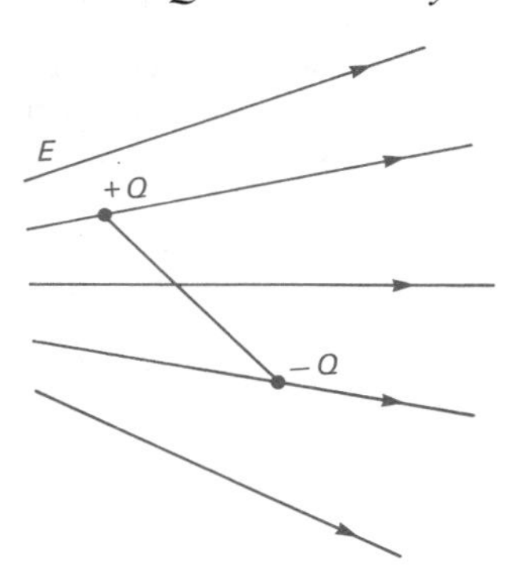

A line of force gives the direction in which a free + ve charge will move. Mark the force direction for each charge. The magnitude of the forces will not necessarily equal one another. Your force diagram should show you that there can be no resultant force normal to the diagram. See p. 19 and F1.8 for further information about couples.

2.9M Two point charges $X = +2\mu C$ and $Y = -3\mu C$ are placed 100 mm apart. The electric potential V due to the two charges (along the line between them) will be zero at a distance, in mm, from X of:
A 30. B 40. C 50. D 60. E 70.

(—: *all other Boards except SEB H*)

Make a simple sketch to start with, then write down the equation for V near a point charge; see E2.3 if necessary. For zero potential due to X and Y, the contribution from X is equal and opposite to that for Y.

2.10M The equipotentials of a pair of point charges $+Q$ and $+Q$ are shown in the diagram. If an electron moves from X to Y:
A its electric pe falls.
B its electric pe is unchanged.
C it experiences a resultant force pulling it back to X.
D it experiences a resultant force pushing it away from X.
E it does not experience a resultant force.

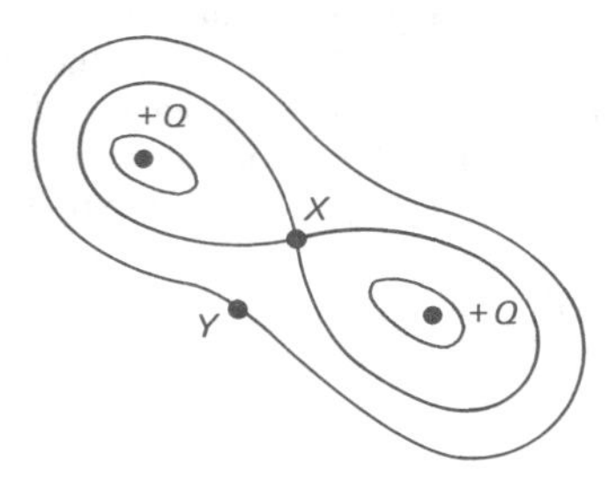

(—: *all Boards except SEB H*)

Remember that an equipotential is a line along which a point charge has constant electric pe. Also, the resultant force on the electron is the vector sum (see p. 19 if necessary) of the two individual forces acting on it.

2.11M X and Y represent two parallel metal plates. X is connected to the cap of a gold-leaf electroscope and Y is connected to earth. The plates are charged and the leaf rises. When an uncharged slab of perspex is inserted between the metal plates, the divergence of the leaf:
A decreases.
B increases, then decreases.
C increases.
D remains unaffected.
E drops to zero immediately.

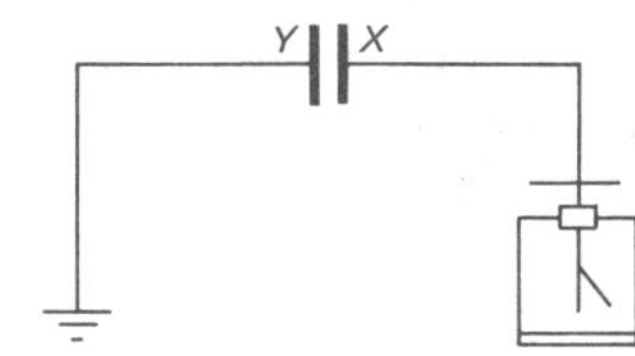

(**SEB H**: *all other Boards*)

The electroscope indicates the pd between the plates. The charge on the plates remains unchanged when the perspex is inserted. What happens to the capacitance of the plates when the perspex is inserted? Remember that perspex is a dielectric; see

p. 36 if necessary. Knowing what happens to the capacitance, and remembering that the charge is fixed, consider what happens to the pd? Use E2.11 if necessary. Then choose from A–E accordingly.

2.12M Which of the following statements about the force between the atoms in a diatomic molecule is/are correct?

1 The force obeys the inverse square law for all separations of the atoms.
2 The force changes from an attraction when the separation is large to a repulsion when the separation is small.
3 The force is zero when the potential energy of the system is a minimum.

A None of these. B **1** only. C **2** only.
D **3** only. E **2** and **3** only.

(**NISEC**: *all other Boards except SEB H, SEB SYS and AEB*)

See p. 33 and, in particular, see F2.5(a) and F2.5(b). Note that the diagram of pe against separation of the molecules provides the answer to **3** above; this should be carefully distinguished from the diagram showing variation of force with separation.

2.13M Two solenoids P and Q, of equal length but different numbers of turns, are arranged coaxially as shown in the diagram. P has 200 turns and Q has 300 turns.

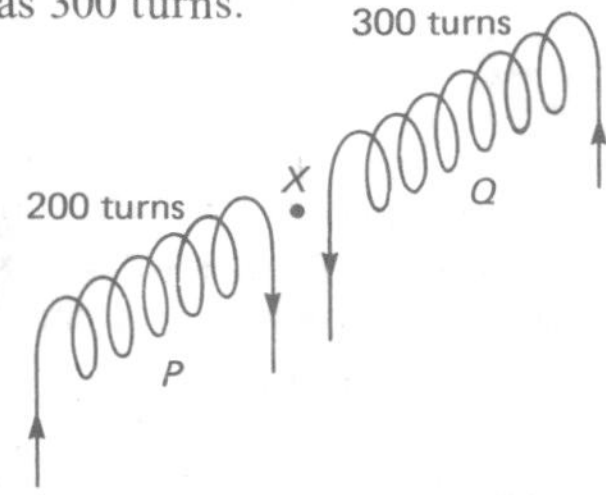

There is a current of 1 A in Q. What must be the current in P in order that there is no resultant field at X, midway between the coils?

A 2/3 A. B 3/4 A. C 1 A. D 4/3 A. E 3/2 A.

(**London**: *all other Boards except SEB H*)

Refer to E2.18 (although this formula is for the field inside the **centre** of a solenoid, the field at X, along the axis, will be proportional to the field at the centre of each of P and Q since the coils are of equal length and X is midway between them). Equate: field due to P = field due to Q, and solve to find the current in P.

2.14M Two moving-coil galvanometers X and Y are connected in series and a current passed through. Their readings are found to be full scale for X and half scale for Y. Given that their scales are identical and their restoring springs of the same strength, but that X's coil is twice as large in area and has twice the number of turns compared with Y's coil, the magnetic field strength of X's magnet compared with Y's magnet is:

A $\frac{1}{4}$. B $\frac{1}{2}$. C 1. D 2. E 4.

(—: *JMB, AEB, Oxford, Cambridge*)

Current sensitivity (ie, deflection per unit current) is equal to $\frac{BAn}{k}$ where B = magnetic field strength, A = coil area, n = number of turns, and k is the 'spring constant' of the restoring couple. With the same current and the same spring strength, the field strength ratio is equal to the ratio of $\frac{\text{deflection}}{An}$ for X to that for Y. Take Y's values as deflection = 1, area = 1, turns = 1 and then write down the X values, and hence the value of $\frac{\text{deflection}}{An}$ for X.

2.15M The diagram represents a current-carrying conductor RS of length 2 m placed at right angles to a magnetic field of flux density 0.5 T. The conductor experiences a force of 1 N as shown in the diagram. The magnitude and direction of the current is:

A 1 A from R to S. B 1 A from S to R.
C 2 A from R to S. D 2 A from S to R.
E 4 A from R to S.

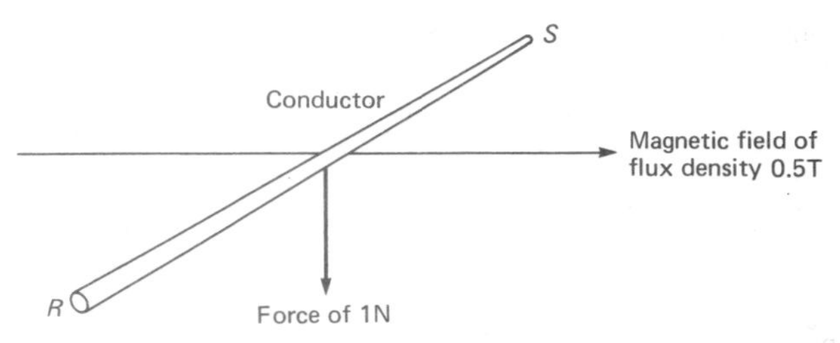

(**AEB** Nov 81: *all other Boards except SEB H*)

The magnitude of the current may be obtained using E2.15; the direction of the current is given by the left-hand rule (see p. 36 if necessary).

2.16M A simple solenoid X is made by winding a 1 m length of insulated wire around a hollow cardboard tube of diameter 1 cm and length 10 cm so as to produce uniform spacing of turns in a single layer along the whole length of the tube. A similar solenoid Y is then made by winding a 2 m length of wire around another tube of diameter 2 cm and length 20 cm, again so as to produce a uniform single layer of turns along the whole length. How many times must be the current in Y compared with that in X to produce the same magnetic flux density inside each tube?

A $\frac{1}{2}$. B 1. C 2. D 4. E 8.

(—: *all Boards except SEB H*)

Remember that the magnetic flux density along the axis of a solenoid is proportional to (i) the number of turns per unit length, (ii) the current. Work out what the turns ratio of Y to X will be, not forgetting that the circumference of Y is twice that of X. Then consider the turns per unit length for Y compared to X. Finally remember that the product of turns per unit length and current must be the same for X as for Y if they are to have equal field strengths.

2.17M The coil of a ballistic galvanometer is wound on a non-metallic frame because the moving system should have the least possible:

1 moment of inertia. **3** damping.
2 electrical resistance.

A **1, 2, 3** correct. B **1, 2** only correct. C **2, 3** only correct.
D **1** only correct. E **3** only correct.

(**AEB** June 80: *and Oxford, Cambridge, SEB SYS*)

The galvanometer is said to be 'ballistic' because the idea is that a sudden brief flow of charge through the coil sets the coil off into oscillating motion that should ideally be of constant amplitude. The charge must pass through in a pulse of short duration compared with the time period of the coil, so a heavy coil is useful since it would have a large time period. With low electrical resistance, problems would arise in 'search coil' experiments where the total circuit resistance was low, allowing induced currents to be generated by the coil motion so damping the oscillations. Damped oscillations would make the 'first throw' of the galvanometer difficult to measure. Refer to your textbook for further details.

2.18M Two long parallel wires X and Y carry currents of 3 A and 5 A respectively. The force experienced per unit length by X is 5×10^{-5} N to the right as shown in the diagram. The force per unit length experienced by wire Y is:

A 2×10^{-5} N to the left.
B 3×10^{-5} N to the right.
C 3×10^{-5} N to the left.
D 5×10^{-5} N to the right.
E 5×10^{-5} N to the left.

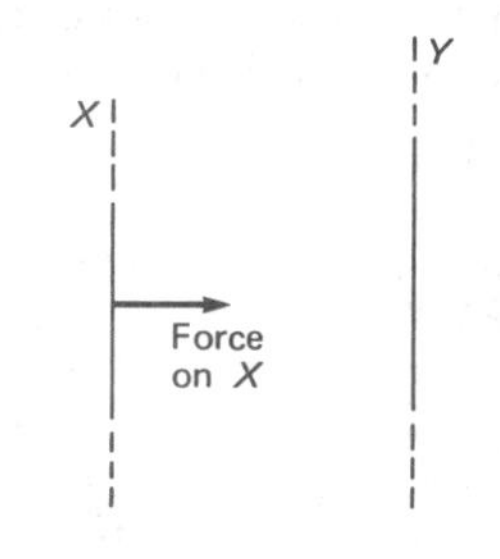

(**Cambridge**: *all other Boards except SEB H*)

Remember Newton's third law: see p. 17 if necessary.

2.19M Two moving-coil galvanometers, 1 and 2, are constructed with exactly similar permanent magnets and springs, but with coils of different numbers of turns (n_1,n_2), effective areas (A_1,A_2) and resistances (R_1,R_2). When they are connected in series in the same circuit, their deflections are respectively θ_1, and θ_2. The ratio θ_1/θ_2 equals:

A A_1n_2/A_2n_1. B A_1R_2/A_2R_1.
C n_1R_2/n_2R_1. D A_1n_1/A_2n_2.
E A_2R_1/A_1R_2.

(**Cambridge:** *and JMB, AEB and Oxford*)

The deflection of the coil is given by $\theta = \frac{BAn}{k}I$ where I = current, k is the 'strength' of the hairsprings, and B is the magnetic field strength. You will find a proof of the equation in your textbook. Since the magnets and springs are exactly similar, then B and k are the same for the two galvanometers. Also, because they are connected in series, then I is the same. Use the above expression to derive an expression for θ_1/θ_2 in terms of the factors of the expression that are not the same.

2.20M Coil P is connected to a 50 Hz ac source and lies close to a separate coil Q which is connected to the Y-input terminals of an oscilloscope. A wave-shaped trace appears on the screen of the oscilloscope.

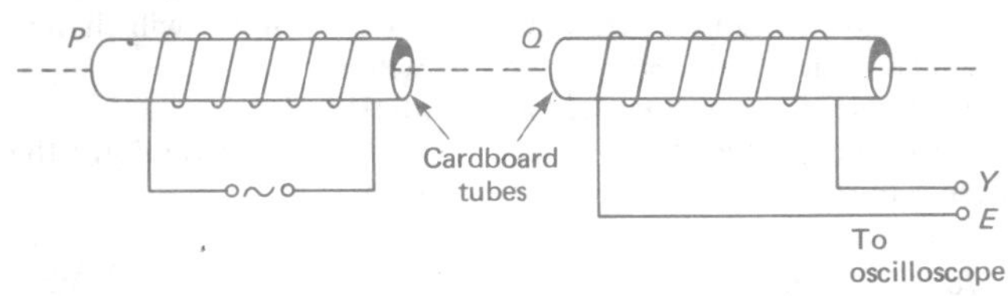

What would be the effect upon the oscilloscope trace of linking the coils by a common soft-iron core

	Height of trace	Number of waves on screen
A	increases	increases
B	decreases	increases
C	stays the same	increases
D	increases	stays the same
E	stays the same	stays the same

(**SEB H:** *all other Boards*)

The introduction of the iron core will have two effects: (i) it will increase the magnetic flux in the primary coil P (see p. 39); (ii) it will increase the flux linkage between coils P and Q, thus increasing the amplitude of the induced emf in coil Q without increasing its frequency.

2.21M The magnetic flux in the iron core of diagram X is proportional to (i) the current in the windings, (ii) the overall number of windings. Five identical iron cores with windings and currents as shown are represented in diagrams A – E. Which iron core has the greatest flux?

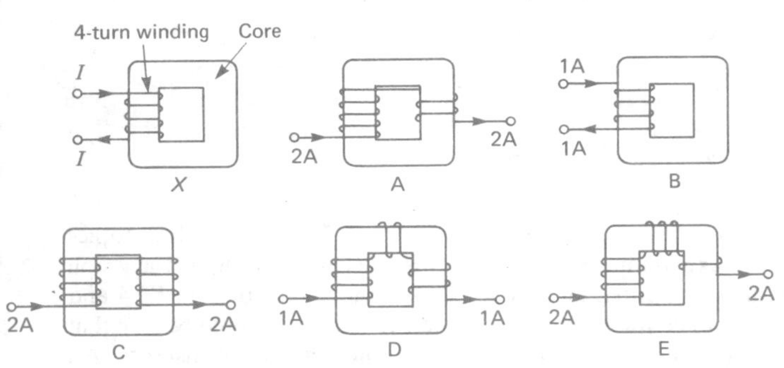

(—: *all Boards except SEB H*)

You must take account of the 'sense' of the windings. For example, C has two sets of windings that 'oppose' one another so there will be little flux in its core. Also, the position of the windings on the core makes no difference to the total flux.

2.22M The thick copper disc in the diagram (a) has its plane at right-angles to a magnetic field. The disc is then rotated until the plane is parallel to the magnetic field, as in diagram (b). As a

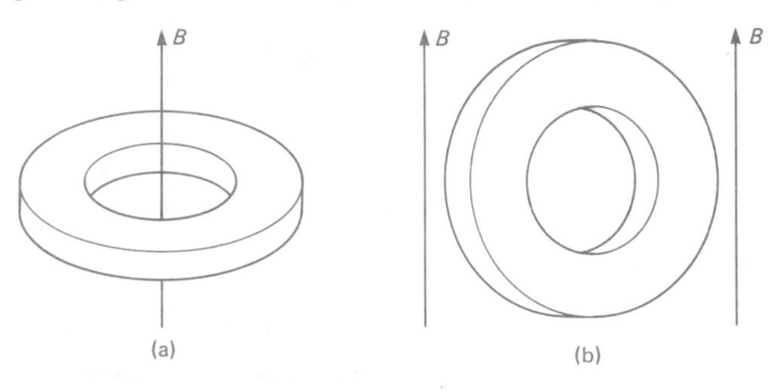

result of this movement an induced emf, and hence an induced current, is produced in the disc. This induced current could be increased by:

1 increasing the size of the central hole.
2 increasing the speed of rotation of the disc.
3 using a metal of lower resistivity.

A **1, 2, 3** all correct. B **1, 2** only correct. C **2, 3** only correct.
D **1** only correct. E **3** only correct.

(**AEB** June 81: *all other Boards*)

The induced current depends upon the induced emf and upon the resistance of the disc. For 2, you must consider how the speed affects the induced emf (and thus the current). For 3, you must consider how the resistivity affects the disc resistance (and thus the current). For 1, consider the relationship between hole size and initial flux, and so consider the change of flux and hence the induced emf: a larger hole will also increase the disc resistance, but does the effect of extra resistance act alone or does the emf differ as well?

2.23M Which one of the following effects observed in the study of electromagnetism is a consequence of self-induction?

A A large voltage may occur when the current in an iron-cored coil is interrupted.
B A magnet plunged into, or removed from, a solenoid in a circuit encounters forces opposing its motion.
C Galvanometer coils are often wound on light metal frames.
D The introduction of an iron-core increases the magnetic field of a solenoid.
E The magnetic flux density well inside a long solenoid is independent of the cross-sectional area of the solenoid.

(**London:** *and all other Boards except SEB H*)

Observe that the question refers to self-induction, so that consideration of E2.29 should provide the answer.

2.24M X, Y and Z represent three wires perpendicular to the paper. Currents in X and Y are into the paper while that of Z is

out of the paper. The resultant force on Y due to the currents in X and Z is:

A zero.
B perpendicular to the line joining XYZ.
C in a direction from Y to Z.
D in a direction from Y to X.
E in a direction dependent on the current strengths.

(**AEB** June 80: *all other Boards except SEB H*)

Remember that two parallel wires carrying current in the same direction tend to attract, whereas they tend to repel if the currents are in opposite directions. Thus, consider first the force on Y due to the interaction of Y with X; then consider the force on Y due to its interaction with Z. Then consider the combined effect of the two forces in terms of direction, and hence choose your answer.

2.25S **(i)** The magnitude of the force on an electron when it is at a point A, 5.0 mm from a point charge of 2.0×10^{-6} C, is 1.2×10^{-10} N.

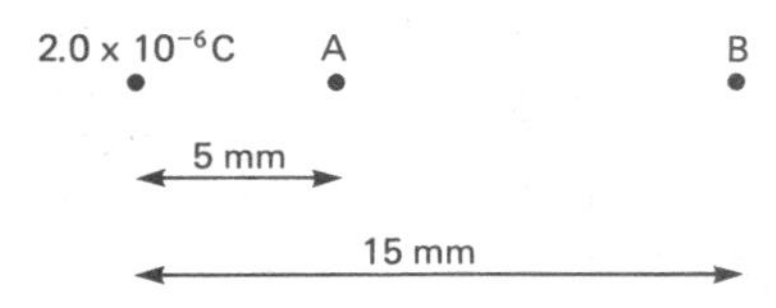

What is the magnitude of the force on an electron when it is at point *B*, a distance of 15 mm from the same point charge?
(ii) Calculate the electric field strength due to the 2.0×10^{-6}C charge at point *A*.

(**SEB H**: *all other Boards*)

(i) Remember that the force between two point charges is an **inverse-square** law, ie, $F \propto \frac{1}{d^2}$ which can be written as $F \times d^2 =$ constant. Therefore, $F_1 d_1^2 = F_2 d_2^2$ and since F_1, d_1, and d_2 are given, then F_2 can be calculated.

(ii) Remember that electric field strength = force per unit (+) charge. Assume that the charge on an electron is 1.6×10^{-19}C.

2.26S

(a) (i) An *electric field of force* exists around an isolated point charge. Explain briefly the meaning of the term in italics. (ii) The units for electric field intensity may be stated as volts per metre or newtons per coulomb. Show that these are equivalent.

(b) Sketch the pattern of electric lines of force for the following: (i) an isolated negative charge; (ii) an isolated positive charge.

(c) (i) Two equal charges of opposite sign are placed at *A* and *B*, two of the vertices of an equilateral triangle *ABC*. The field due to each charge may be considered separately, and the results combined to give the resultant field. Using this idea, indicate the direction of the electric field at C.

(ii) Gravitational fields may be combined in the same way as electric fields. If the charges at *A* and *B* were replaced by equal point masses *m*, indicate on the diagram the direction of the gravitational field at *C* due to the masses at *A* and *B*.

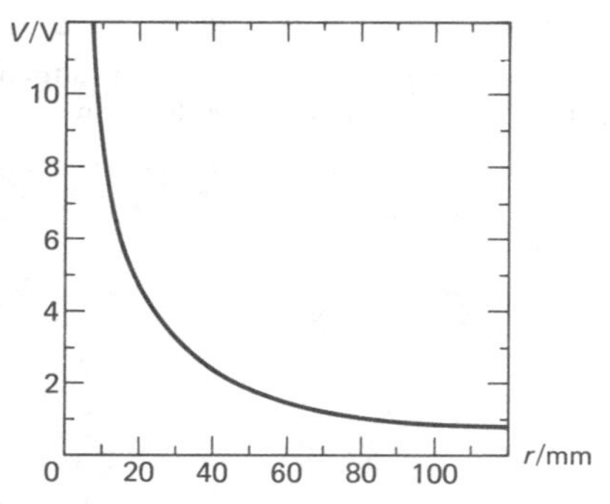

(d) The graph shows the variation of electric potential *V* with distance *r* from an isolated positive point charge. (i) Making use of the graph, calculate the force on a point charge of + 2 C placed 30 mm from the isolated point charge. (ii) The + 2 C charge is moved from the point 30 mm from the positive charge to a point 10 mm from the positive charge. How much energy is required to do this?

(**NISEC**: *all other Boards except SEB H*)

(a) (i) Remember that a field is a region around an object in which it can exert a force on a second object. Apply this to an electric field of force. (ii) Use the definition of the volt as 'joule per coulomb'.

(b) (i) and (ii) Imagine a **positive** test charge placed in the field.

(c) (i) Imagine a positive test charge at *C*. Apply the parallelogram rule, made easier in this case because the force on the test charge due to *A* has the same magnitude as the force on the test charge due to *B*. Take care with directions! (ii) Remember that the force on a test object is always attractive in a gravitational field.

(d) (i) Since $E = -\frac{dV}{dr}$ (see E2.1), the value of *E* in V m^{-1} can be deduced from the gradient of the graph at the point corresponding to $r = 30$ mm (care with units) V m^{-1} is equivalent to NC^{-1}, hence the force on the +2 C charge can be calculated. (ii) Find the pd between the two points from the graph. This gives the energy required to move *unit* charge between two points. Apply this to a charge of + 2 C.

2.27S Define electric field strength and electric potential at a point. Derive the relationship between the electric field strength and the electric potential gradient at a given point.

(**AEB** June 81: *all other Boards except SEB H*)

Express your definition in terms of the force per unit + ve charge acting on a **small** positive charge; a unit charge of one coulomb is so large that in most circumstances distortion of the charge producing the field will occur, so don't give your definition as force acting on a unit charge, since the unit charge will change the field – but a small positive charge will not!

Take similar care for electric potential!

For the relationship, see E2.1. Consult your textbook for the derivation.

2.28S The diagram shows the variation of the Earth's gravitational potential *V* with distance *r* from the centre of the Earth.

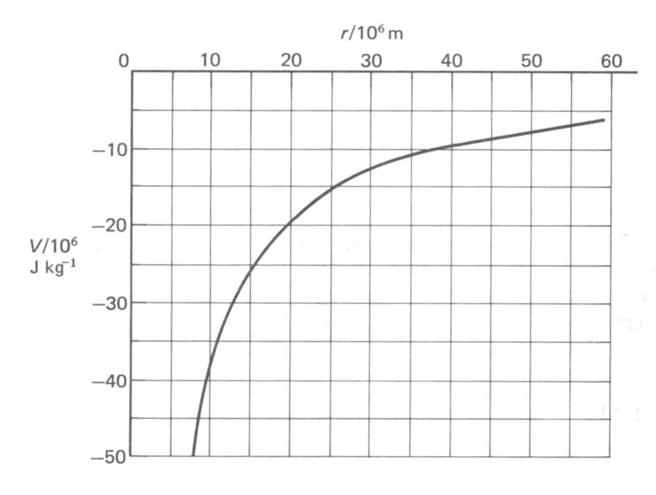

(a) (i) What is the potential at an infinite distance from the Earth? (ii) Why is the potential on the graph negative?

(b) What is the value of the gravitational field at 25×10^6m from the centre of the Earth? Show how you arrive at your answer.

(c) How much energy is required to raise a mass of 70 kg from 10×10^6 m to 50×10^6 m above the Earth's centre? Show how you arrive at your answer.

(**O and C Nuffield**: *all other Boards except SEB H*)

Those in difficulties can refer to 2.2, 2.3 and 2.5. These topics will explain most of this question. Be careful in **(b)** as the graph shows potential and the 'field' is being asked for; use E2.4 and E2.5 to help solve this. From these formulae it should be seen that it is only necessary to divide the potential by the distance. An alternative method is to use the negative value of the slope of the potential graph at 25×10^6 m from the centre of the Earth (see E2.1). Part **(c)** is more straightforward; simply read the potential difference between the two points off the graph and multiply by the mass involved to find the energy needed.

2.29S A closed wire loop in the form of a square of side 4.0 cm is mounted with its plane horizontal. The loop has a resistance of $2.0 \times 10^{-3}\Omega$, and negligible self-inductance. The loop is situated in a magnetic field of strength 0.70 T directed vertically downwards. When the field is switched off, it decreases to zero at a uniform rate in 0.80 s. What is (a) the current induced in the loop, (b) the energy dissipated in the loop during the change in the magnetic field? Show on a diagram, justifying your statement, the direction of the induced current.

(**SUJB**: *all other Boards except SEB H*)

(a) The change of magnetic flux through the loop induces an emf in the loop. The induced emf then drives a current round the loop since the loop is a closed circuit. Calculate the initial flux through the loop; see E2.24 if necessary. Then use E2.25 to calculate the induced emf. Finally, use the value of induced emf and the given value of resistance to calculate the current. Remember that the area of the loop must be in units of m^2.

(b) Since the emf is constant (because the flux changes at a steady rate), then the current is constant. To calculate the energy dissipated, calculate the power involved from E7.4, and then calculate the energy dissipated in 0.8 s.

Draw the loop as seen from above, and indicate that the magnetic field lines pass into the plane of the diagram. The decrease of applied flux induces a current in such a direction as to try to maintain the flux through the loop; use your knowledge of magnetic field patterns to work out the current direction in the loop that would give flux in the same direction as the applied flux. Note that if the applied flux had been increased instead of decreased, the induced current flow would have been in the opposite direction so as to try to keep the flux at its initial level. See p.38 if necessary.

2.30L This question is about gravity and the similarities between gravity and electric and magnetic effects.

The passage below sets out three sets of ideas about gravity. For each of the sections (i) to (iii) you are asked to write a more complete explanation of the ideas: your explanations may include

- quantitative calculations to illustrate the ideas,
- fuller explanations of the theoretical ideas,
- discussion of possible experiments.

You should pay particular attention to the words and phrases that are underlined in each section.

Passage

(i) There is something peculiar about gravity: it is such a small force that if we didn't live on a big lump of matter called the Earth we might not notice that it affected man-size objects at all. In fact the simplest calculations can show that it is very hard to demonstrate that the effect exists between all pieces of matter.

(ii) There is a close analogy between the theoretical ideas involved in electricity and in gravity, and this can be of great value in discussing such abstract ideas as field and potential. Thus problems such as the scattering of alpha-particles by a nucleus and the path of a spaceship round the moon have many similarities though there are also important differences.

(iii) However, electrical and magnetic effects are so much bigger, for man-size experiments, that they swamp all effects of gravity. The fact that when we come to matter on an astronomical scale, gravity is by far the most important force is then hard to explain – it must be due to electrical neutrality of big objects.

(**O and C Nuffield**: *all other Boards except SEB H*)

(i) Clearly you are expected to provide a calculation of the force (E2.4) between man-sized objects, and should compare this with forces that are large enough to be easily measured (eg, with a spring balance). The experimental techniques needed to measure these gravitational forces should be discussed, eg, by reference to experiments to measure G (as described in your textbook). Concentrate on describing how the experiment copes with measurement of the tiny forces involved, but do not start answering the wrong question (ie, how do you measure G?)!

(ii) An analogy is said to exist between two physical situations if they can be described by the same mathematics. This can be discussed for electric and gravitational fields using E2.2–E2.5. In discussing these equations the terms 'field' and 'potential' can be explained more fully (2.1, 2.2, 2.3). Alpha-particle scattering by the nucleus is an electric field effect and spaceship paths round the moon are controlled by gravitational fields; hence the 'similarities' of the analogy must apply; however differences exist in the scale of distances involved and the nature (repulsive or attractive) and magnitude of the forces. Calculations can back up discussion of the magnitude of the forces. A mention may be made of the experimental model of both situations that can be created by rolling a ball on a '$1/r$' hill.

(iii) Calculate some electric and magnetic forces for the man-sized objects of part **(i)** using typical values of charge and current; use ratios to show how much bigger these forces are. Explain that electrical neutrality means 'uncharged' and the effect of magnetic and electric forces in astronomy becomes negligible. This can be confirmed by experiments on planets/moons showing that their motion is entirely controlled by gravitational forces.

2.31L

(a) A conducting sphere of radius 0.10 m is suspended by a long insulating thread. The sphere is charged to an electric potential of 2000 V.

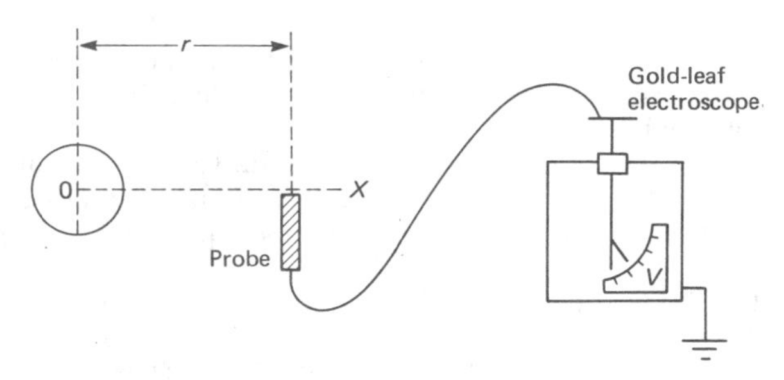

The approximate values for the electric potential at different distances r from the centre of the charged sphere are measured using a probe connected to a gold-leaf electroscope. This gold-leaf electroscope is calibrated in volts. The values obtained are given below.

Electric potential (V)	2000	1000	700	500	400
r (m)	0.10	0.20	0.30	0.40	0.50

(i) Draw a graph to show the variation of electric potential with distance along OX from the centre of the sphere. (ii) Plot the corresponding graph to show how the electric field strength varies with the distance along OX.

(b) Write down an expression for the electric potential at the surface of an isolated conducting sphere of radius r which carries a charge Q. Show that the capacitance of the sphere C is given by $C = 4\pi\varepsilon_0 r$ where ε_0 is the electric permittivity of free space.

(c) The spherical dome of a Van de Graaff generator has a radius of 0.10 m and when operating steadily gives two spark discharges every second to a nearby earthed sphere. With the generator operating at the same rate the dome is connected to earth through a meter which indicates a steady current of $10\,\mu A$. (i) Calculate the average charge transferred during a discharge. (ii) Using the relationship $C = 4\pi\varepsilon_0 r$ given in **(b)**, calculate the change in potential difference between the dome and earth which occurs as a result of each of these spark discharges. (iii) Discuss the validity of using the given expression $C = 4\pi\varepsilon_0 r$ in this situation.

(**SEB SYS**; *and O and C Nuffield, Oxford, JMB*)

(a) See F2.4(a) and F2.4(b) if necessary.

(b) See E2.3. For the capacitance formula, remember that $C = Q/V$.

(c) (i) See E7.1 if necessary, to give the charge flow per second. Then with two discharges per second, calculate the charge flow per discharge. (ii) If Q is known per discharge, and capacitance C is known, then V can be found. (iii) On an isolated sphere the charge distribution will be uniform. If a second earthed sphere is placed nearby, it will affect the charge distribution considerably, hence affecting the validity of the expression. In general the capacitance of a conductor is increased by the presence of any nearby earthed conductor. Furthermore consider the shape of the Van de Graaff dome and compare it with the shape of conductor for which the formula applies.

2.32L Write equations for Coulomb's law in electrostatics and Newton's law of gravitation. Identify the symbols in the equations.
Explain what is meant by the 'electric potential' of a conductor. Deduce, from first principles, an expression for the potential of an isolated, charged conducting sphere.
Electrons 'leak' easily from the surface of many stars so that such stars acquire a positive charge. This charging stops when the charge on the star is so large that protons in the surface also begin to be repelled. This occurs when the sum of the gravitational potential energy and the electrical potential energy of a proton near the surface is zero.

(a) Write down the equation relating these two energies.

(b) Show that, in the steady state, the maximum charge carried by a star of given mass is independent of its radius.

(c) Calculate the maximum charge for the Sun which has a mass of 2.0×10^{30} kg.

(**Cambridge**: *all other Boards except SEB H, SEB SYS*)

For Coulomb's law of force between two point charges, see E2.2. For Newton's law of gravitation for the force between two masses, see E2.4.

To explain electric potential, see p. 32. E2.3 gives the formula for the electric potential at the surface of an isolated, charged conducting sphere. For the proof of the equation, see your textbook.

(a) The field potential at the surface of an object such as a charged star can be written as k/r; for the gravitational field, $k = -GM$, and for the electric potential $k = +Q/4\pi\varepsilon_0$, where M is the mass and Q is the charge of the star. If the mass and charge of a proton is represented by m_p and $+e$, then the pe of the proton due to the gravitational field can be written down, and also the pe of the proton due to the electric field can be written down. At the surface, these two expressions are equal in size although of opposite sign (ie, electric pe $= +X$ joules, gravitational pe $= -X$ joules).

(b) Use your expressions from **(a)**. Make Q the subject of the equation.

(c) Substitute values for G, e, m_p, ε_0 and M. Consult the chart on page 186 for these values if necessary.

2.33L Three parallel metal plates A, B, C each of area $2.0 \times 10^{-2}\,\text{m}^2$, are separated by distances $AB = 1.0$ cm, $BC = 2.0$ cm. The plates A and C are connected to earth, and the plate B is maintained at a potential of $+500$ V. What is the charge on (a) the left side, (b) the right side of the plate B? What is the energy of the system?

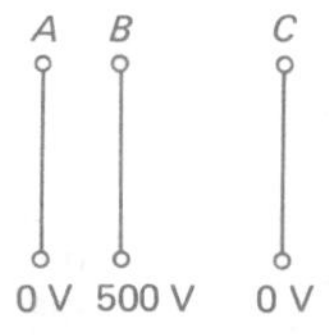

How would you use a proof plane and electrometer to measure the charge density on each surface of the plate B?
What does the potential of B become if it is first isolated and then moved, at right angles to its plane, to a position midway between A and C?
What is the energy of the system when B is midway between A and C, and how do you account for the change of energy which has occurred?

(**SUJB**: *all other Boards except SEB H*)

Plate B will have more +ve charge on its left-hand side (ie opposite A) than on its right-hand side, because the electric field is stronger from B to A than from B to C. Use E2.9 to calculate the strength of the electric field across (i) AB, (ii) BC. The charge per unit area on either the right-hand side of A or the left-hand side of B is then given by E2.10; hence calculate the charge on the left-hand side of B. The same equation can be used to calculate the charge on the right-hand side of B, but remember this time that the electric field strength corresponds to that between B and C.

To calculate the energy of the system, use E2.14 in the form $\frac{1}{2}QV$. Treat AB as a pair of parallel plates with $V = 500$ V and Q as calculated above, and so determine its energy. Then, repeat the method for BC. Add the two energy values to give the total energy of the system.

For the proof plane and electrometer experiment, consult your textbook if necessary.

When B is isolated, its total charge cannot be changed. Movement of B, after isolation, to the midpoint means that its total charge is unaltered but is redistributed from one side of B to the other side of B. At the midpoint, the charge on either side must be equal, so you can calculate its value. You should find that the left-hand side of B loses charge to the right-hand side. Once you have calculated the new charge on each side of B, then use E2.10 to calculate the electric field strength in each gap, and then use E2.9 to calculate the potential of B.

For the new energy of the system, you can follow the same steps as before, but take care to use the new charge and the new potential values. Alternatively, since all the charge is at the same new potential, you can use '$\frac{1}{2}$ × total charge × new potential'. To account for the change of energy, remember that charge redistribution means current flow through resistance. Don't forget that the negative charge on the two earthed plates is also redistributed. (assume A and C are joined by a wire) and that work (= force × distance) can be done by moving the charged plate B in the electric field.

2.34L

(a) (i) State Newton's Law of gravitation. (ii) Find the dimensions of the gravitational constant G. (iii) Calculate the work done against gravitational attraction when two particles of masses m_1 and m_2, initially at distance r apart, are moved to an infinite separation.

(b) (i) Show that the mass M of the Earth may be expressed in terms of the radius R of the Earth, the gravitational constant G, and the acceleration of free fall, g, by the formula $M = gR^2/G$. (ii) What approximations, made about the Earth in the foregoing derivation, restrict the validity of the formula? (iii) Calculate the radius of the Earth, assuming the mean density of the Earth's crust, $3.0 \times 10^3\,\text{kg m}^{-3}$, is representative of the Earth as a whole (iv) The mean radius of the Earth is known to be 6360 km. What does your result suggest about the mean density of the Earth's crust compared with the mean density of the rest of the Earth? Explain your answer.

(c) An astronomer has found the mass of the Earth to be 6.00×10^{24} kg. His observations of the Moon show that its period of rotation round the Earth is 27.3 days. Find the radius of the Moon's orbit, assuming it to be circular and concentric with the Earth.
(Gravitational constant $G = 6.67 \times 10^{-11}$ SI units; acceleration of free fall $= 9.81\,\text{m s}^{-2}$; volume of a sphere $= \frac{4}{3}\pi r^3$.)

(**NISEC**: *all other Boards except SEB H, SEB SYS*)

(a) (i) See E2.4 if necessary. (ii) Make G the subject of the equation of **(a)** (i), then substitute the dimensions of each term into the rearranged equation. Remember that force can be represented by mass × acceleration. Thus prove the dimensions of G are $\text{M}^{-1}\text{L}^3\text{T}^{-2}$. (iii) A proof of E2.5 is required. Use your textbook.

(b) (i) Use Newton's law of gravitation to show that $g = GM/R^2$. Hence . . . etc. (ii) What are the assumptions underlying the law of gravitation that do not apply to the Earth? Is g a universal constant? (iii) From mass = volume × density, obtain an expression for the mass of the Earth in terms of R. Use this in the equation $M = gR^2/G$. (iv) The calculated value of R is too high, so check from your equation in (iii) how the value of density would need to be changed to give a lower value for R.

(c) See p. 35 and E2.8: $v = \sqrt{\frac{GM}{R+h}}$. Since speed $v = 2\pi(R+h)/T$ then combine the two expressions and then solve for $(R+h)$. Take care with units!

2.35L

(a) What characteristics of solid materials suggest that the forces between molecules in such substances have both an attractive and a repulsive component? Explain your answer.

Figure (1) shows how these components individually might vary with separation for a pair of molecules. Construct on Figure (1), using the same axes, a graph showing how the **resultant** force between the molecules varies with their separation. Use the graph

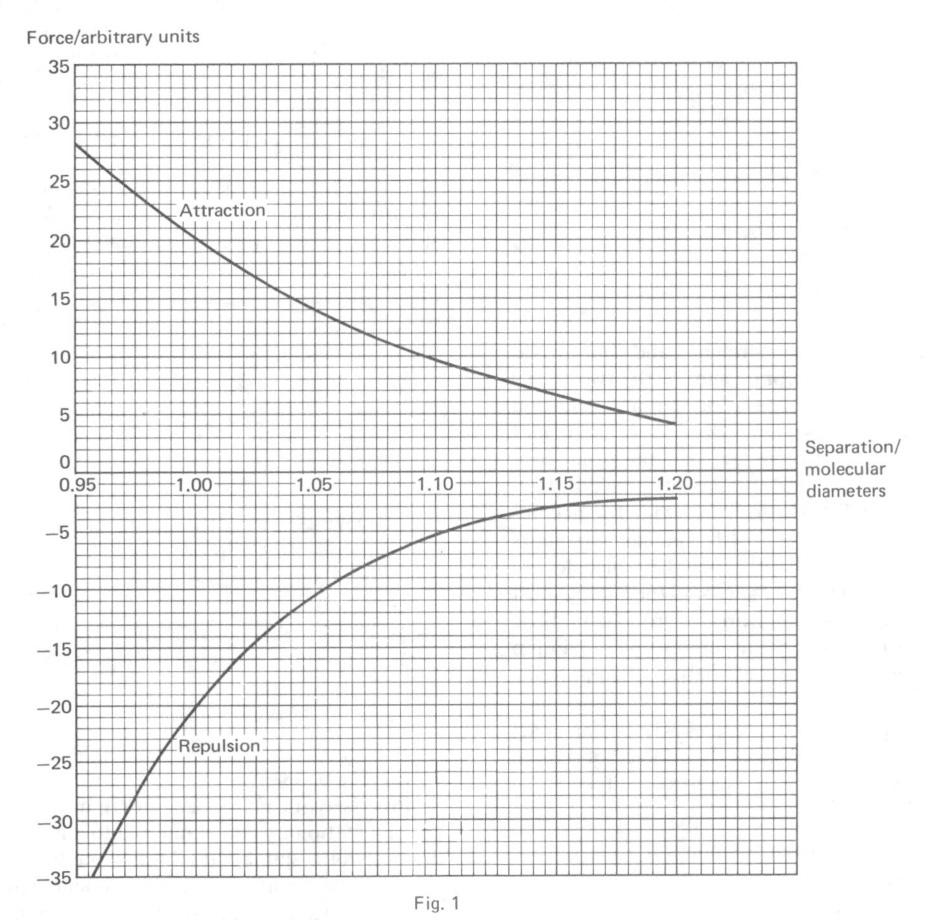

Fig. 1

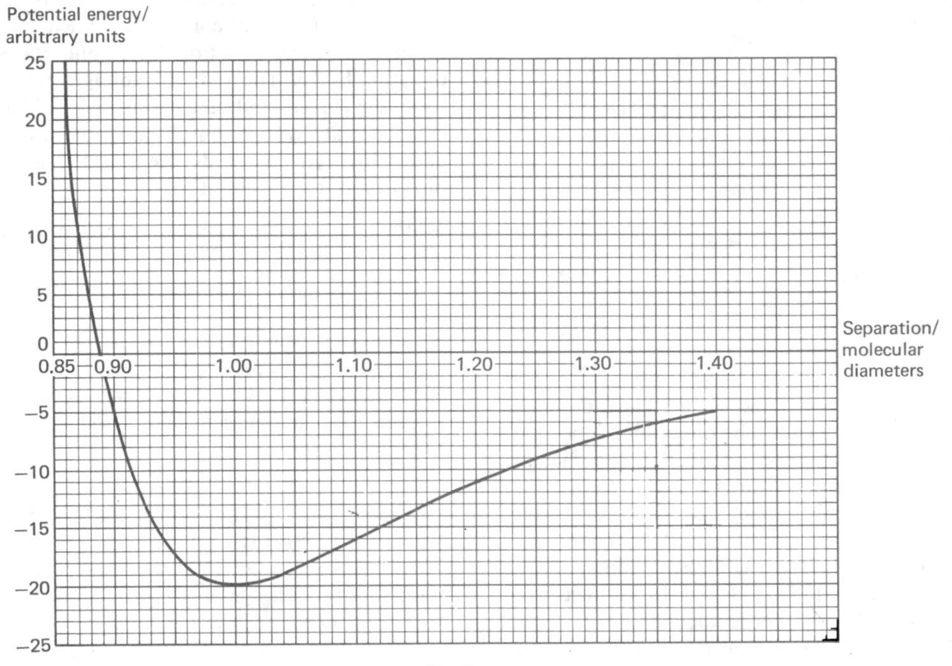

Potential energy/ arbitrary units

Separation/ molecular diameters

Fig. 2

you have drawn to determine (i) the equilibrium separation of the molecules, and (ii) the separation when the intermolecular attraction is a maximum.

(b) Figure (2) shows how the potential energy of this pair of molecules depends on their separation. Account for the shape of the curve by referring to the graph you have drawn on Figure (1). Use the graph in Figure (2) to find (i) the energy, in arbitrary units, required to squeeze the molecules together from a separation of 1.0 molecular diameter to a separation of 0.9 molecular diameter, and (ii) the energy, in arbitrary units, required to free the molecules from each other. If the pair of molecules, initially at the equilibrium separation found in (a), are given 10 units of vibrational energy find, using the graph, their new mean separation.

(c) In the gaseous state, the forces between molecules appear to be negligible except at very high pressures. Show that this behaviour could be explained if the forces between the gas molecules were of the form shown in Figure (1).

(**London**: *all other Boards except SEB H, SEB SYS, AEB and NISEC*)

(a) Think how solids behave when stretched and compressed (particularly rubber!). Find the algebraic sum of the two forces at convenient values of molecular separation (eg, at separation $= 0.975$, resultant force $= (+24) + (-28) = -4$) and plot several values. Don't be misled by the fact that the usual convention of making repulsive forces positive has not been adopted.

(b) Refer to F2.5(a) and F2.5(b) together with the accompanying explanation. (i) The energy must be increased from its value at 1.00 to the new value at 0.90. (ii) When the molecules are free from each other their pe is defined as zero, so the energy must be increased from the minimum value up to zero. To find the mean separation, take the average of the minimum and maximum displacements at the level which is 10 energy units above minimum pe.

(c) At normal pressures the gas molecules will be far apart (so force = ?) and at very high pressures the molecules will be only a few molecular diameters apart (so force will be . . . ?).

2.36L Describe an experiment by which the magnetic flux density, B, at different points between the poles of a horse-shoe magnet could be compared.

(b) (i) Draw a diagram showing the magnetic field pattern due to a long, straight wire carrying current. Mark clearly the relative directions of the current, I (conventional), and the magnetic flux density, B. (ii) At a distance r from the wire, the magnitude of B is given by

$$B = \frac{\mu_0 I}{2\pi r}$$

where μ_0 is the permeability of free space. Calculate the value of B at a distance 50 cm from a long straight cable carrying a direct current of 1000 A ($\mu_0 = 4\pi \times 10^{-7}$ H m^{-1}). (iii) Ions in the atmosphere will be affected by the magnetic field produced by such a cable. Calculate the force acting on an ion travelling directly towards the cable at a speed of 10^4 m s^{-1} when it is 1.0 m away from the cable. Assume the ion carries a charge of $+1.6 \times 10^{-19}$ C. Explain, with the aid of a careful diagram, how the path of the ion is affected by the field as it approaches the cable.

(**London**: *all other Boards except SEB H*)

(a) Either: use a **small** search coil and ballistic galvanometer (draw a diagram of the arrangement). The direction of the field must first be determined so that the coil is initially perpendicular to the field. The coil should have many turns and long connecting leads. If θ_{max} is the maximum galvanometer deflection when the coil is rapidly removed from the field, $\theta_{max} \propto B$. Refer to your textbook for further details. **Or**: Use a Hall probe (again, draw a diagram of the circuit). The Hall pd (V_H) must be measured when the face of the probe is perpendicular to B, then $V_H \propto B$ provided the current is kept constant. Refer to your textbook if necessary.

(b) (i) See p. 37. (ii) Take care with units! (iii) Find B as above, then use $F = Bqv$ (q = charge, v = speed). Draw a diagram showing the direction of the magnetic field round the wire and the direction of the ion. From Fleming's left-hand rule, deduce the direction of the force on the ion, and so its effect on the motion of the ion. Remember that as the ion approaches the wire, the field (and thus the force) will increase **considerably** ($\propto 1/r$).

2.37L

(a) State Faraday's law of electromagnetic induction in words and in the form of an equation.

(b) Describe two methods of investigating Faraday's law, (i) qualitatively, by a method which does not involve the movement of a conductor in a magnetic field, and (ii) quantitatively, by a method in which a conductor moves in a magnetic field.

(c) As a physics project, two students investigate the strength and direction of the Earth's magnetic field by moving a long wire connected to a sensitive microammeter in the field of the Earth. See the diagram.

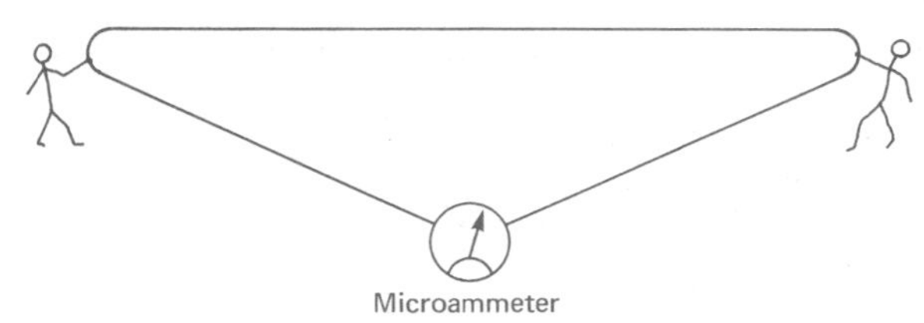

The straight section of the wire is of length 3.3 m and is at right angles to the direction of magnetic north. While the straight section is moving vertically downwards at 6.0 m s^{-1}, there is a current of 4.0×10^{-6} A in the circuit. While it is moving horizontally at 3.6 m s^{-1} the current is 7.0×10^{-6} A. Calculate the strength and direction of the magnetic field at the location of the experiment. The resistance of the microammeter is 100 ohms and the resistance of the wire is negligible.

(**SEB SYS**: *all other Boards except SEB H*)

(a) See p. 38.

(b) Use your textbook. For (i), movement of magnet in and out of a stationary coil connected to a galvanometer will provide the basis of a suitable qualitative method for your description. For (ii), the 'rotating disc' experiment described in most textbooks will suffice, but you must state what measurements are made (and with what instruments) and how Faraday's law is verified from your measurements.

(c) The Earth's magnetic field has both a horizontal component and a vertical component. Each can be found from the given data using equation E2.25 and the ideas illustrated by diagram F2.15. The induced voltages must first be calculated using the current and meter resistance values. Then use E2.25 to calculate each component. Finally, the two components must be added by vector methods (see p. 19) to give the magnitude and direction of the resultant.

2.38L

(a) The defining equation for self inductance is $E = -L\frac{dI}{dt}$. Explain what is meant by 'self-inductance', and define the SI unit of self-inductance.

(b) The apparatus represented in the diagram is to be used to demonstrate self-induction in dc circuits. In circuit 1, the value of the resistor R is 3Ω, and the lamp (of constant resistance 9Ω) is correctly lit when the switch is closed. In circuit 2, the resistance

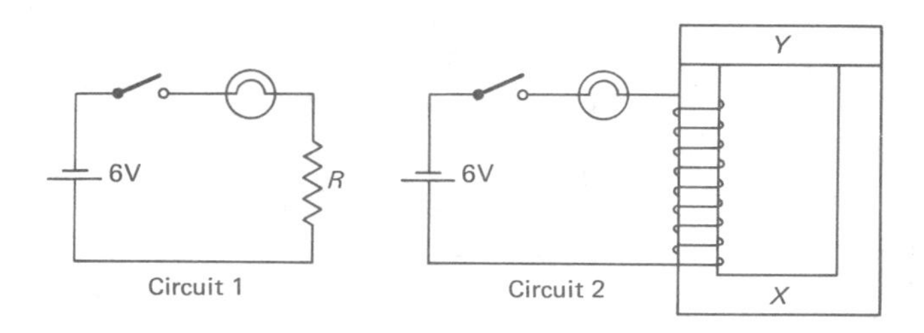

of the coil L (of about 2000 turns) is 3Ω and its self-inductance (when the two parts of the core are together as shown) is 4.0 H, and the lamp (of constant resistance 9.0 Ω) is similar to that of the first circuit.
(i) State and explain what difference would be observed in the behaviour of the lamps when the switches are closed in each of the circuits. (ii) In circuit 2 what is the greatest current and the greatest rate of change of current after the switch is closed? (iii) No difference is observed in the behaviour of the lamps in the two circuits when the switches are opened after having been closed for some time. State any real difference between the behaviour of the circuits, and explain why it is not shown by the lamps. (iv) With the switch in circuit 2 closed, the yoke Y is removed and then replaced. State and explain what is observed in the behaviour of the lamp if this is carried out (A) slowly and (B) quickly.

(**Oxford:** *all other Boards except London and SEB H*)

(a) See p. 39 and E2.29. You should explain why a changing current causes a self-induced emf in a coil, and to do this you must bring in the basic principle of electromagnetic induction (ie, change of magnetic flux in a coil causes an induced emf across the coil terminals) as the link between **changing** current and the self-induced emf. To define the SI unit of L (ie, the henry), use the defining equation given; if necessary, consult your textbook.

(b) (i) The observed difference in the lamps is required, so you should state if their individual brightnesses grow at the same rate, and if their final brightnesses are the same. Remember that self-inductance opposes the change of current (which is growth here) but does not affect the value of the final steady current; give your explanation in terms of current, but remember to state that the lamp gets brighter as the current gets larger. (ii) For the greatest **current**, the current must have grown to its 'final' steady value when $\frac{dI}{dt} = 0$. Then, all the battery emf will be dropped across R since the self-induced emf due to L will have fallen to 0. So use $I = V/R$. For the greatest **rate of change** of current, this is when the self-induced emf is equal and opposite to the battery emf, and there is no pd across R. This is the situation when $I = 0$ (ie, initially). Calculate $\frac{dI}{dt}$ from the defining equation. (iii) In this situation, the effect of L is to try to maintain the current, so giving a big emf and a spark across the switch gap. (iv) Removal of the yoke causes the flux in the core to drop, and decreasing flux gives an induced emf in the same direction as the battery emf. Remember the **rate** of change determines the size of the induced emf. Answer in terms of lamp brightness.

2.39L This question asks you to say more about **four** brief statements about magnetic flux.

(a) The passage below consists of **four** main statements, numbered **(i)**, **(ii)**, **(iii)** and **(iv)**. For each of these statements you are asked to give arguments which explain and support the statement; these might refer to theoretical principles, or to experimental evidence, or both. Your arguments should include explanation of the principles and description of the evidence.

(b) Statement **(iii)** could be described in general terms as a statement about practical applications based on an experimental result (the value of μ_r). For each of statements **(i)**, **(ii)** and **(iv)** give a similar description to say what sort of ideas are in the statement, eg, it is a definition based on previous ideas or it is a summary of experimental results or some other description you think appropriate; give a **brief** justification for each description.

Passage

(i) We can show from experiments with solenoids in air that magnetic flux φ is related to the current I and number of turns N by an equation of the form

$$NI = \frac{\varphi L}{\mu_0 A}$$

where L is the length and A the cross-section area; however, it takes a whole set of careful experiments to sort out the various factors correctly.

(ii) The equation is very like the equation which relates the flow of electric current in a circuit to the various factors which determine it: we can compare the two equations term by term, so that it seems reasonable to say that magnetic flux is like the flow of something.

(iii) Useful practical applications nearly always use solenoids filled with iron because when iron is used, μ_0 is replaced by $\mu_0\mu_r$ and the value of μ_r can be as much as 1000.

(iv) The idea of flux as a flow around a magnetic circuit is often helpful. For example, it can easily be shown that if an iron ring electromagnet has a small air gap cut in it there is a large reduction in φ – just like the effects on the current when a large resistor is inserted in series in an electrical circuit.

(**O and C Nuffield**: *all other Boards except SEB H*)

The question asks for a description of the principles and evidence of theoretical and experimental support for each of the four statements and a description of the nature of each statement. You are not asked to describe experiments in any detail, but more to concentrate on the results of these experiments and how they support the statements. You are not asked to actually prove any formulae but more to explain the theoretical ideas on which a proof of formulae (or statements) could be derived. The following comments on each of the four statements may help.

(i) Rough experiments show that φ is proportional to N and I, but far more careful detailed experiments are needed to deduce the full formula. Refer to the nature of the experiments (with coils) and the results that enable one to draw the conclusions that lead to the statement (eg, that the field in a solenoid is independent of its area). Show how the results link into the equation given.

(ii) An analogy is drawn between the equation of part **(i)** and $V = IR$ (see 2.13); this results in magnetic flux being analogous to electric current, which in turn produces the concept of magnetic flux having a 'flow' property. Compare the equations term by term showing how the analogous quantities all have similar properties in their own type of system (be it electrical or magnetic).

(iii) This is an application of the full version of the formula from part **(i)**; ie, a theoretical application of $NI = \varphi L/\mu_r\mu_0 A$. You may also argue that it is a consequence of carrying out experiments with iron (rather than air) cores to the coils. However the full formula is derived from such experiments anyway! The large flux that can be obtained from iron because of its value of μ_r should be explained, together with the need for large flux in practical applications (eg, electromagnets, motors).

(iv) The analogy between magnetic and electric circuits enables the behaviour of magnetic systems to be deduced from studying the more familiar electric circuits that are analogous to them. Draw diagrams of the magnetic and electric circuits described in this case and explain how the reluctance of the air gap is analogous to the high value resistor; show how the formula for resistances in series in electrical circuits ties in with the magnetic reluctances in series in the magnetic circuit. Hence even the smallest air gap can produce a dramatic reduction in flux, just like a break in an electric circuit can produce a dramatic drop in current.

2.40L A straight solenoid of length L and circular cross-section of radius r is uniformly wound with a single layer of N turns and carries a current I. The solenoid is so long that end-effects may be neglected. Write down expressions for the flux density B inside the solenoid and the total flux through the solenoid.

(a) The current in the solenoid is increased at a uniform rate from zero to I in time t. Find an expression for the back-emf induced in the solenoid during the change.

(b) Find the electrical work done against this back-emf in establishing the current.

(c) The work found in **(b)** is the energy stored in the magnetic field of the solenoid. Show that this energy is $B^2/2\mu_0$ per unit volume.

(d) What becomes of this energy when the circuit is broken?

(**O and C**: *all other Boards except SEB H*)

For B inside a long solenoid, see E2.18. Magnetic flux is defined in E2.24; remember that $B \times A$ (A = area of cross-section) gives the flux per turn, so the total flux is BAN. You must give A in terms of the radius r.

(a) Your expression for the total flux ought to be of the form Φ = constant × I (where Φ = total flux). The back-emf is given by the rate of change of flux which equals constant × rate of change of current. See E2.25 and E2.29 for further details if necessary.

(b) The constant in **(a)** is the self-inductance of the solenoid. The electrical work done is given by E2.30. You must write the equation in terms of the factors that make up the self-inductance.

(c) Assume the volume of the magnetic field is the same as the volume of the inside of the solenoid (ie, length × area of cross-section). You will need to rewrite your expression for **(b)** in terms of B rather than I, and then 'divide' that expression by the volume expression.

(d) The sudden collapse of the field in the solenoid induces a large emf across the switch gap, and the emf tries to maintain the current flow. A spark may result.

Unit 3 Oscillations and waves

Energy carried by waves appears in several familiar forms; eg, sound, heat, light and X-rays. There are not many subjects in Physics that do not directly feature a study of some form of wave motion and even such subjects invariably include important applications of wave behaviour. A typical example is materials science, where vital information about solids is obtained by using the wave properties of X-rays to unlock the secrets of molecular structure. All waves are carried by some form of oscillation; eg, sound waves in air are carried by oscillating gas molecules, and electromagnetic waves are carried by oscillating electric and magnetic fields. Hence an understanding of oscillations would be fundamental to any theory about waves. However, oscillations are important in themselves; eg, atoms oscillate in solids, and currents can be made to oscillate in electric circuits. This Unit examines principles, theories and examples of waves and oscillations; the applications of this Unit can be found throughout the rest of the text, in particular in Unit 4 (Optics) where interference and diffraction of light is studied in detail, in Unit 5 (Materials) as outlined above, and in Unit 7 (Electricity and electronics) in the treatment of ac circuits.

Before commencing your studies on this Unit, it may be helpful to look back at your GCSE study notes on Waves. Although such studies are generally descriptive, the basic ideas are developed further in this unit. The treatment of oscillations in the unit involves using the language of shm, so re-reading the relevant part of Unit 1 ought to be valuable at this stage. Also, before tackling 3.7 (Electromagnetic waves), a look back at Electromagnetic fields in Unit 2 may prove helpful.

3.1 GENERAL INFORMATION ABOUT OSCILLATIONS

If any quantity changes regularly with time about a fixed value it can be said to be oscillating. At A-level most examples feature the change in displacement of an oscillating mass; it should always be remembered that other systems 'oscillate' (eg, electric and magnetic fields oscillate to create electromagnetic waves, and currents can oscillate in electric circuits). It is possible to represent oscillatory motion by plotting graphs of the oscillating quantity against time. F3.1 shows three different possible modes of oscillation for a body vibrating about a fixed point.

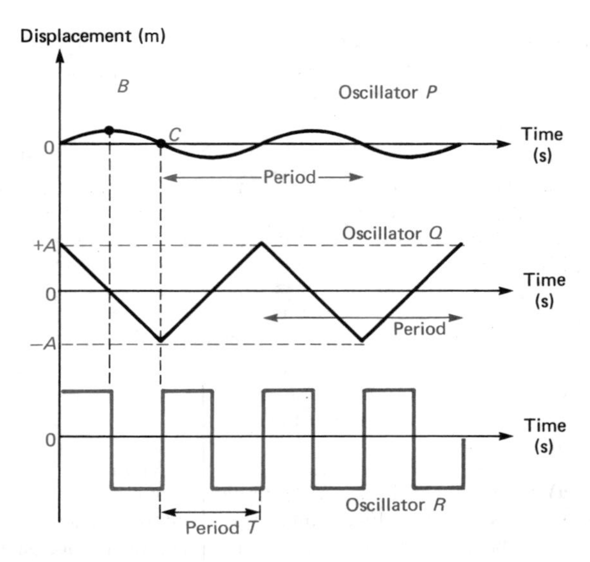

F3.1 Displacement – time graphs for three oscillators

In F3.1: slope, velocity of oscillator; area, no great physical significance; intercepts, etc., the oscillations pass through zero displacement twice every cycle.

It is important to remember that a vector quantity (displacement) is being plotted on the vertical axis and the purpose of the positive or negative sign is purely to indicate direction; so the 'smallest displacement' occurs at zero displacement, NOT at a negative amplitude ($-A$). Of the three systems shown in F3.1, oscillator P is by far the most important; any oscillator such as this which has a displacement/time graph whose shape happens to be sinusoidal (sine wave form) is said to be oscillating with shm (See 1.6). Any systems performing shm will also have velocity/time and acceleration/time graphs that are sinusoidal; the three graphs will not be in phase. These three graphs are shown as F1.10.

The **phase** of a particular point in an oscillation is a measure of what fraction of an oscillation has been completed compared with a chosen 'starting point'. A complete oscillation is normally expressed either as 360° or 2π radians (as in 'once round' a circle). For example, point B on oscillator P is ¼ of a cycle BEFORE point C, so we say in 'phase terms' that B LEADS C by 90° (or ½π); alternatively, C LAGS B by 90°. Two other important quantities can be measured directly from the graphs of F3.1. The **amplitude** of an oscillation is the largest displacement encountered during the oscillation. The amplitude of oscillator Q is A.

The **period** of an oscillator is the time taken to complete one cycle. The period of oscillator R in F3.1 is T. The frequency of oscillation is defined by:

E3.1 $$f = \frac{1}{T}$$

where f = frequency of an oscillation (Hz or s^{-1}), T = period of the oscillation (s).

The **frequency** of any oscillator is the number of cycles completed every second. A comparison of oscillators P and Q in F3.1 reveals that they have different amplitudes and 'motions', but the same period (and hence the same frequency). As the oscillations are continuous (without any discontinuities, ie, breaks or jumps, in the 'wave' pattern of the graph) no matter which point on the time axis is used the same phase relationship between P and Q is always observed; eg, at point B oscillator P is a positive amplitude whereas at the same moment in time, oscillator Q is already ¼ of a cycle AFTER its positive amplitude, so P LAGS behind Q by 90°. Only phase measurements up to 180° are used, so it is NOT usual to say instead that P leads Q by 270°. Note that there is NOT a constant phase relationship between P and R (or Q and R) as they do not have matching frequencies; eg, P lags R by $\frac{1}{2}\pi$ at B, P lags R by $\frac{1}{2}\pi$ at C. Note that systems completely out of phase (in antiphase) can be said to lead OR lag by 180°; there is zero phase difference between systems in phase.

3.2 ENERGY STORED IN OSCILLATORS

Oscillators such as those illustrated in F3.1 store a fixed amount of total energy in two ways; as ke and as pe. At maximum displacement ($x = \pm A$) there is zero ke (oscillator instantaneously has zero velocity) so the pe stored equals the total energy; at the equilibrium position the pe stored is zero and the ke equals the total energy. For an shm oscillator this can be shown as in F3.2

In F3.2: slope of pe graph, negative value of the force exerted on oscillator (E2.1); slope of ke graph, no physical significance; areas of both; no physical significance; intercepts, etc., ke zero at amplitudes.

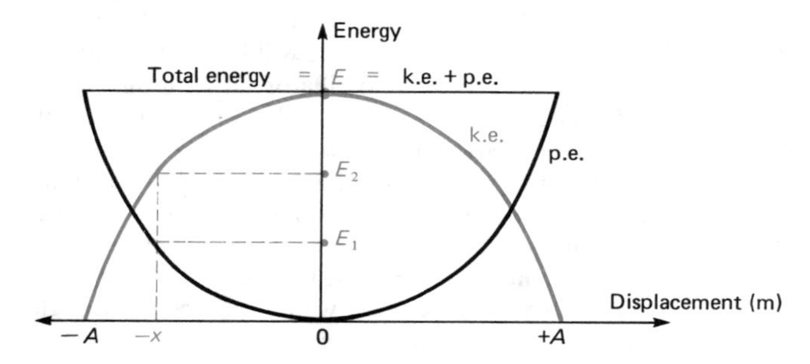

F3.2 Energy stored in a simple harmonic oscillator

Both ke and pe when added together will always give the total energy stored; eg, at a displacement of $-x$, pe $= E_1$ and ke $= E_2$ so total energy $= E = E_1 + E_2$. The total energy stored by a simple harmonic oscillator is given by:

E3.2 $E = \frac{1}{2}m(2\pi f)^2 A^2$

where E = total energy stored by shm oscillator (J), m = mass of oscillator (kg), f = frequency of oscillator (Hz), A = amplitude of oscillations (m).

Note that the pe and ke both depend upon the square of the displacement giving the parabolic shape of the curves in F3.2.

However oscillators normally lose their stored energy due to damping effects when energy from the oscillating system is passed to the surroundings (eg, by heating up the air as a result of doing work against air-resistance forces).

Oscillations of constant amplitude as a result of zero damping are called **free oscillations** (as in F3.1).

Oscillations of reducing amplitude owing to energy from the system being absorbed by the surroundings are called **damped oscillations** (see F3.3). If a system is so damped that it JUST fails to oscillate it is said to be **critically damped**. (JUST failing to oscillate means that if released from a positive displacement owing to the damping effects, see F3.3). More damping than critical damping is referred to as **overdamping** (see F3.3).

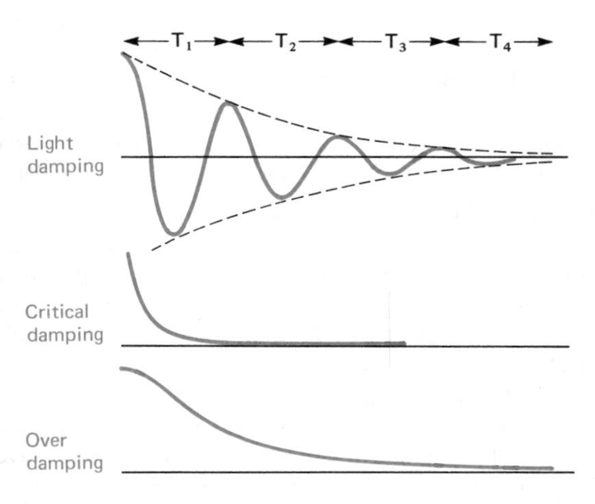

F3.3 Damped oscillations

It is a common mistake to think that the period changes in a damped system, for until the oscillations have completely died away the period stays the same (eg, T_1, T_2, T_3 and T_4 in F3.3 are all the same). In any system of damped oscillations it is possible to draw a locus showing the changing amplitudes (the dotted lines in F3.3); this locus takes the shape of an exponential decay (or rise).

If a system is made to oscillate by periodic impulses from a **driver** system, then these oscillations are referred to as **forced oscillations.** Forced oscillations will have the same (or almost the same) frequency as the driver system. It is important to realize that forced oscillations can maintain a constant amplitude despite the presence of damping; the energy needed to prevent the amplitude decaying is provided by the driver.

If the **natural frequency** of an oscillator is equal to the frequency of a driver in a forced oscillations system then **resonance** will occur (see p. 108). The forced oscillations will acquire a large amplitude limited only by the damping losses. The phase of the displacement of the driver will be found to lead that of the oscillator by 90°.

Several useful resonance effects can be encountered in A-level physics; eg, tuning circuits for radios, studies of absorption spectra, 3 cm waves are produced in cavity resonators, loudspeaker designs can benefit from bass resonant cavities, etc. However resonance can have unpleasant effects; eg, causing distortion in hi-fi, unpleasant vibrations in cars at certain engine speeds, even collapsing bridges and buildings purely from effects of winds. Careful damping of oscillations can be used to control room acoustics, car suspensions, servo mechanisms for robots etc.

3.3 FREQUENCIES OF DIFFERENT SIMPLE HARMONIC OSCILLATORS

An shm can be fully described from two items of information; the amplitude A and the frequency f. 1.6 explains how to analyse the motion; the following equations define the frequency of different simple harmonic oscillations:

E3.3 $f = \frac{1}{2\pi}\sqrt{\frac{g}{2l}}$

where f = frequency of oscillation of simple pendulum (Hz), g = acceleration due to gravity (m s^{-2}), l = length of the pendulum (m).

E3.4 $f = \frac{1}{2\pi}\sqrt{\frac{g}{l}}$

where f = frequency of oscillation of water column in a U-tube (HZ), g = acceleration due to gravity (ms^{-2}), l = total length of the water column (m).

E3.5 $f = \frac{1}{2\pi}\sqrt{\frac{k}{m}}$

where f = frequency of oscillation of mass on spring (Hz) k = spring constant of spring (Nm^{-1}), m = mass attached to spring (kg).

E3.6 $f = \frac{1}{2\pi}\sqrt{\frac{1}{LC}}$

where f = frequency of oscillation of electric current (Hz), L = inductance in circuit (H), C = capacitance in circuit (F).

Proof of these equations may be found in textbooks. It is possible to derive E3.6 from E3.5 using an electrical/mechanical analogy.

3.4 GENERAL INFORMATION ABOUT WAVES

Imagine a line of particles that can oscillate up and down as shown in F3.4(a). In F3.4(a) just the left-hand oscillator (A) has energy, the two amplitudes of its oscillation being shown by the arrows. A little time later A has passed some of its energy onto B and a little has travelled onto C(F3.4(b)). Later still, A has stopped oscillating completely (F3.4(c)) and the energy has been passed further to the right down the line of oscillators. We have

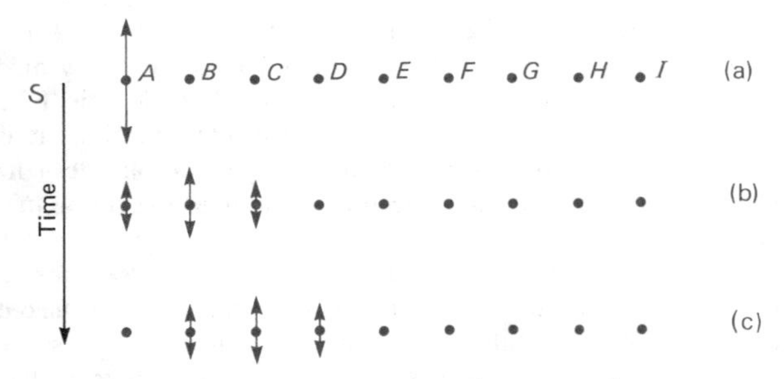

F3.4 A transverse progressive wave pulse

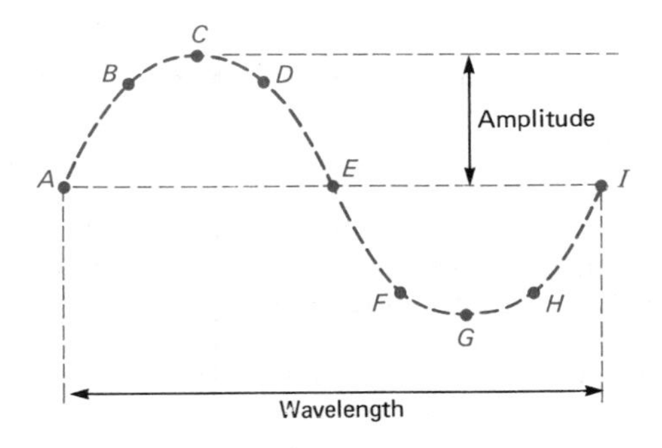

F3.6 'Instant' picture of 1 cycle of transverse progressive or standing wave

just described what is referred to in Physics as a **'transverse wave pulse'**. It is possible to repeat the description but talking about oscillations to the left and right (ie, in the same direction as the energy will travel instead of at right angles – or transversely – as before): in this new case we have a **'longitudinal wave pulse'**. If instead of feeding a limited amount of energy along the line of oscillators we continually feed in energy at the left-hand end, energy will continually emerge at the right-hand end; now we say that a **progressive wave** is travelling from left to right. The progressive wave can be transverse or longitudinal. If the motion of the oscillators is at all damped, then the medium will absorb some of the oscillator energy, and hence the wave energy will be gradually absorbed as the progressive waves travel through the medium. Sound wave energy being absorbed as it travels through the air is an obvious example; the oscillator energy (wave energy) will be converted into heat energy of the air molecules.

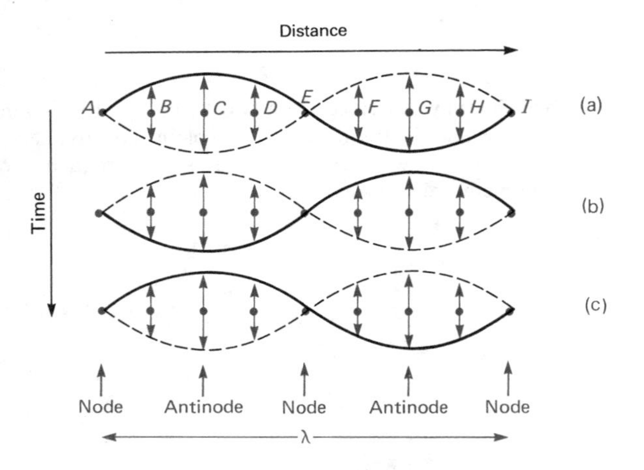

F3.5 A transverse standing wave

F3.5 shows a system in which the amplitude of each oscillator stays the same. Hence no energy is being passed down the line and this situation is referred to as **a 'standing wave'** or a **'stationary wave'**. The example shown is transverse, but again it could be longitudinal. Points on the standing wave where the amplitude of the oscillations is zero are called **'nodes'** (no disturbance); points on the standing wave where the oscillations have the greatest amplitude are called **'antinodes'**. If the oscillator's motion is at all damped (the medium is absorbing energy) then it will be necessary to feed in energy continually to maintain the amplitude of the standing wave (see resonance in 3.2).

F3.4 and F3.5 do not show the actual vertical positions of the oscillators; the oscillating particle may be found anywhere along the arrowed line. In nearly all important wave motions the oscillators perform shm and a photograph of the position of the oscillators carrying a progressive or standing transverse wave might look like F3.6.

F3.6 defines the wavelength and amplitude of the wave. Of course if F3.6 is representing a standing wave then *A*, *E* and *I* will never oscillate, whereas for a progressive wave all particles oscillate. It is interesting to note that on a standing wave *B*, *C* and *D* are in phase with one another, but exactly out of phase with *F*, *G* and *H*. On a progressive wave there will be the same phase difference between each successive oscillator; in F3.6 this phase difference is 45°. The sinusoidal shape of the wave is important but must not be confused with F3.1 where the horizontal axis represents time and not distance as in F3.6. It is vital to note that this introduction to waves has talked about particles oscillating in the medium; but any oscillating quantity can produce waves and it may not need a **medium** to propagate (carry) it. An obvious example is the electromagnetic wave which can travel through free space and is carried by oscillating electric (and magnetic) fields.

Standing waves are created when two waves of equal frequency and amplitude travelling in opposite directions combine with one another when they are exactly out of phase. This situation occurs when a wave is reflected and the point of reflection serves either as a node or antinode depending on the exact situation. The consequence of this is that it is possible to use the same formulae and measurements for progressive and standing waves.

E3.7 $v = f\lambda$

where v = speed of progressive wave (m s^{-1}), f = frequency of the wave (Hz), λ = wavelength of the wave (m).

When using this formula with standing waves, it is vital to remember that v refers to the speed at which the energy is carried by a progressive wave; so the information refers to the speed of the progressive waves that created the standing wave. A further error can occur when measuring wavelength of standing waves; looking at F3.5 it is often wrongly assumed that the wavelength is the distance between two nodes (eg, from *A* to *E*). F3.6 should make it clear that the wavelength is TWICE the distance between nodes.

The energy carried by oscillators is proportional to the square of their amplitude (E3.2); hence there is a proportionality between the square of the amplitude of a wave and its intensity. The **intensity** at a given point is related to the energy carried by the wave at that point.

3.5 PROPERTIES OF WAVES

Reflection The **laws of reflection** that apply to smooth surfaces are:

1 The **incident ray, reflected ray** and **normal** to the reflector at the point of incidence all lie in the same plane

2 The **angle of incidence** equals the **angle of reflection**.

Reflection can be explained using a particle model, or using a wave model and Huygens constructions (see 4.3). It can be important to know that waves undergo a 180° phase shift on reflection; eg, a trough travelling up a slinky spring is reflected as a peak.

Refraction **Refraction** means bending; waves are 'bent' (refracted) only if there is a change in the medium through which they travel. This may be a sudden change (eg, waves travelling from air to glass) or a gradual change (eg, waves travelling through layers of air at different temperatures). Refraction occurs as a result of the difference in velocity of the wave in different media.

Principle of superposition The **principle of superposition** can be applied to any waves that are caused by the same type of oscillation; eg, to two (or more) sound waves. It cannot be applied to fundamentally different waves; eg, a water wave and a light wave. The principle is summarized by F3.7.

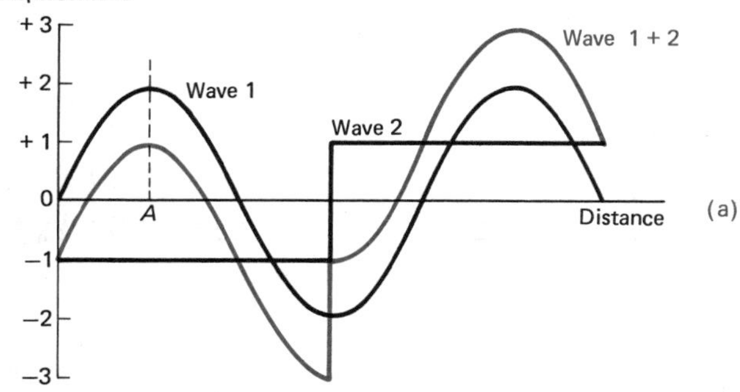

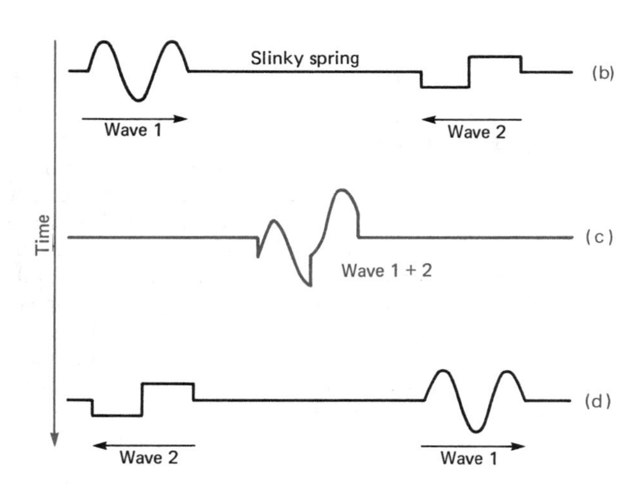

F3.7 Superposition of waves

The first part of the principle of superposition states that where two waves of the same type meet in space, the resulting disturbance is the sum of the disturbances due to the individual waves. This simply means that they can be added together graphically as in F3.7(a); eg, at position *A*, wave 1 has a displacement of +2, wave 2 has a displacement of −1, adding these means that the combined wave has a displacement of +1. Note that in superposition the waves do not need have a constant phase relationship, or be at the same frequency or amplitude. It is possible to apply the principle to an intense beam of X-rays mingling with a weak light source.

The second part of the principle is illustrated in F3.7(b), (c) and (d). These show successive moments in time as wave pulses as in F3.7(a) pass along a slinky spring, superpose, and then continue on their way. Although when superposed a new wave shape is formed, a little time later both pulses emerge unscathed from the encounter (exactly as they were originally before they met!). There has been no lasting interaction between the waves. Although reflection and refraction are properties of both waves and beams of particles, any property that can only be explained by superposition is a 'waves only' property. Interference is such a property, as is diffraction.

Interference Interference of waves (longitudinal or transverse) takes place whenever two or more waves meet in space in such a way that there is a constant phase relationship between them (hence they must also have the same frequency, wavelength and speed) and that the waves are of similar amplitudes. Waves that have these properties are said to be **'coherent'**. When coherent waves meet, **interference** takes place. Interference results in the amplitude of the waves being constant at the point where they meet. If two waves of identical amplitude interfere, it is possible that zero amplitude will result (**destructive interference**); alternatively it is possible to have a maximum amplitude equal to that of the two waves added together (**constructive interference**). When interference takes place, wave energy does not go missing, it is redistributed. Hence the 'energy missing' at points of destructive interference is found as 'extra energy' at points of constructive interference.

Diffraction In a uniform medium (no refraction) waves travel in straight lines (**rectilinear propagation**). The arrowed lines (**rays**) in F3.8 show the direction of travel. The waves in the diagram are

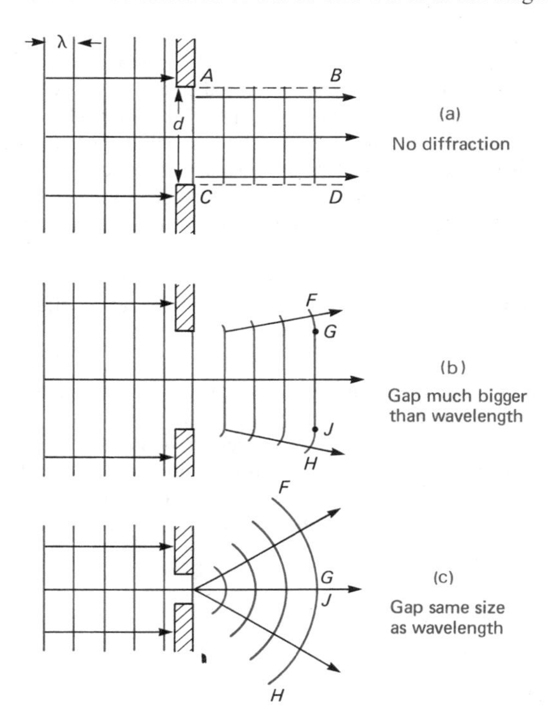

F3.8 Diffraction by a 'gap'

plane waves (parallel rays), and the waves encounter two obstacles (forming a gap). If there was no diffraction the situation of F3.8(a) would occur; common knowledge about waves suggests that this situation is impossible. Imagine F3.8 happening in a ripple tank; along the lines *AB* and *CD* there would be complete calm on one side of the line and large waves on the other side, yet such discontinuities are NOT characteristic of waves. The waves must build up gradually to their amplitude. The principle of superposition and Huygens principle (see p. 63) show that bending occurs at the edges as in F3.8(b), with the waves building up from zero amplitude at *F* (or *H*) to their full amplitude at *G* (or *J*). If the points *G* and *J* can be made to meet, then plane waves (parallel rays) have been converted into **circular waves** (rays spreading from a point): this effect occurs when the gap width (d) becomes as small as the wavelength (λ). Hence **diffraction effects** become particularly noticeable when the size of gaps (or obstacles) becomes so small that they are comparable to the wavelength of the waves. Full diffraction theory shows that there must also be places of constructive and destructive interference in the regions of diffraction (ie, on the right-hand waves this happens between *F* and *G*, and also between *H* and *J*) – see your textbook and 4.2.

Polarization Waves and the media carrying them are often invisible; this means that the only way a transverse wave can be successfully distinguished from a longitudinal wave is by a demonstration of polarization. **Polarization** is a property of transverse waves only. If F3.4 were viewed from the right-hand side of the page, then you would be looking down the line of oscillators with the wave travelling towards you: the oscillators

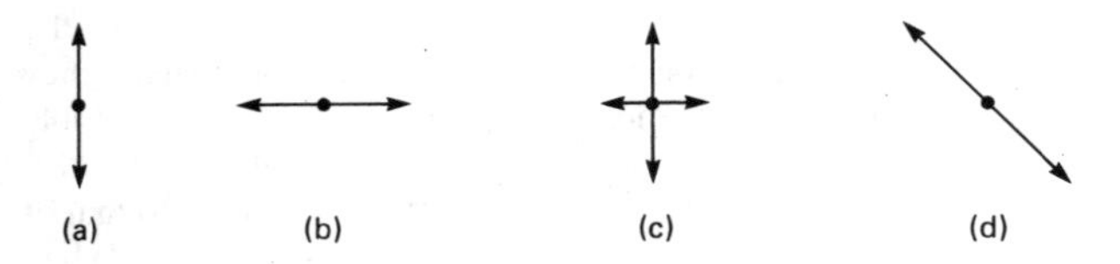

F3.9 Modes of oscillation

would be seen vibrating along the line defined in F3.9(a). However the requirement for a transverse wave to travel is that the oscillations should be at right angles to the direction of wave travel: this still happens in F3.9(b), which would mean in F3.4 that the oscillators would have to move 'in and out' of the paper rather than 'up and down' as they are shown. If the oscillations only take place in one plane as in F3.9(a), (b) or (d), then the waves are plane polarized. Other special polarization effects can be achieved according to specified modes of oscillation. Oscillations may occur in two directions simultaneously, (F3.9(c)), and in unpolarized waves the oscillation is random. Surface water waves, most radio waves, and 3 cm microwaves from a Klystron are all plane polarized. Reflection can polarize by cutting out much of one mode of oscillation, a fact exploited as the 'glare reducing' property of Polaroid sunglasses. Longitudinal waves are always produced by oscillations along the line of propagation and hence cannot exhibit polarization effects.

3.6 SOUND

Sound waves are **longitudinal** pressure waves travelling through a medium (ie, they cannot travel through a vacuum). In air, macroscopically this can be considered to generate regions of higher pressure (called **compressions**) and regions of lower pressure (called **rarefactions**). In air, microscopically the air molecules perform shm along the line of travel of the wave.

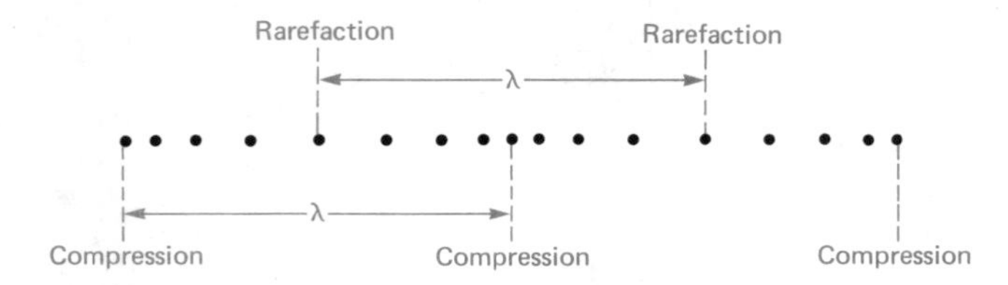

F3.10 'Instant' picture of a longitudinal progressive or standing wave

F3.10 is an instantaneous 'picture' of a longitudinal wave produced in the same way as F3.6 was for a transverse wave. Longitudinal waves have all the same properties as transverse waves (reflection, refraction, interference, diffraction, superposition, standing waves, etc.) with the one notable exception of polarization.

Beats The effect of beats can occur with ALL types of waves, although it is most often encountered with sound. Two wave forms of slightly different frequencies (f_P,f_Q) but similar amplitudes meet at a point in space.

F3.11 has been drawn for transverse waves (the diagram is easier to draw and understand), but the theory is the same for longitudinal waves. It is important to realize that the diagram shows two wave forms, P and Q, meeting at a point, that the horizontal axis represents time and that the waves are added together using the principle of superposition to produce the resultant wave form $P+Q$. At points A and C waves P and Q are in phase, so the resultant wave form has maximum amplitude; at points B and D waves P and Q are exactly out of phase (phase difference of π or 180°), so the resultant wave form has zero amplitude. So at the point in space where the wave forms meet, at one instant a 'loud' signal will be heard (at time A), gradually reducing to no signal (at time B) then increasing to another loud signal (at time C), and so on. This 'loud' and 'soft' variation is the effect called **beats**; the period of the beats is T (see F3.11) and the beat frequency ($1/T$) is given by:

E3.8 $f = f_P - f_Q$

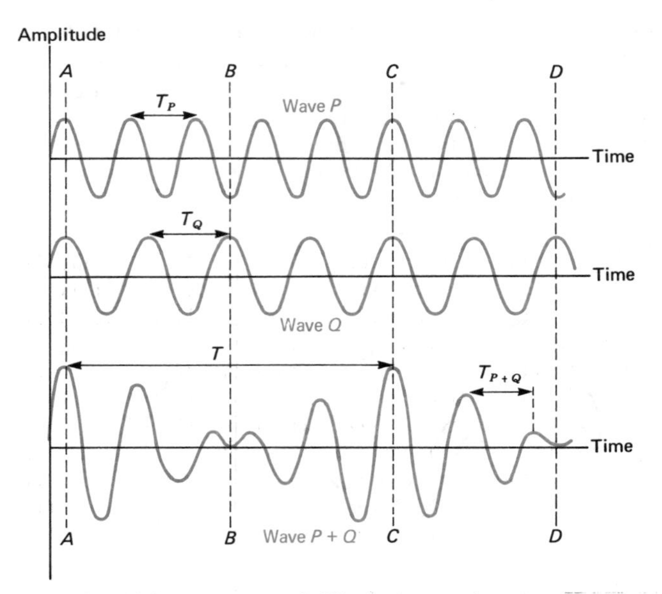

F3.11 Combining two waves of different frequencies to produce beats

where f = frequency of the beats (Hz), f_P = frequency of one of the wave forms producing the beats (Hz), f_Q = frequency of the other wave form producing the beats (Hz).

It is important to realize that the beat frequency (defined by T) is not the frequency of wave $P+Q$ (defined by T_{P+Q}) (see F3.11); the beat frequency is the frequency of the amplitude variation of the wave $P+Q$. Using sound waves, beats become clearly audible when frequencies are nearly matched; this leads to a useful application in the tuning of musical instruments. The 'beating' effect between two electromagnetic waves of slightly different frequency is used in police radar traps (see Doppler effect) and modern (superheterodyne) radio receivers. The effect of 'beats' is that of 'superposition' of waves that are not quite 'coherent' because there is a small difference in their frequencies. This means that an observer at a fixed position will receive a signal that will interfere destructively and constructively as time passes; this time varying interference is the effect called 'beats'.

Doppler effect An observer who stands still will receive waves (speed v) from a stationary transmitter at the same frequency at which they were emitted (f). However, motion of either the observer or the transmitter along the line between them, will cause the observed frequency (f') to be different from the basic frequency of the transmitter (f). Similar effects can also be observed if the medium is caused to 'move'.

For sound suppose the transmitter and observer are moving in the *same* direction at speeds u_s and u_0 respectively.

(a) In 1 second, the transmitter emits f waves which occupy a distance $(v - u_s)$. Hence the wavelength λ is given by the equation:

E3.9 $\lambda = \dfrac{(v - u_s)}{f}$ where v = wavespeed and f = transmitter frequency

(b) The observer receives waves of this wavelength. In 1 second, a stationary observer would receive v/λ waves; since the observer here is moving away from the transmitter, the number of waves received each second is reduced by u_0/λ. Hence the observed frequency f' is given by:

E3.10 $f' = \dfrac{(v - u_0)}{\lambda} = \dfrac{(v - u_0)}{(v - u_s)} \times f$

Notes: (i) If either velocity is reversed, change the appropriate sign in the above formula.
(ii) If the direction of either velocity is not along the line between the transmitter and the observer, the velocity component along the line should be used.

If electromagnetic waves are involved then the equations above may not apply. It is simplest to deal with the motion of the transmitter **relative** to the observer; then since the speed of light is the same for any observer, the observed frequency is given by E3.11:

E3.11 $f' = f\dfrac{c}{(c - u_r)}$

where f' = observed frequency (Hz), f = transmitted frequency (Hz), c = speed of light (m s^{-1}), u_r = speed of transmitter relative to the observer (m s^{-1}).

The Doppler effect in sound is well known for altering the pitch (frequency) of train whistles, car horns, etc. In a police radar speed trap the radar beam reflected from the moving vehicle has a frequency shift dependent upon the vehicle velocity; the frequency shift is detected by mixing the 'transmitted' and 'reflected' beams and measuring the 'beat frequency' that results. The Doppler shift of light frequencies (**red shift/blue shift**) has applications for astronomers involved in measuring the motion of stars and planets. Problems arise for spectroscopists from the way in which the Doppler effect (due to the motion of atoms) causes the lines of a line spectrum to be broadened.

Standing waves on strings and in pipes A stretched string may be made to vibrate transversely or longitudinally; but most applications involve transverse waves on the string, which can then set up longitudinal sound waves in the surrounding medium. The ends of the string must be fixed, and reflections at these points (nodes) create standing waves; in a pipe, longitudinal sound waves can be reflected at open (antinode) or closed (node) ends to create standing waves. The lowest frequency that will generate the standing wave (f) is known as the fundamental and other modes of vibration that create standing waves at higher frequencies are called overtones.

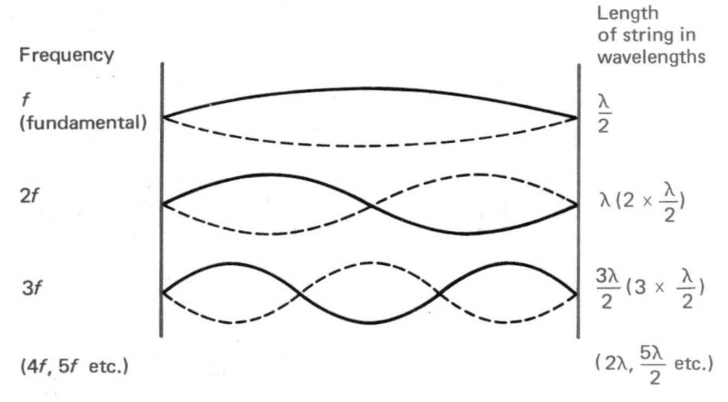

F3.12 Transverse standing waves on a stretched string

It is important to realize that the standing waves in the pipe in F3.13 are longitudinal, but diagrams invariably use a transverse standing wave graph to show how the amplitude of the longitudinal wave varies. The transverse standing wave lines drawn in F3.12 and 13 are the solid and dotted black lines in F3.5 and they show the maximum displacement (or amplitude) of the standing wave. Although the wavelengths have all been marked as λ, the size of these wavelengths of course gets smaller as higher overtones are excited and the frequency rises (speed of the waves staying the same). A study of pipes and stretched strings is obviously of great importance in a study of the behaviour of musical instruments. Care is needed with pipes in that their physical length is NOT the length of the standing wave set up in them, there is an **end correction** that must be taken into consideration at open ends.

3.7 ELECTROMAGNETIC WAVES

The transverse oscillations that carry these waves are of electric fields and magnetic fields (see 2.15) which need no medium to support them; hence electromagnetic waves can travel through a vacuum. The waves are transverse and always travel at the same speed (c – the speed of light) in a specified medium (regardless of wavelength); though of course if this specified medium is changed then the speed changes. Electromagnetic waves are

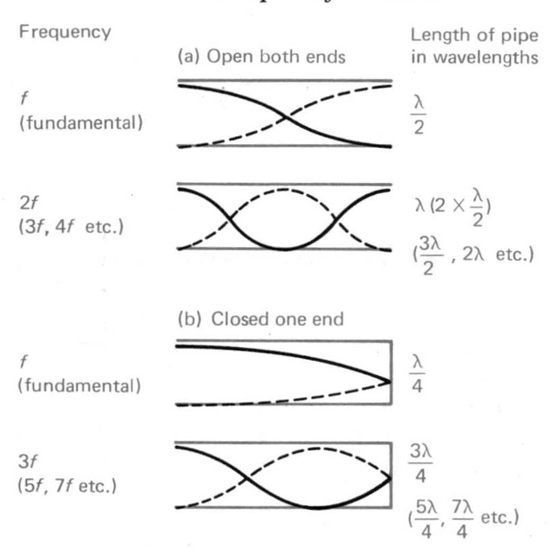

F3.13 Standing waves in pipes

created by the acceleration of charged particles.

A confusion sometimes arises because the categories of radiation can sometimes overlap; eg, it is possible to have γ rays and X-rays at 10^{-11} m wavelength. These radiations are of course absolutely identical; they are only given different names because they have been produced in different ways. Quantum theory says that the energy of any oscillator is quantized in multiples of hf (1.10); often hf is such a small quantity of energy compared with the total energy of the oscillator that the quantum effect can be ignored. However individual quanta of electromagnetic radiation (photons) can carry so much energy (because of their high frequencies) that individual **photons** can become important (particle model of light). See F3.14.

Radio and TV broadcasts These are carried by radio waves and microwaves. Signals carrying information modulate the carrier waves. A radio or a TV channel is a band of radio or microwave frequencies allocated to carry signals from a particular station.

The **bandwidth** of a channel is the width of the band of frequencies allocated to that channel. For example, the bandwidth of a Medium Wave radio channel is 4 kHz which is sufficient to carry audio signals without significant loss of quality. VHF and TV channels have much wider bandwidths because much more information needs to be carried.

The range of a broadcasting station is determined by the wavelength of the carrier waves:

(a) Low frequency radio waves follow the Earth's curvature and are therefore used for inter-continental broadcasting;

(b) Medium frequency radio waves (MW) reflect from the ionosphere of the atmosphere in suitable weather conditions. This is why MW programmes can have a long range;

(c) High frequency (HF), VHF and UHF radio waves carry radio and TV channels and have 'line of sight' range only since they pass straight through the atmosphere;

(d) Microwaves are used for satellite broadcasting. A communications satellite must be in a **geostationary** orbit. Receiver dishes on the ground need to point to the satellite. The **beamwidth** of a dish or aerial is the angle from the direction of maximum signal strength to the direction giving 50 per cent signal strength. See page 170.

3.8 WAVE SPEED FORMULAE

General formula E3.7 $v = f\lambda$

where v = speed of wave (m s^{-1}), f = frequency of wave (Hz), λ = wavelength of wave (m).

Water waves E3.12 $v = \sqrt{gh}$

where v = speed of 'ripple tank type' waves whose $\lambda \gg h$ yet amplitude of waves $\ll h$ (m s^{-1}), g = acceleration due to gravity

NAME OF RADIATION	λ WAVELENGTH m	f FREQUENCY Hz	E ENERGY OF A PHOTON eV	METHOD OF PRODUCTION	METHOD OF DETECTION	USES
	10^{-14}, 10^{-13}	10^{22}	10^{8}, 10^{7}	Mainly from space	PHOTOGRAPHIC FILM; GM TUBES, SCINTILLATORS	
γ-RAYS	10^{-12}	10^{21}, 10^{20}	10^{6}	Radioactive nuclei, accelerators	PHOTOGRAPHIC FILM; GM TUBES, SCINTILLATORS	MEDICAL DIAGNOSIS + TREATMENT; CHECKING METAL
x-RAYS	10^{-11}, 10^{-10}, 10^{-9}	10^{19}, 10^{18}	10^{5}, 10^{4}	X-ray tubes	PHOTOGRAPHIC FILM; GM TUBES, SCINTILLATORS	MEDICAL DIAGNOSIS + TREATMENT
ULTRAVIOLET	10^{-8}, 10^{-7}	10^{17}, 10^{16}	10^{3}, 10^{2}	Very hot bodies, excited atoms in discharge tubes	PHOTOGRAPHIC FILM; PHOTO CELLS; PHOTO MULTIPLIERS	MINERAL ANALYSIS; ATOMIC SPECTROSCOPY
LIGHT (VIOLET – RED)		10^{15}	10^{1}	Very hot bodies, excited atoms in discharge tubes	EYE; PHOTO CELLS; PHOTO MULTIPLIERS	ATOMIC SPECTROSCOPY; AERIAL PHOTOGRAPHY
INFRARED RADIATION (Heat)	10^{-6}, 10^{-5}, 10^{-4}, 10^{-3}	10^{14}, 10^{13}, 10^{12}	10^{0} (1), 10^{-1}, 10^{-2}	Hot bodies	THERMOPILES; BOLOMETERS	AERIAL PHOTOGRAPHY; HEATING; ATOMIC SPECTROSCOPY; COOKING
MICROWAVES: EHF, SHF, UHF	10^{-2}, 10^{-1}, 1	10^{11}, 10^{10}, 10^{9}	10^{-3}, 10^{-4}, 10^{-5}	Klystrons	SOLID STATE DIODES; CRYSTAL DETECTORS	COMMUNICATION; NAVIGATION; RADAR; COOKING
RADIO FREQUENCIES (RF): VHF, HF, MF, LF, VLF	10, 10^{2}, 10^{3}, 10^{4}, 10^{5}	10^{8}, 10^{7}, 10^{6}, 10^{5}, 10^{4}	10^{-6}, 10^{-7}, 10^{-8}, 10^{-9}, 10^{-10}	Electronic oscillators (valves transistors ICs' etc.)	TUNED CIRCUITS	COMMUNICATION; NAVIGATION

F3.14 The Electromagnetic spectrum

($m\,s^{-2}$), h = depth of the shallow water (m).

Mechanical waves E3.13 $v = \sqrt{\dfrac{F}{\mu}}$

where v = speed of waves (longitudinal or transverse) along a stretched string or spring ($m\,s^{-1}$), F = force used (tension) to stretch the string (N), μ = mass per unit length of string/spring ($kg\,m^{-1}$).

E3.14 $v = x\sqrt{\dfrac{k}{m}}$

where v = velocity of wave along a line of masses joined together by springs ($m\,s^{-1}$), x = distance between centres of the masses (m), k = spring constant of the springs ($N\,m^{-1}$), m = mass of each mass (kg).

Sound Waves E3.15 $v = \sqrt{\dfrac{E}{\rho}}$

where v = velocity of sound in a solid ($m\,s^{-1}$), E = Young's modulus for the solid (Pa or $N\,m^{-2}$), ρ = density of the solid ($kg\,m^{-3}$).

E3.16 $v = \sqrt{\dfrac{\gamma P}{\rho}}$

where v = velocity of sound in a gas ($m\,s^{-1}$), γ = ratio of molar heat capacities of the gas (no units), P = pressure of the gas (Pa or $N\,m^{-2}$), ρ = density of the gas ($kg\,m^{-3}$).

Electromagnetic waves E2.34 $c = \dfrac{1}{\sqrt{\varepsilon_0\mu_0}}$

where c = velocity of light in free space ($m\,s^{-1}$), ε_0 = permittivity of free space ($C\,V^{-1}\,m^{-1}$), μ_0 = permeability of free space ($N\,A^{-2}$).

E3.17 $c_m = \dfrac{c}{n_m}$

where c_m = velocity of light in medium 'm' ($m\,s^{-1}$), c = velocity of light in a vacuum/air ($m\,s^{-1}$), n_m = refractive index for light passing from air/vacuum into medium 'm' (no units).

In the two electromagnetic wave speed equations the word light can be replaced by 'electromagnetic waves' as of course all electromagnetic waves travel at the speed of light.

UNIT 3 QUESTIONS

3.1M A wave travelling in the positive direction of the x-axis produces displacements of the particles in its path which at one instant are as shown:

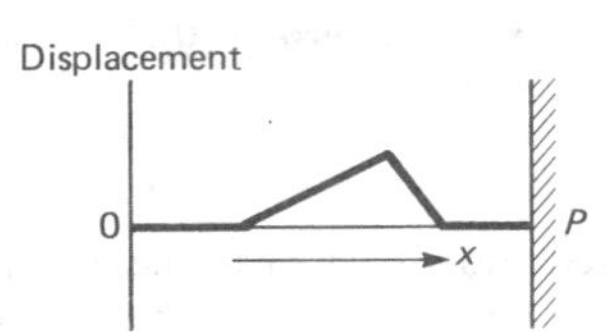

The wave is reflected from a boundary P which does not permit particle movement. Which graph below best represents the displacement of the particles at some instant during the passage of the reflected wave?

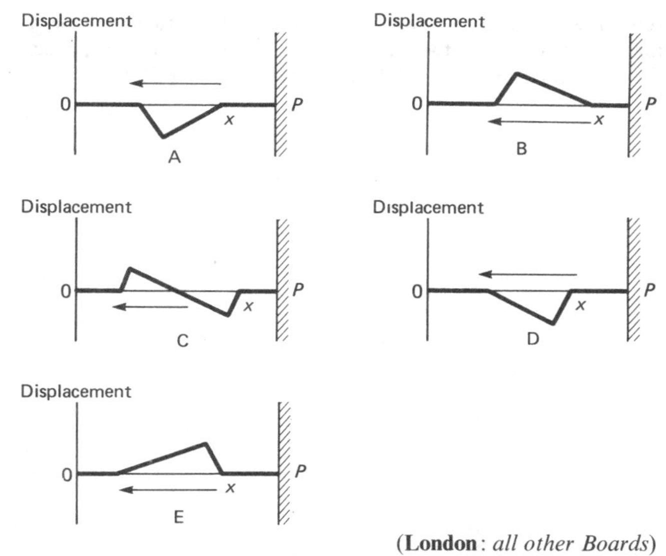

(**London**: *all other Boards*)

The pulse will undergo a 180° phase change on reflection (see p. 52), which eliminates two answers. It will keep the same shape, so eliminating another answer. Which of the two remaining has the same shape **in the direction of travel** as the original?

3.2M Physical properties shared by both radio waves and sound waves are:
1 both can be polarized.
2 both can be reflected.
3 both can be diffracted.
A if **1**, **2**, **3** all correct B if **1**, **2** only correct.
C if **2**, **3** only correct. D if **1** only is correct.
E if **3** only is correct.

(—: *all Boards except SEB H, SEB SYS*)

Which of the above properties are common to all types of waves? Remember that only transverse wave forms can be polarized, so are the two wave forms both transverse? See p. 54 if necessary.

3.3M Which of the following graphs represents the variation about their mean values of air pressure and particle displacement with distance from a given point (at a particular instant) for a progressive sound wave in air? (The wave is moving from left to right.)

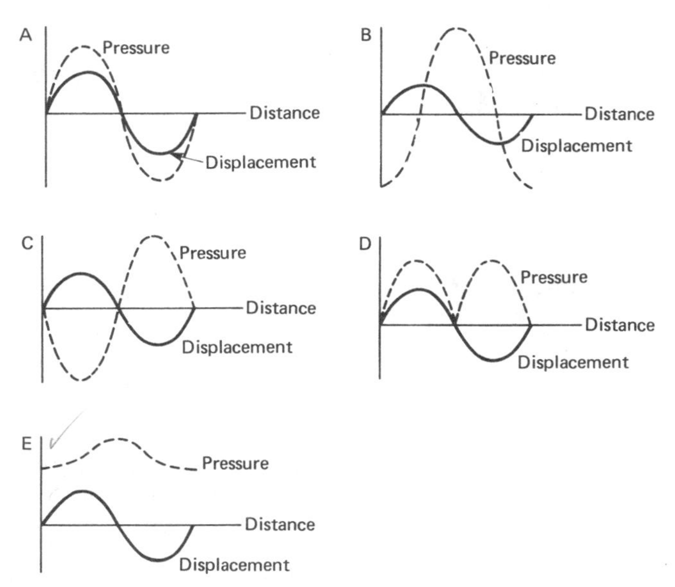

(**London**: *all other Boards*)

See F3.10. Note that the question asks for the **variation about the mean values.**

3.4M The following are all capable of vibrating. Which ones are critically damped as an essential part of the design?
1 The coil of a moving coil ammeter.
2 The coil of a ballistic galvanometer.
3 The suspension system of a car.
A **1** only. B **2** only. C **3** only. D **1** and **2**.
E **1** and **3**.

(—: *AEB, Oxford, Cambridge, SEB SYS*)

Damping is discussed on p. 51. Remember that a ballistic galvanometer is a charge-measuring device; the 'first throw' is proportional to the charge passed, so energy loss is undesirable.

3.5M The diagrams show a 'snapshot' of a horizontal 'slinky' when in equilibrium (ie, when at rest) and when carrying a longitudinal progressive wave.

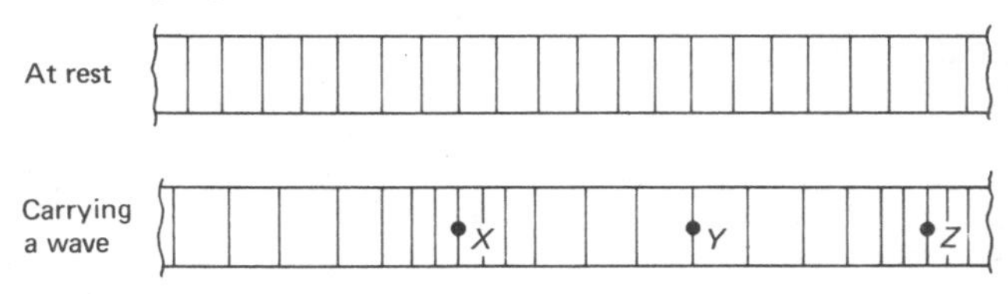

Which of the following statements is/are correct?
1 X is a point of zero displacement at the instant shown.
2 The distance from X to Z is one full wavelength.
3 X is in phase with Y.
A **1** only. B **2** only. C **3** only. D **1** and **2**.
E **1** and **3**.

(—: *all Boards*)

For **1**, compare the coil spacing of the slinky at rest with the spacing at X. For **2**, one full wavelength is the distance from one point to the next point with the same displacement. For **3**, consider the distance XY in terms of the wavelength.

3.6M Here is a list of things involving oscillations:
1 visible light.
2 the 'hum' of a mains transformer.
3 a pendulum of length 10 m.
4 '1500 m wavelength' radio waves.
5 the ebb and flow of the tide.
Assume that the speed of light is 3×10^8 m s^{-1}, which sequence A – E correctly places them in order of increasing frequency:
A 32154. B 32451. C 53241. D 53124.
E 52143.

(—: *all Boards except SEB H, SEB SYS*)

Try to pick out the highest and the lowest frequencies, and see if you can make your choice from those two. Assume that the wavelength of visible light is approximately 600 nm.

3.7M The diagram shows the variation of displacement with time for which **one** of the following objects and situations:

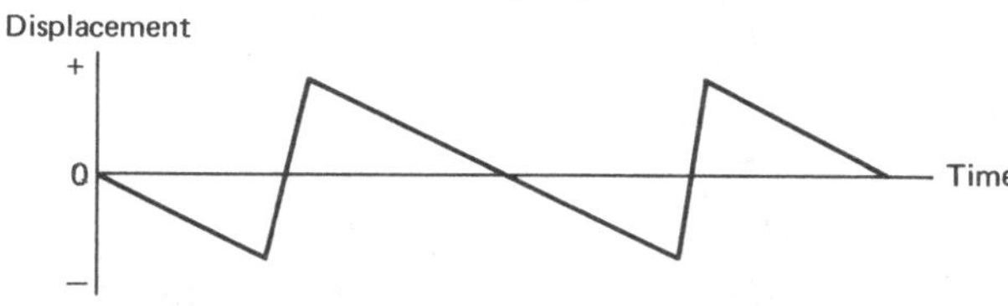

A A rubber ball dropped onto a sponge mat.
B A rubber ball dropped onto a concrete floor.
C The central point of a wire vibrating in its fundamental mode.
D The spot on a cro screen with a linear time base on the X-plates, and no Y-input.
E The 'bob' of a simple pendulum.

(—: *all other Boards except SEB H*)

The graph above is clearly not sinusoidal, so you can eliminate those choices that involve shm and sine waves. Before making your final choice from the remaining alternatives, remember that the graph is of **displacement** against time, not velocity against time.

3.8M Two sources of waves, S_1 and S_2, a distance a apart, vibrating in phase, produce an interference pattern of nodes and

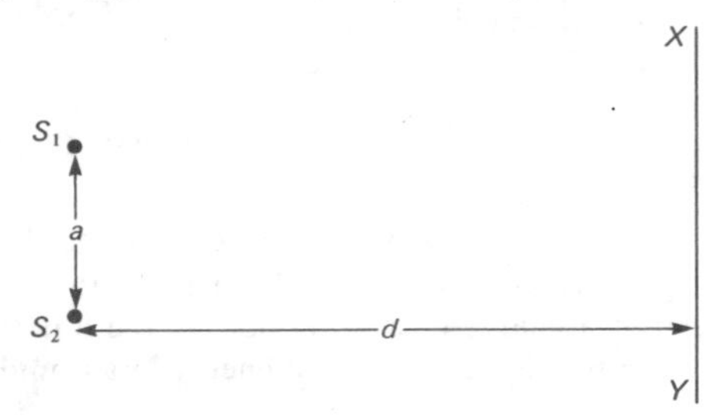

antinodes. This pattern is observed on a line XY parallel to the line joining the sources a distance d from it. The separation between adjacent nodes increases when:
1 the source separation a is increased.
2 the distance d is increased.
3 the frequency of vibration of the sources is increased.
Which of these statements is/are correct?
A **2** only. B **1** and **2** only. C **1** and **3** only.
D **2** and **3** only. E **1, 2** and **3**.

(**NISEC**: *all other Boards*)

See E4.3, which becomes $\lambda = \frac{ya}{d}$ in the notation of the question. Remember that the separation between nodes is y in the equation.

For **3**, assume the wave speed is constant, so change of frequency changes the wavelength.

3.9M Which of the following statements about electromagnetic waves is/are correct?
1 X-rays in a vacuum travel faster than light waves in a vacuum.
2 The energy of an X-ray photon is greater than that of a light photon.
3 Light can be polarized but X-rays cannot.
A **1** and **2**. B **2** and **3**. C **1, 2** and **3**. D **2** only.
E **3** only.

(—: *all other Boards except SEB H, SEB SYS and NISEC*)

Remember that light and X-rays are both electromagnetic radiations. Is electromagnetic radiation a transverse or a longitudinal waveform?

3.10M A parallel beam of white light is passed through a piece of Polaroid into a transparent specimen. When viewed through a second piece of Polaroid crossed with respect to the first, a number of coloured bands is observed in the specimen. Possible reasons for this are the specimen is
1 being subjected to a non-uniform stress distribution.
2 causing interference in multiple slits in the Polaroid.
3 rotating the plane of polarization of the incident light.
Answer: A if **1, 2, 3** correct.
B if **1, 2** only.
C if **2, 3** only.
D if **1** only.
E if **3** only.

(**London**: *all other Boards except SEB H and NISEC*)

Refer to your textbook to distinguish between **photoelasticity** (ie, double refraction under stress) and **optical activity** (ie, rotation of the plane of polarization). Polaroid does not produce interference of light.

3.11M The diagram shows the overhead view of a ripple tank in which two dippers X and Y are just touching the water surface. Dipper X is made to vibrate at 4.5 Hz and dipper Y is made to vibrate at 5.5 Hz. A cork at Z bobs up and down with a displacement which varies with time as given by which graph A–E below?

(—: *all Boards*)

At Z, the phase difference between the waves received from X and those received from Y will change steadily. When the two sets of waves are in phase, the resultant amplitude will be a maximum; when the two sets of waves are exactly out of phase, the resultant amplitude will be zero, See p. 54 and E3.8.

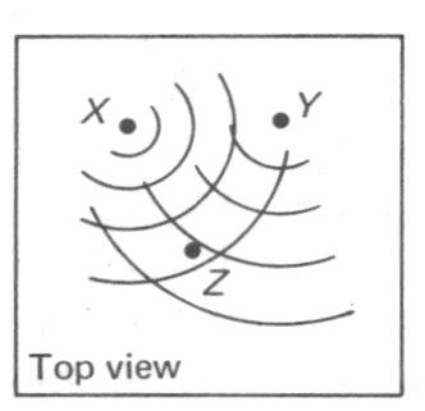

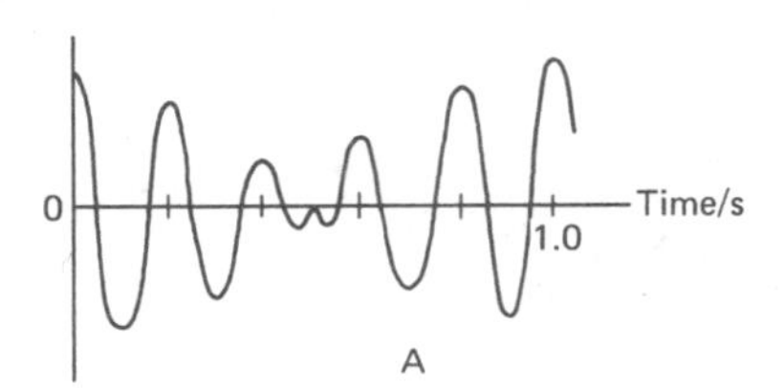

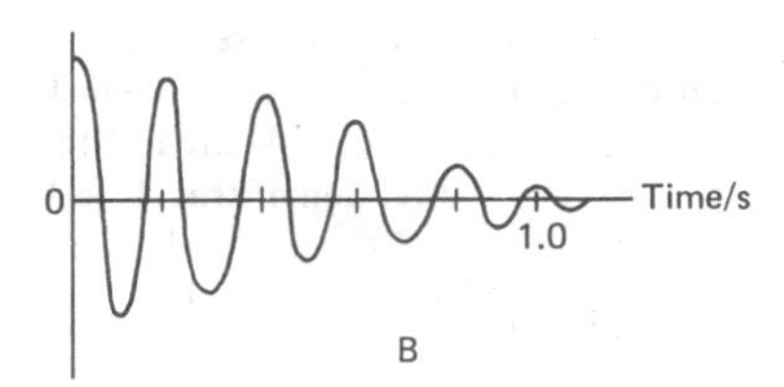

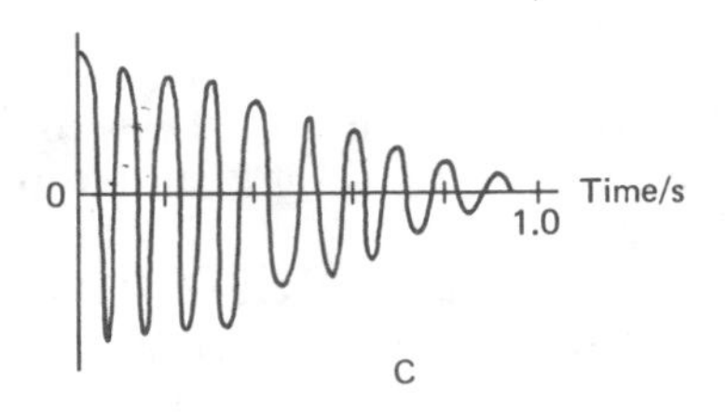

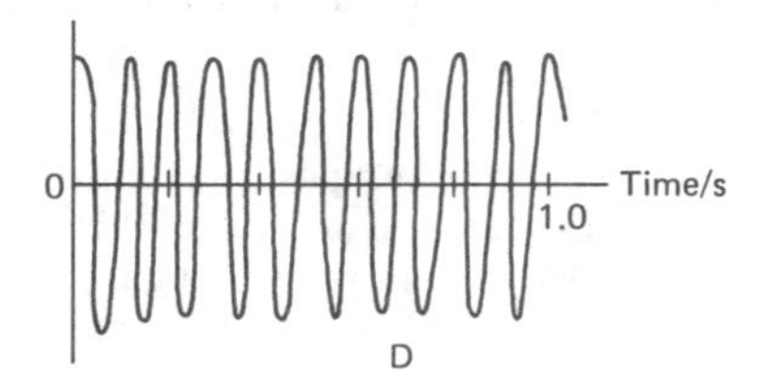

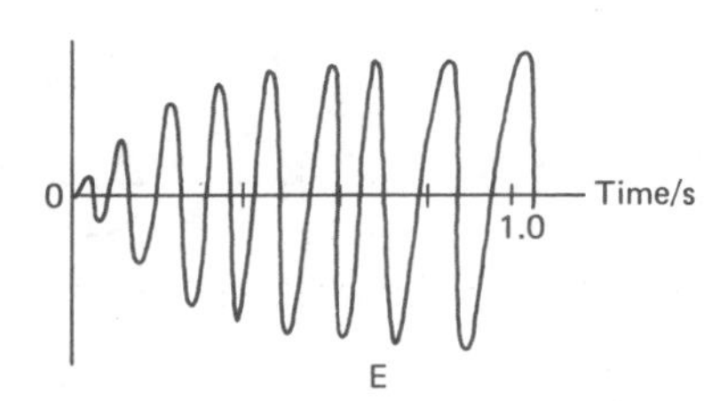

3.12M A transmitter of 3 cm electromagnetic waves and a small metal plate are set up as shown. A receiving aerial is connected

to a suitable amplifier and meter. (The speed of electromagnetic radiation is 3.0×10^8 m s^{-1}). Which of the following is then true?
1 When the receiving aerial is moved along the line XY, a maximum response is noted every 1.5 cm.
2 The frequency of the waves is 10^{10} Hz.
3 When the receiving aerial is placed at a suitable point behind the plate but on the line XY produced, a response can again be noted because diffraction occurs at the edges of the plate.

Answer: A if **1**, **2**, **3** correct.
B if **1**, **2** only.
C if **2**, **3** only.
D if **1** only.
E if **3** only.

(**London**: *all other Boards*)

Refer to F3.6 and E3.7 (noting the warning regarding standing waves following E3.7). Diffraction is discussed on p. 53; note that the wavelength is 3 cm and the plate is small.

3.13M When a sonometer wire is set up as shown, the length XY vibrates with a fundamental frequency of 50 Hz. The fundamental frequency could be doubled by:
A halving the load M. B doubling the load M.
C doubling the length XY. D halving the length XY.
E doubling the diameter of the wire.

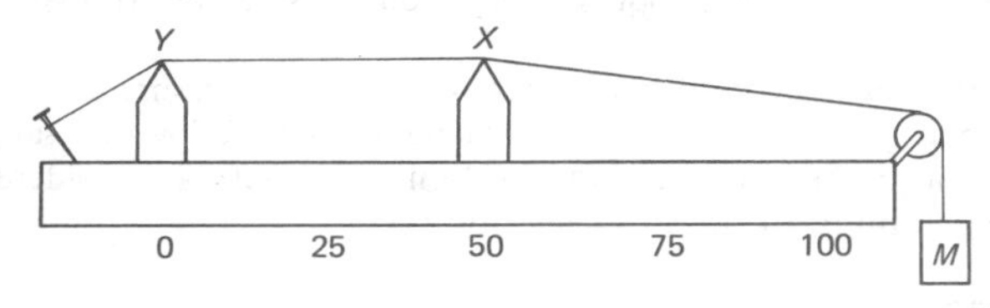

(**AEB** Nov 80: *all other Boards except SEB H*)

The formula for the fundamental frequency of a standing wave on a string can be derived from E3.7 and E3.13. Remember that the tension is Mg, and that the wavelength of the fundamental mode is twice the distance XY.

3.14M Stationary waves are set up in five tubes containing air, three of which are closed at one end. The diagrams represent the displacements of the particles of air from their mean positions at each point along the tube axis. Each tube has a length L, in metres, and an end correction e, in metres. The speed of sound in each of the air columns is c, in m s^{-1}. Which diagram represents a stationary wave of frequency, in Hz, $\frac{c}{4(L+e)}$.

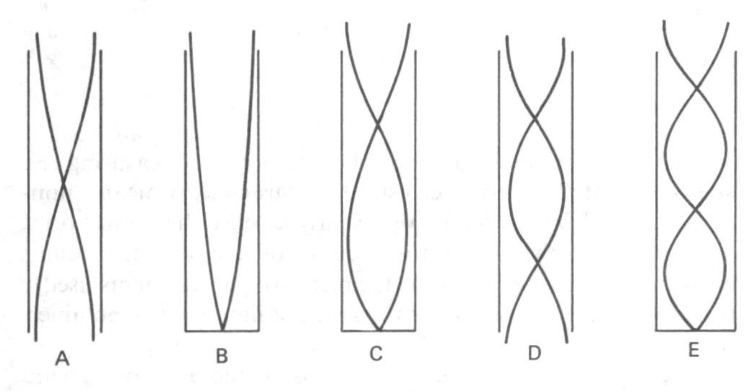

(—: *all other Boards except SEB H*)

See p. 55 and F3.13. The diagram clearly illustrates the end-errors so you must remember an end correction e must be added to L for each open end.

3.15M A vibrator sets up standing waves in a string as shown in the diagram. If the frequency of the vibrator is 20 Hz, then the speed of the waves is:
A 3.0 m s^{-1}. B 10.0 m s^{-1}. C 15.0 m s^{-1}.
D 22.5 m s^{-1}. E 43.0 m s^{-1}.

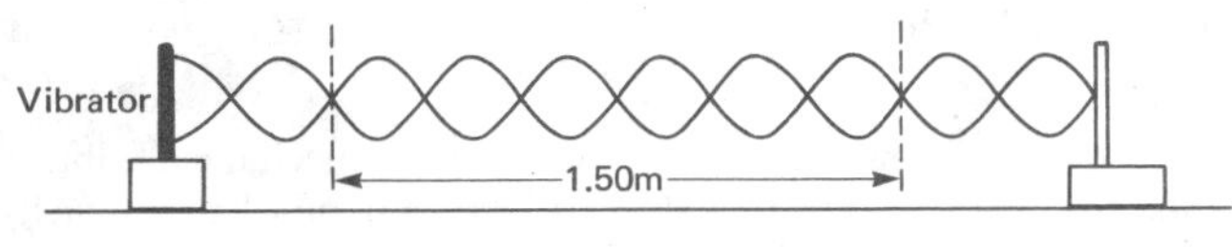

(**SEB H**: *all other Boards*)

You must firstly determine the wavelength by considering the diagram. See F3.5 if necessary. Then calculate the wave speed using E3.7.

3.16M A string is stretched under constant tension between fixed points X and Y. The solid line in the diagram shows a standing

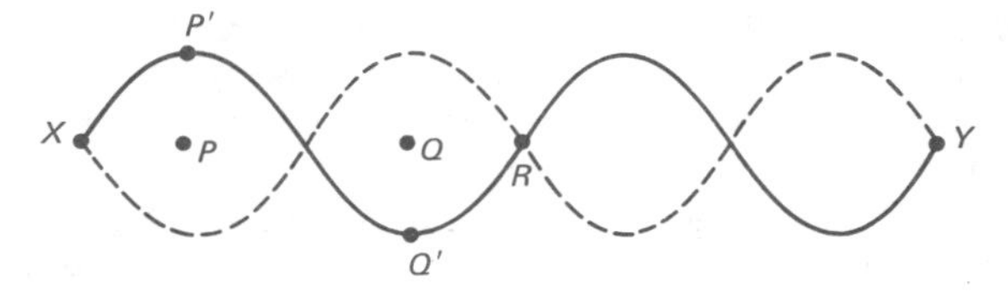

(stationary) wave at an instant of greatest displacement. The broken line shows the other extreme displacement. Which one of the following statements is **correct**?
A The distance between P and Q is one wavelength.
B A short time later, the string at R will be displaced.
C The string at P' and the string at Q' will next move in opposite directions to one another.
D At the moment shown, the energy of the standing wave is all in the form of kinetic energy.
E The standing wave shown has the least possible frequency for this string stretched between X and Y under this tension.

(**Cambridge**: *all other Boards except SEB H*)

See p. 55 and F3.5. Remember that displacement nodes (points of zero displacement) are spaced out at half-wavelength intervals. At the moment shown, the displacement of each particle is at its maximum value so the velocity of each particle is zero.

3.17M Which one of the following groups of electromagnetic waves is in order of increasing frequency?
A gamma-rays, ultraviolet rays, radio waves.
B gamma-rays, visible light, ultraviolet rays.
C visible light, infrared radiation, microwaves.
D microwaves, ultraviolet rays, X-rays.
E radiowaves, visible light, infrared radiation.

(**Cambridge**: *all other Boards except SEB H and SEB SYS*)

See F3.14.

3.18S A 3 cm wave transmitter was set up in front of a metal barrier in which there were two narrow, parallel gaps P and Q.

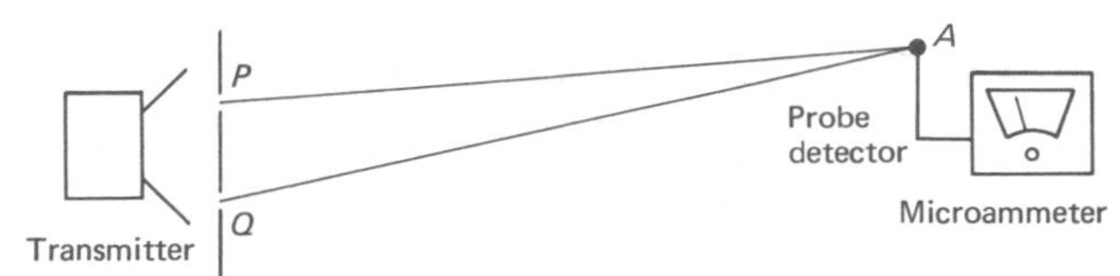

If the distance $AP = 88.1$ cm and the distance $AQ = 92.6$ cm, comment on the intensity of radiation detected by the probe detector at A. Explain clearly your reasoning.

(**SEB H**: *all other Boards*)

The path difference $AQ - AP = 4.5$ cm. Does this mean that waves from P and Q will interfere constructively or destructively at A? See p. 63 if necessary.

3.19S
(a) The sketch shows a long wooden rod clamped at its centre. Longitudinal vibrations of period T may be excited in the rod using a resined cloth gripped tightly to the rod and drawn sharply in the direction of the arrow. Sketch, in exaggerated form, the outline of the rod at the times indicated:

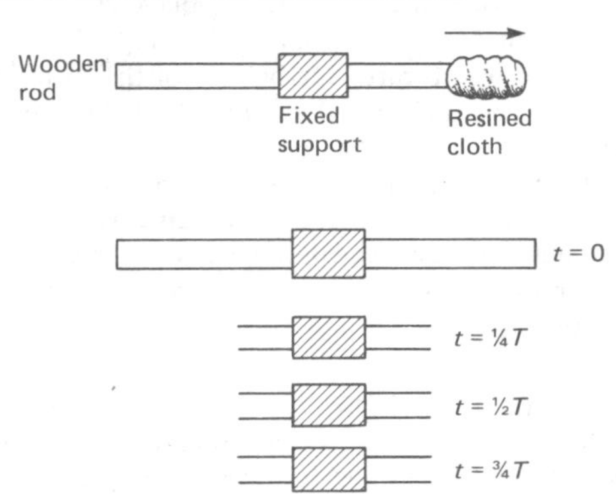

(b) (i) Explain the combination of simple travelling waves which gives the vibration in the rod. (ii) The waves in the gas in a sounding resonance tube have some features in common with the waves in the rod. Name two of them.

(c) The shortest length of the air column in a tube, closed at one end, resonating with a tuning fork of frequency 384 Hz was 206 mm. Another resonance was found at 624 mm. Find the speed of sound in the air inside the tube.

(**NISEC**; *all other Boards except SEB H*)

(a) A stationary wave is set up in the rod as a result of the initial vibration being reflected at each end. Where will the nodes and antinodes occur in the simplest mode of vibration? Look at F3.6 and consider *C* and *G* as the ends of the rod. Remember that you are dealing with longitudinal vibrations, so that, in a displacement diagram, movement to the right is plotted above the axis and movement to the left below it.

(b) (i) See p. 52. (ii) See F3.5 and F3.13 (first part), and p. 55, which deals with standing waves in pipes.

(c) See F3.13 (third and fourth parts) and use E3.7, remembering to allow for the 'end correction'.

3.20S

(a) The diagram shows a standing wave in a long narrow spring which is vibrating from side to side on a smooth horizontal surface. What factors together determine the frequency of the vibrations?

(b) The diagram shows a standing wave in the same spring when it is arranged vertically between two supports; the speed of propagation of the wave, and hence the wavelength, is now decreasing towards the lower end. Why does the speed decrease towards the lower end?

(c) Suggest one similarity and one difference between the standing waves shown in the diagram for part (b) and electron standing waves in an atom.

(**O and C Nuffield**: *all other Boards (a & b only) except SEB H*)

(a) The frequency of any wave form is defined by its wavelength and its speed of propagation. What defines the wavelength in this system, and what controls the velocity of transverse waves down a stretched spring (E3.13)?

(b) E3.13 provides the clue. The role played by the tension in the spring is crucial. If the spring can be regarded as 'heavy', how will the tension towards the bottom of the spring compare with the tension at the top? Does this affect the wave velocity in the right way to explain the diagram?

(c) The electron in an atom can be modelled as a standing wave in a box. The diagram shows a standing wave in a box (the similarity), but does the electron have a wavelength that varies along its path as shown in this diagram? (See Unit 10 of Nuffield Student Books.)

3.21S What is the frequency of the sound emitted by an open-ended organ pipe 1.7 m long when sounding its fundamental frequency? (the speed of sound in air = 340 m s^{-1}).

What would be the effect (if any) on the frequency of the sound emitted of an increase of **(a)** the atmospheric pressure, **(b)** the temperature of the air?

(**SUJB**: *all other Boards except SEB H*)

To calculate the fundamental frequency of an open-ended pipe, you must first calculate the wavelength corresponding to that frequency. See p. 55 and F3.13. Neglect end-errors. Then use E3.7 to calculate the frequency.

(a) If the pressure changes, then the density will also change (assuming constant temperature). E3.17 indicates the factors that determine the speed of sound; at constant temperature, consider if the (pressure/density) changes when pressure changes. Hence decide if the speed of sound alters. Since the wavelength depends only on the tube length, then if there is a change of the speed of sound, there must be a change of frequency. Your answer will depend upon whether you consider the (pressure/density) changes.

(b) Again, consider whether or not the (pressure/density) changes when the temperature changes. Then follow the same steps as in the last part of (a). E6.14 may be useful if considered in the form P = constant × density × T.

3.22S With the aid of suitable diagrams, explain the difference between transverse waves which are polarized and those which are unpolarized.
State one useful application of plane polarized waves, specifying the type of wave involved (eg, electromagnetic).

(**London**: *all other Boards except SEB H, SEB SYS*)

Polarization is discussed on p. 53 and two applications are mentioned. For the transverse wave, indicate clearly on your diagram the plane in which the transverse wave vibrations occur.

3.23L Describe a terrestrial method of measuring the speed of light in air or vacuo. List the principal regions of the electromagnetic spectrum, and mention some of the important properties of the radiation in each part. Indicate the evidence that light and one other kind of radiation are both electromagnetic waves.

(**SUJB**: *O and C Nuffield, Cambridge, NISEC and WJEC*)

Use your textbook for a suitable method for measuring the speed of light. The term 'terrestrial' in this context means 'non-astronomical'. You should give a clearly labelled diagram, a brief description of the apparatus, details of the procedure and measurements made (and state the measuring instruments used); then, indicate how a value for the speed of light may be obtained from the measurements made.

See E3.14 for the principal regions of the electromagnetic spectrum. See p. 55 and your textbook for the important properties of each part – you might find it helpful to draw up a list under the following headings: speed, wavelength, means of production, means of detection, absorption by matter, ionizing effects, photoelectric effects, etc.

Use your textbook to find details of evidence for light, etc., as electromagnetic waves. You might like to consider the link between the speed of the waves (c) and the field constants ε_0 and μ_0 as given in E2.34 and related comments. See your textbook for information about the 'spark gap' transmitter of radio waves discovered by Hertz.

3.24L

(a) A device for measuring the frequency of mechanical vibrations consists of eleven thin steel strips of different lengths each fixed at one end to a block. The frequency of the fundamental transverse oscillations of the strips ranges from 35 Hz to 45 Hz in 1 Hz steps. The device is attached to a machine which is vibrating, and it is observed that the 40 Hz strip is set into strong oscillation while all the others show little movement. (i) Explain the principle on which the device operates. (ii) What conclusions can be drawn about the frequency of vibration of the machine?

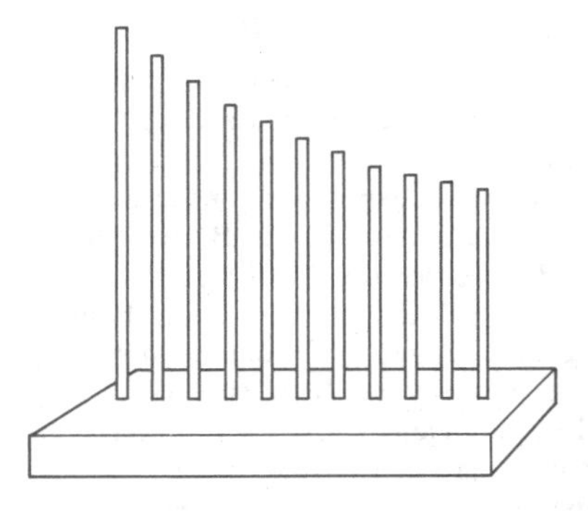

(iii) What would be observed if the experiment were repeated with the strips immersed in a bath of light oil?

(b) A quartz crystal is in the form of a rectangular bar clamped at the mid-point and oscillates longitudinally in its fundamental mode. The crystal is designed to oscillate at a frequency of 100 kHz. Calculate: (i) the length of the crystal; (ii) the limits between which the length of the crystal must lie if the frequency is to be correct to within $\pm$ 10 Hz. (Take the speed of longitudinal waves in quartz to be 5.72×10^3 m s^{-1}.)

(**Oxford**: *all other Boards except SEB H, SEB SYS*)

(a) (i) The vibrations of the machine are transmitted through the base to each rod. Each rod undergoes forced oscillations. See p. 51. The key point is resonance. (ii) At resonance, the frequency of natural oscillations (ie, fundamental) of the resonant system is equal to the frequency of the 'forcing' system (ie, of the machine vibrations). Use this fact to state the machine vibration frequency. Also, since none of the other rods show much movement, you can comment on the 'spread' of the machine vibration frequency spectrum (ie, the presence of other frequencies in the range 35 to 45 Hz in the range of frequencies produced by the machine). (iii) Oil immersion would have a considerable damping effect on the response. See p. 51. You should give your answer in terms of a comparison of observations with and without oil.

(b) (i) At the midpoint there must be a node (ie, zero disturbance), so in the fundamental mode, there would be an antinode at either end with no intervening antinodes or nodes between the centre and the end. You must consider if this means that the length from end to end is $\frac{1}{4}$ × wavelength or $\frac{1}{2}$ × wavelength, etc., so that you can write down the length in terms of the wavelength. Then use the given values of wavespeed and frequency to calculate the wavelength. Once you have the wavelength value, then you can calculate the bar length. (ii) The percentage limits for the length must be the same as the percentage limits for the frequency, so work out the % limits for length first, then calculate the actual limits, using the length value from **(b)** (i).

3.25L

(a) Distinguish between transverse and longitudinal waves. State into which of these categories you would put (i) light waves, (ii) sound waves and indicate how you would demonstrate experimentally that your statements are correct.

(b) In terms of the motion of the particles of a medium in which there are sound waves, describe **one** similarity and **two** differences between a progressive sound wave and a stationary sound wave.

(c) By means of suitable apparatus stationary waves of sound are set up in air. Describe how you would attempt to measure the wavelength of these waves.

(**JMB**: *all other Boards except SEB H and SEB SYS*)

(a) See p. 52. State the difference in terms of the direction of vibration related to the direction of travel of the wave. For light waves, describe a simple polarization experiment to show that the wave vibrations are in one direction only. Give a simple diagram. For sound waves, describe a simple experiment in which sound waves are directed at a diaphragm. With a magnet attached to the diaphragm (must be a small magnet) and a coil near the magnet, vibrations of the diaphragm will induce a voltage across the coil terminals. An alternative experiment would be to pass sound waves down a horizontal hollow glass tube which has some light powder sprinkled along its length; with one end of the tube closed and with a suitable source of sound waves at the other end, a stationary wave pattern can be established. The vibrations of the sound waves at the antinodes cause the powder to move to the nodes. See Kundt's tube experiment in your textbook.

(b) The motion of the particles in each case is described on p. 52. You must give your answer in terms of the particles' motion, so you must think about the amplitude of particle vibrations at different points, and about the phase difference between vibrations at different points. Remember that all particles which vibrate (which means each and every particle in the progressive case, but excludes particles at the nodes in the stationary case) do so at the same frequency.

(c) Refer to your textbook for the 'resonance tube' experiment.

3.26L

(a) State the necessary conditions for the establishment of a stationary wave. Explain how these conditions are fulfilled when a wire stretched between two fixed points is plucked in the middle.

(b) Describe how the amplitude and phase of the vibrations vary with position along the length of the wire when it is vibrating in (i) its fundamental mode, (ii) its first overtone.

(c) Describe an experiment using a sonometer to investigate how the fundamental frequency of transverse vibration of a fixed length of wire varies with diameter. You may assume that you are provided with a set of wires of the same material, but of different diameters, and any other essential apparatus. Show how you would establish the relationship by graphical method and state the relationship you would expect to obtain.

(d) Two steel violin strings of the same length and which are subjected to the same tension have fundamental frequencies of 440 Hz and 660 Hz, respectively. Calculate the ratio of the diameter of the two strings.

(**JMB**: *all other Boards except SEB H*)

(a) The conditions for stationary waves are discussed on p. 52. When the wire is plucked in the middle, progressive waves travel in either direction away from the centre and reflect off the ends. You should describe how this leads to standing waves.

(b) Give a diagram showing how the wire would appear at maximum displacement for each mode to aid your explanation of how the amplitude (ie, maximum displacement) varies with position. For the phase, remember that all points between adjacent nodes are in phase, whilst points on either side of a given node (as far away as the next node) are out of phase by 180°.

(c) For a suitable sonometer experiment, refer to your textbook. To establish the fundamental frequency of each wire (under the same tension, and of the same length), the ac resonance method as drawn in the diagram of question 7M is the best method. You must describe how you would determine the fundamental frequency for each wire, how you would ensure each wire is under the same tension and of the same length, and how you would measure the diameter of each wire. The expected relationship can be derived from the equation for the fundamental frequency: $f_0 = \frac{1}{2L}\sqrt{\frac{T}{m}}$ where m = mass per unit length. Since m = (volume × density)/length = $\pi d^2 \rho$ where d = diameter, ρ = density then you can show how f_0 depends upon d. You must consider how you would present the relationship graphically so as to obtain a straight line.

(d) Use the relationship between f_0 and d which you have already derived in **(c)**.

3.27L Explain briefly how you would display a sound wave form on a cathode ray oscilloscope.

Two sources of sinusoidal sound waves of equal amplitude but of frequencies 500 Hz and 1000 Hz, respectively, are placed 7 m apart. The sources are in phase and the velocity of sound is 350 m s^{-1}. Sketch a graph showing how the resultant disturbance varies with time at a test point half way between the sources. State any physical principle involved, and explain clearly how you arrived at your graph. What is the period of the resulting disturbance? What is the shortest distance the test point may be moved in order to obtain a resulting disturbance of maximum amplitude?

(**WJEC**: *all other Boards except SEB H*)

Your explanation should include a diagram showing a means of producing sound waves and a means of converting a sound wave signal into an electrical signal which can then be displayed on an oscilloscope. Refer to your textbook if necessary.

To determine the resultant disturbance at the midpoint of the two sources, start by making a sketch showing the two sources (with frequencies) and the midpoint. Then calculate the wavelength of the sound waves emitted by each source (see E3.7) and put that information on your diagram. Next, consider the distance from each source to the midpoint in terms of wavelengths appropriate to the source being considered; hence, determine the phase difference (if any) between the source and the waves at the midpoint from that source, for each source in turn.

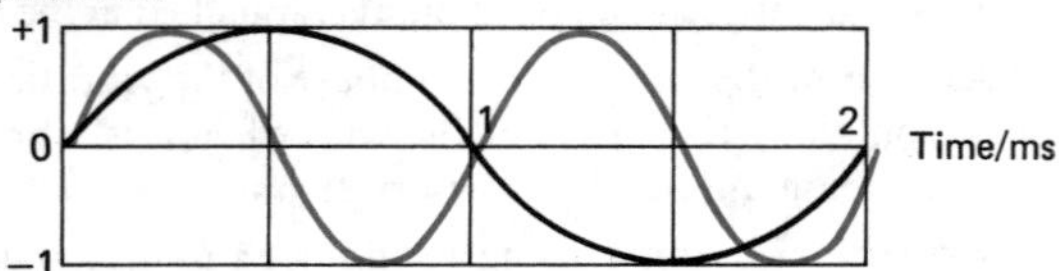

The next stage is to consider what 'in phase' means for the two sources; take it to mean that each 500 Hz cycle starts at the same time as a 1000 Hz cycle starts, as in the diagram. Then, bearing in mind any phase differences that exist between midpoint and source, sketch on the same axes displacement/time graphs for each set of waves at the midpoint. Your graphs ought to be similar to the diagram above because the midpoint is in phase with the source for both sources. To obtain the resultant disturbance, mark regular points on the time axis of your graph, and at each point, add the two displacements to give the resultant displacement. Remember to take account of signs; for example, if the displacement at a given time due to one source is $+0.6$ and the displacement at the same time due to the other source is -0.8, then the resultant is -0.2. The principle involved here is that of 'superposition' (see p. 53).

Now consider moving the test point towards one of the sources; waves from the nearer source will arrive earlier than from the other source, so you must consider how far the test point must be moved in order that wavepeaks from one source will arrive at the test point at the same time as wave peaks from the other source arrive. Coincident wavepeaks will then give maximum amplitude.

3.28L

(a) Give an expression for the shift in wavelength of a spectral line as emitted by a distant star which is moving with a velocity V_1 through space when it is observed on the Earth which is moving through space with a velocity V_2, both velocities being in the same direction in the line of sight, and measured with respect to some third body. Verify that your expression would remain unchanged if both the Earth and the star are supposed to be moving through space with an additional velocity V_3 in the line of sight.

(b) A particular spectral line has a wavelength of 500 nm when observed in the laboratory. The same line when emitted by a distant star and observed from the Earth has an apparent wavelength of 550 nm. What can you deduce about the velocity of the distant star?

(c) Calculate the change in the observed wavelength of the same spectral line if it is viewed from a spacecraft leaving the Earth in the direction of the star if the speed of the spacecraft were 3 km s^{-1} relative to (i) the Earth, (ii) the distant star.

The speed of light, $c = 3 \times 10^8$ m s^{-1}.

(**JMB***: *and Oxford, Cambridge*)

(a) Start by making a sketch showing Earth, the distant star and the third body along the same straight line. Then indicate the velocities (including direction) of Earth and distant star, taking the third body to be at rest.

Now write down an expression for the relative velocity of the distant star with respect to Earth. By treating the situation in terms of fixed observer (ie, Earth) and moving source (with velocity = relative velocity as above), you should appreciate that the wavelength is changed by the source motion. For the shift of wavelength (ie, the difference between the wavelength with source moving and the wavelength from a stationary source), see p.54 and E3.9 (this equation is for stationary observer and source moving towards the observer, so take care).

To show that the expression is unchanged with an additional velocity V_3 for both Earth and star, add V_3 onto the individual velocities. Then, write down an expression for the relative velocity of the star with respect to the Earth; you should find that V_3 does not appear in the relative velocity expression.

(b) Use your expression from (a) for the wavelength shift. Hence deduce the velocity of the distant star with respect to Earth.

(c) (i) Make a new sketch. Indicate the spacecraft velocity and that of the distant star relative to Earth. Write down the relative velocity of the star with respect to the spacecraft, and then use the expression from (a). (ii) Treat the observer (in the spacecraft) as stationary and the distant star as moving towards the observer at 3 km s^{-1}. Then use E3.9.

3.29L Describe as fully as you can the nature of linearly polarized light.

When unpolarized light falls on the surface of a block of glass, the reflected light is partially polarized. If the angle of incidence is $\tan^{-1} n$, where n is the refractive index of the glass, the reflected light is completely linearly polarized. Describe the apparatus you would use and the experiments you would perform in order to verify these statements for a sample of glass of known refractive index.

Why would it be necessary, if very accurate results were required, to use monochromatic light to verify the second statement?

Show that, when the condition for completely polarized light is satisfied, the reflected and refracted beams are at right angles to one another.

(**O and C**: *all other Boards except SEB H, SEB SYS, NISEC*)

General ideas about polarization are discussed on p. 52 and p. 53. See also F2.18. You will need to describe light in terms of electric and magnetic fields, best done with the aid of a diagram, and to explain what polarized light means in terms of your description.

See your textbook; as part of your answer, give a diagram showing how a narrow beam of light may be produced. Also mark the angle of incidence on your diagram and state how you would measure the angle. Show the means of detecting whether or not the reflected beam has been completely polarized by reflection; a simple Polaroid filter would suffice, but you should explain its use. Remember that there are two statements to verify.

The key point here is to consider how (in general terms) refractive index varies with wavelength. See p. 67 if necessary. Then consider how the use of non-monochromatic light (eg, white light) would affect the accuracy of locating the angle of reflection corresponding to total polarization.

The condition for total polarization of the reflected beam is given by the equation $\tan i = n$, which may be rewritten as $\sin i = n \times \cos i$. Snell's Law provides the link between angle of incidence (i) and angle of refraction r, $\sin i = n \times \sin r$. It follows that for this particular angle only, $\sin r = \cos i$ which means that $i + r = 90°$. Now sketch the paths of the incident, reflected and refracted rays; simple geometrical considerations ought to enable you to prove that the angle between the reflected and refracted rays is 90°, using the fact that $i + r = 90°$ here.

3.30L

(a) The strings of a guitar responsible for producing the lower notes are usually of metal, but those for the upper notes are often of a non-conducting material. (i) Describe an experimental method of measuring the fundamental frequency of oscillation of each of the strings of a tuned guitar. (ii) Suggest an approximate value for the fundamental frequency of the string responsible for the lowest note.

(b) A cine film of a certain wave motion on a section of rope is taken against a scale in the immediate background. The film is found to show an unchanging rope pattern, as in the diagram.

There is no record of whether the film is of a travelling wave or of a standing wave on the rope, only that the cine camera operated at 24 frames per second and the length of the scale was one metre. (i) Can you tell from the information given whether the film is of a travelling wave or of a standing wave? Explain your reasoning fully. (ii) What velocity can be deduced from the information above for the waves on the rope? If the information given does not define a unique velocity, explain how other possibilities arise. (iii) Sketch what would be recorded on a sequence of three consecutive frames if a similar cine film is taken with a **slightly** reduced frame rate, (A) of a wave travelling from left to right, and (B) of a standing wave.

(**SEB SYS**: *all other Boards except SEB H*)

(a) (i) The metal wire can be made to resonate if a variable ac source is connected across it. Arrange the wire to pass between the poles of a horseshoe magnet so that the wire is perpendicular to the magnetic field, as in the diagram of question 7M. Gradually increase the frequency of the ac source from a low value and observe the wire; it will resonate when the ac frequency equals the fundamental frequency. For the non-conducting strings, a stroboscope of variable frequency could be used, but care would need to be taken to ensure that the string vibrates in its fundamental mode when plucked. (ii) 125 Hz perhaps, which is about one octave lower than C (256 Hz).

(b) (i) See F3.5 and 3.6. A standing wave of frequency 24 Hz (or any multiple of this) would occupy the same position every 1/24 s. This would also be true of a travelling wave. Thus... (carry on the argument). (ii) From the diagram $\lambda = 0.3$ m and $f = 24$ Hz, hence find v; see E3.7 if necessary. (iii) In each case the wave would have travelled beyond its 'stationary' position shown in the diagram. Remember that if a rotating spoked wheel is 'frozen' stroboscopically, then when the strobe rate is slightly reduced, the wheel appears to advance slowly. The same idea applies here.

Unit 4 Optics

This Unit deals firstly with the experimental evidence for the wave theory of light, and then shows the usefulness of the theory. The link between waves and rays is then established, so leading into a study of optical instruments. Treatment of the unit in this way reflects the division of the subject of light into physical optics, dealing with the wave nature of light, and geometrical optics, which involves using ray theory. The subject reaches into other units, since light is a form of electromagnetic radiation, and a fuller discussion of electromagnetic waves is given in Unit 3. Electromagnetic radiation is quantized in the form of photons, as discussed in Unit 1, and the implications of this fact are discussed in Unit 8.

Before tackling the sections on interference and diffraction, it is worthwhile re-reading the section of Unit 3 dealing with the principle of superposition, since interference and diffraction involve application of that principle. GCSE studies of optics vary widely from one course to another; most courses contain an introduction to lenses and mirrors, so it ought to prove useful to re-read your notes on these topics before you begin the section of this Unit on ray optics.

4.1 WAVE NATURE OF LIGHT

The idea that light consists of waves was first put forward by **Huygens**. He assumed that a point source of light emits spherical wave fronts, and that every point on a wave front acts as a secondary emitter. Using his theory, the propagation of light can be explained, as in F4.1. We need to assume that the new wave front in the diagram, formed by the envelope of secondary 'wavelets', is formed ahead (and only ahead) of the old wave front.

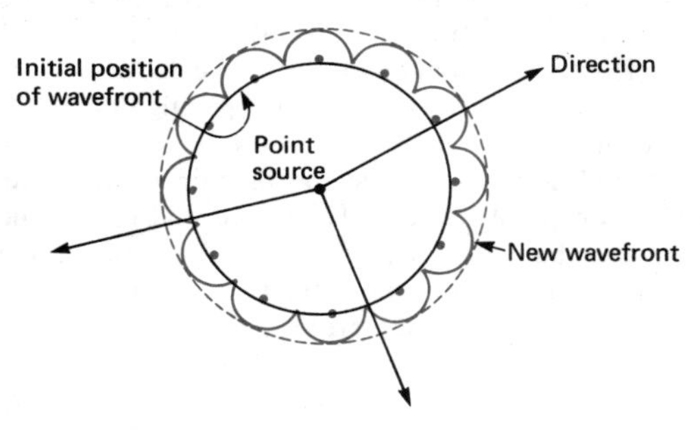

F4.1 Huygens wavelets producing a spherical wave front

Interference of light: Young's fringes Direct evidence for the wave nature of light was first obtained by Young, more than a century after Huygens' time. What Young did was to observe the superposition of two sets of light waves derived from the same source, and he arranged this by allowing light from a point source to pass through two very close slits, as shown in F4.2.

The pattern on the screen consists of alternate bright and dark fringes, parallel to the slits. The pattern can only be explained by using Huygens' wave theory; each slit acts as a source of waves, so that where two sets of waves overlap, an interference pattern can be produced.

The formation of a bright fringe on the screen is because the waves from one slit **reinforce** the waves from the other slit. In other words, a wave crest from one slit arrives at the bright fringe at the same time as a wavecrest from the other slit. Referring to the diagram, the condition for a bright fringe to be formed at a given point P on the screen is:

E4.1 $S_1P - S_2P = m\lambda$

where $m = 0$ or 1 or 2 or 3 etc, λ = wavelength of the light used (m).

For a dark fringe to be formed, the light waves from one slit must **cancel** the waves from the other slit. In other words, a wave crest from one slit must arrive at the dark fringe at the same time as a wave trough from the other slit.

E4.2 $S_1P - S_2P = (m + \frac{1}{2})\lambda$ (for a dark fringe)

By making suitable measurements, the wavelength of light can be calculated from the equation below (you should consult your textbook for a proof of this equation):

E4.3 $\lambda = \dfrac{Yd}{X}$

where Y = fringe spacing (ie, distance between centres of two adjacent fringes) (m), λ = wavelength (m), d = slit spacing (centre to centre) (m), X = slits screen distance (m).

A simple test to check that the pattern is caused by interference between light from the two slits is to block off one of the slits. The fringes will then disappear.

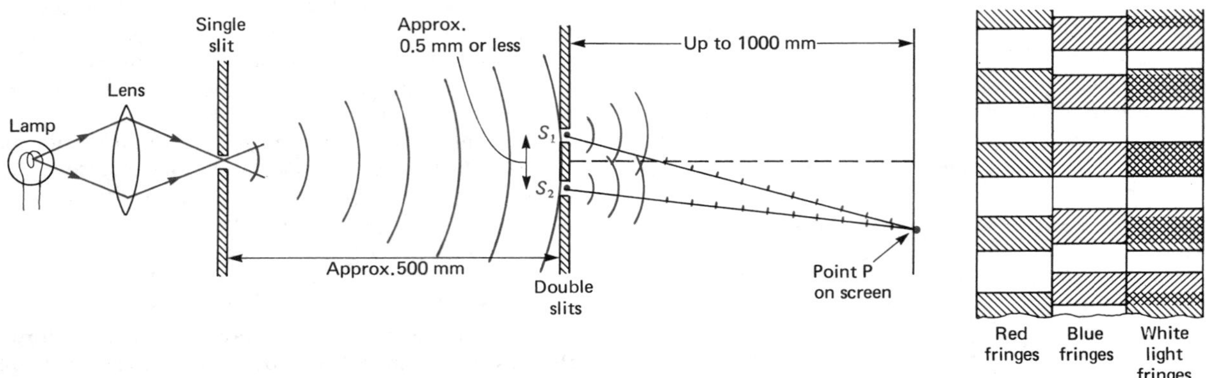

F4.2 Interference from Young's double slits experiment

If a white light source is used, then the central fringes will be white and black, but the outer fringes will be tinged with colours. The reason is that white light is composed of all the colours of the spectrum from red ($\lambda = 700\,\text{nm}$) to blue ($\lambda = 400\,\text{nm}$). Each wavelength component of the white light gives its own interference pattern on the screen, with spacing in proportion to the wavelength. Thus, each non-central bright fringe will be tinged with red on its outer side and blue on its inner side.

Coherence Two point sources of light are said to be **coherent** if they each emit light waves of the same frequency and with a **constant phase difference** between the two sets of waves. For an interference pattern to be observable in the double slits experiment, the sources must be coherent. When two sets of waves overlap, cancellation and reinforcement takes place; however, if the phase difference between the two sources changes, then the points of interference move. Thus, an overlap of light from two separate light bulbs will not produce an observable interference pattern because the two bulbs emit light waves with a randomly changing phase difference. However, in the double slits experiment, waves are emitted by the two slits only when a wave front arrives from the initial light source. Therefore, the two slits act as coherent sources, and an observable interference pattern can be produced. See also p. 53.

Because atoms emit light in short 'bursts' only, lasting of the order of nanoseconds (ie, 10^{-9}s) only, light from a given source will consist of 'wave trains' of length up to 0.3 m or so (= speed of light × time taken to emit). In Young's double slits experiment, the two sets of light waves, arriving at the screen, will only be coherent (ie, have a constant phase relationship) if the path difference $S_1P - S_2P$ is less than the length of a typical wave train. If the path difference exceeds the length of a typical wave train, then the coherence will be lost because the end of the wave train from one slit will arrive before the start of the simultaneously emitted wave train from the other slit. From a sodium light source, a typical wave train length is 1 cm, whereas for a laser source it is typically 0.40 m; hence, the laser produces an interference pattern with many more fringes visible.

4.2 EFFECTS DUE TO FILMS AND GAPS

Thin-film interference: near-normal incidence When monochromatic light is incident upon a thin film, interference takes place between light reflected from one boundary and the light reflected from the other boundary. F4.3 shows a parallel monochromatic beam of light at near-normal incidence upon a thin film of thickness t. The two reflected beams are produced as a result of partial reflection and transmission at each boundary. If the two reflected beams are in phase, then they reinforce one another so that observer O_1 sees the film as a single colour corresponding to the incident light. The path difference, $2t \cos r$, must not exceed the length of a typical wave train in the film; hence thick films do not produce such effects.

The condition for **reinforcement** of the two reflected beams is:

E4.4 $2nt = (m + \frac{1}{2})\lambda_0$

where $m = 0$ or 1 or 2, etc, n = refractive index of the film, t = thickness of film (m), λ_0 = wavelength **in air** (m), assuming $\hat{r}$ is small so that $\cos r \simeq 1$.

Given this condition, the two reflected beams will be in phase because:

1 **external reflection** always causes a phase reversal of 180° (see 3.5, p. 52), so the reflection of R_1 causes the extra half-wavelength;
2 R_2 travels an extra distance $2t$ compared with R_1;
3 the wavelength **in the film** is λ_0/n.

In this situation, the two **transmitted beams** will be exactly out of phase (neither T_1 nor T_2 undergo phase reversal due to reflection) so that an observer at O_2 will see the film as entirely dark.

If the film thickness is such that the following condition holds:

E4.5 $2nt = m\lambda_0$

then the **film will appear dark by reflection** and bright by transmission. The reason is that the two reflected beams will be exactly out of phase because the path difference $2t$ is a whole number of wavelengths in the film, **and** R_1 undergoes phase reversal upon reflection. The two transmitted beams will be exactly in phase because $2t$ = whole number of 'film' wavelengths **but** neither T_1 or T_2 undergo phase reversal.

Thin-film interference: near-normal incidence for films with non-parallel sides The simplest example is **the wedge.** The conditions for interference given by equations E4.4 and E4.5 still apply, but the film thickness t varies from one position to another. An observer will therefore see dark and bright fringes running along positions of equal thickness. Each fringe will therefore follow a contour of equal film thickness; for a wedge, the fringes will be alternate bright and dark lines parallel to the edge of the wedge.

A more interesting example is provided by the film formed between a convex lens and a flat glass plate when the lens and plate are in contact. Whether the film is an air film or a liquid film, the same general pattern, known as **Newton's rings** is produced. The pattern consists of alternate, concentric bright and dark rings, as in F4.4, centred upon the point of contact. The ring pattern may be seen by transmission or reflection. The explanation is essentially the same as for the wedge and the parallel-sided film. In this case, the contours of equal thickness are rings centred upon the point of contact. The same conditions for bright

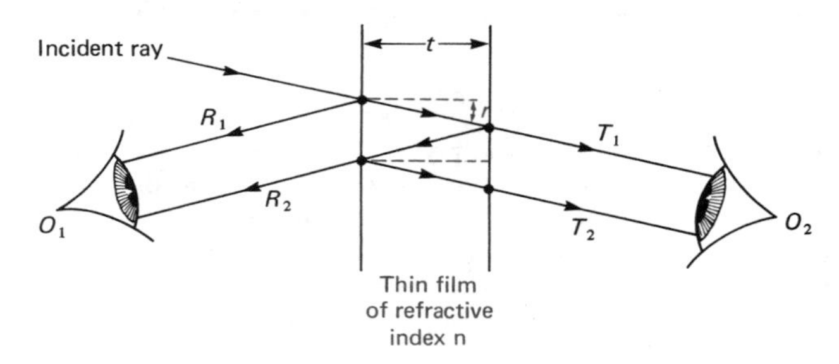

F4.3 Interference from a parallel thin film

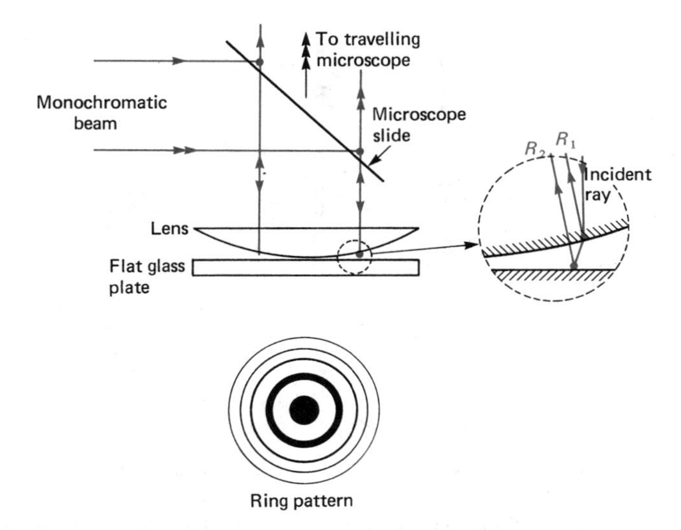

F4.4 Interference from lens and flat surface – Newtons rings

and dark fringes as before hold (ie, E4.4 and E4.5 apply); thus for a **dark ring by reflection** $2nt = m\lambda_0$. However, the diameter of the mth dark ring can be expressed by the following equation derived by geometrical considerations: $D_m^2/4 = 2Rt$, which gives:

E4.6 $D_m^2 = \dfrac{4mR\lambda_0}{n}$

where D_m = diameter of mth dark ring (m), R = radius of curvature of lens face (m), t = film thickness at ring position (m), $m = 0$ or 1 or 2 or etc, λ_0 = wavelength (m) in air, n = refractive index of medium between lens and plate.

E4.6 is known as Newton's rings equation.

At the centre of the rings ($m = 0$), a dark spot will be seen because the reflected beams will be out of phase by 180° owing to phase reversal upon external reflection. If the lens is slowly lifted off the plate, the central spot will first change from dark to bright (at spacing = $\frac{1}{4}\lambda_0/n$ then back to dark (at spacing = $\frac{1}{2}\lambda_0/n$ etc. When using Newton's rings to calculate the wavelength of light, a graph of $y = D_m^2$ against $x = m$ is usually plotted. The gradient (ie, slope) of the graph is then measured, and since $4R\lambda_0/n$ equals the gradient, then λ_0 can be calculated provided values for R and n are known. The line may not pass through the origin since the lens may not be in actual contact with the plate; hence, it is better to use the graph method rather than to insert values directly into E4.6.

Single slit diffraction When a plane wave front is incident upon a narrow slit, each point of the transmitted wave front acts as a secondary emitter of waves. The net effect is to produce a diffraction pattern as shown in F4.5, with a central bright band, and bright and dark fringes parallel on either side. The formation of the fringes and central band is due to interference of the secondary wavelets emitted by each point on the plane wave front at the gap. The condition for a **minimum** in the diffraction pattern is:

E4.7 $d \sin \theta_m = m\lambda$

where d = slit width (m), m = 1 or 2 or 3 or etc, θ_m = angular position of the mth minimum from the centre, λ = wavelength (m).

The reason for this condition can be appreciated by considering point P on the screen at the first dark fringe. Each point on the wavefront XY acts as a secondary emitter, so that waves from point 1 arrive at P out of phase with waves from point 2, which arrive out of phase by the same amount compared with the waves from point 3 etc. Provided P is distant, the contributions from each point on XY are of the same amplitude; for P to be a minimum, the resultant of all these contributions will be zero. This will be so if corresponding points in each half give contributions out of phase by exactly 180°. The path difference from two such points, $\frac{1}{2}d \sin \theta_1$, must therefore be equal to one half-wavelength. The single-slit diffraction pattern causes intensity 'modulation' of the diffracted orders of a diffraction grating. See p. 66 and F4.9.

Diffraction gratings With the aid of a spectrometer, diffraction gratings are used to study light spectra. A diffraction grating consists of many parallel, close slits which are ruled either on glass (transmission grating) or on metal (reflection grating). The effect of a diffraction grating upon a parallel beam of monochromatic light at normal incidence is shown in F4.6.

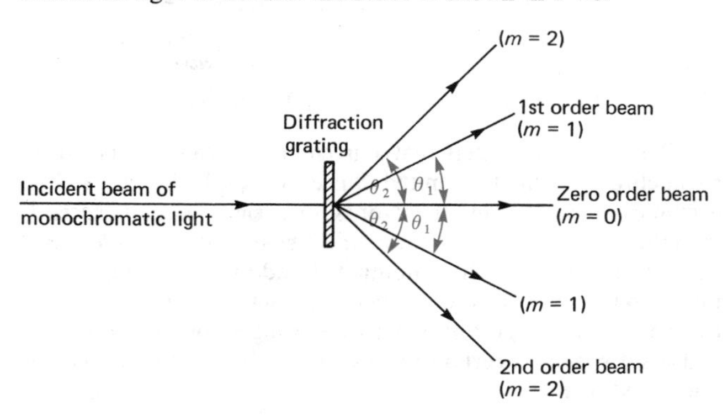

F4.6 Interference from a diffraction grating – monochromatic light

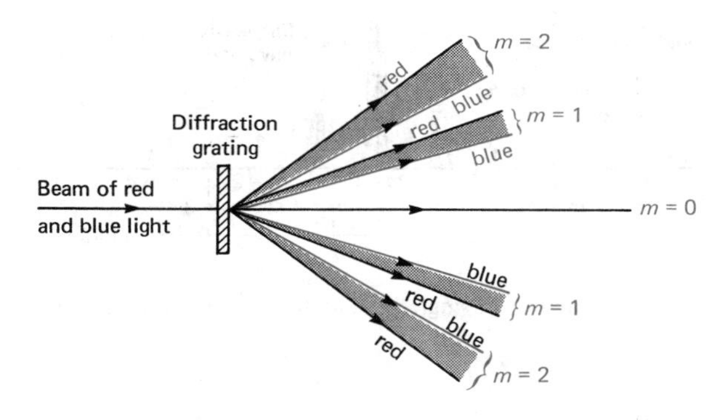

F4.7 Interference from a diffraction grating—red and blue light

By measuring the angle of diffraction, θ_m, the wavelength of the light can be calculated from the diffraction grating equation:

E4.8 $m\lambda = d \sin \theta_m$

where d = distance from the centre of one slit to the centre of the next adjacent slit (m), m = 0 or 1 or 2 or etc, λ = wavelength (m), θ_m = angle of diffraction.

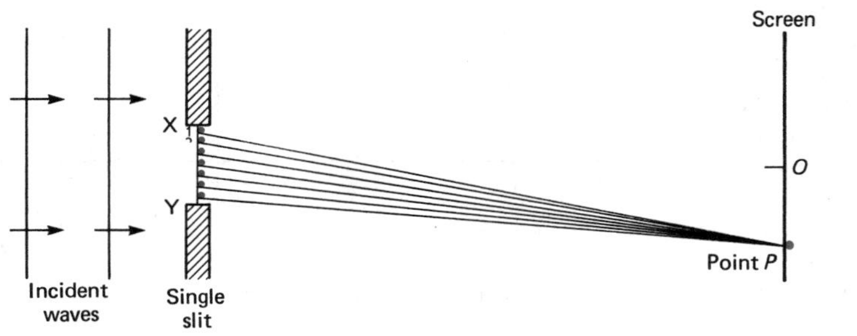

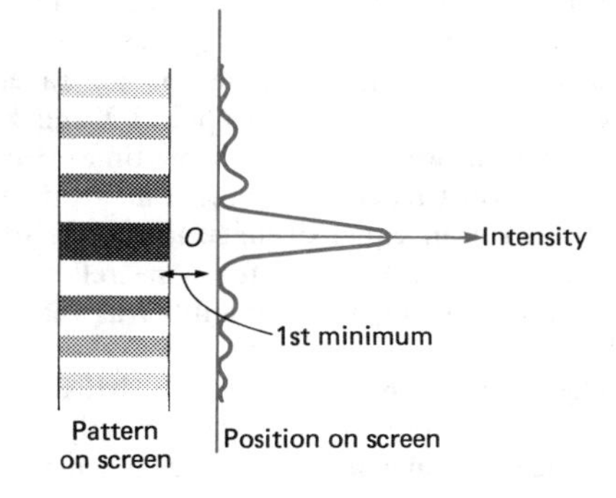

F4.5 Diffraction from a single slit

When the incident beam contains several different values of wavelength (ie, non-monochromatic), the diffraction grating will split the beam so that the longer the wavelength, the greater will be the angle of diffraction, as in F4.7. If white light is used, then each order will contain the full spectrum with blue light diffracted least and red most.

The operation of a diffraction grating relies upon the property of diffraction of a plane wave front incident upon a narrow slit; secondly, a series of narrow slits upon which plane waves are incident will produce sets of diffracted wavefronts, one set from each slit. Thirdly, the sets of wavefronts interfere constructively in certain directions only, as given by E4.8. and as illustrated by F4.8.

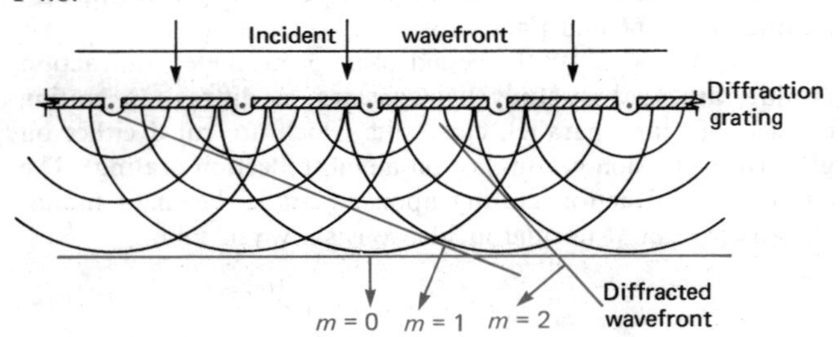

F4.8 Huygen's theory for a diffraction grating

The number of orders that a given grating can produce from monochromatic light is limited. Since $\sin\theta_m \leqslant 1$, then $m\lambda \leqslant d$, so m cannot exceed d/λ; in other words, since $\sin\theta_m$ cannot exceed 1, then the maximum order number must be the ratio (d/λ) rounded down to the nearest whole number. In addition, missing orders may also be caused by the single-slit diffraction pattern produced by the individual grating slits; depending upon the ratio (slit width/slit spacing), certain orders will be fainter than others, as illustrated by F4.9.

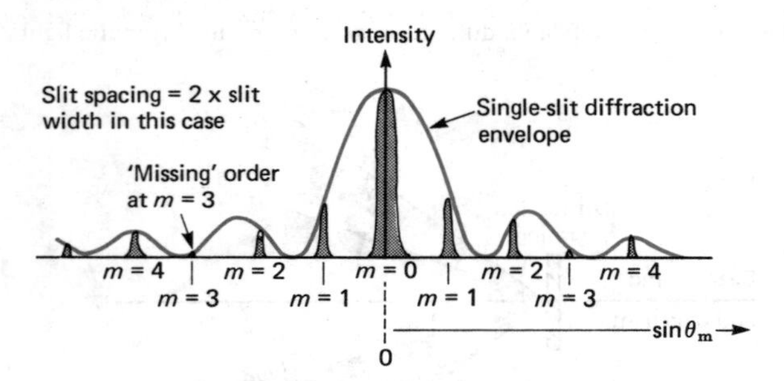

F4.9 Combination of single-slit and diffraction-grating patterns

4.3 RAY OPTICS

A light ray indicates the direction in which a wave front travels. The rays from a point source spread out radially because the wave fronts are expanding spheres. Provided diffraction of wave fronts by gaps and obstacles is negligible, then the rays will be straight lines. Ray optics is based upon the assumption that diffraction can be neglected. Many aspects of optics can be more easily dealt with by considering the rays rather than the waves.

Reflection The simple law of reflection from **a plane mirror** can easily be explained by considering light as a wave form, as in F4.10. Incident wave fronts, assumed plane, travel forward by producing secondary wavelets. Thus points X and Y on the same wave front send out wavelets at the same time. However, by the time that X's wavelet reaches the mirror at X', then Y's wavelet will have travelled an equal distance to form part of a plane, reflected wave front $X'Y'$. Therefore, the reflected wave front $X'Y'$ makes the same angle to the mirror as the incident wave front XY.

An image seen in a plane mirror is a **virtual image**, because rays reflected off the mirror to the viewer only appear to come from the image (**a real image** is one from which rays actually come **direct** to the viewer). The image formed by a plane mirror is

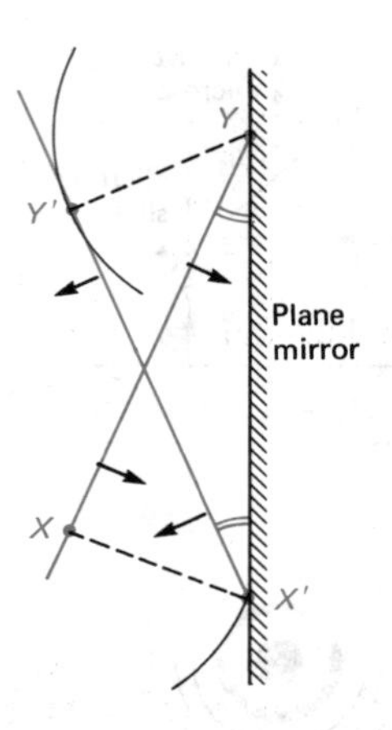

F4.10 Huygen's theory for reflection at a plane mirror

always the same size as the object, and is as far behind the mirror as the object is in front of the mirror.

A convex mirror always forms a virtual image from a real object, although the image is diminished in size compared with the object. **A concave mirror** can either form a virtual image or a real image, depending upon how far the object is from the mirror. To calculate image positions, the following equation is usually used:

E4.9 $$\frac{1}{u}+\frac{1}{v}=\frac{1}{f}$$

where u = object – mirror distance (m), v = image – mirror distance (m), f = focal length of mirror (m).

Also, the **linear magnification**, defined as the ratio

$$\frac{\text{image height}}{\text{object height}}=\frac{v}{u}.$$

A sign convention must be employed to use the equation correctly. The most popular convention is: **real is positive, virtual is negative**; for real images and objects, numerical values of u and v are positive, and negative values of u and v indicate virtual images or objects. Also, concave mirrors are given + values of f, and convex mirrors are given – values of f. Remember that the focal length of a curved mirror is the distance from the pole (ie, mirror centre) to the point where rays parallel to the axis (ie, parallel to the normal through the pole):

1 converge, after reflection, if the mirror is concave;
2 appear to diverge from, after reflection, if the mirror is convex.

Scale ray diagrams are an alternative way of locating image positions, and some examples of these diagrams are shown in F4.11 and F4.12.

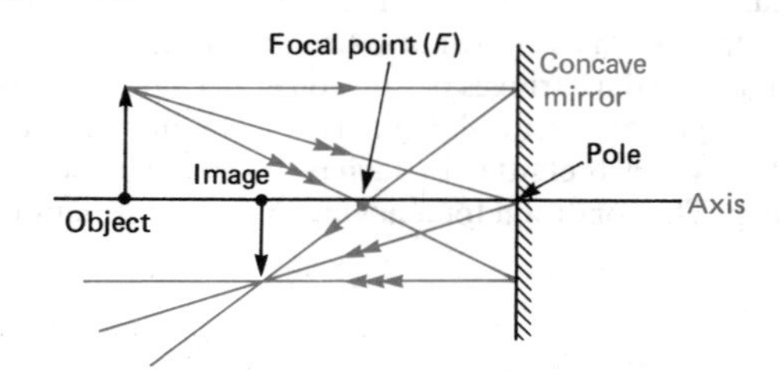

F4.11 Real diminished inverted image from a concave mirror

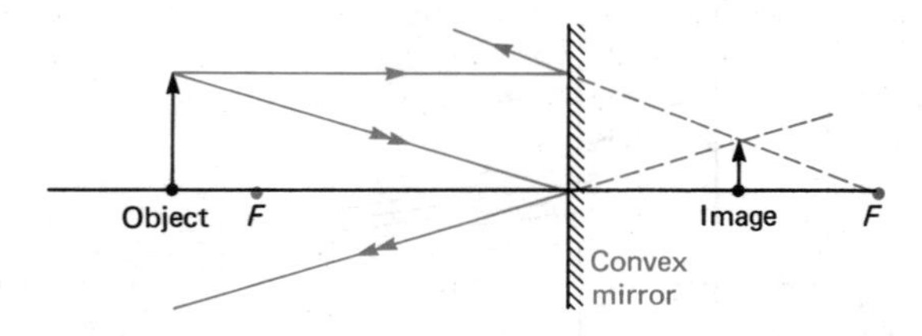

F4.12 Virtual diminished upright image from a convex mirror

Refraction Huygens' wave theory can be used to explain refraction. In F4.13, wave fronts are incident on the boundary as shown. Points X and Y on the same wave front emit secondary wavelets at the same time, but because the wavelet from X is in the more dense optical medium, then it travels more slowly than that from Y. Thus the wavelet from Y reaches Y' in the same time as that from X reaches X'. The refracted wavefront $X'Y'$ travels therefore in a direction closer to the normal than the incident wave front. From the diagram, the following equation can be proved:

E4.10 $$\frac{\sin i}{\sin r} = \frac{\lambda_0}{\lambda_m} = \frac{c_0}{c_m}$$

where i = angle of incidence, r = angle of refraction, λ_0 = wavelength in air (m), c_0 = wave speed in air (m s^{-1}), λ_m = wavelength in the medium, c_m = wave speed in medium (m s^{-1}).

This equation is, therefore, in agreement with **Snell's law**, an experimental law, which states that the ratio $\frac{\sin i}{\sin r}$ is a constant. The constant is known as **the refractive index** (n), and by considering E4.9, it follows that the refractive index is equal to $\frac{\text{wavelength in air}}{\text{wavelength in medium}} = \frac{\text{wave speed in air } (c_0)}{\text{wave speed in medium } (c_m)}$.

If the light ray is travelling from an optically more dense medium to a less dense medium, then **total internal reflection** will take place if the incident angle (i) exceeds the critical angle, c, where $\sin c = 1/n$ from Snell's law.

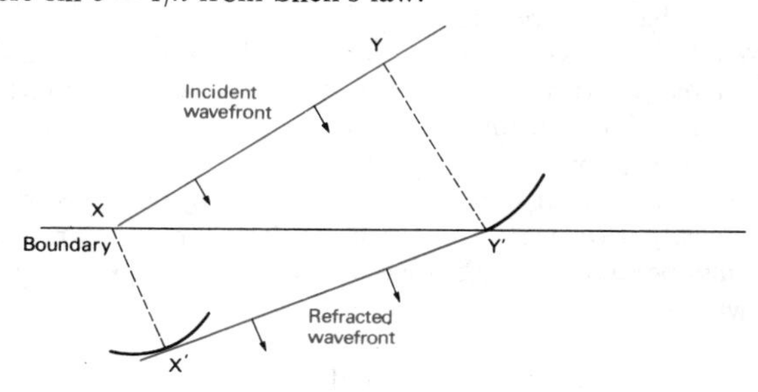

F4.13 Refraction of waves

The passage of a light ray through a prism involves use of Snell's law if the path is to be calculated, given the values of the refractive index and the incident angle. F4.14 shows the light path for a monochromatic beam; the angle of refraction at the first boundary is given by $\frac{\sin a_1}{\sin g_1} = n$. For the second boundary, take care because the path is from glass to air, so the incident ray is in the medium (eg, glass); the angle of refraction at the second boundary is therefore given by $\frac{\sin g_2}{\sin a_2} = \frac{1}{n}$. It can be shown that the

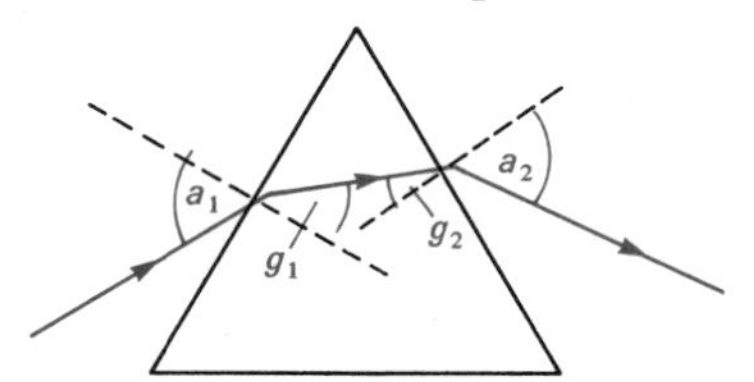

F4.14 Refraction through a prism

angle of deviation of the light beam from its original path is a minimum when the light passes symmetrically through the prism.

When white light passes through a prism, the beam is split into the colours of the spectrum, as shown in F4.15. This effect is known as **dispersion**. The cause of dispersion is that the refractive index (of the prism) is greater for blue light than for red light. Hence, red light is deviated by the least amount.

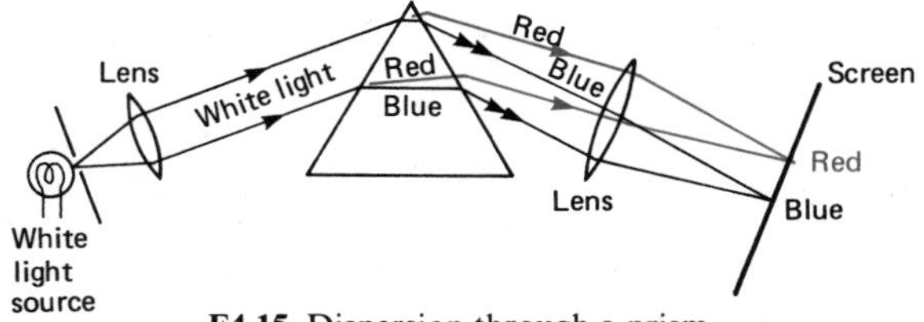

F4.15 Dispersion through a prism

Lenses Image formation by thin lenses can either be treated by ray diagrams, or by use of E4.9. If the equation is used, then the sign convention should be employed; also, the focal length value for a convex lens is taken as +, and for a concave lens is taken as −. Remember that the **focal length** (f) is the distance from the lens centre to the point at which rays parallel to the axis;

1 converge after refraction, if the lens is convex.

2 appear to diverge from after refraction, if the lens is concave.

Single-lens problems can usually be dealt with using the lens formula E4.9, although a sketch ray diagram is often valuable as well. In two-lens problems, either make an accurate ray diagram or do a 'rough' ray diagram **before** use of E4.9 so that you can check on approximate positions etc. A two-lens problem may sometimes involve dealing with a 'virtual object', as in F4.16.

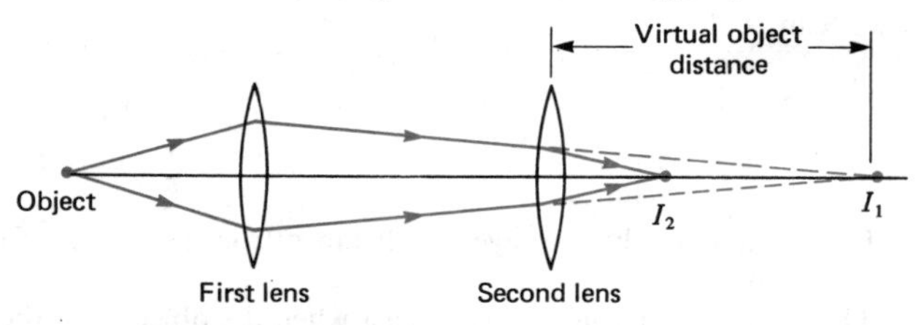

F4.16 Virtual object in a two-lens system

The first lens would form a real image at I_1 if the second lens was absent; refraction by the second lens takes place, so giving its image at I_2. The final image position can be calculated by treating I_1 as a virtual object for the second lens. For example, if the second lens is positioned such that its distance from I_1 is 10 cm, then the object distance for the second lens, u_2, is −10 cm.

The use of scale ray diagrams is shown in F4.17 and F4.18. The three 'key' rays in each diagram are shown clearly. Given the object distance and the focal length of the lens, the three key rays should be drawn in from the tip of the object so as to give the tip of the image.

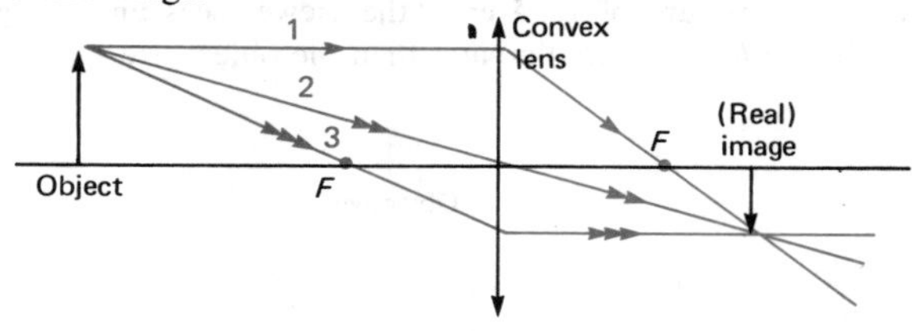

F4.17 The three key rays of geometric optics

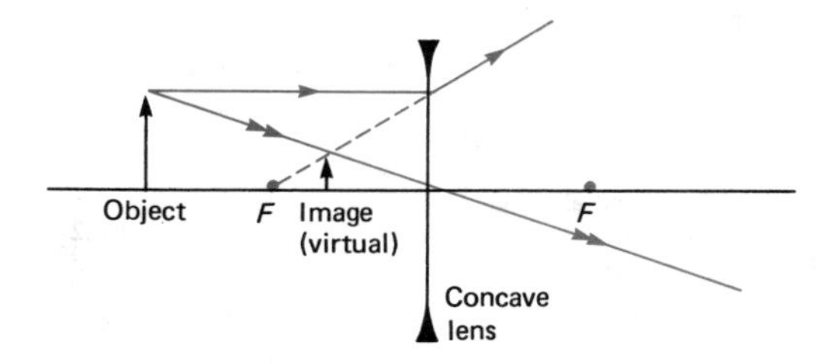

F4.18 Virtual upright diminished image – concave lens

4.4 OPTICAL INSTRUMENTS

The simple microscope A convex lens may be used as a simple microscope (ie, a magnifying glass) by placing the object to be viewed between the focal point and the lens. The viewer, looking from the other side of the lens, will see an enlarged virtual image, as shown in F4.19. If the object is brought closer to the lens, the apparent size of image is increased because the viewer judges apparent size in terms of the angle subtended at the eye. However, if the object is brought too close to the lens, the image position will be closer than the viewer's near point of vision, so the image will appear blurred.

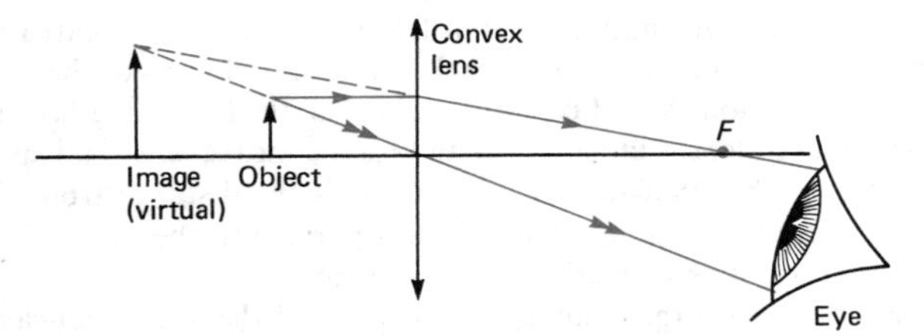

F4.19 Virtual upright magnified image – simple microscope

The **magnifying power** (or angular magnification) is defined as

$$\frac{\text{angle subtended by the image to the eye}}{\text{angle subtended by the object at the near point to the eye}},$$

and for small angles, the following equation for magnifying power can be derived:

E4.11 $\text{mp} = \dfrac{D}{u}$

where D = least distance of distinct vision (ie, to near point) (m), u = object – lens distance (m).

For the average human eye, D (distance from near point to eye) is 25 cm.

Thus, the magnifying power is least when the object is at the focal point; the image will then be at infinity. The magnifying power will be greatest when the image lies at the near point (ie, $v = -D$), so that the object position must be given by $\dfrac{1}{u} + \dfrac{1}{-D} = \dfrac{1}{f}$. The magnifying power is then $\left(\dfrac{D}{f}\right) + 1$.

The compound microscope Essentially, a compound microscope consists of two convex lenses, the objective and the eyepiece. The object to be viewed is placed just beyond the focal point of the objective, so that the objective lens forms a real image (I_1) of the object (O), as shown in F4.20. The eyepiece lens is then arranged so that I_1 lies between it and its focal point F_e. Consequently, the eyepiece lens then acts as a simple microscope to give the viewer a magnified picture of I_1. Hence the viewer sees an enlarged virtual image I_2 of size (h_2) greater than the object size (h_0).

In normal adjustment, the eyepiece lens is positioned so that image I_2 lies at the near point of the viewer. In this situation, the magnifying power is equal to the linear magnification (h_2/h_0) (assuming small angles) because the object subtends an angle h_0/D when at the near point, and image I_2 subtends an angle h_2/D when at the near point.

The refracting telescope A simple refracting telescope can be constructed from two suitable convex lenses. The less powerful lens, the objective, forms a real image I_1 of the distant object being viewed. The eyepiece lens is then positioned so that, in **normal adjustment**, I_1 lies in its focal plane; then the viewer sees the final image I_2 at infinity. In this situation, the magnifying power is given by:

E4.12 $\text{mp} = \dfrac{\text{Angle subtended by final image at infinity}}{\text{Angle subtended by distant object}} = \dfrac{f_0}{f_e}$

where f_0 = focal length of the objective (m), f_e = focal length of the eyepiece (m).

The best position for the eye is at the **eyering**, as shown in F4.21, since all the light that enters the objective from the object must pass out through the eyering. It can be shown that the ratio f_0/f_e is equal to the ratio (objective diameter)/(eyering diameter). Usually, the telescope is designed to make the eyering diameter no larger than the eye-pupil width.

When used to view a point object, the brightness is considerably increased with the aid of the telescope than without; the larger diameter objective collects much more light than the unaided eye pupil. Large objective diameters also enable the viewer to see greater detail of an extended object, or to resolve more easily two very close objects (eg, binary stars). The **resolving power** of a telescope is theoretically given by the following equation:

E4.13 $\text{rp} = \dfrac{1.22\lambda}{D_0}$

where λ is the wavelength of the light (m), D_0 is the objective diameter (m).

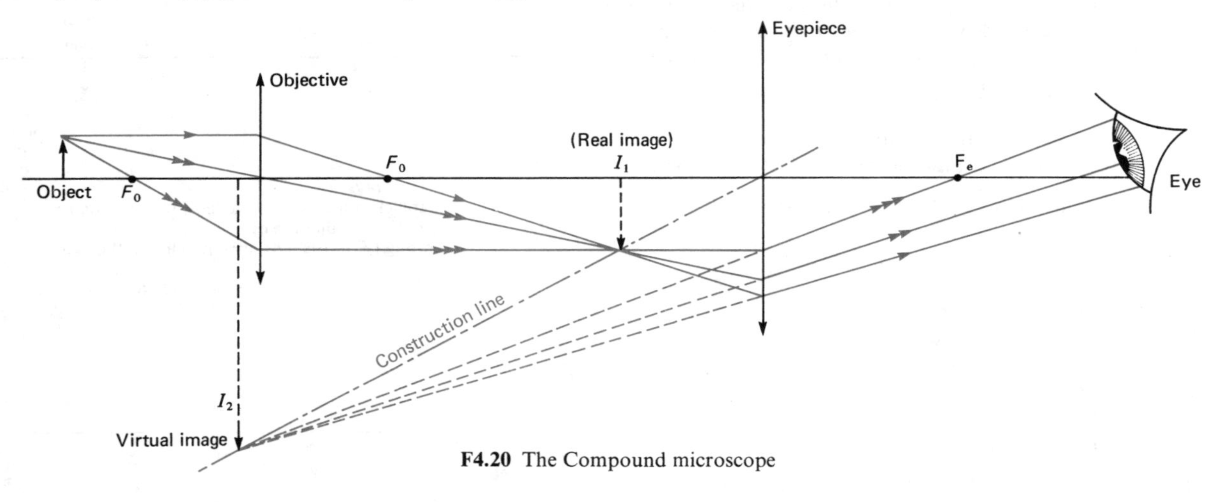

F4.20 The Compound microscope

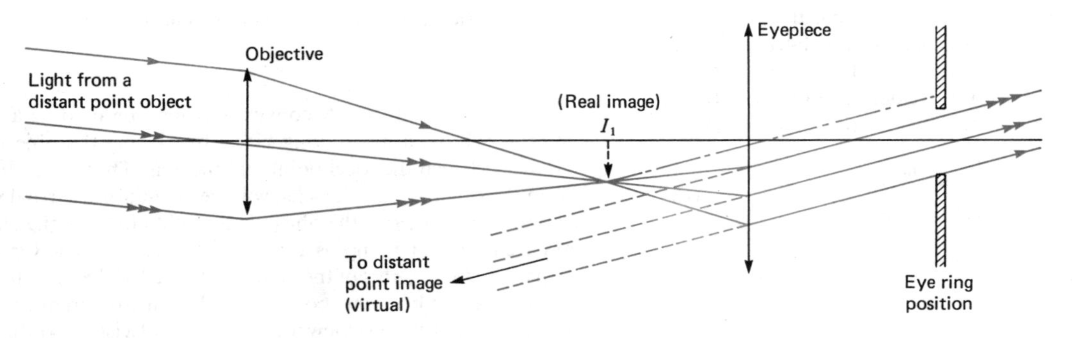

F4.21 Astronomical telescope in normal adjustment

This equation gives the angular separation, in radians, of two point objects at infinity which can just be resolved (ie, the resolving power, rp). If any closer, then the viewer sees one 'merged' image only. The basis of the formula is the diffraction of light as it passes through a gap, as explained on p. 65. For a circular gap (eg, eye pupil or telescope objective), E4.13 applies; if the gap is the eye pupil, then D_0 is the eye-pupil diameter. Because the objective of a telescope is usually several times wider than an average eye pupil, then more detail can be seen with a telescope than without. Each point object gives a diffraction pattern of rings, as shown in F4.22. The rings from one point object will overlap the rings from the other point object if the two objects are too close; only one image will then be seen.

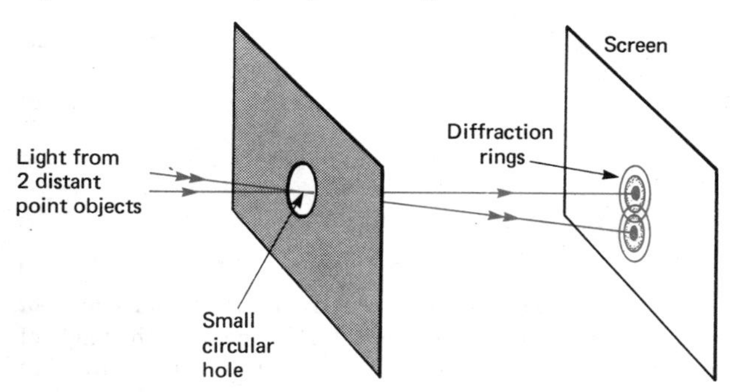

F4.22 Resolving distant sources

A **reflecting telescope** uses a concave mirror as its objective. Since increased objective diameter increases brightness of point images, and also increases the ability to resolve detail, etc, then large-diameter objectives are desirable. It is much easier to make large diameter concave mirrors than large diameter lenses.

The spectrometer The use of a spectrometer with a diffraction grating is shown in F4.23. The essential parts of a spectrometer

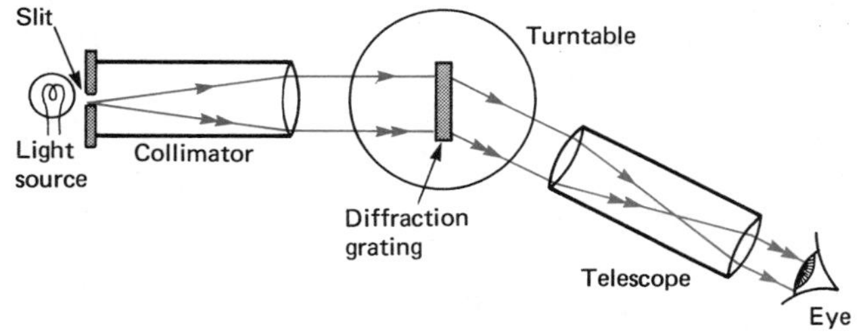

F4.23 The spectrometer

are the collimator tube, the turntable and the telescope. Before use, the telescope must be set to receive parallel light (by focusing on a distant object), and the collimator must be set to produce parallel light. This is done by viewing the slit through the correctly adjusted telescope via the collimator. The collimator is adjusted to bring the slit into focus. The turntable must be level, and the slit is then narrowed to a fine line.

UNIT 4 QUESTIONS

4.1M When light enters glass from air:

A its wavelength increases and its frequency increases.
B its velocity decreases and its frequency increases.
C its velocity decreases and its wavelength increases.
D its wavelength decreases and its frequency decreases.
E its wavelength decreases and its velocity decreases.

(**AEB** June 80: *NISEC* and all other Boards except O and C, WJEC, O and C Nuffield*)

It is important to understand that the frequency of light on refraction cannot change; the same number of waves per second must pass a given point in the glass as pass a corresponding point in the air. This knowledge reduces the number of possible options and a final choice can be made using the knowledge that refractive index is equal to the ratio of velocities; see E4.10 if necessary.

4.2M In a double slits experiment, light of wavelength 600 nm gave fringe separation of 0.4 mm on a suitably placed screen. Use of a different light source, also monochromatic, gave ten fringes across a distance of 3.3 mm on the same screen at the same position. The wavelength of the second source, in nm, was:

A 50. B 495. C 660. D 727. E 7300.

(—: *all Boards*)

Use E4.3; since slit spacing and slit–screen distance are constant, the wavelength ratio of the two sources is the same as the ratio of fringe separation. Remember that the fringe separation of the second pattern is **not** 3.3 mm.

4.3M Under which set of conditions will the bright fringes of a double-slit interference pattern be furthest apart?

	Distance between slits	Distance from slits to screen	Wavelength of source
A	small	large	long
B	small	large	short
C	small	small	short
D	large	small	long
E	large	small	short

(**SEB H**: *all other Boards*)

See p. 63 and E4.3 if necessary. Write down the equation for fringe spacing (Y) in terms of wavelength (λ), slit spacing (d) and slit–screen distance (X). Then choose from A–E so that Y will be as large as possible; you want 'large' terms on the top of your equation for Y and 'small' terms on the bottom.

4.4M Which one of the following effects provides direct experimental evidence that light is a transverse, rather than a longitudinal, wave motion?

A Light is diffracted by a narrow slit.
B Two coherent light waves can be made to interfere.
C The intensity of light from a point source falls off inversely as the square of the distance from the source.
D Light is refracted by a glass prism.
E Light can be polarised by reflection at a water surface.

(**Cambridge**: *all other Boards except SEB H, SEB SYS, NISEC*)

See p. 53 if necessary.

4.5M Monochromatic light of wavelength 600 nm is used in a spectrometer to illuminate a diffraction grating set normally to the collimator. The grating has 3×10^5 lines per metre. The telescope is used to scan the field to one side of the straight-through position. Not counting the 'straight-through' image, the maximum number of diffracted images of the slit visible to the observer will be:

A 2. B 5. C 8. D 10. E 11.

(**Cambridge**: *all other Boards except SEB H*)

See p. 65 if necessary.

4.6M The surface of a glass lens may be made non-reflecting for light of a particular wavelength by coating the lens with a thin layer of transparent material of such thickness that the light reflected from the front surface of the film destructively interferes with light reflected from the rear surface. The refractive index of the material used is intermediate between that of air and that of glass. If λ is the wavelength of the light in the coating material, the 'minimum' thickness of the film would be:

A $\lambda/4$. B $\lambda/2$. C $3\lambda/4$. D λ. E $3\lambda/2$.

(**Cambridge:** *and JMB*, SEB H*)

Consider if there is a phase change of 180° upon reflection at the front surface, and if there is a similar phase change upon reflection at the rear surface. Remember that phase change at both surfaces means that there is no phase difference on account of reflection between the two reflected beams. Taking any such phase change into account, consider the extra distance travelled by the light reflected at the rear surface ($= 2 \times$ film thickness) in terms of wavelength so that destructive interference is produced. See p. 64 if necessary.

4.7M Sun shining on a wet road reveals coloured patterns where oil has been spilled. The varied colours arise because:

A some wavelengths, after being reflected at the lower surface of the thin oil film, are totally internally reflected at the upper surface, whereas other wavelengths are not.
B the natural colour of the oil shows up better when the layer is thin whereas a thicker layer absorbs more light.
C the oil selectively absorbs some wavelengths and reflects other wavelengths.
D the phase difference between light reflected at the upper surface of the oil film and that emerging after reflection at the lower surface and refraction at the other depends on the wavelength of the light.
E water forms small droplets on the oily surface so that an effect similar to that of a rainbow is produced.

(**London:** *JMB*, Cambridge and SEB H*)

Consider E4.5. Wavelengths in phase will reinforce and those in anti-phase will be removed.

4.8M A parallel beam of microwaves is directed at a single slit as shown. A detector D is moved from the central maximum X to the first minimum Y. Then, a plate is placed in the single slit gap to make two equal-width narrow slits. The detector signal will be zero:

A at X. B at Y.
C between X and Y. D along the line OX.
E at Z.

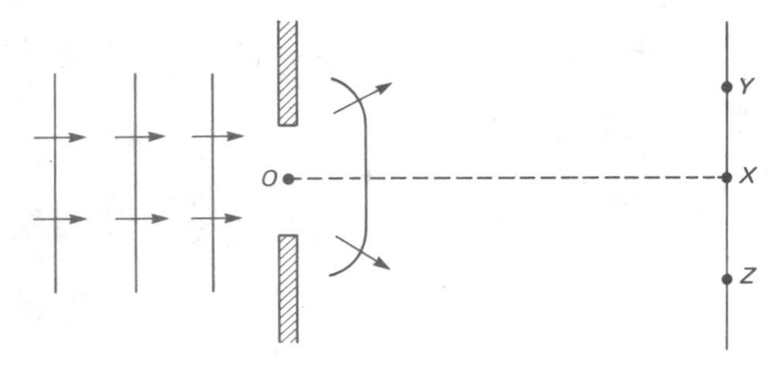

(—: *JMB*, and all other Boards except WJEC, SEB H*)

Compare the equation for single-slit diffraction (E4.7) with that for the first minimum of Young's slits equation (derived from E4.2). Assume single slit width d becomes the double slit spacing when the plate is placed in the gap.

Alternatively, consider the explanation of single-slit diffraction after E4.7 to determine the path difference from Y to opposite ends of the gap in terms of wavelength. Then, with the introduction of the plate, consider whether or not waves from the two 'slits' reinforce or cancel one another at Y.

4.9M A convex lens of focal length 0.20 m is placed at a distance of 0.3 m from an illuminated object. A screen is moved until a clear image of the object is seen on the screen. Then, without moving object or image, the lens is moved towards the screen to a new position where a smaller clear image is seen on the screen. The distance through which the lens has been moved, in m, is:

A 0.9. B 0.6. C 0.5. D 0.4. E 0.3.

(—: *London, AEB, O and C, Oxford, Cambridge, JMB, NISEC*)

For the first position, use the lens formula ($1/u + 1/v$ etc.) to calculate the image distance. The second position corresponds to object and image distances interchanged; make a simple rough sketch to show the two positions, and then you ought to be able to calculate the distance between the two lens positions.

4.10M A diverging lens of focal length 15 cm is placed in contact with a converging lens of focal length 20 cm. The focal length of the combination will be, in cm:

A + 10. B + 5. C − 5. D − 35.
E − 60.

(—: *AEB, NISEC*)

The combined focal length F is given by $1/F = 1/f_1 + 1/f_2$, but remember to use the sign convention. Refer to your textbook if necessary.

4.11M A narrow monochromatic beam of light passes through a triangular glass prism, being refracted at Q and R. Which of the following is correct?

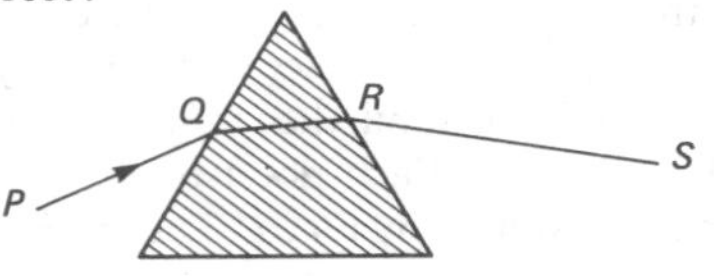

1 The angle between ray PQ and ray QR is known as the angle of deviation of the prism.
2 If the angle of incidence of ray PQ increases, the angle that ray RS makes with the normal at R decreases.
3 When the angle of deviation is a minimum, ray QR makes equal angles with the two refracting surfaces.

Answer: A if **1**, **2**, **3** correct. D if **1** only.
B if **1**, **2** only. E if **3** only.
C if **2**, **3** only.

(**London:** *and AEB, Oxford, NISEC**)

Deviation is the angle between the incident and emergent rays. As the angle of incidence of ray PQ increases, the angle of refraction at Q must also increase. By geometry, it follows that the angle QR makes with the normal must decrease, so... For minimum deviation, see p. 67.

4.12M A lens forms a real image of an object. The distance from the object to the lens is x cm and that from the lens to the image is y cm. The graph shows the variation of y with x. It can be deduced that the lens is:

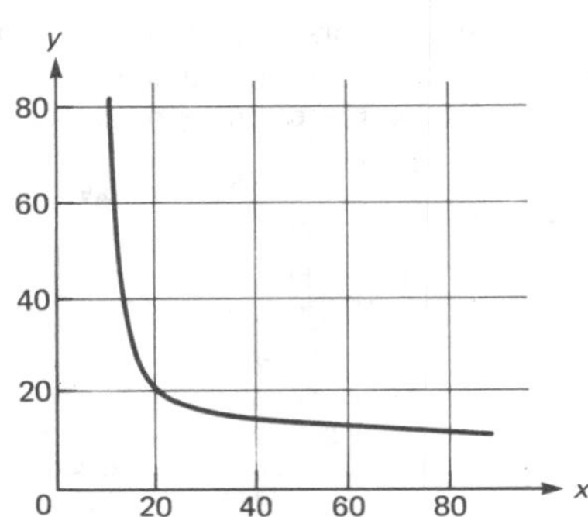

A converging and of focal length 10 cm.
B converging and of focal length 20 cm.
C converging and of focal length 40 cm.
D diverging and of focal length 20 cm.
E diverging and of focal length 10 cm.

(**Cambridge:** *London, AEB, O and C, Oxford, NISEC, JMB*)

Use E4.9.

4.13M An illuminated object and screen are placed 90 cm apart. To produce a clear image on the screen twice the size of the object, the required lens must be:

A diverging, f = − 60 cm. B diverging, f = − 10 cm.
C converging, f = + 20 cm. D converging, f = + 30 cm.
E converging, f = + 60 cm.

(—: *London, AEB, Oxford, Cambridge, NISEC*, JMB*)

Use the lens formula, having first worked out u and v from the given facts that $u + v = 90$ cm, and $v/u = +2$ (from the magnification).

4.14M The diagram represents two thin lenses L_1 and L_2 placed coaxially 30 cm apart. A beam of light parallel to the axis is incident on L_1. The final image formed by refraction through both lenses is:

A real and between L_1 and L_2.
B real and on the right of L_2.
C virtual and on the left of L_1.

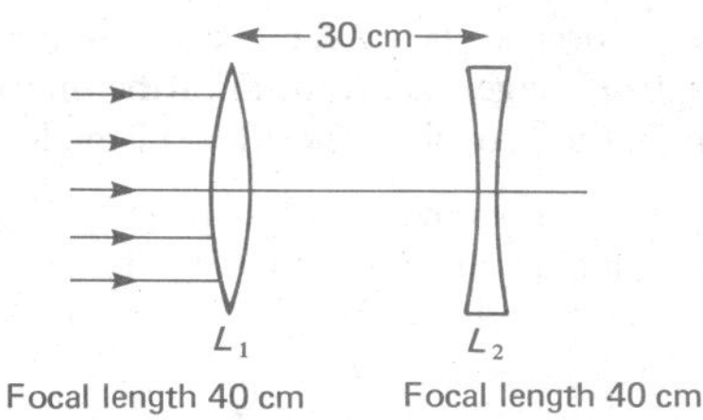

D virtual and on the right of L_2.
E at infinity.

(**London:** *AEB, NISEC, Oxford, Cambridge*)

Draw the continuation of the rays to where they would meet in the absence of L_2. Bearing in mind that L_2 will diverge the rays, you may be able to deduce the answer by elimination. If not, consider that L_1 will form a virtual object 10 cm beyond L_2 and apply $1/u + 1/v = 1/f$ to L_2 (watch those signs).

4.15M An astronomical telescope has an eye lens of focal length 20 mm. In normal adjustment, when the final image of a distant object is at infinity, the separation of the lenses is 500 mm. The angular magnification of the telescope under these conditions is:

A 22. B 23. C 24. D 25. E 26.

(**London:** *JMB, Oxford, SEB H*)

Refer to F4.18. Find the focal length of the objective and then use E4.12

4.16M For the telescope in 4.15M, to obtain the best view, the distance between the observer's eye and the eye lens should be:

A zero. B 19.2 mm. C 20.0mm. D 20.8 mm.
E 24.0 mm.

(**London:** *JMB, Oxford, SEB H*

The position of the eye-ring is discussed on p. 68. Apply $1/u + 1/v = 1/f$ to the eyepiece, with the objective acting as the object.

4.17S

(a) The wave theory of light provides an explanation of many aspects of reflection, refraction and interference. (i) The three parallel lines in the sketch represent sections of a wave front of a parallel beam of light at times $t = 0$, $t = t_1$ and $t = 2t_1$. Using the principle of secondary wavelets, show how to construct one further wave front, for $t = 3t_1$. (A written explanation is *not* required). (ii) The sketch shows a plane light wave encountering a glass block of refractive index 1.5. The four parallel lines represent the position of a single wave front at a series of equal time intervals. Draw six further wave fronts at the same equal time intervals showing how the light wave continues beyond *AB*. (Your drawing should be to scale.)

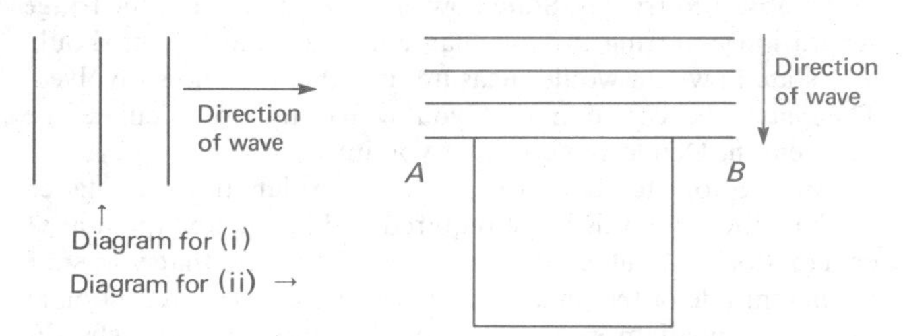

(b) Three parallel narrow slits P, P_1 and P_2 are illuminated by the source S of monochromatic light of wavelength λ as shown in the diagram (which is not to scale). The distances of P_1 and P_2 from O are equal. A series of bright and dark fringes are formed on the screen at T, with a **bright** fringe at O', on the axis of the system, and a **dark** fringe at R. (i) The beams emerging from P_1 and P_2 are 'coherent'. What does this mean? (ii) What relationship must exist between the lengths P_1R and P_2R? (iii) When the wavelength of the source is 500 nm, the centre of the 120th **dark** fringe, counting from O', lies at R. Upon replacement of the source by one of unknown wavelength, R is found to be the location of the 90th **bright** fringe (counting from O' as zero). Find the wavelength of the unknown source.

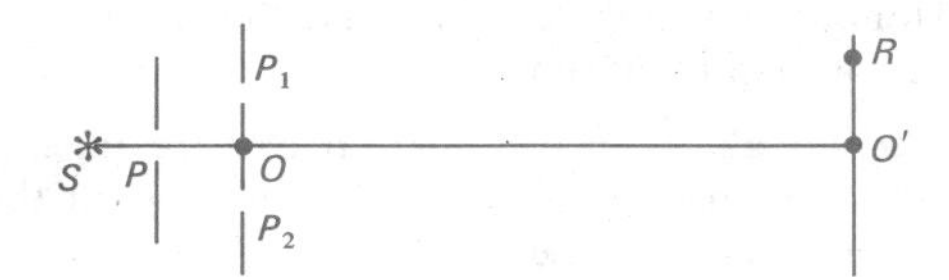

(**NISEC:** *all other Boards except O and C Nuffield, SEB H, SEB SYS*)

(a) (i) Use Huygens' construction. See p. 63 and F4.1. (ii) Note that a scale drawing is required. Measure the distance between two positions of the incident wave front. Calculate the separation of the wavefront positions in the glass from the comments after E4.10. Show the new positions of the wave front at each edge of the block as well as inside it.
(b) (i) See the section on Interference on p. 64. (ii) See E4.2. (iii) See your textbook for the derivation of E4.3. This will indicate that for the nth dark fringe $O'R = (n + \frac{1}{2})\lambda X/d$, and for the nth bright fringe $O'R = n\lambda X/d$. Substitute the data given and equate the two expressions for $O'R$.

4.18S A refracting telescope has an objective of focal length 1.0 m and an eyepiece of focal length 2.0 cm. A real image of the sun, 10 cm in diameter, is formed on a screen 24 cm from the eyepiece. What angle does the Sun subtend at the objective?

(**London:** *JMB, Oxford, SEB H*)

If the final image is real, it must be beyond the eyepiece (ie, on the opposite side to the objective). Apply $1/u + 1/v = 1/f$ to the eyepiece to find where the intermediate image must be (this acts as an object for the eyepiece to give the final image). Knowing that the linear magnification $= v/u$, you can now find the height of this intermediate image. As the Sun is distant, this image I_1 will be in the focal plane of the objective, as shown in the diagram.

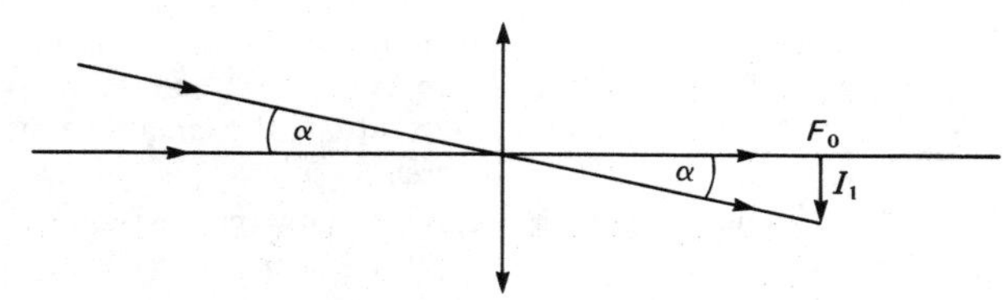

Using the approximation that $\tan \alpha = \alpha$, α can now be found.

4.19S A parallel beam of monochromatic light of wavelength 580 nm is incident normally on a diffraction grating having a large number of regular slits, each of width 0.70×10^{-6}m, as shown in the diagram. After passing through the grating the light will have, as a result of interference, intensity maxima in certain directions. Calculations predict that for the first, second, and third order interference maxima, the values of the angle θ between the direction of the incident light and the directions of these maxima should be approximately 16°, 34° and 56°, respectively.

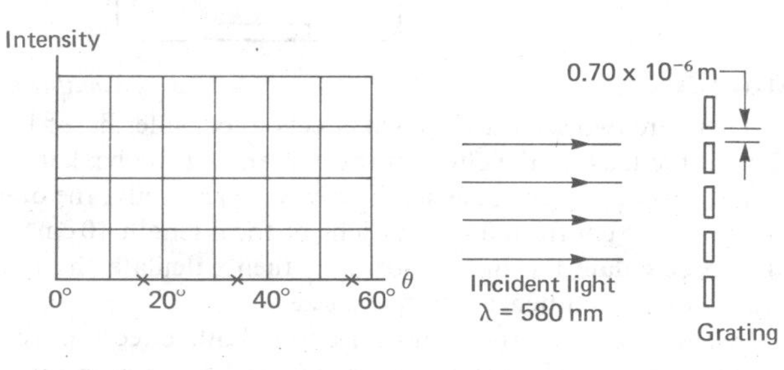

(a) At what value of θ would the light diffracted by a SINGLE slit of width 0.70×10^{-6}m have its first intensity MINIMUM? Show the steps in your calculation.

(b) Produce a sketch graph showing how the intensity of light varies when plotted against θ. Make the range of θ vary from 0° to

60° and mark the 16°, 34° and 54° points on the axis. Plot the intensity on the vertical axis.

(**O and C Nuffield**: *JMB* and all other Boards except SEB H and WJEC*)

(a) The question refers to the single slit diffraction pattern; use E4.7 with $m = 1$. The angle involved is measured in the same way as the angle in the diffraction grating formula.

(b) Two effects have to be allowed for in sketching the graph, diffraction from the individual slits and the interference pattern produced by the grating. The single-slit diffraction pattern should be drawn to form an 'envelope' to the interference pattern produced by the grating. The minima of the single-slit pattern are given by E4.7 and the maxima of the interference pattern are given in the question. If still stuck you can refer to F4.9; in this question a comparison of the answer to part (a) with the 54° third order interference maximum should indicate that it will be very difficult to observe this maximum.

4.20S* The phenomenon of Fraunhofer diffraction may be demonstrated by illuminating a wide slit by a parallel beam of monochromatic light and focusing the light that passes through the slit on to a white screen. A diffraction pattern may then be observed on the screen.

(a) Sketch the intensity variation in the diffraction pattern as a function of distance across it.

(b) What would happen to the intensity variation if the width of the slit were halved?

(**Cambridge**: *all other boards except WJEC and SEB SYS*)

(a) See p. 65 for a discussion of single-slit diffraction. For the intensity distribution, see F4.5.

(b) E4.7 can be used here. For fixed wavelength, consider the effect on the position of the first minimum caused by reducing the slit width. Give your answer by sketching the new intensity distribution, and comment on it and the initial distribution.

4.21S An opaque disc P, 3 mm diameter, lies at the bottom of a glass beaker, and is illuminated from below by a source S. A converging lens L of focal length 10 cm, situated 15 cm above the disc, forms an image of this at Q. Where is Q situated, and what is the size of the image? Explain qualitatively how the position and size of the image of the disc is changed when the beaker is filled with water.

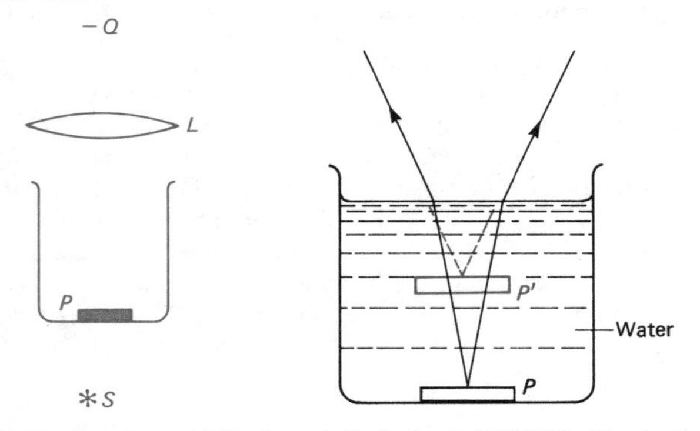

(**SUJB**: *London, AEB, O and C, Oxford, NISEC*, Cambridge*)

There are two separate optical effects to consider here: 1, the effect of the lens; 2, the effect of the contents of the beaker.

The first part of the question involves the lens only; the object P is placed 15 cm from a convex lens of focal length 10 cm. Use E4.9 to calculate the image position, then calculate the linear magnification and hence the image size.

In the second part of the question, both effects must be considered. The effect of the water in the beaker is to give an image nearer the water surface than the object. The reason for the formation of this image is shown in the black diagram. The image P', which would be seen if the lens was removed and a view taken from above, is treated as an object when the effect of the lens is considered. Thus the effect of filling the beaker with water is to move the 'object' closer to the lens. Hence, state the effect on the image position. For image size, consider if the image P' is larger or smaller than P; then consider the effect of the lens on P'.

4.22S In an experiment to find the focal length of a convex lens, measurements of image distance (v) were made for different values of object distance (u). The following set of results was obtained:

u (cm)	25	30	40	50	60
v (cm)	99	61	40	33	29

(a) Plot a graph of $\frac{1}{v}$ (y-axis) against $\frac{1}{u}$ (x-axis). **(b)** Determine from the graph the focal length of the lens.

(**SEB H**: *London, AEB, O and C, Oxford, Cambridge, NISEC, JMB*)

Because $1/u + 1/v = 1/f$, then the graph should give a straight line of slope (gradient) -1. This is because $1/v = -1/u + 1/f$, so that the intercept on either axis will be equal to $1/f$. Use the average of the two values.

4.23L

(a) Explain what is meant by (i) interference and (ii) diffraction of waves. Describe experiments by which each of these phenomena may be demonstrated.

(b) The interference of light waves may be used to determine the wavelength of light from a monochromatic source. Describe Young's double-slit experiment by which such a determination may be made. The theory of the experiment is not required.

(c) Explain the colours which are produced when a soap film is illuminated with white light. Why might the film appear black just before it breaks?

(d) When a diffraction grating is illuminated by white light at normal incidence the violet in the third order spectrum may overlap the red in the second order spectrum. Explain why this happens giving figures to support your answer.

(**AEB** Nov 81: *all other boards a, b, d only*)

(a) (i) See p. 53. You should mention superposition of waves, coherent sources, constructive and destructive interference; link these points together for your explanation. (ii) See p. 53 for single-slit diffraction, which is just one example of the effect of obstacles and gaps on wave fronts. You should make it clear that diffraction results from superposing secondary wavelets which originate from coherent sources on the same wave front after the wave has been distorted by some obstacle. Use single-slit diffraction to support your explanation, but the detailed theory is not required. For experiments, give brief details with a diagram for each one. Refer to your textbook if necessary. For (ii), single-slit diffraction is suitable.

(b) Use your textbook if necessary. Indicate approximate values for the slit spacing, and the distance from the double slits to the observed fringes. State how you would measure the fringe separation, referring to your diagram if necessary. You should also state how you would measure the other distances involved. Then give the equation that you would use to calculate the wavelength. Define symbols that you introduce!

(c) See your textbook for a detailed explanation. A detailed mathematical analysis is not required: a diagram and qualitative explanation will suffice. It is essential, however, that you state the magnitude of the phase change which occurs on reflection at the denser medium since this is the key to explaining why the film might appear black just before it breaks.

(d) F4.7 and the paragraph following E4.8 contain the essential information to explain this effect. For support in terms of figures, show that the term $d \sin \theta$ in E4.8 has a greater value for $m = 2$ with red light of suitable wavelength compared with the value for $m = 3$ with violet light. Take the range of the visible spectrum as 400 nm to 700 nm.

4.24L Both Young's double-slit interference experiment and the diffraction grating may be used to obtain a value for the wavelength of light. For each experiment

(a) give a diagram of the experimental arrangement,
(b) state what measurements need to be made, and
(c) show how a value of the wavelength of light used is obtained from the measurements.

In Young's double-slit experiment a sodium lamp emitting light of wavelength 589 nm illuminates two slits with centres 0.60 mm apart. Interference fringes are observed on a screen 150 mm from the slits. Calculate the spacing of the maxima in the interference pattern.

Describe qualitatively the changes in the pattern that would be observed if each of the following changes were separately made to the Young's double-slit experiment. (i) One slit only is covered with a thin sheet of mica which delays the light through the slit by one half of a period. (ii) The slits are made narrower without changing the separation of the centres of the slits. (iii) One slit is closed. (**O and C**: *all other Boards except SEB H*)

(a) The arrangement for Young's double-slit experiment is shown in F4.2. For the diffraction-grating experiment, see the spectrometer diagram F4.23.

(b) For Young's double-slits experiment, E4.3 will tell you which measurements need to be made. For the fringe spacing, you would need to measure across several fringes to obtain an accurate value of spacing between two adjacent fringes. For the diffraction-grating experiment, E4.8 will tell you which measurements need to be made.

(c) See E4.3 and E4.8. For the calculation of the spacing between maxima in the double-slits experiment, use E4.3. (i) Before insertion of the mica over one slit, each bright fringe on the screen is at a point where arrival of a 'wave crest' from one slit coincides with arrival of a wave crest from the other slit. At one of these points, the effect of the mica is to make wave crests from one slit arrive at the same time as wave troughs from the other slit. You must describe how this will cause the observed pattern to change. (ii) See F4.9. The key point here is about single-slit diffraction. If single-slit diffraction could be ignored, the bright fringes would all be of the same intensity; however, the effect of single-slit diffraction is to cause 'missing orders' (as discussed for the diffraction grating, p. 66). You must describe how the observed pattern will change if the slits are made narrower but kept at the same spacing. (iii) See the comments after E4.3 if necessary.

4.25L **(a)** (i) State the conditions necessary for optical interference to be observed when waves emerge from two sources.
(ii) With the aid of a labelled diagram, showing the necessary apparatus, describe how the wavelength of monochromatic light can be measured by a two-source interference method (eg, using Young's slits). Show how the result is obtained from the measurements.
(ii) How would the interference pattern differ if white light was used instead of monochromatic light?
(b) (i) A parallel beam of monochromatic light of wavelength λ falls normally on a plane diffraction grating where the slit separation (grating element) is d. Diffracted beams emerge from the grating at various angles to the grating. With the aid of a sketch, derive an equation relating d, λ and the angle θ between each beam and the normal to the grating. Define any other symbols you use.
(ii) A parallel beam containing light of two different wavelengths is incident normally on a grating which has 600 lines per mm. Diffracted beams are observed at angles 16°57′, 20°21′, 35°41′, 44°6′ and 61°1′. Work out the two wavelengths from these measurements and explain why the two wavelengths do not produce the same number of beams.
(c) (i) The solar spectrum is a *continuous spectrum* crossed by a series of thin dark vertical lines which make up the *absorption spectrum* of sunlight. Outline in general terms the cause of these lines.

(—: *all Boards*)

(a) (i) and (ii) See 4.1 if necessary. Proof of E4.3 is **not** required but you must state what measurements are made and how these measurements are used in conjuction with E4.3 to obtain the wavelength.
(iii) See p. 64 if necessary for the effect of white light.
(b) If necessary, refer to your textbook for a proof of E4.8.
(ii) From E4.8, the lesser wavelength gives smaller angles of diffraction. A reasonable assumption would be to try m = 1 for 16°51′ and work out λ then see if this value of λ with m = 2 etc. gives any of the other angles. Then try m = 1 for the next largest angle etc. to work out the other wavelength. Work out the average value for each wavelength. Your calculations ought to show that 'm = 3' for one of the wavelengths is not possible.
(c) See pp. 125-126 and your textbook if necessary. The solar corona is a gaseous envelope surrounding the Sun. Light from the Sun's interior contains all wavelengths. Hence explain why light reaching the Earth has some wavelengths missing.

4.26L This question is about diffraction.
The passage below presents three sets of ideas about diffraction. For each of the sections **(i)** to **(iii)** you are asked to write a more complete explanation of the ideas. Your explanation may include:
fuller explanations of the theory;
quantitative calculations to illustrate the ideas;
discussion of possible experiments.

Passage

(i) Light from a point source appears to cast sharp shadows and this leads to the familiar idea that it travels in straight lines. However, this is not exactly true: the shadows are not perfectly sharp, although special experiments are needed to show the effect because it is so small. This unfamiliar property is called diffraction and is explained by a wave model of light.
(ii) The consequence is that the eye, or a camera, or even the best possible telescope, doesn't produce a perfect image. Instead it gives an image which is slightly blurred. When we try to make a telescope magnify more to show finer details of the stars this blurring effect can become an obstacle.
(iii) But diffraction can also be put to good effect. Diffraction gratings are made to enhance the effect and make use of it to give a powerful method of investigating spectra.

(**O and C Nuffield**: *all except SEB H*)

This question is in three parts **(i)** **(ii)** and **(iii)** and each part refers to a passage and should contain three sections; fuller explanation, quantitative calculations, possible experiments. However these three sections need not be written separately and will often end up 'interwoven' as you answer each part of the question. However it is a good idea to produce a rough plan based on the three sections and such a rough plan (deliberately slightly incomplete!) provides the rest of this commentary.

(i) Fuller explanation – diagram and discussion of plane waves passing obstacle and being bent at the edges.
Quantitative calculations – diffraction only really noticeable when obstacle of comparable size to wavelength, but wavelength of light makes it difficult to get suitable objects.
Possible experiments – ripple tanks for wave properties, lasers (intense parallel beam) to project shadow over large enough distance to make diffraction noticeable (diagram).

(ii) Fuller explanation – 'obstacles' behave like 'holes', optical instruments have 'hole' (objective lens) at front and hence must diffract. Two stars very close together can blur into one (resolution diagrams).

Quantitative calculations – use resolving power formula (E4.13) applied to eye or telescope (diagram to illustrate the angle involved).
Possible experiments – eye looking at lines on paper, telescope with variable aperture focused on adjacent point sources.

(iii) Fuller explanation – each slit diffracts light from a parallel beam so it spreads over a wide angle interfering with beams from

other slits. Constructive interference only at certain specific angles (diagram).
Quantitative calculations – show that it is possible to get light at different wavelengths to interfere constructively at measurably different angles, use formula (E4.8) to show this.
Possible experiments – describe how to use gratings to measure wavelengths, remember need for parallel beam, importance of source of light, mention reflection gratings used on X-rays.

Other ideas would be equally acceptable, but the above plan should provide an adequate base for getting full marks on this question. Two major faults occur when students answer questions of this nature; 1, they tend to write too much, putting down all scraps of information on the subject they can glean whether asked for in the question or not; 2, they tend to produce unplanned answers which often fail to cover the points asked for (explanation, calculation, experiments). On this question much time was wasted by students discussing in depth the effect of different wavelengths; this is only of major significance in part **(iii)**.

4.27L Describe the steps in outline you would take in setting up a diffraction grating in combination with a spectrometer to examine the spectrum of visible light. Explain the function of each part of the system and show on a clear diagram the passage of light rays from the source, through the system, to the observer's eye.
When the spectrum of light containing violet and red components only is examined with a diffraction grating, it is found that the fourth line from the centre (not counting the zero-order line) is a mixture of red and violet. Explain this. If the grating has 500 lines per mm, and the diffraction angle for the composite line is 43.6°, find the wavelengths of the violet and red components.
What will be the fifth line in the spectrum and at what diffraction angle will it occur?

(**WJEC**: *all other Boards except SEB H*)

See p. 69 and F4.23 for the spectrometer used with a diffraction grating. You should explain the function of (i) the collimator, (ii) the turntable and levelling screws, (iii) the diffraction grating, (iv) the telescope. You should include the following 'setting up' steps, (i) adjusting the telescope to receive parallel light, (ii) adjusting the collimator to give parallel light, (iii) levelling the turntable. Refer to your textbook if necessary.

To explain the overlap of red and violet in the fourth line, remember that for a given order, the longer wavelength is diffracted most, so you need to find out if red light has a longer wavelength than violet light. Then decide if the red part of the fourth line is second or third order, and if the violet part is second or third order; to overlap, red and violet must belong to different orders. See F4.7 if necessary. Once you have determined which order each of the two colours belongs to, then use E4.8 to determine the wavelength of each colour.

You should now know which order the red part of the fourth line belongs to, so calculate the angle of diffraction for the next order red line. Then do likewise for the violet (ie, calculate the next order diffraction angle for violet), and then you should be in a position to state if the fifth line is red or violet.

4.28L
(a) A narrow parallel-sided beam of white light is dispersed by a diffraction grating which is placed perpendicular to the beam, ie, parallel to the incident wavefronts. Explain this effect in detail.

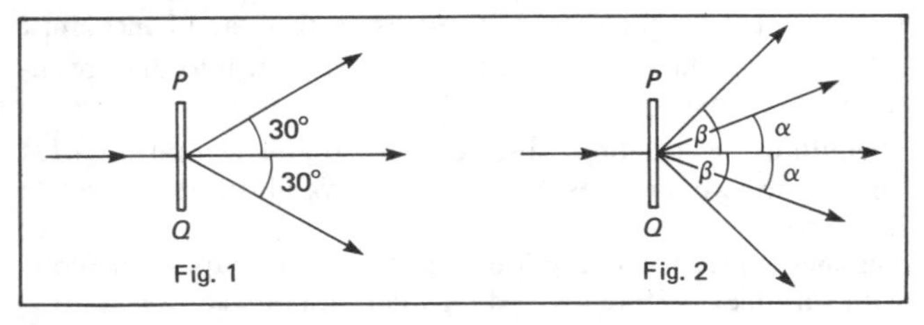

(b) Figure 1 shows the action, in air, of a plane diffraction grating PQ on a monochromatic beam of light which falls normally on the grating. Figure 2 shows the whole arrangement immersed in water, of refractive index 1.33. The beam is now diffracted as shown, making angles a (< 30°) and β(> 30°) with the normal. Explain these changes produced by the water.
Given that the grating has 8.0×10^5 lines per metre, calculate (i) the wavelength of the light in air, (ii) the wavelength of the light in water, (iii) the angle α, and (iv) the angle β.

(**London**: *all other Boards*)

(a) The question asks you to **explain** the effect in **detail**, so you need to explain diffraction at each slit producing secondary wavelets, which reinforce in certain directions. With the aid of a diagram show $m\lambda = d \sin \theta_m$, from which it can be deduced that different wavelengths of the white light beam will reinforce at different angles (ie, white light will be dispersed). Describe what will be seen – see F4.7.

(b) As the refractive index of water is 1.33, the speed of light in the water will be reduced by a factor of 1.33. As $c = f\lambda$, what will happen to λ? (and hence θ, from the diffraction formula?). This explains the magnitude of a and also why β is possible (see p. 66 for the number of orders possible). (i) Use E4.8 where $d = 1/$(No. of lines per metre). (ii) See p. 67 if necessary. (iii) Use the wavelength value from (ii) in E4.8, assuming $m = 1$. (iv) What is the value of m for the beam at angle β? Use E4.8 again.

N.B. This is a good example of where reading through the complete question is of value. The distinction between λ_a and λ_w in (i) and (ii) should give you a clue as to how to explain the changes produced by the water.

4.29L
(a) (i) Draw a diagram showing the path of rays (from a non-axial point on a distant object) through a simple refracting telescope to the eye. (ii) Explain what is meant by the 'magnifying power' of the telescope, and show how it is related to the focal lengths f_o of the objective and f_e of the eyepiece.

(b) A refracting telescope has an objective of focal length 900 mm and an eyepiece of focal length 12 mm. (i) What is the distance apart of the lenses when the telescope is in normal adjustment and what is its magnifying power? (ii) If the telescope is used to project a sharp image of the Sun onto a screen that is at a fixed distance of 150 mm from the eyepiece, what adjustment to the telescope will be needed? (iii) Given that the Sun's diameter subtends an angle of 9×10^{-3} rad, calculate the diameter of the image on the screen.

(**Oxford:** *JMB, SEB H*)

(a) (i) See F4.21 if necessary. The rays from a non-axial point on a distant object enter the objective lens as a set of parallel rays inclined at a small angle to the axis. (ii) See E4.12. You can use your diagram by marking the angle which the rays entering the objective lens make with the axis (a) and the angle which the rays leaving the eyepiece make with the axis (β). Then give the magnifying power in terms of these angles. To show how the m.p. is related to the focal lengths, you need to state and prove E4.12. You will find a proof in your textbook.

(b) (i) In normal adjustment, the eyepiece is adjusted so that the intermediate image (formed in the focal plane of the objective) lies in the focal plane of the eyepiece. It should be clear from your diagram how the distance apart of the lenses is related to the focal lengths. The magnifying power is given by E4.11. (ii) In this situation, the eyepiece lens is further from the objective lens than in normal adjustment. The intermediate image formed by the objective still lies in the objective's focal plane, but because the eyepiece lens has been moved away, the intermediate image now lies outside the focal point of the eyepiece. Since the intermediate image acts as an object for the eyepiece lens, calculate the object distance (for the eyepiece lens) given the image distance (+150 mm) and the focal length of the eyepiece. Then you can state how far the eyepiece lens must be moved from its 'normal adjustment' position. (iii) Proceed in two stages: A, calculate the height of the intermediate image; B, given object and image distances for the eyepiece lens, calculate the linear magnification of the eyepiece (see E4.9). Then you can calculate the height of the final image (= lin. mag. × height of intermediate image). To calculate the height of the intermediate image, use of the small angle approximation, $\tan a = a$, and reference to the diagram below may be helpful.

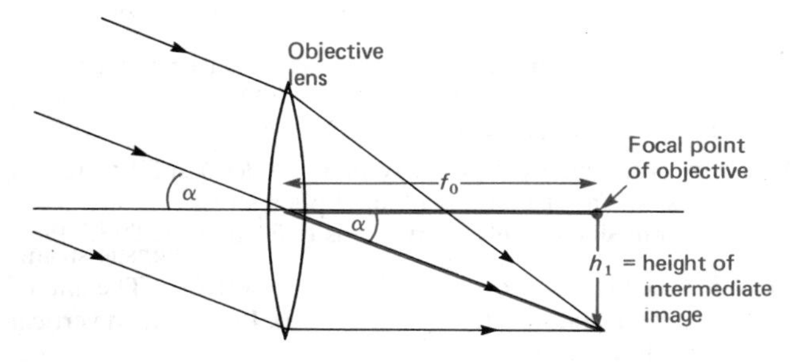

4.30L

(a) A compound microscope consisting of two thin converging lenses is adjusted to view an object so that the final image is at infinity. Draw a diagram showing the passage of **three** rays from a non-axial point on the object, through the instrument, and into the observer's eye.

(b) By reference to your diagram (i) indicate how the instrument can be adjusted so that the final image is at the least distance of distinct vision, (ii) identify the lens which acts as a magnifying glass and state and explain an advantage of the compound microscope over the magnifying glass in viewing small objects.

(c) A laboratory microscope has cross-wires. Show on your diagram where these would be located and hence explain their use.

(d) The objective of the microscope for which you have drawn a diagram is replaced by a lens of shorter focal length in the same position, the object position being unchanged. Explain (i) how you would adjust the eyepiece to produce again a final image at infinity, (ii) in what way this final image differs from the original one.

(**JMB:** *Oxford, Cambridge*)

(a) See F4.20. Start your diagram by drawing the axis, objective lens, and small object just beyond the focal point of the objective, as in F4.20. Then draw three rays from the tip of the object to the objective lens, and then locate the tip of the intermediate image. Since the final image is at infinity, the eyepiece lens must be placed so that its focal point lies at the same position as the foot of the intermediate image. Thus, draw the eyepiece lens in, and mark its focal point clearly. Then continue the three rays on to the eyepiece lens from the tip of the intermediate image, and complete the diagram.

(b) (i) Compare your diagram with F4.20. (ii) See p. 67. A magnifying glass is sometimes called a simple microscope. The difference between the compound microscope as in F4.20 and a magnifying glass (as in F4.19) is the presence and use of the objective lens in the compound microscope. So you should explain why it is advantageous to have an objective, rather than to place the object directly between the eyepiece lens and its focal length.

(c) The crosswires must be seen in focus with the final image. Any wire placed in the same plane as the intermediate image will therefore be seen alongside the final image. See your textbook for further details.

(d) (i) With the new objective, decide if the intermediate image will be formed nearer to or further from the eyepiece lens than before. For the final image at infinity, the eyepiece focal point must be in the same plane as the intermediate image. Hence, decide which way to move the eyepiece lens from its initial position to put the final image at infinity again. (ii) The key lies in the size of the intermediate image. Consider if the intermediate image becomes larger by using a shorter focal-length, objective lens.

Unit 5 Materials

This Unit is about those properties of materials that distinguish between the three main states (phases) of matter. Such properties are essentially related to shape and surfaces; a liquid takes the shape of its container, but has its own surface. A gas has neither shape nor surface. Thus, a solid has rigidity whereas liquids and gases can flow. Consequently, the study of shapes and surfaces in this Unit means solids and liquids only.

Other branches of physics contribute extensively to this Unit. From Unit 1, basic mechanics is used in the sections dealing with strength of solids and with elasticity. Understanding X-ray diffraction requires ideas from Unit 3 and the treatment of surface tension draws on the section on intermolecular fields of Unit 2. From GCSE studies, revision of Hooke's law and of pressure ought to be valuable for the earlier sections of the unit whilst a careful re-reading of the experiment 'Estimate of the size of an oil molecule' helps to set the scale for crystal structures.

5.1 STRENGTH OF SOLIDS

Stress is defined as force per unit area applied to a surface. The unit of stress is $N\,m^{-2}$ (or the Pascal, Pa). Note that the unit of pressure is the same.

Strain is defined as change of length per unit length. Since strain is a ratio, it has no unit.

When stress is applied to a solid, the solid will change shape, even if only by a small amount. Consequently, the applied stress will cause a strain within the solid. Applied stress is treated by considering its normal and tangential (ie, perpendicular and parallel to the surface) components separately. Normal stress can either be **tensile**, if the solid is stretched, or it can be **compressive**. Tangential stress, if excessive, will cause the solid to shear. Knowledge of stress–strain characteristics is vital to ensure correct use of materials, either alone or in composite form.

Stress–strain curves Various methods are available for obtaining information. The equipment must be able to apply regular increments of stress or strain to the sample, and it must enable accurate measurement to be made of both stress and strain. For metals, the sample is often in the form of a thin wire; in this form, large stresses may be applied by means of suspending the wire vertically and then hanging weights up to 100 N from its lower end.

Consider the test wire shown in F5.1. The applied stress is given by the (load weight/initial cross-sectional area); the strain is calculated from the ratio (increase of length/initial length). Results are usually displayed graphically, with y = stress and x = strain. F5.2 shows how the strain in a wire varies with stress. With the form of the test apparatus as outlined above, the wire would snap at B when the applied stress from the load exceeds the breaking stress.

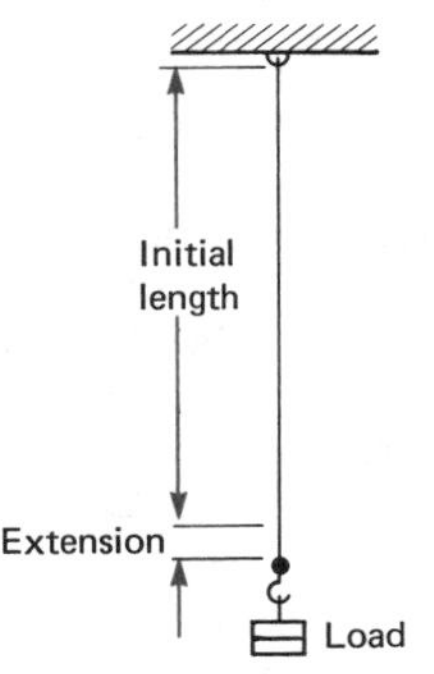

F5.1 Loading a wire

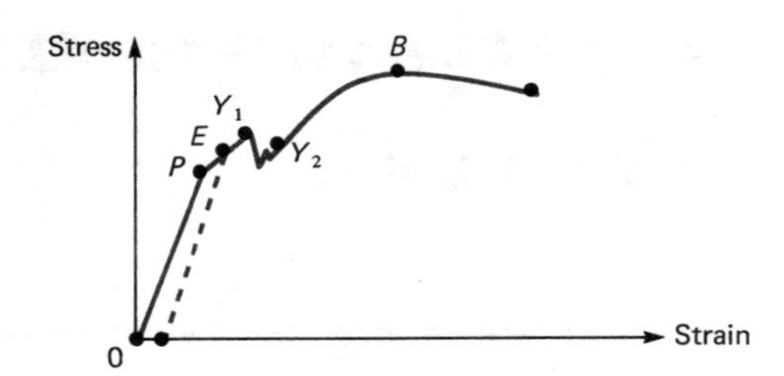

F5.2 Stress – strain curve for a typical metal

An alternative method of obtaining stress – strain information is to use a more specialized piece of apparatus known as a 'tensometer'. This equipment acts like a 'rack' because it extends the test material by measured amounts (ie, the strain is increased in steps), allowing the tension to be measured at each step. Consequently, the material can be taken beyond the strain corresponding to the 'breaking stress'; the stress will fall as the material loses its strength beyond *B*.

The essential features of the stress – strain curves for metals are shared by most metals, and are as follows (see F5.2):

1 *OP* is a straight line, thus **stress is proportional to strain** (ie, load is in proportion to extension) in agreement with Hooke's law.

2 *P* is the **limit of proportionality**, beyond which the ratio (stress/strain) is no longer constant.

3 *E* is the **elastic limit**. If the elastic limit is not exceeded when the material is stretched, then the material will regain its initial shape; otherwise, it will suffer permanent distortion.

4 Y_1 and Y_2 is the range over which **plastic flow** begins. From *E* to the lower yield point Y_1, removal of stress will reduce the strain, although not completely. If unloaded between *E* and Y_1, stress and strain would follow the dotted line of (F5.2). Beyond Y_2, the strain would be permanent. For some metals, Y_1 and Y_2 are very close (ie, a single yield point).

5 *B* is the point of maximum stress, referred to as the **breaking stress** (or sometimes as the 'ultimate tensile strength', UTS).

For **non-metals**, stress – strain curves vary widely from one material to another. F5.3 shows curves for rubber and polythene. The curves show that both materials readily depart from Hooke's law, but rubber regains its initial shape and so behaves elastically; polythene becomes permanently distorted from its initial shape, so it behaves in a 'plastic' manner. The term **hysteresis** is used to describe the difference between the 'loading' and 'unloading' curves. This means that when the stress is reduced from point *A* to point *B* on the diagram, the strain remains greater than if there was no hysteresis effect (ie, strain 'lags' behind stress).

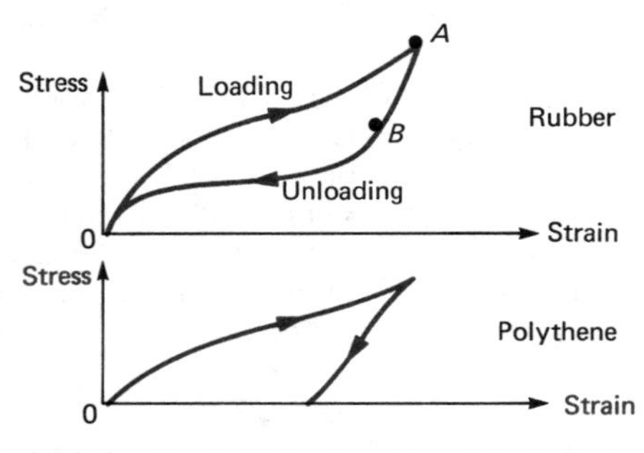

F5.3 Elastic loading (with hysteresis) and Inelastic loading

Knowledge of stress – strain behaviour is essential in giving scientific meaning to common terms used to describe strength of materials. For example, a **ductile** material is one which can easily be lengthened and worked into 'shape' without fracture. This is because the stress – strain relationship has an extensive plastic section, as shown by F5.4. Lead is a good example of a ductile material.

A **brittle** material, such as glass, will easily snap after reaching the elastic limit. Plastic behaviour of brittle materials is almost non-existent, as the stress–strain curve of F5.4 shows.

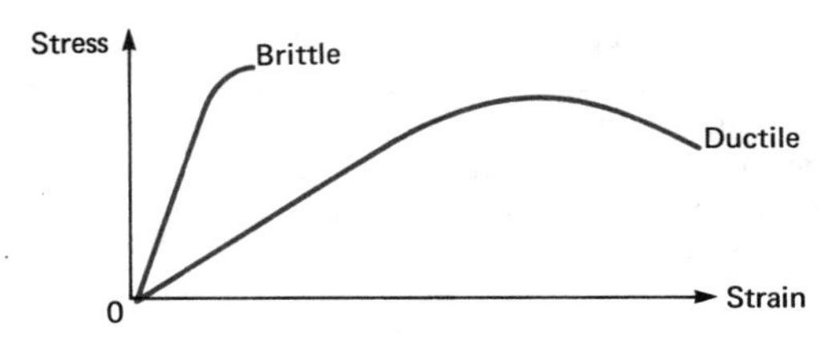

F5.4 Stress – strain curves for brittle and ductile materials

5.2 ELASTIC BEHAVIOUR

Elasticity is the physical property that enables a solid to regain its initial shape after removal of an applied stress.

Young's modulus of elasticity (E) is defined as $\frac{\text{tensile stress}}{\text{tensile strain}}$, provided the limit of proportionality is not exceeded. The unit of E is the same as that of stress (N m^{-2} or Pa). For a loaded vertical wire, the definition gives:

E5.1 $E = \frac{Tl}{Ae}$

where E = Young's modulus (N m^{-2}), T = tension due to load, (N), l = initial length (m), A = cross-sectional area (m^2) = $\frac{\pi d^2}{4}$ (d = diameter), e = extension from initial length (m).

The value of E is obtained from the stress – strain curve by determining the gradient *OP* in F5.2. Also, from E5.1, the **spring constant** (k) may be defined by $T = ke$, where k is given by $(\frac{AE}{l})$ for a wire.

The **work done** to stretch a wire can be determined from a graph of y = tension, x = extension, as in F5.5. Since work done = force × distance, then the area under the line gives the work done.

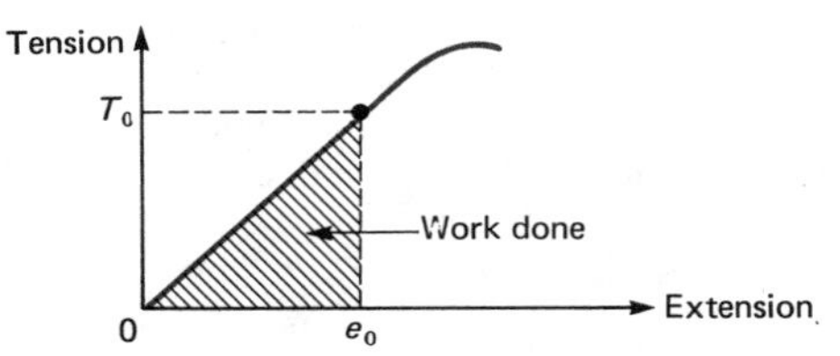

F5.5 Work done during Hooke's law loading

Provided that the limit of proportionality is not exceeded, the work done to stretch the wire to extension e_0 is therefore given by $\frac{1}{2}T_0 e_0$, where T_0 $(= \frac{EAe_0}{l})$ is the corresponding tension.

E5.2 Work done $= \frac{1}{2}(\frac{EA}{l})e_0^2 = \frac{1}{2}ke_0^2$

The area under a stress – strain curve gives the work done per unit volume; so the area of the hysteresis 'loop' for rubber in F5.3 gives the energy absorbed (per unit volume) by the material in one cycle of stretching and relaxing the material. **Resilient** materials are those which can be stretched and relaxed without dangerous overheating; they must therefore have as small a hysteresis area as possible.

5.3 STRUCTURE OF SOLIDS

Solids can be classified in terms of their structure as either **crystalline** or **amorphous**. The atoms of a crystalline solid are arranged in a regular pattern, known as a crystal lattice. Some materials are obviously crystalline, but others, known as polycrystalline materials, are composed of many tiny crystals (eg, metals). Amorphous solids, such as glass, totally lack any regularity of atomic arrangement in the sense of a lattice. Materials such as rubber are composed of molecules in the form of long chains of atoms, and the chains are tangled together in a haphazard way.

Composite materials such as fibreglass are a physical mixture of two separate materials, chosen so that the weaknesses of each separate material are reduced by the presence of the other material.

Inter-atomic forces The forces which bond atoms together are essentially electrostatic in origin, although there are several different varieties. The more important are:

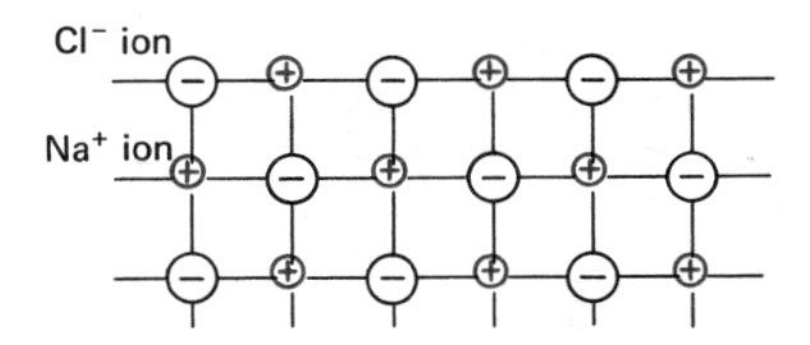

F5.6 An ionic solid – sodium chloride

1 Ionic bonds, formed in crystals such as sodium chloride. When formed, each sodium atom gives up an electron to a chlorine atom; the resulting structure is a lattice of alternate sodium ions (Na^+) and chlorine ions (Cl^-), as in F5.6.
2 Covalent bonds, formed in many organic compounds and in semiconductors such as silicon. Each bond involves sharing of two electrons, one from each atom, so that each atom's electron shells are full. For example, the outer shell of a silicon atom takes 8 electrons when full; an isolated atom of silicon has only 4 electrons in its outer shell. In the solid state, each silicon atom forms 4 covalent bonds, one bond with each of its four neighbours, so satisfying the 'full shell' requirement (see F5.7).

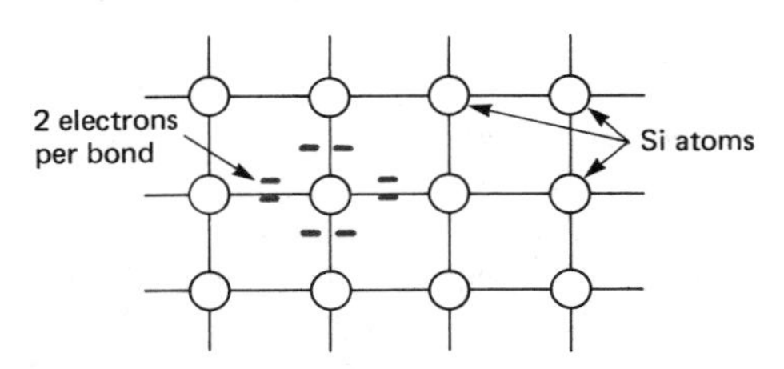

F5.7 A covalent solid – silicon

3 Metallic bonds, in which metal atoms give up their outermost electrons, so as to form a lattice of positive metal ions in a 'sea' of electrons.
4 Van der Waals' bonds exist between neutral atoms, and are thought to be due to the pull on the electrons of one atom by the nucleus of the other atom.
Atoms and elasticity The various types of bonds share several common features, including long range attraction, short range repulsion, and an equilibrium separation, as illustrated by F2.5(a). At small displacements from equilibrium, the force is in proportion to the displacement (ie, the curve is linear near equilibrium), but this is not so at greater displacements. Thus, Hooke's Law and the limit of proportionality can be explained.

At the elastic limit, crystal planes begin to slip over one another, so that the solid will not return to its initial shape when freed from stress.

Atomic spacing can be estimated if values are given for density (ρ), relative atomic mass (M) and Avogadro's number (N_a). Since the mass of a single atom of a solid element is given by M/N_a, then the volume occupied by one atom is $M/\rho N_a$. Hence the appropriate spacing between atoms of a solid element is given by $D^3 = \dfrac{M}{\rho N_a}$ where D is the approximate distance between the centres of two adjacent atoms. Typically, values of the order of 0.1 nm are obtained.
X-ray crystallography This is a widely used method for determining the structure of crystals. By directing a monochromatic (ie, single wavelength value) beam of X-rays at a crystal and then measuring the X-ray reflections from the atomic planes, the arrangement of atoms in the crystal can be deduced. The essential action is that:
1 X-rays are diffracted by each individual atom;
2 the combined effect of all the atoms in a single plane is to produce a reflected beam at the same angle to the crystal plane as the incident beam;
3 the reflections from similar planes should reinforce one another at certain angles. If this third condition is met, the following equation, known as **Bragg's law**, gives the angle of 'reinforced reflection':

E5.3 $m\lambda = 2d \sin \theta_m$

where d = crystal plane spacing (m), m = order number (ie. 1 or 2 or 3 or etc), λ = X-ray wavelength (m), θ_m = angle of 'reinforced reflection', usually known as the 'Bragg angle'.

Note that θ_m is the 'grazing' angle of the X-rays to the crystal planes, not the incident angle. The equation shows that wavelengths in the X-ray range are essential, since crystal plane spacing is of that order.

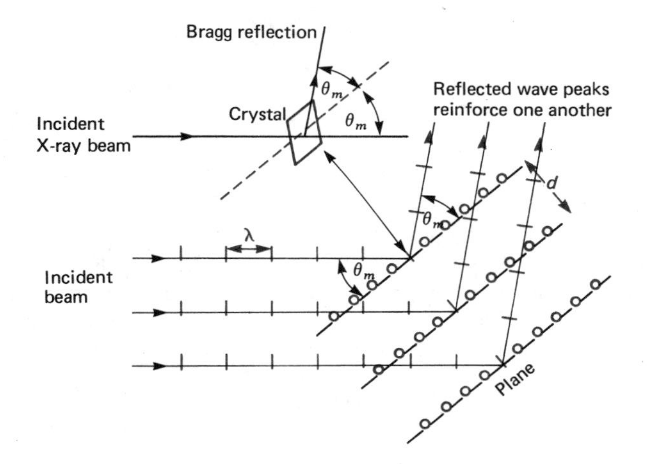

F5.8 Bragg reflection

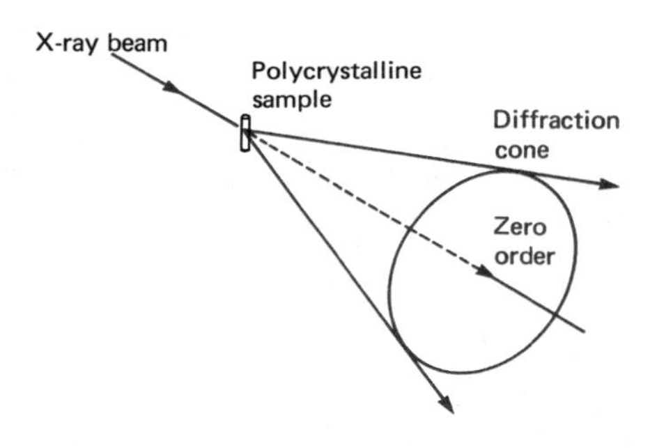

F5.9 Bragg reflection from a polycrystalline sample

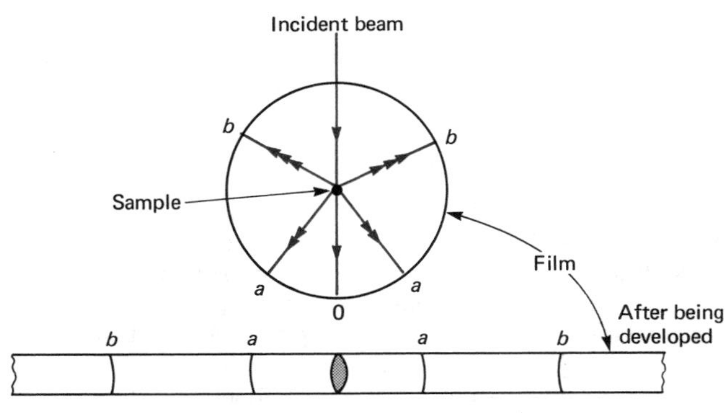

F5.10 The film in a Bragg 'powder' camera

Each reinforced reflected beam corresponds to a particular set of crystal planes at a suitable angle to the incident beam, as in F5.8. For a polycrystalline sample, since there are crystals in all directions, each reflected beam will form a cone, as in F5.9. A circular strip of photographic film is usually used to record the positions of the reflected beams (see F5.10). Use of Bragg's law then enables the different crystal plane spacings to be calculated.

A model of X-ray 'diffraction' (ie, Bragg reflection) by crystals can be devised using a structure of polystyrene balls as the 'crystal' and with microwaves as the incident radiation. With the balls in the arrangement known as hexagonal close-packing (hcp), shown in F5.11, a beam of microwave radiation of suitable wavelength is directed at the model. Strong reflections can be detected at the same angle as is observed for X-rays directed at the copper. It is then reasonable to assume that the actual arrangement of copper atoms in a single crystal is the same as that of the polystyrene spheres (ie, hcp).

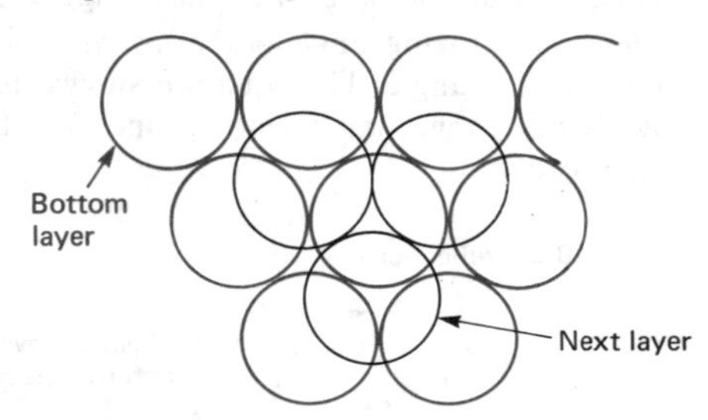

F5.11 Hexagonal close-packed Structure

There is a wide variety of crystal structures, of which hcp is just one. To deduce crystal structure from X-ray experiments by the 'analogue' approach above would involve unrealistic trial and error methods; in practice, Bragg's law is used to calculate crystal plane spacing values. Then the structure can be determined. Nevertheless, the analogue experiment shows very directly how crystal structures could be worked out.

Estimate of Avogadro's number (N_a) Avogadro's number is defined as **the number of atoms in exactly 12 grams of carbon-12.** Note that the mole is the amount of substance that contains as many elementary units as there are carbon atoms in 12 grams of carbon-12 (the elementary unit might be an atom, a molecule, an ion, etc).

The microwave analogue experiment, as above, can be used to estimate N_a by using a scale factor equal to the ratio (microwave wavelength/X-ray wavelength) to determine the number of copper atoms in 1 mole of copper. Since the microwave results are the same as the X-ray results as regards angle of Bragg reflection, then the polystyrene model of balls is a scaled-up version of a copper crystal. The scale factor is the ratio of wavelengths as above.

The geometry of the HCP crystal structure gives the number of balls per unit volume as $\sqrt{2}/d^3$ where d is the ball diameter. Therefore, scaling down by a factor Z (the microwave wavelength/the X-ray wavelength) gives the number of copper atoms per unit volume as $Z^3\sqrt{2}/d^3$. Given the density of copper, ρ, the number of copper atoms per unit mass is therefore $(Z^3\sqrt{2}/d^3)/\rho$.

The number of copper atoms per mole, N_a, is then worked out from the number of copper atoms per unit mass × the relative atomic mass of copper.

5.4 FLUIDS

Fluids at rest The **pressure** in a static fluid acts equally in all directions and increases with depth. At a certain depth beneath the surface of a liquid, the pressure due to the liquid is given by depth × density × g. This equation is used to work out atmospheric pressure from the reading of a mercury barometer. For example, if the height of the mercury column is 0.760 m, the equation gives 0.760 m × 13600 kg m^{-3} (density of mercury) × 9.8 Nkg^{-1} = 101 kPa for atmospheric pressure.

Any object immersed in a fluid experiences an **upthrust** due to upward pressure on its base. The upthrust is equal to the weight of fluid displaced. This is known as **Archimedes' Principle.**

Fluid flow A description of fluid flow, whether for a liquid or gas, is usually made by considering the motion of a small element of the fluid. The size of the fluid element must be small enough to enable it to follow the flow accurately, but if the fluid element is too small, then molecular motion becomes significant.

A **flowline** is a line along which a fluid element moves. A **streamline** is a stable flow line, along which every element follows the same path. Thus, a fluid element that passes through a given point will describe a flowline. If every subsequent fluid element which passes through that same point takes the same path as the first fluid element, then the flowline is stable, and is therefore a streamline.

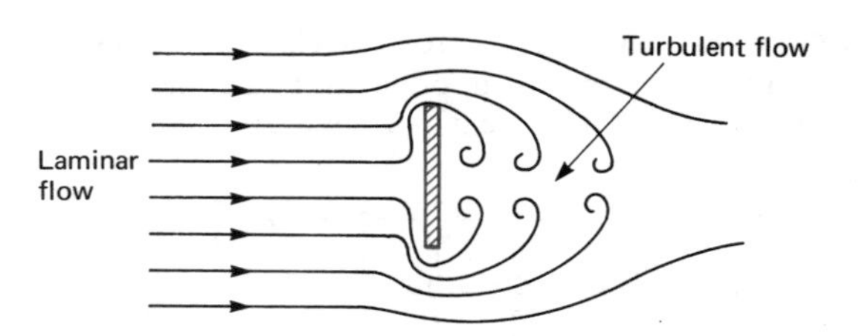

F5.12 Laminar and turbulent flow

A flow pattern can exist with both stable and unstable flow in different parts of the fluid, as in F5.12. Where the flow is stable, it is described as **laminar** (or streamlined) flow; unstable flow is referred to as **turbulent** flow. Thus streamlines cannot exist in areas of turbulence; in streamlined flow, fluid elements cannot cross streamlines.

Continuity Flow of fluid along a stream-tube (ie, a tube bounded by streamlines) will be fastest where the tube is narrowest. The mathematical statement of this fact is known as the **continuity equation**:

E5.4 $\rho A v$ = constant (= mass flow per unit time)

where ρ = fluid density (kg m^{-3}), A = area of cross-section of the tube (m^2), v = fluid speed (m s^{-1}).

For two points X and Y along a stream-tube, the mass per second flowing past X is equal to the mass per second flowing past Y. Since mass flow per second can be proved to be given by $\rho A v$ (see your textbook), then $\rho_x A_x v_x = \rho_y A_y v_y$. If the fluid is incompressible, then its density is constant (ie, $\rho_x = \rho_y$) so the product vA is then constant (ie, $v_x A_x = v_y A_y$).

5.5 NON-VISCOUS FLOW

Bernouilli's equation In the absence of viscous effects, laminar flow of a fluid can be described by Bernouilli's equation:

E5.5 $P + \rho g H + \frac{1}{2}\rho v^2$ = constant

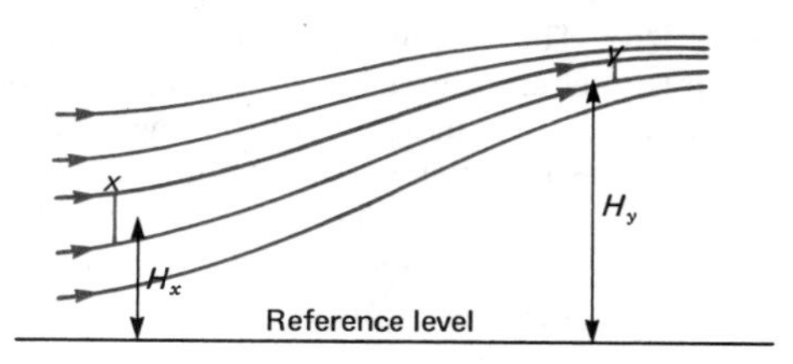

F5.13 Laminar flow

where P = pressure (N m^{-2}), ρ = density (kg m^{-3}), H = height (m), measured from a fixed level, v = fluid speed (m s^{-1}).

For two points X and Y along a stream-tube, as in F5.13, Bernouilli's equation gives:

$$P_x + \rho_x g H_x + \tfrac{1}{2}\rho_x v_x^2 = P_y + \rho_y g H_y + \tfrac{1}{2}\rho_y v_y^2$$

Each of the three terms can be seen in relation to the total energy of a fluid element, as follows:

1 $\frac{1}{2}\rho v^2$ is the ke per unit volume.
2 $\rho g H$ is the gravitational pe per unit volume.
3 P is the work done per unit volume from work done by pressure.

Thus, E5.5 is an expression of the fact that the ke + gravitational pe + energy due to pressure of a fluid element is constant. It follows that frictional energy loss due to viscous effects must be negligible for the equation to apply.

An example of the application of E5.5 is provided by laminar flow along a horizontal pipe which has a narrow section and a wide section, as in F5.14. For points X and Y as in the diagram, $H_x = H_y$; assuming the fluid is incompressible (ie, $\rho_x = \rho_y$), then E5.4 gives $v_x A_x = v_y A_y$. Thus, $P_x - P_y = \frac{1}{2}\rho_y v_y^2 \left(1 - \frac{A_y^2}{A_x^2}\right)$; since

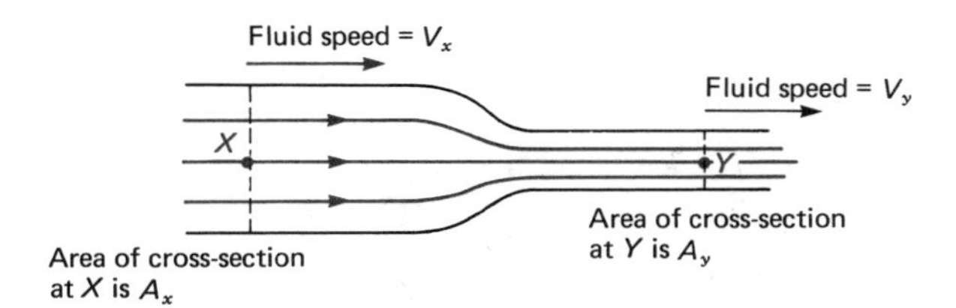

F5.14 Flow in a narrowing pipe

the fluid moves fastest in the narrow section, it follows that the pressure in the narrow section is less than in the wide section.

Pitot tubes These tubes are used to measure the flow speed of laminar flow which may be in a pipe or in the 'open' fluid. The pitot tube itself consists of a central narrow tube inside a second coaxial tube, as in F5.15. The tube is turned into the flow so that

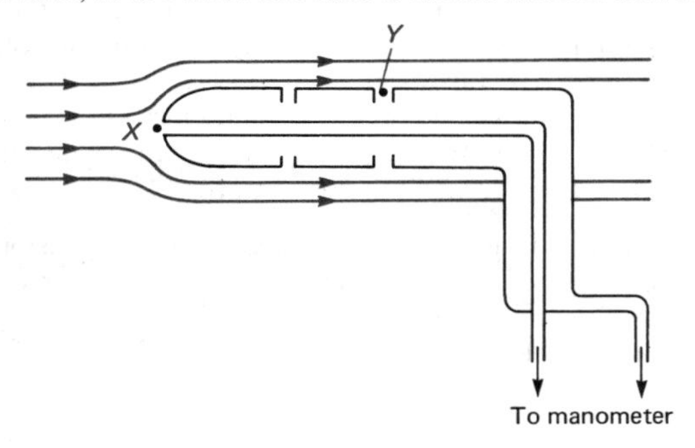

F5.15 Pitot-static tube

the opening X of the inner tube is a **stagnation point** (ie, fluid speed is zero at X). Point Y, at one of the holes into the outer tube, is at a point where the fluid speed is the same as the upstream speed (U). A manometer is used to measure the pressure difference between points X and Y. Use of Bernouilli's equation then gives $P_x - P_y = \frac{1}{2}\rho U^2$ (assuming $H_x = H_y$), so that the fluid speed U can be calculated if the fluid density ρ is known.

5.6 VISCOUS FLOW

Fluid friction exists between layers of a fluid moving at different speeds. Consider laminar flow in a uniform channel, as in F5.16.

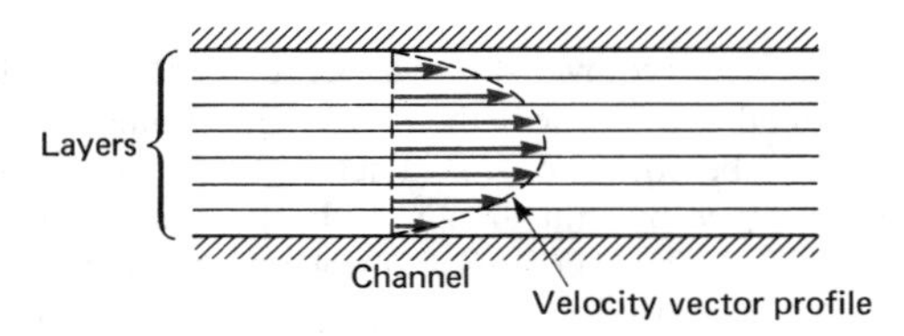

F5.16 Viscous flow in a pipe

The fluid in the centre moves fastest, whereas at the side the fluid will be stationary. Therefore, a **velocity gradient** will exist between the centre and the sides. In the absence of viscosity, the velocity at the centre would be the same as at the sides; the effect of viscosity is to cause the slow moving surface layers to drag on fast moving inner layers which in turn drag on the fastest moving central layers. The viscosity of a fluid is a measure of its resistance to flow.

The **coefficient of viscosity** (η) is defined as the tangential stress (T), in a fluid, required to produce unit velocity gradient across the fluid:

E5.6 $T = \eta \frac{dv}{dx}$

where T = tangential stress ($N\,m^{-2}$), η = coefficient of viscosity ($N\,s\,m^{-2}$), $\frac{dv}{dx}$ = velocity gradient (s^{-1}).

F5.17 illustrates the meaning of this definition.

Poiseuille's law For viscous flow through a horizontal uniform pipe, Poiseuille derived the following equation:

E5.7 $Q = (\frac{\pi r^4}{8L\eta})P$

where η = coefficient of viscosity ($N\,s\,m^{-2}$), P = pressure difference across the ends of the pipe ($N\,m^{-2}$), r = internal radius of the pipe (m), L = length of the pipe (m), Q = volume per unit time ($m^3\,s^{-1}$).

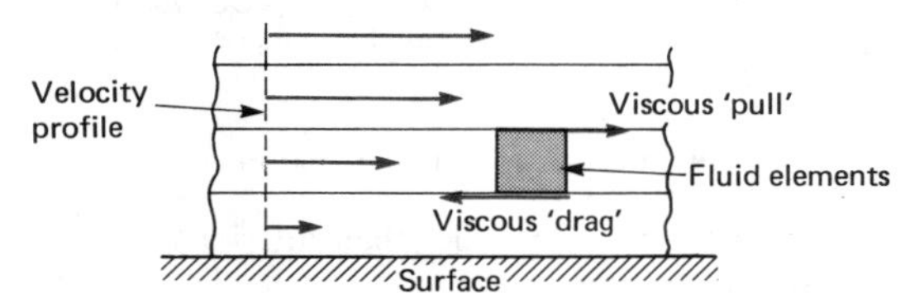

F5.17 Liquid–surface interface

Stokes' law When an object passes through a viscous fluid, the object experiences a frictional drag. For a uniform sphere moving at constant velocity through a viscous fluid, Stokes derived the following equation for the drag force:

E5.8 $F = 6\pi\eta r v$

where F = viscous force (N), η = coefficient of viscosity ($N\,s\,m^{-2}$), r = sphere radius (m), v = speed of sphere relative to the undisturbed fluid ($m\,s^{-1}$).

Molecular explanation of viscosity In any fluid, molecules move in random directions with a wide range of speeds. Consider the interchange of molecules between fluid layers moving at different speeds, as in F5.18. The number of molecules that transfer each

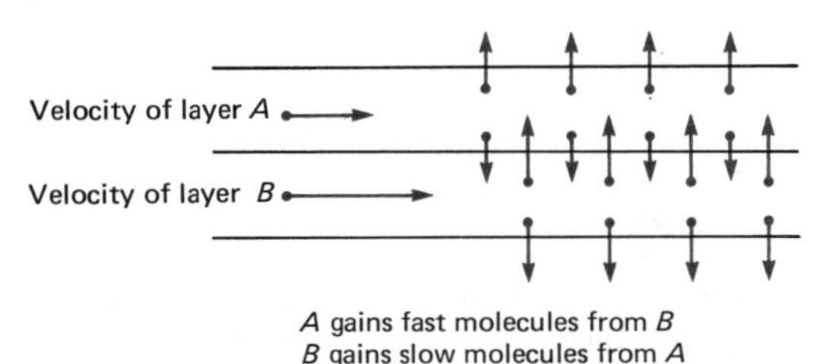

F5.18 Molecular motion in a fluid

second from one layer to an adjacent layer will be unchanged from one part of the fluid to any other part. Thus, the number of molecules per second that transfer from layer A to layer B will be the same as the number that transfer in the reverse direction in the same time. However, since layer B moves faster than layer A, then B gains slow-moving molecules from A while A gains fast-moving molecules from B. Since exchange of fast- for slow-moving molecules involves change of momentum, then the loss of momentum from layer B to layer A means that A drags on B whereas B pulls A forward. To maintain the flow at a steady rate, a pressure gradient is required along the direction of flow, but the work done by the pressure force is dissipated as heat (ie, increased molecular motion). Hence, viscosity corresponds to friction forces.

5.7 SURFACES

A surface is characteristic of either a solid or a liquid, and is the interface between the solid (or liquid) and the surrounding fluid. For a liquid/liquid surface, such as an oil/water surface, the two liquids must not mix with one another. For a solid, the surface is well defined, with little change in the atomic arrangement at the surface compared with the interior. However, solid surfaces easily become contaminated with 'foreign' atoms. Also, the state of the surface is an important factor in the strength of materials such as glass; in such materials, small cracks easily spread causing fracture.

For a liquid, its molecules move in a random manner although in contact with one another much more frequently than in a gas. There is a difference between the molecular arrangement at the surface compared with the interior; surface molecules are at

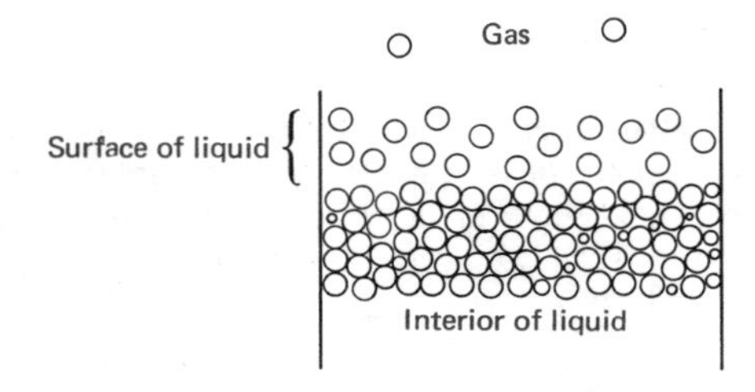

F5.19 Molecules at a liquid–gas interface

greater spacing from one another than interior molecules, and this increased separation gives rise to a state of tension between surface molecules (see F5.19).

Surface-tension effects are evident in simple experiments such as floating a needle on a clean water surface. To explain such effects, it is necessary to assume that the surface is in a state of tension, like a stretched drumskin.

The **coefficient of surface tension** (γ) is defined as the surface tension force per unit length acting along the surface, perpendicular to a line in the surface. The unit of γ is $N\,m^{-1}$. Using the above definition and F5.20:

E5.9 $F = \gamma L$

where L = the length in the surface (m), F = surface tension force (N) acting on length L.

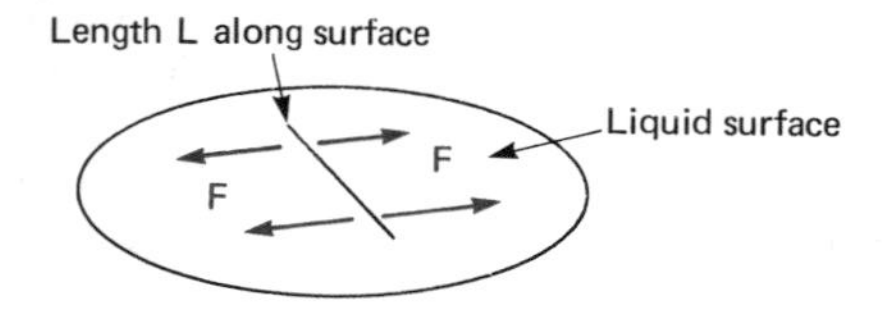

F5.20 Coefficient of surface tension

Capillary action The rise of water up the inside of a narrow vertical capillary tube, with its lower end in a beaker of water, is because of surface tension forces acting on the water surface where it is in contact with the glass, as in F5.21:

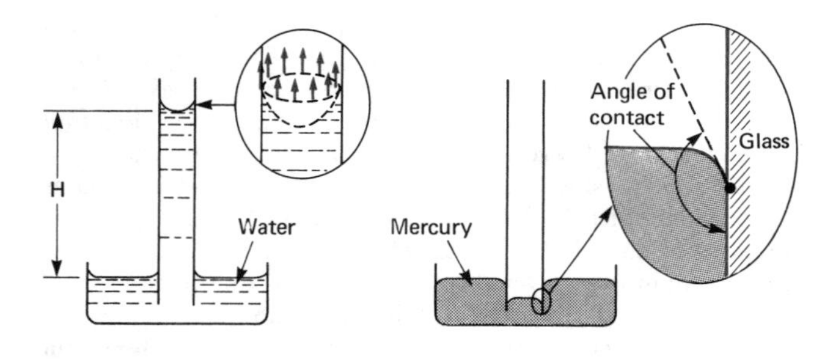

F5.21 Meniscuses of mercury and water

E5.10 $\gamma = \frac{1}{2}Hr\rho g$

where H = height as shown (m), r = internal radius (m), ρ = density of the liquid ($kg\,m^{-3}$).

The equation is arrived at by equating the weight of the liquid in the capillary thread ($\rho H\pi r^2 g$) to the surface tension force acting upon the perimeter of the meniscus at the top of the thread ($\gamma.2\pi r$).

Angle of contact This is the angle (in the liquid) that a liquid surface makes with a solid surface at the point of contact. For water, the angle of contact is zero when in contact with glass; thus the surface tension forces in the above example (ie, capillary action), where water is in contact with glass, act upwards parallel to the tube axis hence the water has to rise. For mercury on clean glass, the angle of contact exceeds 90°, so mercury in a capillary tube with its lower end in a beaker of mercury, will be pushed downwards, as in F5.21.

Bubbles The internal pressure of a bubble must be greater than

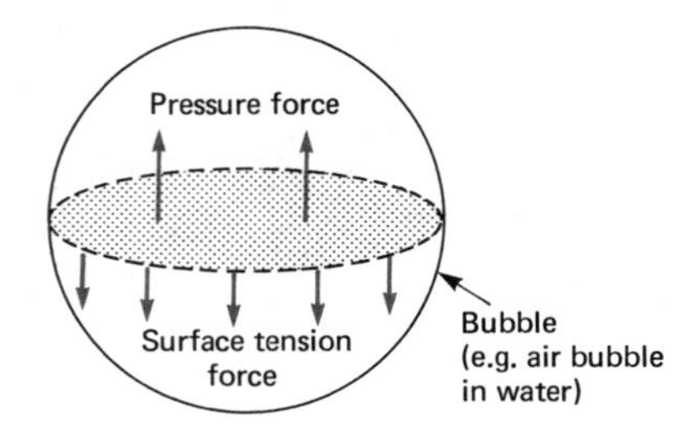

F5.22 Forces in a bubble

the external pressure, so that the pressure difference will be sufficient to enable the surface tension forces (which try to collapse the bubble) to be withstood. By considering the forces on half the bubble, as in F5.22, the excess pressure is given by:

E5.11 $P = \frac{2\gamma}{r}$

where P = excess pressure ($N\,m^{-2}$), γ = coefficient of surface tension ($N\,m^{-1}$), r = bubble radius (m).

The equation is derived by equating the surface tension force on the perimeter of the hemisphere to the pressure force on the flat area of the hemisphere (ie, $\gamma.2\pi r = P\pi r^2$). For a soap bubble or any other bubble with *two* surfaces, the excess pressure is given by $\frac{4\gamma}{r}$.

Surface energy When a surface is extended, work must be done against the surface tension forces. Consider a surface, as in F5.23, which is extended by moving one of its edges. The surface tension force on the edge is γL, where L is the edge length.

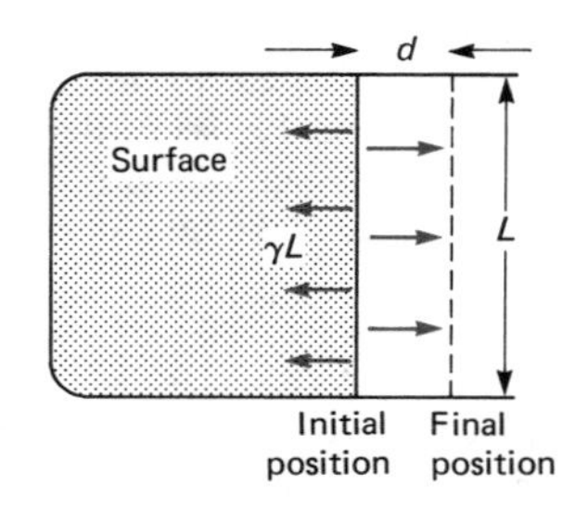

F5.23 Stretching a liquid film

If the edge is moved by distance d, perpendicular to its length, so as to extend the surface, then the work done (= force × distance) must be equal to γLd. Therefore, the work done to extend the surface by unit area is γ. If there is no change of temperature, then the increase of surface energy per unit area, σ, is equal to the work done per unit area, γ. In other words, the work done to extend a surface under isothermal conditions (ie, constant temperature) can be considered as surface energy.

Molecular explanation of surface tension

The variation of pe with separation for a pair of molecules is shown in F5.24. In the interior of a liquid, molecules will be at equilibrium separation, so there will not be an overall force on any given 'interior' molecule; their pe will be a minimum. At the surface, molecules will be at slightly greater separation so weak

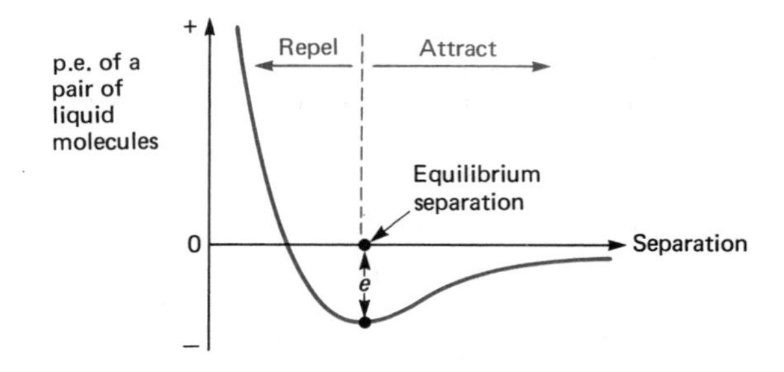

F5.24 Potential plotted against separation for two molecules

attractive forces will link surface molecules, thus preventing escape unless a surface molecule has sufficient ke.

When a surface is extended, molecules will be brought from the interior to the surface. Each such molecule will lose half its number of neighbours, so that if n is its number of neighbours (on average) in the interior, then at the surface it has neighbours on one side only, giving it (on average) $\frac{1}{2}n$ neighbours. Removal of the molecule from one of its neighbours requires energy, e, to be supplied, so the removed molecule and the neighbour each require energy $\frac{1}{2}e$. Therefore, if the removed molecule loses half its number of neighbours, then it requires energy $\frac{1}{2}n.\frac{1}{2}e$. Finally, if the number of molecules per unit area of surface is A, then the energy required to increase the surface by unit area must be $\frac{1}{4}neA$. Since the surface energy per unit area is equal to γ, then it follows that:

E5.12 $\gamma = \frac{1}{4}neA$

where A = number of molecules per unit area (m^{-2}), n = number of neighbours of each interior molecule, e = equilibrium pe between two molecules (J), γ = coefficient of surface tension ($N\ m^{-1}$).

UNIT 5 QUESTIONS

5.1M The diagram shows stress/strain curves for three different materials taken to 'fracture'. Which curve best shows the behaviour of a copper wire, and which one best shows the behaviour of a 'glass fibre'?

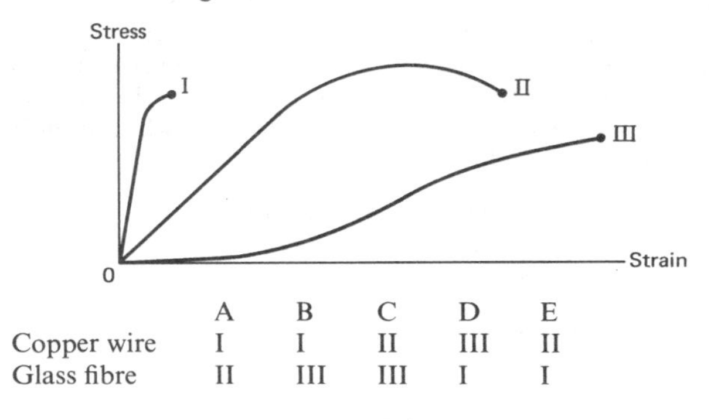

	A	B	C	D	E
Copper wire	I	I	II	III	II
Glass fibre	II	III	III	I	I

(—: *NISEC and all other Boards except SEB H, SEB SYS*)

Remember that copper is ductile, and glass is brittle. See p. 76 if necessary.

5.2M The following data were obtained when a wire was stretched within the elastic region:
Force applied to wire 100 N.
Area of cross-section of wire $10^{-6}\ m^2$.
Extension of wire 2×10^{-3} m.
Original length of wire 2 m.
Which of the following deductions can be correctly made from this data?
1 The value of Young's Modulus is $10^{11}\ Nm^{-2}$.
2 The strain is 10^{-3}.
3 The energy stored in the wire when the load is applied is 10 J.

Answer: A if **1**, **2**, **3** correct. D if **1** only.
B if **1**, **2** correct. E if **3** only.
C if **2**, **3** correct.

(**London**: *NISEC and all other Boards except SEB H, SEB SYS*)

For **1**, use E5.1. For **2**, see p. 75 if necessary. For **3**, see F5.5 and the subsequent discussion if necessary.

5.3M A helical spring extends 40 mm when stretched by a force of 10 N, and for tensions up to this value the extension is proportional to the stretching force. Two such springs are joined end-to-end and the double-length spring is stretched 40 mm beyond its natural length. The total strain energy, in J, stored in the double spring is:

A 0.05. B 0.10. C 0.20. D 0.40. E 0.80.

(**NISEC**: *all other Boards except SEB SYS, SEB H*)

Since stress/strain remains constant for a given material, within the limits of proportionality (see p. 76), it follows that a double-length spring will give double the extension for the same applied force. The energy stored in the stretched spring will equal the work done on it in stretching it. Remember that the average force will be half the applied force.

5.4M A certain wire which obeys Hooke's law is extended by a force F. If the wire is replaced by a similar piece of wire which is twice as long as before, and which is subjected to the same force F, which of the following statements is/are true?
1 The stress in the longer wire is the same as that in the shorter.
2 The strain of the longer wire is the same as that of the shorter.
3 The work done in extending the longer wire by the force F is the same as for the shorter.

A **1** only. B **2** only.
C **1** and **2** only. D **1** and **3** only.
E **2** and **3** only.

(**NISEC**: *all other Boards except SEB SYS, SEB H*)

For **1**, assume that the two wires are of the same diameter. See p. 75 for the definition of stress if necessary.

For **2**, remember that strain = stress/Young's modulus.

For **3**, use E5.2 but take care since the extensions are not equal.

5.5M The diagram below shows how the extension of 5 wires, each of different material, changes with tension. The diagram also shows the initial length and diameter of each wire. Which wire exhibits least strain for a given stress (ie, for the same stress applied to each wire)?

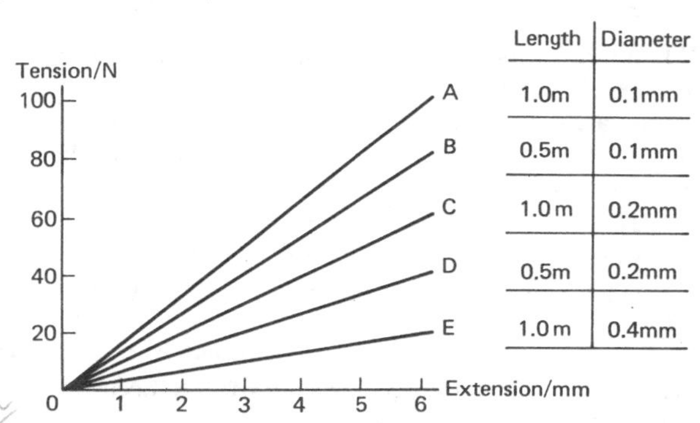

(—: *all other Boards except SEB SYS, SEB H*)

You must pick the wire with the greatest value of Young's modulus of elasticity. See p. 76 for the definition of Young's modulus if necessary.

5.6M Which of the wires A – E of the previous question has most energy stored in it when a load of 20 N is applied to each wire in turn?

(—: *all other Boards except Oxford, SEB SYS, SEB H*)

Remember that energy stored in this situation can be calculated from the area under the line of the tension – extension graph. Take care when you compare areas because you must consider the area under each line up to the point where tension = 20 N.

5.7M The force F required to extend a length of rubber by a length x was found to vary as in the diagram. The energy stored in the rubber for an extension of 5 m was:

A less than 100 J.
B 100 J.
C between 100 J and 200 J.
D 200 J.
E 500 J.

(**Cambridge**: *and all other Boards except Oxford, SEB H, SEB SYS*)

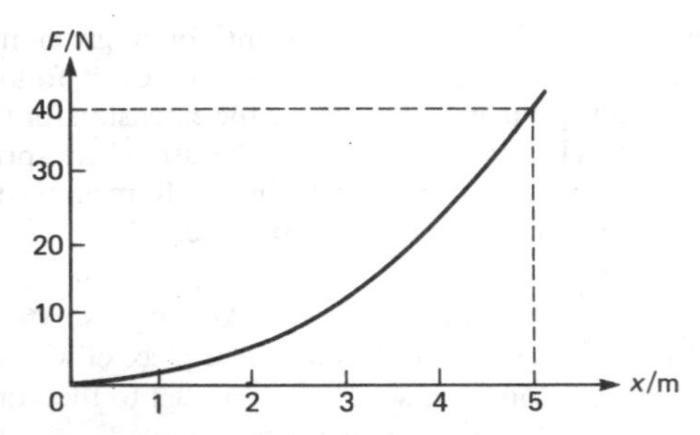

You must consider the area under the curve up to $x = 5$ m. It might be helpful to mark the point on the curve at $x = 5$ m, and then draw a line from that point to the origin. Compare the area under the curve with the area under the straight line which you have drawn in; you ought to be able to easily calculate the area under the straight line.

5.8M A uniform wire, fixed at its upper end, hangs vertically and supports a weight at its lower end. If its radius is r, its length L and the Young's modulus for the material of the wire is E, the extension is:

1 directly proportional to E.
2 inversely proportional to r.
3 directly proportional to L.

Answer: A if **1**, **2**, **3** correct D if **1** only
B if **1**, **2** correct E if **3** only
C if **2**, **3** correct

(London: *NISEC and all other Boards except SEB H and SEB SYS*)

Rearrange E5.1 with the extension e as the subject. Remember area $= \pi r^2$.

5.9M Which of the following statements about Bragg 'diffraction' is/are correct?

1 X-rays are used because they pass straight through crystals.
2 X-rays reflected from successive layers of atoms in the crystal are always in phase with one another.
3 X-rays are used because their wavelength is of the same order of size as the spacing between atoms in crystals.

A **1** only B **2** only C **3** only D **1** and **2**
E **2** and **3**

(—: *London, O and C only*)

Remember that reinforced reflections only occur at **certain** 'grazing' angles. See p. 77 if necessary.

5.10M A sealed U-tube contains nitrogen in one arm and helium (at pressure p) in the other arm. The gases are separated by mercury of density ρ with dimensions as shown below.

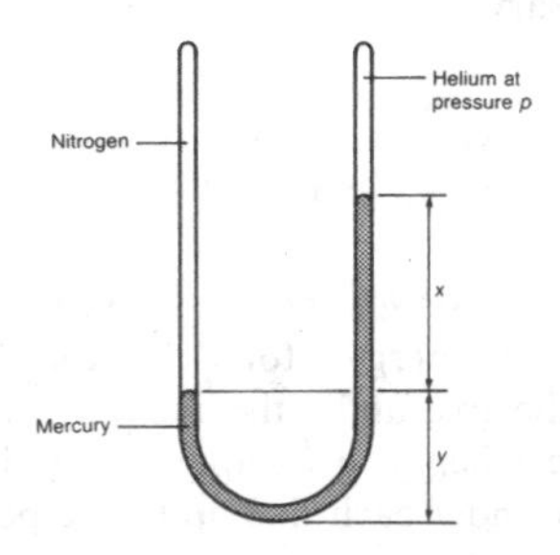

Which one of the following gives the pressure of the nitrogen?

A p
B $x\rho g$
C $p - x\rho g$
D $p + x\rho g$
E $p + (x+y)\rho g$

(Oxford: *and all other Boards*)

See p. 78 if necessary

5.11M The diagram shows the capillary rise of water in a capillary tube. If a small hole is made at point P:

A water will run out of the hole because the hole is below Q.
B water will run out of the hole because water is drawn up the tube by capillary action.
C air will bubble through the hole because the hole is above R.
D neither water nor air will flow through the hole because surface tension will prevent it.
E neither water nor air will flow through the hole because viscosity will prevent it.

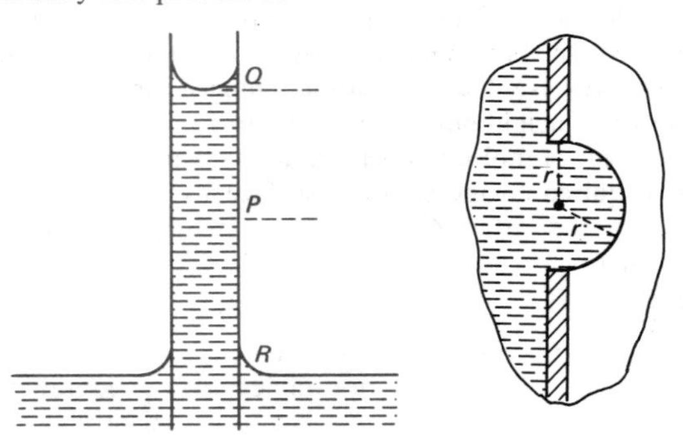

(Cambridge)

The key point is that the hole is small. For a hole of radius r, the force due to hydrostatic pressure (which cannot exceed $\rho g H \pi r^2$) pushing water out (or trying to) must be compared with the surface tension force trying to keep the water in. Can the maximum hydrostatic force ever exceed the surface tension force if the hole radius, r, is less than the tube radius? The surface tension force is biggest when the bubble radius equals the hole radius, as in the diagram above. Thus the biggest surface tension force is $\gamma.2\pi r$. Use E5.10 to decide if the hydrostatic force can ever overcome the surface tension force when the hole radius is less than the tube radius.

5.12M A vertical capillary tube, open at both ends, is dipped into a liquid. The height to which the liquid rises in the tube is governed partly by its surface tension and partly by the:

1 angle of contact between the wall of the tube and the liquid.
2 density of the liquid.
3 atmospheric pressure on the liquid surfaces.

Answer: A if **1**, **2**, **3** correct.
B if **1**, **2** correct.
C if **2**, **3** correct.
D if **1** only.
E if **3** only.

(London)

See F5.22 and the discussion that follows. **1** and **2** are straightforward, but **3** needs some thought. Remember that the tube is **open**, so think carefully **where** atmospheric pressure will act on the liquid surfaces (the plural 'surfaces' should give you a clue).

5.13S The average male has a mass of 70 kg and the minimum cross-sectional area of the bone in each leg is approximately $5.0 \times 10^{-4}\,\mathrm{m}^2$. The compressive breaking stress of bone is approximately $1.0 \times 10^7\,\mathrm{N\,m}^{-2}$.

(a) Assuming that a man is standing with his weight equally shared by each leg, showing the steps in your working, calculate (i) the maximum stress in his leg bones and (ii) the ratio (maximum stress in bones)/(breaking stress).

(b) Now suppose that all the linear dimensions of the human body are increased by a factor of nine. (i) What is the new value of the ratio found in (a) (ii)? Show and explain how you reach your answer. (ii) Explain what would happen to this 'giant' if he attempted to chase a normal human being.

(O and C Nuffield: *NISEC and all other Boards except SEB H and SEB SYS*)

(a) (i) Stress is defined as force/area, so maximum stress from a given force will occur when it is applied to the minimum area. Remember that the force on the bones comes from 'weight' (and not 'mass') and that the weight must be divided equally between each leg. (ii) Divide the answer to (a) (i) by the value quoted for the breaking stress; as this is a ratio, remember that it will have no units.

(b) (i) The breaking stress is a property of the material and will be unaffected by the 'size' of the bones. However to determine what has happened to the 'maximum stress' you need to consider the effect on the 'force' and 'area' as a result of the nine times linear growth. The area has increased 81 times but what about the force? The force depends on the mass of the man which is proportional to his volume. (ii) The way in which this question is phrased suggests that the running giant will break his bones whereas the man will not. The answers to (a) (ii) and (b) (i) must be used to justify this; a sensible guess to the extra stress imposed as a result of running rather than standing must be made.

5.14S
(a) Define the Avogadro constant.
(b) The relative atomic mass of aluminium is 27.0 and the density is 2.70×10^3 kg m^{-3}. If the Avogadro constant is 6.02×10^{23} mol^{-1}, calculate the number of atoms in one cubic metre of aluminium. **Estimate** the distance between centres of neighbouring aluminium atoms, stating any assumptions you make.

(**JMB**: *and all other Boards*)

(a) The definition is in terms of carbon-12. Refer to your textbook if necessary.

(b) See p. 77. Calculate the volume of 1 mole of aluminium (ie, 27 g) in m^3 and then determine the number of atoms in 1 m^3. To estimate the distance between centres of neighbouring atoms, you should use your value for the number of atoms per m^3 to determine the volume occupied by a single atom. Then calculate the distance D between centres by assuming that each atom occupies a volume of D^3.

5.15S
(a) The value of the Avogadro constant is 6.0×10^{23} mol^{-1}. Explain what is meant by this constant.

(b) The density of aluminium is 2.7×10^3 kg m^{-3} and its relative atomic mass is 27. (i) How many moles of aluminium atoms does an aluminium cube of side 20 mm contain? (ii) Using the value of the Avogadro constant given above, find the number of aluminium atoms in the cube. (iii) Assuming that a cubical space is available for each aluminium atom, deduce the maximum volume of this space. (iv) Hence make an estimate of the distance between the centres of adjacent aluminium atoms.

(c) Information about the spacing of atoms in crystals may also be obtained by studying the diffraction of X-rays. In such an experiment, X-rays of wavelength 0.154 nm were incident on an aluminium crystal. Strong first-order scattering from a certain set of atomic planes was observed at a Bragg angle of 11.0°. (i) Deduce the separation of these atomic planes. (ii) Find another value of the Bragg angle at which strong scattering would be expected from these planes.

(**NISEC**: *and all other Boards: for (a) and (b)*)

(a) See your textbook. Note that the term 'explain' demands rather more than a definition.

(b) (i) Find the volume of the cube and hence its mass. Take care with units! Then divide the mass in grams by the relative atomic mass to give the number of moles. (ii) Use your answer to (i) and the given value of the Avogadro constant. (iii) Divide the total volume available by the number of atoms. (iv) See p. 78.

(c) Use E5.3 with $m = 1$. for (i). For (ii), repeat but with $m = 2$.

5.16S The diagram shows an X-ray powder camera loaded with a strip of film and having a thin metal wire specimen (S) mounted along the central axis of the camera. A narrow beam of monochromatic X-rays enters the camera and strikes the specimen S.

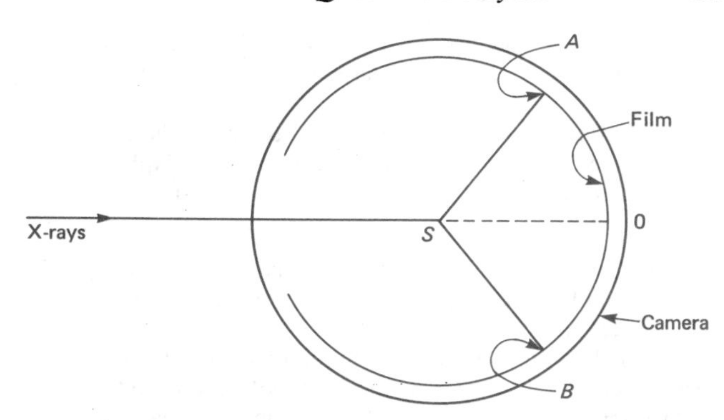

(a) After a suitable exposure the film is developed. Sharply defined lines on the film show that X-rays must have been strongly reflected in certain directions. What can you conclude about the specimen?

(b) A pair of dark lines, resulting from first order interference, occurs on the film at A and B as shown on the diagram (angle ASO = angle BSO). If X-rays of longer wavelength had been used, would this have made the angle ASO (= BSO) bigger or smaller? Show how you arrive at your answer.

(c) Using X-rays of wavelength 0.16 nm and a camera of radius (SA) 30 mm, the length of the film between the lines A and B was found to be 54 mm. Showing clearly the steps in your calculation find: (i) the Bragg angle (θ) to the nearest degree; (ii) the spacing (d) which gives rise to the lines A and B.

(**O and C Nuffield**: *and London*, O and C* only*)

(a) See p. 77.

(b) Explain using E5.3. Remember the general principle that longer wavelengths mean more diffraction or bigger spacings between fringes of interference.

(c) (i) The angle ASO can be calculated by finding what fraction of a circle ($2\pi r$) the length AO is, using the fact that the complete circle is 360°. Note that the angle ASO is the angle between the transmitted beam and the reflected beam; this is TWICE the Bragg angle (see F5.8). (ii) Use the value of the Bragg angle from (c) (i) and E5.3.

5.17S
(a) Calculate the potential energy, in eV, per pair of atoms of a solid for which the latent heat of sublimation is 1.3×10^4 J mol^{-1} and the number of neighbours per atom is 6. The Avogadro constant, $N_a = 6.0 \times 10^{23}$ mol^{-1} and 1 eV $= 1.6 \times 10^{-19}$ J.

(b) For a pair of atoms, sketch a graph showing how the potential energy per atom pair varies with distance between the atoms. Show on your graph (i) the equilibrium separation, r_0, (ii) the value of the energy calculated in (a).

(c) Mark on your sketch graph a point P corresponding to a separation 'other than the equilibrium value' and explain how you would determine, from the graph, the force between the atoms at P. Indicate whether you consider the force at P to be attractive or repulsive.

(**JMB**: *all other Boards except SEB H and SEB SYS*)

(a) Use E6.5 directly. The bond energy is equal to the pe per pair of atoms. The equation gives the bond energy in joules, so you must then convert the value into eV.

(b) See F2.5(b). Beware of confusing this curve with the force–separation curve for two atoms. Remember that equilibrium corresponds to **minimum** pe. On your sketch graph, mark the origin, and indicate clearly the direction of pe > 0 and the direction of pe < 0.

(c) Having marked a suitable point, you must decide if the force at P is repulsive or attractive. To help you to decide, consider the analogous situation of two point charges: (i) which are like charges, so that if moved apart, their electrostatic pe falls; (ii) which are unlike charges, so that if moved apart, their pe rises. Thus consider increasing the separation from P by a **small** amount; if the pe falls, then the atoms must exert repulsive forces on one another, and if the pe rises, then the atoms must exert attractive forces. See the comments with F2.5(b).

5.18S Give a simple argument to show that a molecule in the surface of a liquid has higher energy than one in the body of the liquid.
In capillary rise, energy must be provided to lift the column of liquid in the capillary above the original surface. Explain briefly and in molecular terms where this energy comes from.

(WJEC: *JMB and NISEC*)

To show that surface molecules have more energy than 'interior' molecules, see p. 80. The equation and its proof (E5.12) is not asked for.

F5.21 shows the surface tension forces acting on the meniscus inside a capillary tube. Use your textbook to find out why the surface tension force is upwards for water in a glass capillary. Then complete your answer by considering the work done by the surface tension force when the water rises up to its maximum height in the capillary tube.

5.19L This question is about the way a karate expert breaks blocks of wood or concrete by bringing down the side of his fist vertically at high speed as shown in the diagram.

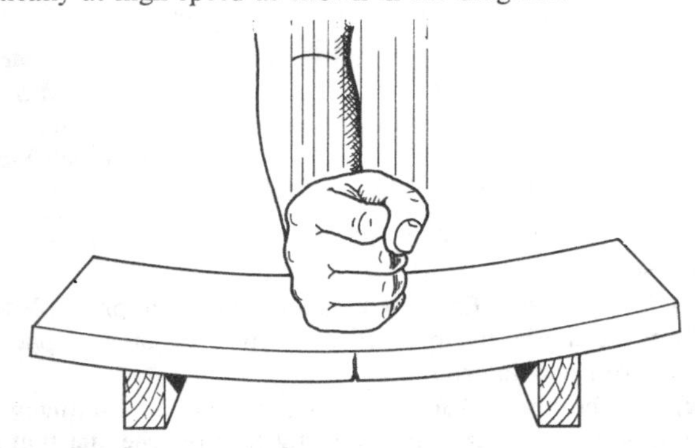

(a) Estimate the kinetic energy a fist and forearm could have on impact, explaining how you arrive at your estimate.

(b) Show how the following results can be derived for a uniformly stretched block of material: (i) energy stored = $\frac{1}{2}$(final force × extension); (ii) energy stored per unit volume = $\frac{1}{2}$(stress × strain); (iii) energy stored per unit volume at breaking point = $\frac{1}{2}\frac{S^2}{E}$, where E is the Young modulus and S the stress at breaking.

(c) Use the results from **(b)** to calculate the energy to break the blocks of wood and of concrete specified in the data given at the end of the question.

(d) The impact is a collision in which a moving mass (the fist and forearm) hits a stationary mass (the block), and the two then move forward together. Using the data given at the end of the question and your estimate in **(a)**, estimate how much of the initial kinetic energy is converted into other forms of energy. Explain how this result might be used in relating the results of parts **(a)** and **(b)** above.

(e) The energies actually needed to break blocks turn out in practice to be smaller than the arguments in **(c)** and **(d)** above would predict. Suggest **two** possible reasons for this and describe in outline how your suggestions could be checked by experiment.

(f) When it strikes a concrete block of the size quoted below, a karate fist slows down in about 3 ms. Estimate the mean force exerted and then the average power involved.

(g) How would you expect the processes of impact and breaking for a wood block to differ from these processes for a concrete block?

Useful Data	**Wood**	**Concrete**
Block size	30 cm × 15 cm × 2 cm	40 cm × 20 cm × 4 cm
Density	350 kg m^{-3}	2150 kg m^{-3}
Young modulus	1.4 × 10^{8} N m^{-2}	2.8 × 10^{9} N m^{-2}
Breaking stress	3.6 × 10^{6} N m^{-2}	4.5 × 10^{6} N m^{-2}

Estimate any quantities that you need and that are not given.

(O and C Nuffield: *NISEC and all other Boards except SEB H and SEB SYS*)

(a) Use $\frac{1}{2}mv^2$ where m is the mass of the fist and forearm and v is the speed at which they are travelling on impact.

(b) (i) Use the explanation of E5.2 in the text; or use Work done = Av. force × distance, where Av. force is half the final force for Hooke's law stretching. (ii) This is easy to prove as long as the definitions of stress and strain (5.1) can be recalled. (iii) This follows from rewriting the formula proved in (b) (ii) as $\frac{1}{2}$(stress)^2 × (strain/stress).

(c) Calculate the volume of each block and use the formula of (b) (iii).

(d) This is a difficult question to follow. It points out that after the collision the hand and block continue to move downward together. If you can estimate the speed at which they move you can calculate the kinetic energy now in the system (using the table of information to help you find the mass of the block). The kinetic energy loss will have been used to break the block; part (a) helps you in the kinetic energy calculations, part (b) supplies an independent method of assessing the energy needed to break the block.

(e) The predictions of part (d) are only correct if all the missing kinetic energy is used to break the block, but is it? However part (b) is based on the properties of the material and can only be wrong if the material were to break below the theoretical value; three reasons could be put forward for this. First one could argue that the calculation is for a 'perfectly' formed slab, but all specimens in practice may be flawed. The other two arguments are better; the calculations of part (b) are based on static tests but the experiment is dynamic (consider the effect of a sudden jerk on a string compared with a steady static pull), secondly the experiment involves bending the material, but the calculation uses Young's modulus which can only be applied to tensile and compressive forces. Whatever two chqices you make you will need to make all the measurements experimentally both in a static and dynamic case. Load the slab with weights until it breaks, taking appropriate measurements to check the theory of (b). Drop a weight (equal to fist and forearm) from the right height to ensure that it hits the block with an appropriate speed, make sure the weight 'sticks' to the block. Stroboscopic photography will aid measurement.

(f) It is easiest to find the average force by halving the breaking force calculated from the breaking stress. The power is found easily from dividing the energy delivered to the block by the time taken to deliver it (3 ms).

(g) Consider the different behaviour of wood and concrete under bending stresses, the different energies needed to break them, the different nature of the fractures involved.

5.20L

(a) Explain what is meant by saying that a material obeys Hooke's law. Describe an experiment which would enable you to investigate the extent to which a given specimen of steel wire obeys this law.

(b) Explain why, in an experiment to measure the Young modulus for a material in the form of a wire (i) the wire chosen is usually long and thin, and (ii) a second wire is suspended alongside the first. Sketch a graph to show the results you would expect to obtain if the extension of the wire were measured for various applied loads below the elastic limit. What other measurements would have to be taken and how would you use your graph to obtain a value for the Young modulus of the material of the wire?

(c) A force is required to compress a body or to extend it. What does this imply about the intermolecular forces?

(d) A load of 35.0 N applied to a wire of cross-sectional area 1.50 × 10^{-7} m^2 and length 2.00 m causes an extension of 1.00 mm. What is the energy stored in the wire per unit volume?

(AEB Nov 81: *NISEC and all other Boards except Oxford SEB H and SEB SYS*)

(a) See your textbook for a statement of Hooke's law. The apparatus required for an investigation of this law is similar to that for the measurement of Young's modulus and is described in your textbook. Your description of the method should include an explanation of how you would check for possible extension of the specimen beyond its elastic limit and for slipping in the clamps.

(b) (i) See p. 75. (ii) The second wire is used to support the vernier scale. This arrangement has the effect of eliminating a particular source of inaccuracy which you should identify. The graph required is that section of F5.2 below the elastic limit. The additional measurements needed to obtain a value for Young's modulus may be deduced from E5.1.

(c) Using the adjectives 'repulsive' and 'attractive', compare the forces acting between neighbouring molecules at distances apart greater or less than their equilibrium separation.

(d) It is useful to remember that whereas the area under the load – extension curve gives the total energy stored in a specimen, the corresponding area for the stress – strain graph gives the energy stored per unit volume.

5.21L Explain what is meant by an elastic solid. Show that the potential energy per unit volume, U, stored in a solid under tensile stress is given by $U = \frac{1}{2}$ (stress $\times$ strain). Estimate the theoretical breaking stress for a certain metal from the data given below. Assume that the metal breaks cleanly between two planes of atoms of interplanar spacing 2×10^{-10} m; any other assumptions should be clearly stated. The measured values for the breaking stress of metals turn out to be very much less than the theoretical values. Give a descriptive account of the reasons for this disagreement.
(Surface energy of metal $= 1\,\text{J m}^{-2}$; Young's modulus $= 2 \times 10^{11}$Pa)

(**WJEC**: *all other Boards except SEB H and SEB SYS*)

For an explanation of the meaning of the term 'elastic solid', see p. 76.

The equation for potential energy per unit volume can readily be proved by considering the solid in the form of a bar of area of cross-section A and length L. The work done to extend it is given by $\frac{1}{2} \times$ tension $\times$ extension. By dividing this expression by the volume of the bar, the expression for U can be obtained; you need to remember the definitions of stress and strain and to apply them here.

To estimate the theoretical breaking stress, consider the metal in the form of a wire of length 1.0 m and of area of cross-section 1 mm^2 ($= 1 \times 10^{-6}$ m^2). The value of surface energy is the amount of energy required to create 1 m^2 of extra surface area; when the wire (as above) breaks, extra surface of area 2 mm^2 is created by the complete separation of a pair of atom planes, so you can calculate the amount of energy required to separate a pair of atom planes of the wire. Then use the given value of interplanar spacing to estimate how many planes there are along the length of the wire; see the diagram. Then calculate how much energy is required to separate all the atom planes, and assume this is equal to the work done on the wire to stretch it to breaking point.

The next stage is to use the estimate of work done with E5.2 to calculate the extension of the wire (assuming proportionality up to breaking point between tension and extension). Then use the calculated value of extension with E5.1 to determine the tension and hence the stress in the wire.

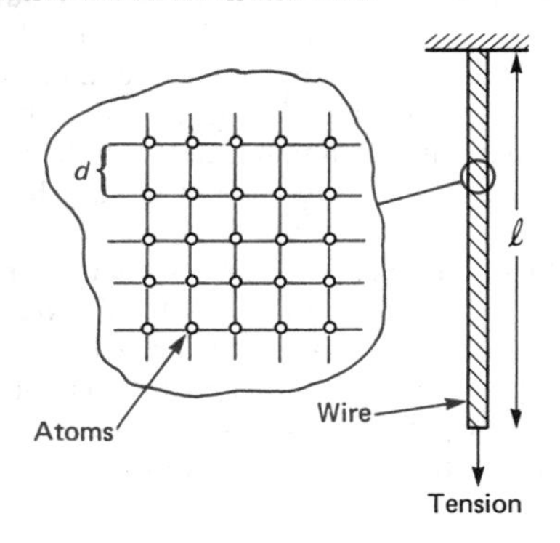

Your explanation should include ideas such as dislocations, slippage of crystal planes and stress build-up at surface cracks. Use your textbook to obtain further information; for each such idea that you choose to write about, give a simple explanation of what it means in terms of atoms, then explain why it helps to account for the disagreement between measured and theoretical stress values.

5.22L

(a) Describe one experimental arrangement for the observation of Brownian motion in 'either' a liquid 'or' a gas. Explain why the effect is only observed for very small particles. Discuss briefly the effects on Brownian motion of cooling the medium in which it is observed.

(b) Sketch a graph that illustrates the variation of force between two atoms with their distance of separation. Use your graph to explain why, for small extensions, the extension of a wire might be expected to be proportional to the applied force. Solids that are difficult to stretch also have a high melting point. Suggest reasons for this in terms of interatomic forces.

(c) A spherical drop of oil of diameter of 0.3 mm placed on a water surface spreads out to an approximately circular shape of diameter 150 mm before breaking up. Estimate the size of an oil molecule from this information. What assumption has been made in the calculation? Discuss briefly the effect on the result if this assumption were not valid.

(Volume of a sphere $= \frac{4}{3}\pi r^3$).

(**NISEC**: *all other Boards except SEB H (a) & (c) only, SEB SYS*)

(a) See your textbook for details of an experiment to demonstrate Brownian motion. To explain why the effect is observed for very small particles only, first explain how Brownian motion is thought to occur, and then consider the effect if the particles were larger. Why is there less likely to be a resultant force on a large particle – think in terms of the surface area of the particle. What effect will cooling the medium have on the mean speed of the bombarding molecules? What effect is this likely to have on the Brownian movement?

(b) See p. 33 and F2.5(a) with particular reference to the interpretation of the slope of the graph. What does stretching do in terms of the separation of atoms? If energy is supplied to a solid in the form of heat what happens to the atoms which eventually leads to the solid melting? Relate these to the strength of the interatomic forces (ie, bonding energy).

(c) Calculate the radius and then the volume of the spherical drop. The volume of oil remains the same when it spreads out on the water surface to form a thin cylindrical slice of thickness h. Write down an expression for the volume of this cylinder in terms of h and its radius, and equate this expression to the volume of the oil; hence find h. On the assumption that the film is a monolayer (ie, one molecule thick), the value of h gives an estimate of the size of an oil molecule.

5.23L

(a) Draw a sketch graph showing how the force between two atoms separated by a distance x varies with x over the range up to about $x = 5$ atomic diameters. Using the same axes, draw a second graph to show how the potential energy of the system varies with x. Use your graphs to explain: (i) why most solids obey Hooke's law for small extensions; (ii) why a solid expands when its temperature is raised.

(b) The density of copper is $8.92 \times 10^3\,\text{kg m}^{-3}$ and one mole of copper atoms has a mass of 6.35×10^{-2} kg. Calculate: (i) the number of atoms in a cubic metre of copper; (ii) the separation (a) of nearest-neighbour atoms, given that the effective volume occupied by an atom is $(1/\sqrt{2})a^3$. The specific latent heat of sublimation of copper at 25°C is $5.34 \times 10^6\,\text{J kg}^{-1}$. (iii) Express this in joules per atom and explain how the result is related to the energy of the bonds between the copper atoms. (Take the value of the Avogadro constant N_a to be $6.02 \times 10^{23}\,\text{mol}^{-1}$.)

(**Oxford**: *and London, JMB, Cambridge, WJEC and NISEC*)

(a) See p. 33 and F2.5(a) and F2.5(b).

(b) (i) Remember that 1 mole of copper contains N_a atoms. As a first step, calculate the volume of 1 mole of copper, given its density and the mass of 1 mole. (ii) Use your answer to (b) (i) to calculate the volume occupied by a single atom of copper, then use the given formula to calculate separation a. (iii) To express the specific latent heat in J atom^{-1}, first of all express the specific latent heat in J mol^{-1} instead of J kg^{-1} by using the fact that each mole of copper has a mass of 6.35×10^{-2} kg. Then use the given value of N_a to convert to J atom^{-1}. See p. 89. Note that the final value for specific latent heat in J atom^{-1} is **not** the bond energy, but it is related to the bond energy. E6.5 shows how to calculate the bond energy, but since you are not given the number of neighbours (n) of each copper atom, then you are not expected to calculate the bond energy. Instead, show how the bond energy (e) and the number of neighbours (n) is related to the latent heat value in J atom^{-1}. Remember that the latent heat value in J atom^{-1} is the energy that must be supplied to a single atom of copper to remove it from the 'lattice' of copper atoms and take it away completely. Remember that energy $\frac{1}{2}e$ is given to the removed atom when each bond between it and a neighbour is broken. One final point; sublimation is change of state from a solid directly to a vapour.

5.24L

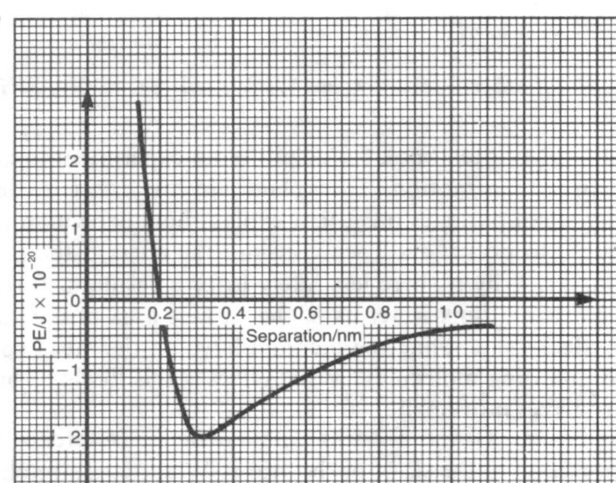

(*a*) The graph shows the variation of potential energy with separation for two atoms of a particular element. Explain how the following quantities can be obtained from the graph and give values for the element concerned:
(i) the equilibrium separation of the atoms;
(ii) the energy required to dissociate the atoms.
(*b*) Using the data on the graph, construct a graph of the variation of the force between the atoms with separation. Put numerical values on your graph where possible.
(*c*) Use your graph or the energy graph to discuss the elastic properties of the element in its solid form.
(*d*) Explain each of the following observations in terms of the microstructure of the solid concerned.
(i) A sheet of glass can be cut by scratching its surface and bending along the scratch.
(ii) Concrete is strong in compression but not in tension.
(iii) An annealed copper wire becomes stiffer when it is bent back and forth.

(**O and C***: *JMB*, London*, O and C Nuffield, Oxford, Cambridge, WJEC*)

(a), (b) and (c) See pp. 33-34

(d) (i) The scratch creates tiny cracks on the surface. When the sheet is bent, stress concentrates at the cracks so the cracks grow and join together. See your textbook for further details.
(ii) In tension, stress builds up at internal boundaries and cracks and fissures develop and grow inside the concrete. In compression, stress is transmitted through internal boundaries. See your textbook for further information.

5.25L This question is about explaining the flow of water from a tank and about designing an investigation of this flow. Approximately one-third of the marks are reserved for the investigation plan in part **(e)**.

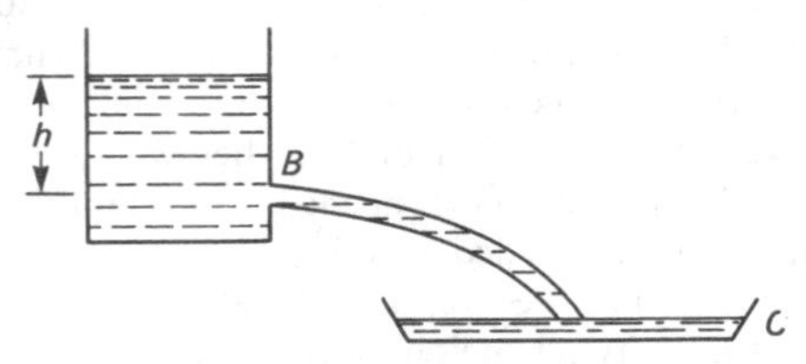

A student is investigating the flow of water through a hole at B in the vessel shown. He arranges a device to keep the head (h) of water constant.

(a) Explain each of the following steps (i), (ii) and (iii) in a theory which the student has worked out:
(i) The mass per second (m) is related to the volume per second (q) and the density (ρ) as follows: $m = \rho q$.
(ii) The value of q should be $q = Av$, where A is the area of cross-section of the stream of water and v is its velocity of flow at B.
(iii) The value of v should be given by $v = \sqrt{2gh}$.

(b) In order to check his theory, the student sets h at 9.0 cm and collects the water for 100 s; the mass collected is 6.91 kg. He knows that the diameter of the circular opening at B is 1.0 cm. Show that, if he assumes that the area A is the same as that of B, his measurements do not agree with his theory.

(c) The student decides that either v or A do not have the expected values. He decides to check on the velocity by making measurements of the drop y in the level of the jet when it is a distance x from the hole and using the equation $v = x\sqrt{g/2y}$. He finds from such measurements that v is within 5% of this theoretical value. Explain the theory of his method to check the velocity.

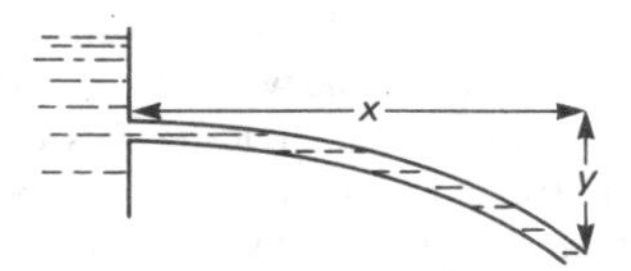

(d) Say what conclusions you could draw from the results so far and discuss the proposal that there *must* be a difference between the assumed and actual values of A. Consider the size of the difference that might be required and whether this could provide a reasonable explanation.

(e) Outline a plan for further experimental investigation designed to check your ideas in (d) and to extend the understanding of them as much as possible. Credit will be given for thinking of a variety of systematic tests, and for stating quite precisely what you would measure and what you would do with the measurements to relate them to your ideas about the situation. You should also consider how you would make the measurements and whether you could obtain adequate accuracy.

(**O and C Nuffield**: *and all other Boards*)

There is normally one long-answer question (for Nuffield) each year that tests your skill at handling algebra. This question is an example, and it follows the usual trend of describing an experiment or system that you will not have encountered during the normal teaching. Some candidates foolishly ignore such questions because they wrongly imagine that they have not covered the subject material. If you are good at manipulating algebra and equations, and found it easy to follow the proofs you have been shown during the course, then you should be able to score well on this type of question.

(a) (i) Mass = density × volume, so mass per second = ...
(ii) Water with velocity v in one second will produce a stream v long with cross-sectional area A from a nozzle of area A; this is a cylinder of length v, cross-sectional area A and volume... (iii) gh normally suggests potential energy in a uniform gravitational field (mgh), equate the pe lost by the water to the kinetic energy it gains and ...

(b) The practically obtained value of v from (a) (ii) you should expect to be less than the theoretical value calculated from (a) (iii); this may help you if you make a mistake in your calculations.

(c) This is a standard projectile problem; the water should have a constant horizontal velocity v and be falling vertically with a constant acceleration g. This enables the formula to be derived; see 1.2.

(d) Part (c) makes it clear that the velocity v is correct (within 5%) so the value of A used in the equation is wrong. Calculate what value of A should be used in (a) (ii) to give the correct value of q and compare with the measured value for A. Is it possible that error in measuring the diameter of the nozzle would account for the discrepancy? Perhaps 'friction' between the water and the tube at the nozzle effectively reduces the value of A which has been measured with good accuracy.

(e) With $\frac{1}{3}$ of the marks available for this part, clearly some depth is needed in this answer. Depending on how you have answered part (d) the emphasis of the answer to this part may vary. Key points in any set of experiments should involve varying both the nozzle diameter and the water height, as well as choosing suitable instruments to measure the diameter. You may choose to try and measure the actual diameter of the water stream as it leaves the nozzle; explain the analysis of the results as well as discussing the practical details and whether you can obtain sufficient accuracy.

5.26L

(a) Explain the terms 'lines of flow' and 'streamlines' when applied to fluid flow and deduce the relationship between them in laminar flow.

(b) State Bernouilli's equation, define the physical quantities which appear in it and the conditions required for its validity.

(c) The depth of water in a tank of large cross-sectional area is maintained at 20 cm and water emerges in a continuous stream out of a hole 5 mm in diameter in the base. Calculate: (i) the speed of efflux of water from the hole; (ii) the rate of mass flow of water from the hole.

Density of water $= 1.00 \times 10^3$ kg m^{-3}.

(**JMB*** *and AEB, Cambridge**)

(a) See p. 78.

(b) See E5.5 and subsequent comments.

(c) Sketch the arrangement, and consider two points as follows; point X on the water surface of the tank, and point Y at the hole. Since both points are open to the atmosphere, then the pressure at $X =$ pressure at $Y =$ atmospheric pressure. Measure levels from a horizontal line through the hole. Apply E5.5 to the situation, with terms for X on one side of the equation and terms for Y on the other side. To calculate the fluid speed at Y (ie, the speed of efflux), you need to make an assumption about the fluid speed at X. Base this assumption on the fact that the level at X would drop very slowly for a large tank.

To calculate the mass flow per second, use E5.4.

5.27L

(a) An air bubble is blown in water. Deduce an expression for the excess pressure inside the bubble.

(b) The bubbles in a bubble chamber are formed with an average radius of 1.0×10^{-6} m. They grow to a radius of 1.0×10^{-5} m in 2.0 μs. Calculate the mean rate at which the pressure in the bubbles is changing during their growth (surface tension of the liquid in the chamber $= 8.0 \times 10^{-3}$ N m^{-1}).

(c) Describe an experiment to measure the surface tension of a liquid which is based on measuring the excess pressure in a bubble. How may this experiment be developed to investigate the way in which the surface tension of the liquid varies with temperature? Explain why such an experiment is likely to be more suitable in measuring the ratio of surface tensions of a liquid at various temperatures than in measuring the value of its surface tension at a particular temperature. Explain qualitatively why it is likely that surface tension will decrease with temperature rise.

(London)

(a) See E5.11; you are asked to derive the expression, not just quote it.

(b) Apply the formula to find the excess pressure in each case, then find the change in pressure, hence the rate of change, stating whether this is an increase or decrease.

(c) Jaeger's method is described in most textbooks. A clearly labelled diagram should be drawn, together with an indication of how each measurement is taken and how the value of surface tension is determined from the measurements. Discuss the uncertainty in taking the bubble radius to be the same as the radius of the tube. At higher temperatures the average molecular separation will be greater, so how will this affect the intermolecular forces?

Unit 6 Heat and gases

Heat energy has many effects upon matter, both at the macroscopic level and at the microscopic (molecular) level. To understand the thermal properties of matter, the effects of energy upon molecules must be studied; such studies often require the use of statistical methods appropriate to handling huge numbers of molecules. Some indication of these methods is given at the end of Unit 1 as well as within this Unit. The study of the kinetic theory of gases also draws on Unit 1 since a thorough understanding of force and momentum is the key to the kinetic theory of gases. Revision of GCSE notes on the gas laws ought to be helpful as well. Finally the link between ideal and real gases can best be appreciated if the section of Unit 2 on intermolecular fields is re-read before tackling real gases in this Unit.

6.1 TEMPERATURE

Temperature is a measure of the degree of hotness of an object, and can be thought of as a measure of the mean ke of the random movements of its molecules. **Heat** is energy transferred on account of temperature difference. Since heat is a form of energy, it is measured in joules. Like work, heat should always be considered in the context of transfer of energy, so is discussed in terms like 'heat gained', etc. However, heat has no meaning for individual molecules, unlike work. In terms of energy, an object can possess either mechanical energy (ie, ke and pe due to bulk movement of the whole mass) or internal energy (ie, ke and pe due to molecular behaviour). An object can do work by using up some of its mechanical energy or some of its internal energy or both. Likewise, its energy can be transformed into heat released by the object.

Two objects in thermal contact will be at the same temperature if there is no resultant flow of heat between the two objects. If there is a resultant flow of heat from one object to the other, then the object that is the 'net' source of heat must be at a higher temperature than the object that accepts the heat energy (ie, the 'net' heat sink).

Temperature scales Every thermometer is based upon a physical property which changes continuously with change of temperature. Such a property is known as the **thermometric property**, and some examples include length (for a liquid-in-glass thermometer) or electrical resistance (for a resistance thermometer). Each thermometer must be calibrated so that readings of the thermometric property give numerical values of temperature on a chosen scale. The calibration requires measurement of the thermometric property at two standard degrees of hotness, known as **fixed points**.

The **Centigrade scale** of temperatures is based upon the melting point of pure ice as the lower fixed point (defined as 0°C), and steam temperature at 1 standard atmosphere of pressure as the upper fixed point (defined as 100°C). For a thermometer with thermometric property X, the following equation gives its temperature θ:

$$\text{E6.1} \quad \theta = \frac{(X_\theta - X_0)}{(X_{100} - X_0)} \times 100$$

where X_θ = value of the thermometric property X at θ°C, X_0 = value of X at ice-point, X_{100} = value of X at steam point.

By definition, all thermometers calibrated on the same scale give the same readings at the fixed points. Between the fixed points, readings will usually differ from one thermometer to another.

The **ideal gas scale** of temperatures serves as a standard scale, against which other temperature scales can be compared. The lower fixed point is **absolute zero** (by definition 0 Kelvin or −273.15°C), and the upper fixed point is the **triple point** of water (point at which ice/water/water vapour are in equilibrium) which is by definition 273.16 Kelvin or 0.01°C. The thermometric property here is the product (pressure × volume) of an ideal gas; in practice, PV at low pressure is used. On the ideal gas scale, temperature is calculated from the equation:

$$\text{E6.2} \quad T = \frac{(PV)}{(PV)_{\text{Tr}}} \times 273.16$$

where $(PV)_{\text{Tr}}$ = value of (PV) at the triple point, T = temperature (K).

The ideal gas scale gives temperatures that are identical to temperatures based on the **thermodynamic scale**, which is outlined on p. 94.

Types of thermometers The following table (F6.1) lists the more important features of 5 types of thermometers commonly used in scientific work. The constant-volume gas thermometer can be used to measure ideal gas temperatures directly, but above 90 K, other thermometers are more suitable in practice. The international practical scale of temperatures (IPTS) gives the type of thermometer which should be used for accurate scientific work between defined fixed points.

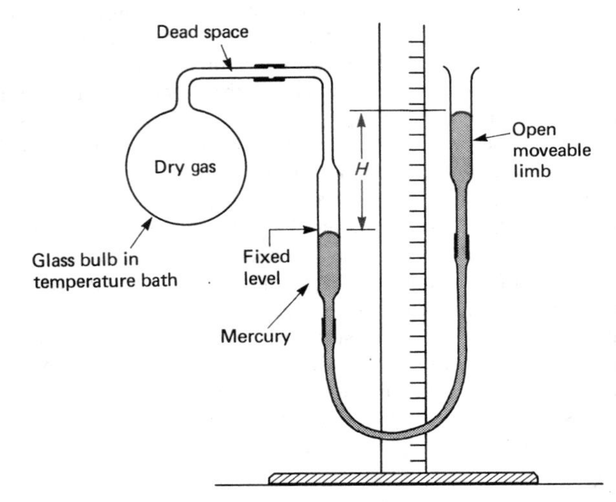

F6.2 Constant-volume gas thermometer

The Celsius scale, defined as T-273.15, is the scale achieved by calibrating a gas thermometer on the Centigrade scale.

F6.1 Types of Thermometers

Type of thermometer and its range	Thermometric property	Advantages	Disadvantages	Particular uses
Mercury-in-glass −39°C to 450°C (234 K to 723 K)	Length of column of mercury in capillary tube	**(i)** Quick and easy to use (direct reading) **(ii)** Easily portable	**(i)** Fragile **(ii)** Small size limits precision **(iii)** Limited range	**(i)** Everyday laboratory use where high accuracy is not required **(ii)** Can be calibrated against constant–volume gas thermometer for more accurate work
Constant-volume gas thermometer −270°C to 1500°C (3 K to 1750 K)	Pressure of a fixed mass of gas at constant volume. See F6.2	**(i)** Very accurate **(ii)** Very sensitive **(iii)** Wide range **(iv)** Easily reproducible	**(i)** Very large volume of bulb **(ii)** Slow to use and inconvenient	**(i)** Standard against which others are calibrated **(ii)** He, H_2 or N_2 used depending on range **(iii)** Can be corrected to the ideal gas scale **(iv)** Used as standard on IPTS below −183°C (90 K)
Platinum resistance −180°C to 1150°C (90 K to 1400 K)	Electrical resistance of a platinum coil. See p. 105 for the Wheatstone bridge method of measuring resistance	**(i)** Accurate **(ii)** Wide range	Not suitable for varying temperatures (ie, is slow to respond to changes)	**(i)** Best thermometer for small steady temperature differences. **(ii)** Used as standard on IPTS between −183°C and 630°C (90 K and 903 K)
Thermocouple −250°C to 1150°C (25 K to 1400 K)	Emf produced between junctions of dissimilar metals at different temperatures. See p. 104 for measurement of 'thermal' emfs	**(i)** Fast response because of low heat capacity **(ii)** Wide range **(iii)** Can be remote readings using long leads	Accuracy is lost if emf is measured using a moving-coil voltmeter (as may be necessary for rapid changes when potentiometer is unsuitable)	**(i)** Best thermometer for varying temperatures **(ii)** Can be made direct reading by calibrating galvanometer **(iii)** Used as standard on IPTS between 630°C and 1063°C (903 K and 1336 K)
Radiation pyrometer Above 1000°C (1250 K)	Colour of radiation emitted by a hot body	Does not come into contact with temperature being measured	**(i)** Cumbersome **(ii)** Not direct reading (needs a trained observer)	**(i)** Only thermometer possible for very high temperatures **(ii)** Used as standard on IPTS above 1063°C (1336 K)

6.2 HEAT CAPACITIES

Specific heat capacity (c) of a material is defined as the heat necessary to raise the temperature of unit mass by one degree, without change of state. The unit of c is usually $J\,kg^{-1}\,K^{-1}$, although mass in moles is frequently encountered so giving the unit of molar heat capacity as $J\,mol^{-1}\,K^{-1}$.

E6.3 $\Delta H = Mc\Delta\theta$

where ΔH = heat energy supplied or removed (J), M = mass (kg), c = specific heat capacity ($J\,kg^{-1}\,K^{-1}$), $\Delta\theta$ = temperature change (K or C°).

Note that the **heat capacity** of an object is the heat energy which will raise its temperature by one degree. The symbol C is used for heat capacity, and its unit is $J\,K^{-1}$. If the object is made from a single material of specific heat capacity c, then $C = Mc$, where M is the mass of the object.

Specific latent heat (l) of vapourization (or fusion or sublimation) is defined as the heat necessary to change the state from liquid to vapour (or from solid to liquid, or from solid to vapour directly) of unit mass of material, without change of temperature. The unit of l is usually $J\,kg^{-1}$, although $J\,mol^{-1}$ is sometimes used.

E6.4 $\Delta H = Ml$

where ΔH = heat energy (J), M = mass (kg) that changes its state, l = specific latent heat ($J\,kg^{-1}$).

Molecular interpretation of latent heat Heat supplied to an object will either change its temperature or change its state. Increased temperature mainly corresponds to increased ke of the molecules, whereas change of state involves energy associated with breaking or forming bonds between molecules. For example, when water freezes, latent heat is given off because bonds are formed; to melt ice, latent heat must be supplied because bonds must be broken. A rough estimate of the energy required to break a single bond can be obtained from E6.5, which relates bond energy to latent heat:

E6.5 $l = \frac{1}{2}Nne$

where l = specific latent heat ($J\,kg^{-1}$), N = number of molecules per unit mass (kg^{-1}), n = number of neighbours per molecule, e = bond energy (J).

The bond energy is equal to the minimum pe of a pair of molecules, as shown in F5.24. Note that if l is given in $J\,kg^{-1}$, then N must be expressed as the number of molecules per kg; if M is the relative molecular mass, and N_a is Avogadro's number, then the number of molecules per unit mass will be (N_a/M).

Experimental methods Most methods for determining specific heat capacities and specific latent heats of materials involve electrical heating nowadays. A typical circuit for supplying electrical power to a heater is shown in F6.3. A rheostat is used to maintain a constant value of current, recorded on the ammeter. The pd across the heater is recorded from the voltmeter, and the time for which electrical power is supplied must be measured. The electrical energy supplied is given by (current (I) × pd(V) × time (t)), where time t must be in seconds.

c of a solid by Nernst's method

The heat capacity of the container is not required, since the solid is suspended in an evacuated container with silvered walls, as in F6.4. The heating coil is also used as a resistance thermometer.

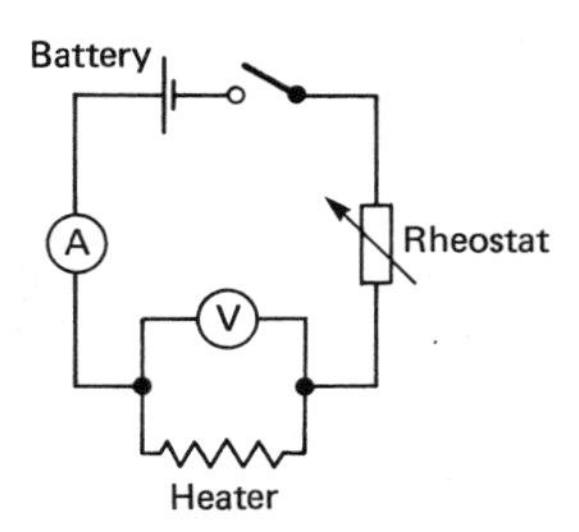

F6.3 Electrical heating

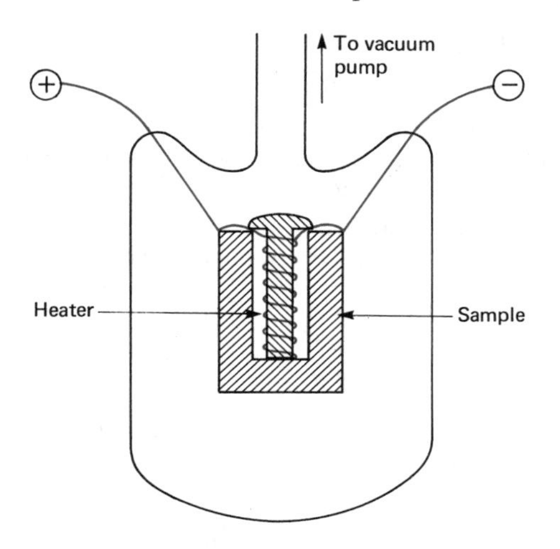

F6.4 Nernst's apparatus

E6.6 $IVt = Mc\Delta\theta$

where I = current (A), V = pd (V), t = time (s), M = mass of solid (kg), c = specific heat capacity of solid ($J\,kg^{-1}\,K^{-1}$), $\Delta\theta$ = rise of temperature of solid (K).

c of a liquid by electrical calorimetry

A measured quantity of liquid is heated electrically in an insulated calorimeter of known heat capacity (C). The temperature rise (θ), current, pd and heating time must all be measured. To calculate c, use:

E6.7 $IVt = Mc\Delta\theta + C\Delta\theta$

where C = heat capacity of calorimeter ($J\,K^{-1}$), M = mass of liquid (kg), c = specific heat capacity of liquid ($J\,kg^{-1}\,K^{-1}$), $\Delta\theta$ = temperature rise of liquid (K), I = current (A), V = pd (V), t = time (s).

The method can be extended to determine c of a solid by putting a known mass of the 'unknown' solid in the calorimeter, and then adding a known mass of liquid (in which the solid is insoluble) of known specific heat capacity.

c of a liquid by continuous flow calorimetry

The liquid is passed at a steady rate through an insulated tube containing an electrical heater, as in F6.5. For a measured rate of flow, electrical power is supplied at a constant rate so that steady temperatures θ_1 and θ_2 are recorded at the inlet and outlet; thus the electrical energy supplied per second (IV) is given by the equation:

$$(IV) = [\frac{m}{t}c(\theta_2 - \theta_1)] + H$$

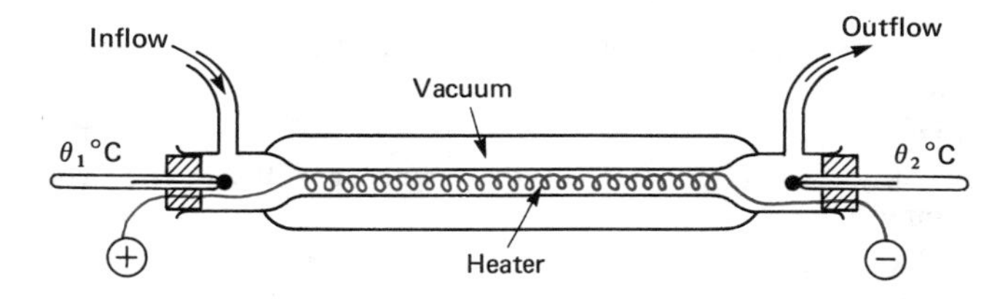

F6.5 Continuous-flow calorimeter

where m/t is the mass flow per second and H is the heat loss per second. The rate of flow is then altered (and re-measured) and the power supply is adjusted so that the inlet and outlet temperatures remain at the **same** values as before. Thus the new rate of supply of electrical energy $(IV)'$ is given by:

$$(IV)' = [(\frac{m}{t})'c(\theta_2 - \theta_1)] + H$$

where $(m/t)'$ is the new rate of flow of mass. Note that H is the same as before because the temperatures are the same as before. By subtraction, the following equation can be used to calculate c:

E6.8 $(IV)' - (IV) = [(\frac{m}{t})' - (\frac{m}{t})]c(\theta_2 - \theta_1)$

where V = pd(V), I = current (A), θ_2 = outlet temperature,

θ_1 = inlet temperature, c = specific heat capacity of liquid, $(\frac{m}{t})$ = mass flow per second ($kg\,s^{-1}$).

Note that the heat capacity of the container is not required, and that heat loss is accounted for by taking the difference between two rates, as in E6.8.

l of a liquid; The liquid is heated electrically in a jacketed container, as shown in F6.6. The vapour in the outer jacket cuts

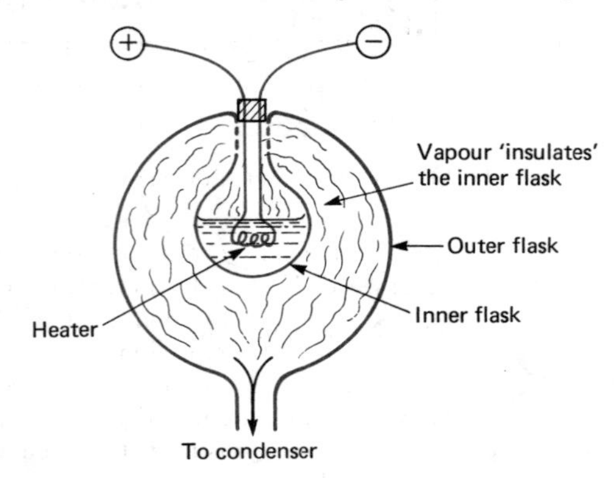

F6.6 Apparatus for latent heat of vapourization

down heat loss from the inner vessel, and condensed vapour is collected as shown in the diagram. The rate of condensation is measured for two different power supply rates; for each rate, the electrical energy supplied per second $(IV) = (\frac{m}{t})l + H$, where (m/t) is the rate of condensation of mass of vapour and H is the heat loss per second. Because H is the same at each rate (remember that H depends only upon the liquid temperature, which is the same at both rates because the liquid is at boiling point in each case), then by subtraction, the following equation gives l:

E6.9 $(IV)' - (IV) = [(\frac{m}{t})' - (\frac{m}{t})]\, l$

where I = current (A), V = pd (V), (m/t) = mass condensed per second ($kg\,s^{-1}$), l = specific latent heat ($J\,kg^{-1}$).

6.3 THERMAL COEFFICIENTS

For a physical property Z, the thermal coefficient of Z is defined by:

E6.10 $a = \dfrac{(Z_\theta - Z_0)}{Z_0\theta}$

where a = thermal coefficient ($°C^{-1}$), Z_0 = value of Z at 0°C, Z_θ = value of Z at θ°C.

Some of the more common examples of thermal coefficients include:

Linear expansivity

$a = (L_\theta - L_0)/(L_0\theta)$ = expansion/(length at 0°C × temp.rise from 0°C).

Cubic expansivity

$\gamma = (V_\theta - V_0)/(V_0\theta)$ where V is the volume, etc. The two coefficients are linked by the relationship $\gamma = 3a$.

Resistance

$a_R = (R_\theta - R_0)/(R_0\theta)$ where R is electrical resistance. For metals, resistance increases with increased temperature so a_R is positive. For semiconductors, resistance falls with increased temperature so a_R is negative.

Gas pressure

$a_P = (P_\theta - P_0)/(P_0\theta)$ where P is gas pressure. For an ideal gas, a_P is equal to 1/273.

6.4 THERMAL CONDUCTION

There are three methods of heat transfer:

1 Convection, which involves bulk motion of a fluid (liquid or gas), and is usually caused by hot fluid (being less dense) rising and displacing cold fluid.

2 Radiation, which involves emission and absorption of electromagnetic radiation.

3 Conduction, which takes place in solids and in fluids, regardless of any bulk motion of the fluid. Conduction, like convection, requires a temperature difference to be established within the material which conducts.

Consider heat conduction along a solid bar of uniform cross-section, as in F6.7, which has one end maintained at temperature θ_1 and the other at θ_2. If the sides of the bar are well insulated, heat flows along straight lines as shown, and the temperature falls uniformly along the axis, as on the accompanying graph.

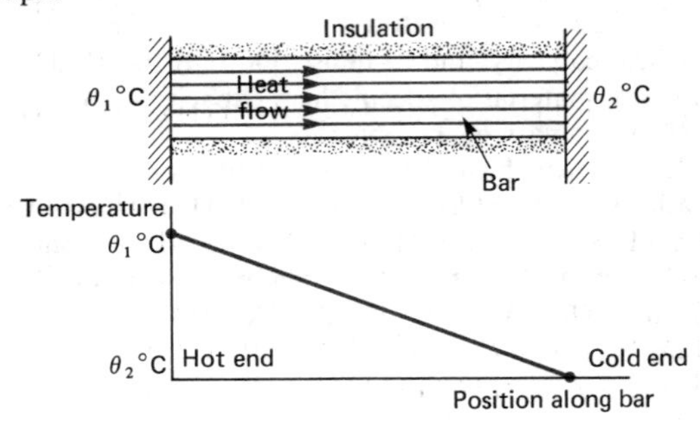

F6.7 Heat flow in a lagged bar

Temperature gradient is defined as the change of temperature per unit change of distance. The unit is $K\,m^{-1}$. For the insulated bar, the temperature gradient is constant and equals $(\theta_1 - \theta_2)/L$, where L is the bar length.

If the bar was not insulated along its sides, then the direction of heat flow would not be constant. The temperature gradient would be different at different positions along the bar, as shown in F6.8.

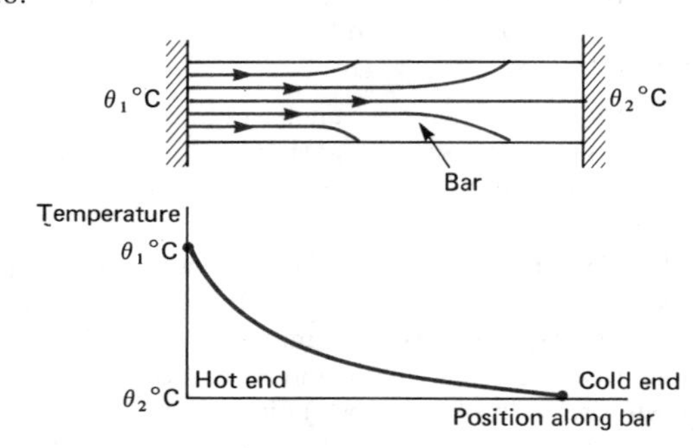

F6.8 Heat flow in an unlagged bar

For the insulated bar, the heat flow per second (dH/dt) is proportional to (i) the temperature gradient and (ii) the area of cross-section (A).

Introduction of a constant of proportionality (k) gives:

E6.11 $\dfrac{dH}{dt} = kA\,\dfrac{(\theta_1 - \theta_2)}{L}$

where (dH/dt) = heat flow/sec ($J\,s^{-1}$), A = area of cross-section (m^2), $\theta_1 - \theta_2$ = temperature difference (K or C°), L = length (m).

The constant k is known as the **coefficient of thermal conductivity** of the material of the bar, and it is defined as the heat flow per second per (metre)^2 of cross-section through which the heat flows normally, per unit temperature gradient along the flow direction. The unit of k is $W m^{-1} K^{-1}$.

Thermal conductors in series When two heat conductors are placed so that all the heat which flows through one then passes through the other, as in F6.9, the interface temperature (x) can be calculated by using E6.11 for each conductor separately, and then equating the heat flow per unit time of one body to the corresponding quantity for the other body, ie:

$$k_pA_p\,\frac{(\theta_1 - x)}{L_p} = k_qA_q\,\frac{(x - \theta_2)}{L_q}$$

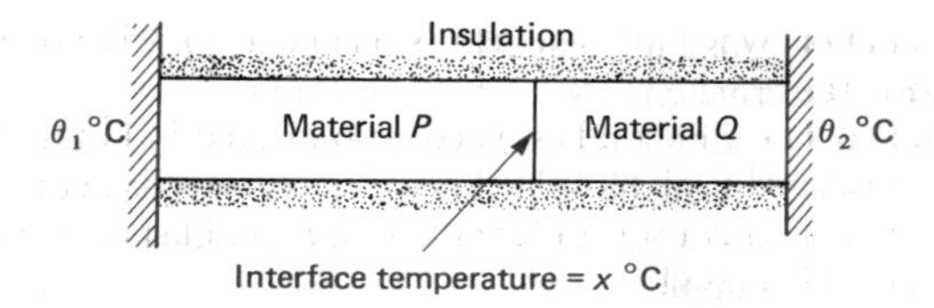

F6.9 Heat flow in two lagged bars in series

This situation is not unlike electrical conductors in series, in that a flow of heat requires a temperature difference, whereas a flow of electrical current requires a potential difference. Thus (dH/dt) is equivalent to current I, and $(\theta_1 - \theta_2)$ is equivalent to pd V; hence the quantity $\frac{L}{kA}$ is equivalent to electrical resistance R. The quantity $\frac{L}{kA}$ is therefore known as the **thermal resistance** of the material.

Comparison of the nature of conduction in good and poor conductors For solids that are poor conductors, heat transfer is due to lattice vibrations, so that when a solid gains heat at one point, the atoms near that point vibrate more. Thus the neighbouring atoms vibrate more, which then makes atoms further away vibrate more, etc., spreading the heat energy gradually through the solid.

Good conductors, such as metals, conduct both heat and electricity well. The reason is the presence of free (conduction) electrons. When one part of a metal is heated, the atoms in that part vibrate more, so free electrons in that part gain extra ke by collisions with atoms. Since the electrons are free to move through the solid, they move away and transfer their ke to other electrons (in other parts) when they collide. Transfer by lattice vibrations still takes place, but energy transfer by movement of free electrons is much greater.

6.5 THERMAL RADIATION

Thermal radiation is electromagnetic radiation, and is emitted by all objects at temperatures greater than absolute zero. The higher the temperature of an object, then the greater is the amount of radiation energy emitted per second. When a hot object is placed in a cold room, the object radiates to the room walls at a greater rate than it receives from the walls; thus, the object loses energy and cools until it is at the same temperature as the walls.

Absorption of radiation by a given surface depends upon:

1 the nature of the surface, in terms of smoothness and colour.
2 the quantity and quality (ie, wavelength) of the incident radiation. **Emission** of radiation by a given surface depends upon:
1 the surface temperature,
2 the nature of the surface.

When radiation is incident upon a surface, it is either absorbed, transmitted or reflected. Surfaces which are good absorbers are also good emitters of thermal radiation.

Black-body radiation A surface that absorbs all wavelengths of incident radiation will appear black. Such a surface can emit all wavelengths of radiation, and is known as a **black-body emitter**. The quality and quantity (sometimes called the 'spectral distribution') of the radiation from a black body emitter changes with temperature, as indicated by F6.10.

The total energy radiated away per second from a surface is given by **Stefan's law**:

E6.12 $W = \sigma e A T^4$

where W = radiated power ($J\,s^{-1}$ or W), A is the surface area (m^2), T = **absolute** temperature of the surface (K), e = surface emissivity, σ = Stefan's constant ($W\,m^{-2}\,K^{-4}$).

The **emissivity** is the ratio of black-body power to actual power for the same area and temperature. Therefore, $e = 1$ for a black body; for non-black bodies, e can vary from 0 to 1.

The net power radiated away from a black body (at temperature T) in cooler surroundings (at temperature T_0) can be written

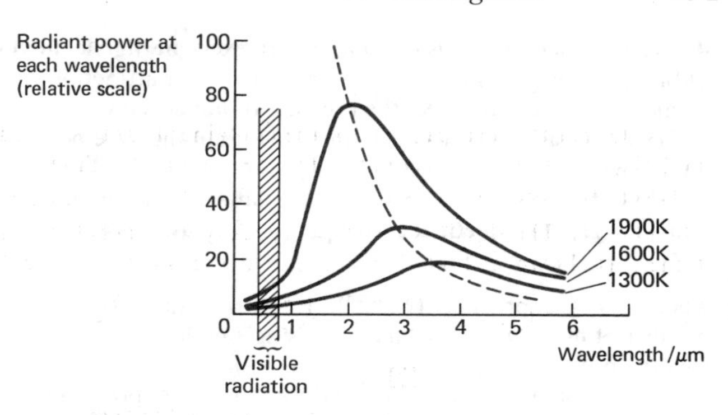

F6.10 Black body radiation curves

as $(\sigma A T^4) - (\sigma A T_0^4)$, where the last term is the power radiated from the surroundings and absorbed by the body. Note that both terms contain the surface area A of the body; the area of the surrounding surfaces do not enter the equation.

It is worth noting that the black-body radiation curves cannot be explained using classical wave theory; the idea of electromagnetic radiation in terms of photons leads to an explanation of the shape of the curves (and of Stefan's law).

6.6 IDEAL GASES

The word 'ideal' used in the study of gases has a special meaning in that it is used to describe gases under conditions such that the experimental gas laws are valid. The gas laws are summarized by the **ideal gas equation:**

E6.13 $PV = nRT$

where P = gas pressure ($N\,m^{-2}$), V = gas volume (m^3), n = number of moles, T = temperature in Kelvins, R = molar gas constant ($J\,mol^{-1}K^{-1}$).

The gas constant R, expressed in terms of moles (ie, $J\,mol^{-1}\,K^{-1}$) has the same value for all ideal gases, and this important point is based on the experimental result that 1 mole of any gas under 'ideal' conditions at stp (ie, 0°C and 1 standard atmosphere of pressure) occupies $0.0224\,m^3$. Thus, substitution of $V = 0.0224\,m^3$, $T = 273$ K, $n = 1$ mole and $P = 1.02 \times 10^5$ $N\,m^{-2}$ (= 1 standard atmosphere) gives the value of R as 8.31 $J\,mol^{-1}\,K^{-1}$.

From the ideal gas equation, you ought to be able to see how the individual gas laws 'fit in':

1 **Boyle's law**; n and T are fixed, hence PV = constant (ie, $P_1V_1 = P_2V_2$).
2 **Charles' law**; n and P are fixed, hence V/T = constant (ie, $V_1/T_1 = V_2/T_2$).
3 **Combined gas law**; n is fixed, hence PV/T = constant (ie, $\frac{P_1V_1}{T_1} = \frac{P_2V_2}{T_2}$)

Brownian motion When smoke particles in air are observed with the aid of a microscope, the path of each individual particle is seen to continually change direction in an unpredictable way. The cause of these unpredictable movements is the random bombardment of the smoke particles by much smaller (hence invisible) fast-moving gas molecules. Any one smoke particle will be subjected to uneven bombardments over its surface, since the gas molecules move about at random.

The kinetic theory of gases The theory is based upon several assumptions:

1 A gas consists of **point molecules**: in other words, the volume of the molecules themselves compared with the container volume is negligible.

2 Gas molecules are in a state of **continual random motion**: this is based upon Brownian motion observations.

3 Molecules collide **elastically** (ie, no loss of total ke) with one another, and with the container walls: if the collisions were inelastic, then the gas would need to be supplied with energy to maintain its motion.

4 Except when in collision, molecules exert **negligible forces** upon one another: in other words, intermolecular attraction is so small that it does not affect the pressure of impacts on the walls.
5 The **duration** of a collision is much less than the time between collisions.

With the above assumptions, the laws of mechanics and statistics give the **kinetic theory** equation for gas pressure:

E6.14 $PV = \frac{1}{3} Nm\overline{c^2}$

where P = gas pressure ($N\,m^{-2}$), V = gas volume (m^3), N = number of molecules, m = mass of a single molecule, $\overline{c^2}$ = mean square speed of the gas molecules ($m^2\,s^{-2}$).

You should consult your textbook for the proof of the equation; in studying the proof in your textbook, you should try to see at which stage each assumption is used. Most proofs consider N molecules in a rectangular box, and treat the proof in two main stages:

1 Use of the laws of mechanics to prove that the pressure on any one face is given by $PV = Nm\overline{u^2}$, where $\overline{u^2}$ is the mean value of (velocity components normal to that face)^2. Assumptions 1,3,4 and 5 are used here. Note that $\overline{u^2}$ is a shorthand way of writing $(u_1^2 + u_2^2 + u_3^2 + \cdots + u_N^2)/N$. Your textbook proof should show you how to deal with this stage in detail.
2 Use of the laws of statistics to prove that $\frac{1}{3}\overline{c^2}$ is equal to $\overline{u^2}$, so giving the equation as stated. The use of statistics is based upon the assumption that the molecules are in random motion. Again, look at your textbook to see the details of this stage.

Note that the **root mean square** (rms) **speed** of the gas molecules is the square root of the mean value of the (molecular speeds)^2; in other words;

$$\text{the rms speed} = \sqrt{\left(\frac{c_1^2 + c_2^2 + c_3^2 + \cdots + c_N^2}{N}\right)}$$

where $c_1, c_2, c_3, \ldots c_N$ are the individual values of molecular speeds.

Kinetic energy and temperature By combining the ideal gas equation (E6.13) with the kinetic theory equation (E6.14), the mean ke of a gas molecule, $\frac{1}{2}m\overline{c^2}$, can be shown to be equal to $\frac{3R}{2N_a}T$, where N_a is Avogadro's number (and is not to be confused with the number of moles, n). Thus the total ke of 1 mole of an ideal gas is $\frac{3}{2}RT$.

6.7 THERMODYNAMICS OF IDEAL GASES

When **heat** is supplied to an isolated system, the amount of heat energy equals the increase of internal energy plus the mechanical work done by the system. This general rule is known as the **First law of thermodynamics**. Note that a gas can only do work when its volume changes; the **work done by a gas** = **pressure** × **change of volume**, since pressure is force/area and volume change is area × distance moved (so that work done = force × distance = pressure × volume change).

E6.15 $\Delta Q = \Delta U + P\Delta V$

where ΔQ = heat supplied (J), ΔU = change of internal energy (J), P = gas pressure ($N\,m^{-2}$), ΔV = change of gas volume (m^3).

The symbol Δ is used to mean a change, and if the change is a +ve value, then that is always taken to mean an increase. If the change is a −ve value, then that is taken to mean a decrease. For example, if ΔV is given or calculated as $-4.0\,m^3$, then the volume has decreased by $4\,m^3$; again, if the gas is supplied with 15 J of heat energy and the gas volume increases by $10^{-4} m^3$ at a constant pressure of $10^5\,N\,m^{-2}$, then it follows by use of E6.15 that the internal energy increases by 5 J.

Constant volume No work is done since ΔV is zero, so heat supplied goes entirely to increase the internal energy of the gas. If c_v is the molar specific heat capacity at constant volume, then a temperature rise of ΔT for n moles of gas will require heat supply $\Delta Q = nc_v\Delta T$. Therefore, the increase of internal energy will be $nc_v\Delta T$.

Constant pressure Since $PV = nRT$, then $P.\Delta V = nR.\Delta T$, since the pressure is constant. Thus, the work done $= nR{\cdot}\Delta T$. Now the increase of internal energy for temperature rise ΔT is $nc_v{\cdot}\Delta T$. It follows that to increase the temperature at constant pressure, the heat supplied must be equal to $(nR{\cdot}\Delta T) + (nc_v{\cdot}\Delta T)$. If c_p represents the molar specific heat capacity at constant pressure, then the heat supplied is $nc_p{\cdot}\Delta T$ so that $nc_p{\cdot}\Delta T = nc_v{\cdot}\Delta T + nR{\cdot}\Delta T$, giving,

E6.16 $c_p - c_v = R$, where c_p = molar specific heat capacity at constant pressure, R = molar gas constant, c_v = molar specific heat capacity at constant volume. In other words, c_p is greater than c_v because heat supplied at constant pressure is used to increase the internal energy **and** to enable the gas to do work when it expands. **Isobaric changes** occur at constant pressure.

Adiabatic changes An **adiabatic change** is one in which no heat enters or leaves the system (ie, $\Delta Q = 0$). For example, the sudden expansion of an ideal gas from a high pressure container into low pressure would not involve heat transfer on account of the rapidness of the expansion. The work done by the gas in expanding comes from a fall in the internal energy of the gas.

Consider an adiabatic change of an ideal gas which involves a small change of temperature ΔT. The work done is $P{\cdot}\Delta V$ and the change of internal energy ΔU is $nc_v{\cdot}\Delta T$. Because the change is adiabatic, then $\Delta Q = 0$ so use of E6.15 gives $P{\cdot}\Delta V = -nc_v{\cdot}\Delta T$. By combining this equation with the ideal gas equation $PV = nRT$ then the following equation for adiabatic changes can be derived:

E6.17 PV^γ = constant, where $\gamma = c_p/c_v$

Adiabatic changes for an ideal gas can be compared with **isothermal changes** ($\Delta T = 0$) by appropriate sets of curves on a $P-V$ diagram, as in F6.11. Note that the adiabatic curves are always

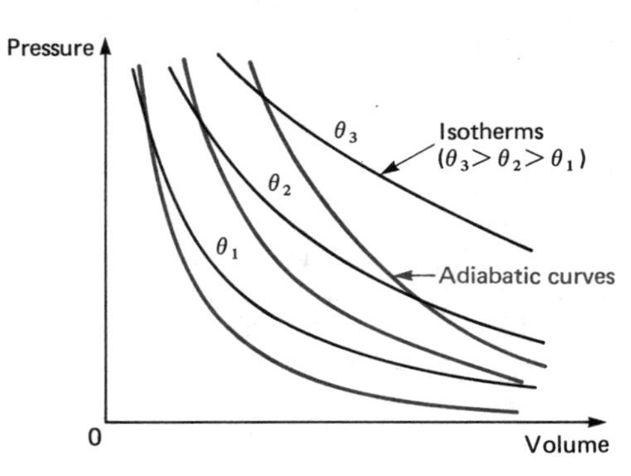

F6.11 Adiabatics and isotherms

steeper than isotherms, since the slope of an isotherm is $(-P/V)$ whereas the slope of an adiabatic curve is $(-\gamma P/V)$, and γ is always greater than 1.

6.8 REAL GASES

The term **'real gas'** is used to describe gases under conditions where the ideal gas equation does **not** hold.

Andrews' experiments The behaviour of gases under a wide range of pressures and temperatures was thoroughly investigated by Andrews. The apparatus used, as in F6.12, was designed to withstand pressures up to 400 atmospheres. The volume of the gas under test (eg, CO_2) could be measured directly from the length it occupies. The pressure was calculated by first measuring the volume of the 'control' gas (eg, dry air), and then using Boyle's law to calculate its pressure. The test gas tube was contained in a water bath so that sets of readings were made at different temperatures.

The results are usually presented as a set of PV isotherms (ie, curves of $y = P$, $x = V$ at constant temperature for each curve) as in F6.13. The important features of the graph are:

1 The **critical isotherm**, above which the gas cannot be liquified, no matter how much pressure is applied.
2 The **critical point**, which is the point of inflexion on the critical isotherm (ie, the point on the critical isotherm at which the slope is zero).

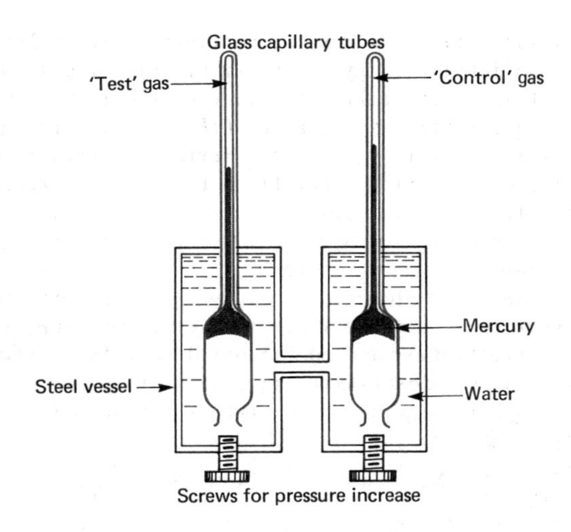

F6.12 Apparatus for testing gases

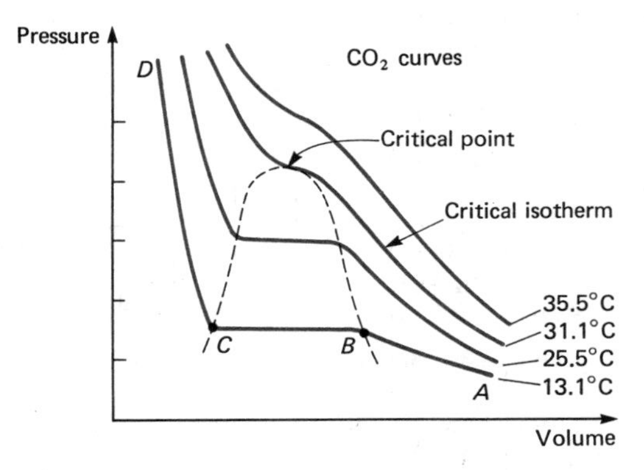

F6.13 P–V curves for carbon dioxide

3 Each isotherm below the critical isotherm exhibits common features. For example, consider the isotherm *ABCD*. (i) Section *AB* is unsaturated vapour; (ii) section *BC* is saturated vapour in equilibrium with liquid. At *B*, the material is 100% vapour; at *C*, it is 100% liquid, (iii) section *CD* corresponds to liquid.

The **critical temperature** is defined as the temperature above which a gas cannot be liquefied by increase of pressure alone. Note that the word 'vapour' is reserved for the gas when its temperature is less than the critical temperature. Each gas has its own critical temperature; for example, the critical temperature of CO_2 is 31°C, as shown in F6.13.

Explanation of real gas behaviour An ideal gas is one that obeys the ideal gas equation. To illustrate the **difference** between ideal gas and real gas behaviour, Andrews' results can be plotted as isotherms of $y = PV$ against $x = P$, as in F6.14.

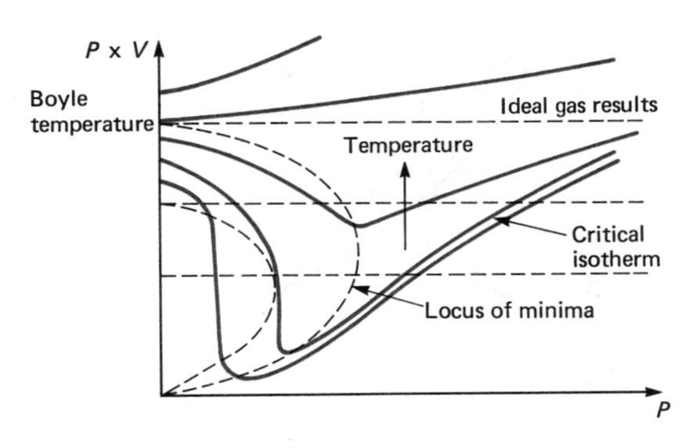

F6.14 *PV* against *P* curves

Ideal gas behaviour would give results as indicated by the dotted lines parallel to the *x*-axis (since $PV = \text{constant} \times T$). Actual gas behaviour only conforms to ideal gas behaviour for a limited temperature range (different for different gases) at low pressure. The temperature for which the curve of *PV* against *P* is initially straight is known as the **Boyle temperature** (T_B).

1 $T > T_B$: *PV* increases as *P* increases.
2 $T < T_B$: *PV* falls then rises as *P* increases.

The explanation of the experimental results lies in reconsidering two of the assumptions of kinetic theory, as follows: (i) point molecules – at high density, the actual volume of the molecules themselves is a significant fraction of the measured gas volume; (ii) intermolecular attraction – at low temperatures, the molecules move more slowly, so the weak forces of attraction (Van der Waal's forces) between molecules have longer to act. Hence the force of impact on the container sides is reduced.

Above T_B, as *P* increases, the density increases and the molecules become closer to one another. Consequently, the measured volume increasingly overestimates the available volume. Hence *PV* rises.

Below T_B, as *P* increases, the molecules again come closer together. However, intermolecular attractions become significant, making the measured pressure less than the 'kinetic theory' pressure. Hence *PV* falls until further increase of pressure causes the volume factor to become more important; then *PV* stops falling and starts to rise.

The **Van der Waals' equation** is an attempt to represent real gas behaviour:

E6.18 $$\left(P + \frac{a}{V^2}\right)(V - b) = RT$$

where *b* depends on the molecular volume, and $\frac{a}{V^2}$ represents the pressure reduction caused by intermolecular forces between molecules hitting the container walls and 'bulk' molecules (ie, those far from any surfaces). The constant *b* is called the **co-volume**, and is about 4 × the actual volume of the molecules.

6.9 VAPOURS

A **saturated vapour** is one that is in equilibrium with liquid. In molecular terms, equilibrium means that the number of molecules per second that transfer from the liquid to the vapour state (evaporation) is equal to the number per second transferring from vapour to liquid (condensation).

Saturated vapour pressure (svp) is: (i) independent of volume (as shown by F6.13), (ii) increased by increase of the temperature, (iii) equal to external pressure when the liquid boils.

In molecular terms, reduction of the volume of a saturated vapour makes the vapour molecules temporarily more concentrated. Therefore, the number per second that return to the liquid state increases. Since the rate of transfer in the opposite direction is unchanged, then the concentration of molecules in the vapour state falls until it returns to its value before the volume reduction. So svp returns to its previous value (ie, value before volume reduction).

The increase of svp with increased temperature is shown in F6.15. When the temperature of the liquid rises, the number of

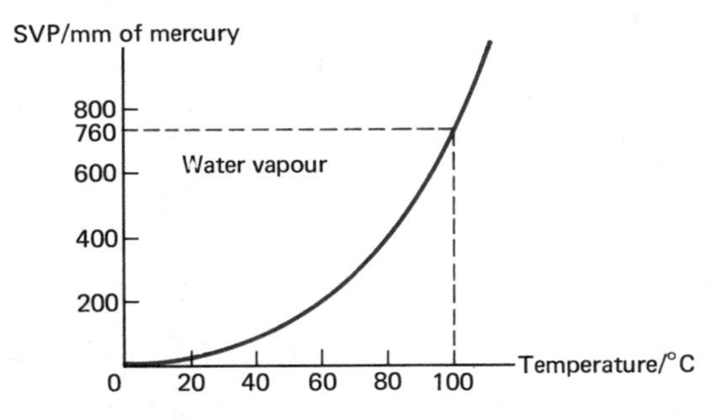

F6.15 Variation of svp with temperature for water

molecules per second which transfer from the liquid to the vapour increases so the concentration of vapour molecules increases. Thus, two effects (increased concentration and increased mean speed) give rise to increased pressure.

6.10 ENTROPY AND EQUILIBRIUM

Energy transformations Conservation of energy is insufficient to explain why certain processes happen whereas others do not. For example, the energy of an oscillating simple pendulum is gradually converted into internal energy of its surroundings; the reverse process (increase of pendulum energy by using some of the internal energy of its surroundings) is most unlikely to happen, even though conservation of energy could be satisfied.

Another example is provided by considering the direction of heat flow between a hot and a cold object. When unaided, heat always flows from hot to cold, never in the reverse direction. Heat flow in either direction could satisfy the requirement of conservation of energy; however, the chance of heat flow from cold to hot is negligibly small when unaided.

Chance The law of chance (in the form that all possibilities are equally likely to occur) applies when dealing with large numbers of objects that behave in a random manner. Its role in the kinetic theory of gases can be extended beyond the development of the kinetic theory equation (E6.14) by considering an ideal gas of N molecules in a rectangular container as shown in F6.16.

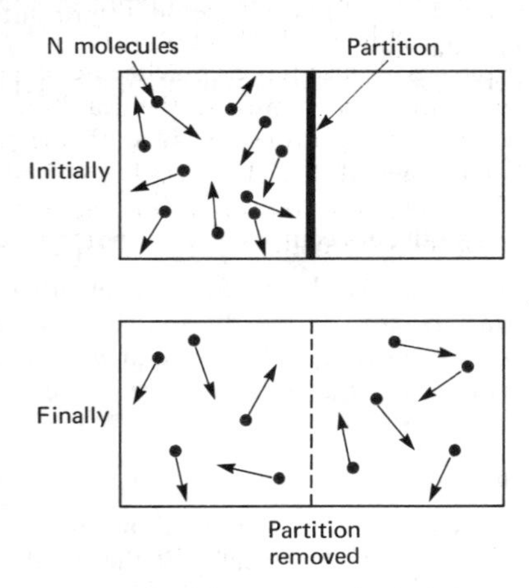

F6.16 Diffusion

Initially, all N molecules are confined to one half of the box; once released the molecules can move to any position of the box, so each molecule can be found in either half with equal likelihood (ie, a 1 in 2 chance of being found in the left half). The probability of **all** N molecules returning to the initial half at the same time is 1 in 2^N, and since N is typically of the order of Avogadro's number, then such a probability is vanishingly small.

Thermal equilibrium Some understanding of the role of chance in thermal equilibrium can be achieved by using the **Einstein model** of a solid. Each atom is considered to have a set of equally spaced energy levels; each energy level can be occupied by a single quantum of energy, or by several quanta of energy. Further, because of random interactions between atoms, the quanta move about at random. The number of quanta of any one atom will be independent of the number of quanta of any neighbouring atom. Therefore, there will be many **ways** of distributing the quanta amongst the atoms so that at any instant, the actual distribution could be any one of these ways (ie, they are all equally probable). For example, the possible ways of distributing 4 quanta amongst 3 atoms are shown in F6.17.

First atom	••••			•••	•••	•		•		••	••		••	•	•
Second atom		••••		•		•••	•••		•	••		••	•	••	•
Third atom			••••		•		•	•••	•••		••	••	•	•	••
	400 pattern			310 pattern						220 pattern			211 pattern		

← Individual ways →

F6.17 Distributing 4 quanta to 3 atoms

In this example, one of the **distribution patterns** occurs in more ways than any other pattern. For a large fixed number of atoms (N) and of quanta Q, the quanta move about at random so that the arrangement of quanta amongst atoms continually changes from one **way** to another. However, as in the simple example, one set of ways (corresponding to a given **distribution pattern**) occurs more frequently (in fact, far more frequently for large N,Q) than any other pattern; the equilibrium distribution will therefore conform to that pattern.

The equilibrium distribution can be represented by a graph of y = number of atoms with n quanta, $x = n$; the shape of the curve is always an **exponential decrease** as in F6.18. Probability theory gives the following equation for curve steepness:

$$\text{E6.19} \quad \frac{\text{No. of atoms with } n \text{ quanta}}{\text{No. of atoms with } (n+1) \text{ quanta}} = 1 + \frac{N}{Q}$$

where N = total No. of atoms, Q = total No. of quanta.

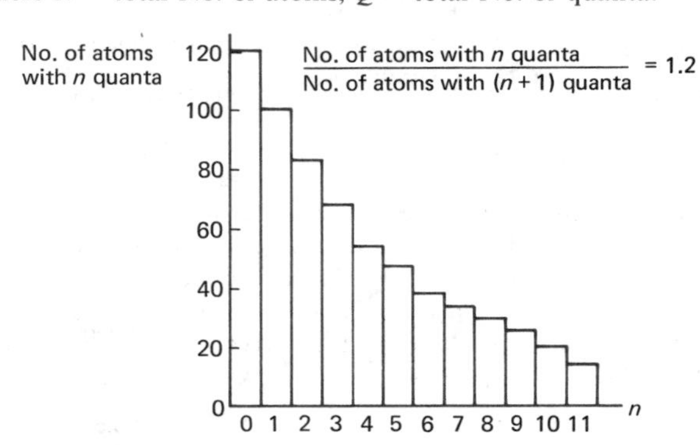

F6.18 Distribution of quanta within a solid

The equation shows that if the total number of atoms is fixed, then when more quanta are supplied (by heat flow into the solid), then the exponential decrease becomes less steep. Thus, high temperature corresponds to a less steep exponential curve; in other words, at high temperature, energy is shared out more evenly, although even sharing is unattainable since this would correspond to infinite temperature.

Achieving equilibrium Consider two Einstein solids A and B in individual thermal equilibrium at different temperatures. Assume equal numbers of atoms for each, and that A is initially hotter than B. Hence A has initially more quanta than B, so that initially there are more ways of distributing quanta in A than in B.

Now suppose that A and B are brought into thermal contact with one another so that quanta can move in either direction (ie, A to B or B to A), but the total number of quanta in A and B must stay the same. The variation of the number of ways of distributing quanta in A and B separately, and together, is shown in F6.19.

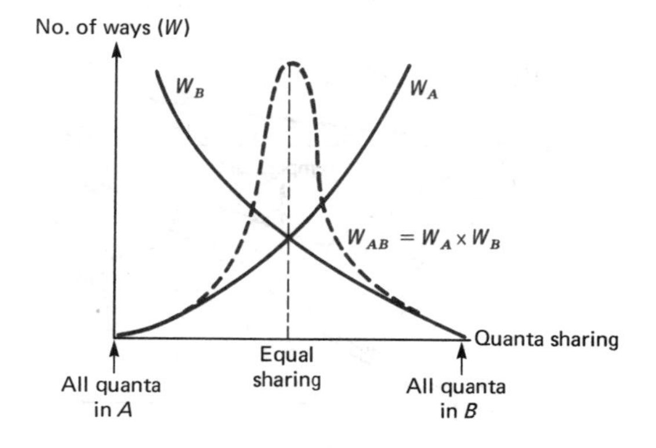

F6.19 Sharing quanta between two solids

Equilibrium between A and B is attained when the number of ways, W, of distributing quanta between A and B together reaches a maximum, and then the distribution pattern for A (and hence temperature) will be the same as for B. In other words, the final distribution pattern is the most frequent of the most ways. Thus heat flows from hot to cold because that is the most likely direction.

Further insight can be achieved by using the result from probability theory that the ratio:

$$\frac{W \text{ for } (Q+1) \text{ quanta amongst } N \text{ atoms}}{W \text{ for } Q \text{ quanta amongst } N \text{ atoms}} = 1 + \frac{N}{Q}$$

so if $N_A = N_B = 400$, and Q_A and Q_B have initial values of 400 and 100, respectively, then the transfer of 1 quantum from A to B makes W_A change by $\times \frac{1}{2}$ (since $1 + \frac{N}{Q} = 2$ for A, so loss of one quantum means that W_A falls by a factor of 2); at the same time B gains a quantum from A so W_B changes by $\times$ 5. Thus W_{AB} changes by $\times \frac{5}{2}$. Now consider the initial situation again, with the transfer of a single quantum from B to A this time; W_{AB} changes by $\times \frac{2}{5}$ by the same method of reasoning as above. Hence, the transfer from A to B of 1 quantum is $6\frac{1}{4}$ times (ie, $\frac{5}{2} \times \frac{5}{2}$) more likely than transfer in the reverse direction.

Entropy Equilibrium involves maximizing the number of ways in which energy can be distributed through a system. In other words, **disorder** always tends to increase. This general result goes far beyond the simple model of a solid used here, and it is a determining factor in deciding the direction of energy flow involving heat. The disorder of a system is measured by its **entropy**, as follows:

E6.20 $S = k \ln W$

where S = the entropy (J K^{-1}), W = the number of ways of arranging the system, ln is the natural log, k = the Boltzmann constant ($= R/N_a$) (J K^{-1}).

Entropy is defined in this way so that the entropy change of a system whose internal energy changes by ΔU is given by $\Delta S = \Delta U/T$. Thus, for A and B in contact, with the temperature of A, T_A, initially greater than the temperature of B, T_B, transfer of internal energy ΔU from A to B causes the entropy of A to fall by $\Delta U/T_A$ and that of B to rise by $\Delta U/T_B$. Since $T_A > T_B$, then the total entropy increases. The flow of energy from A to B continues until the total entropy stops increasing (ie, until the number of ways of distributing the energy has reached its maximum value). The system will then be in thermal equilibrium. Note that the unit of entropy is J K^{-1}.

UNIT 6 QUESTIONS

6.1M A coil of wire has a resistance of 2.00Ω, 2.80Ω and 3.00Ω at temperatures of 0°C, 100°C and t, respectively. What is the value of t on the scale defined by this resistor?
A 20°C. B 25°C. C 80°C. D 105°C.
E 125°C.

(**London**: *all other Boards except O and C Nuffield, Scottish H, Scottish SYS*)

Substitute resistance values into E6.1 to give t.

6.2M A metal cube of length 10.0 mm at 0°C (273 K) is heated to 200°C (473 K). Given its coefficient of linear expansion (a) is 2×10^{-5} K^{-1}, the % change of its volume is:
A 0.1. B 0.2. C 0.4. D 1.2. E 2.0.

(—: *AEB only*)

EITHER calculate the new length using E6.10, then determine the new volume, etc., OR use the fact that the volume coefficient = 3 × the linear coefficient, and use E6.10 to determine the fractional increase of volume directly. Remember that the % change is 100 × the fractional change.

6.3M The heat loss per second from a hot object is proportional to the temperature difference between the object, at temperature θ, and its surroundings, at temperature θ_0. The temperature of the hot object was measured at regular intervals and the results plotted to give the graph. Which of the following statements is/are correct?

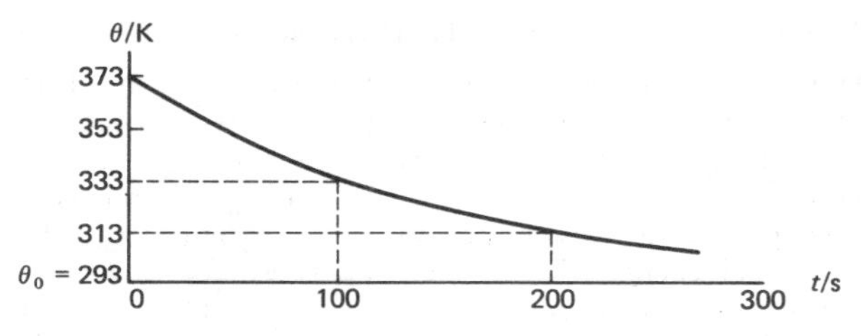

1 The initial rate of loss of heat is twice the rate of heat loss at 100 s.
2 The heat loss in the first 100 s is the same as the heat loss from t = 100 s to t = 200 s.
3 The temperature of the object will become equal to that of the surroundings after 300 s from the start (ie, at t = 300 s).
A **1** only. B **1** and **2**. C **1**, **2** and **3**. D **1** and **3**.
E **3** only.

(—: *all Boards*)

For **1**, remember that the y-scale also represents rate of loss of heat (since the rate of loss of heat is proportional to $(\theta - \theta_0)$).

For **2**, since the y-scale also represents the rate of loss of heat, then the area under the curve represents the heat loss.

For **3**, the temperature drops by 40 K in the first 100 s, then by 20 K in the next 100 s. How much will it drop by in the next 100 s? It would need to drop by another 20 K to fall to the temperature of the surroundings, so is a 20 K drop likely?

6.4M Which of the following quantities must be determined in order that the thermal capacity of a body may be calculated when the specific heat capacity of the body is known?
A Emissivity. B Latent heat. C Mass.
D Temperature. E Thermal conductivity.

(**London**: *and all other Boards except SEB H, SEB SYS, NISEC and WJEC*)

Thermal capacity is the heat needed to raise the temperature of the whole (mass) of a body by one degree.

6.5M When an ideal gas undergoes an adiabatic change causing a temperature change δT:
1 there is no heat gained or lost by the gas.
2 the work done is equal to the change in internal energy.
3 the change in internal energy per mole of the gas is $C_v \delta T$ where C_v is the molar specific heat capacity at constant volume.
A If **1**, **2**, **3** correct. B If **1**, **2** correct.
C If **2**, **3** correct. D If **1** only.
E If **3** only.

(**AEB** June 80: *JMB and all other Boards except O and C, Nuffield, SEB H, SEB SYS, NISEC*)

What is an adiabatic change? See p. 92 if necessary.

6.6M An unlagged cubic tank containing hot water loses heat to its surroundings at a rate of 900 W. This loss is reduced to 60 W if all the faces of the tank are covered with a layer of lagging. What will be the rate of loss of heat if one face is left unlagged? (The temperatures of the water and surroundings are unaltered. You may assume that heat is lost only from the faces and that the rate of loss of heat from a face is unaffected by whether it is vertical or horizontal, top or bottom.)
A 210 W. B 200 W. C 190 W. D 160 W.
E 150 W.

(**Cambridge**: *and all other Boards*)

Since the unlagged tank loses heat at a rate of 900 W, calculate the heat loss per second from each unlagged face.

Since the lagged tank loses heat at a rate of 60 W, calculate the heat loss per second from each lagged face.

Use your calculations to determine the heat loss per second from 5 lagged and 1 unlagged face.

6.7M A composite rod of uniform cross-section has aluminium and copper sections of the same length in good thermal contact.

The ends of the rod, which is well-lagged, are maintained at 100°C and at 0°C, as shown in the diagram. The thermal conductivity of copper is twice that of aluminium.

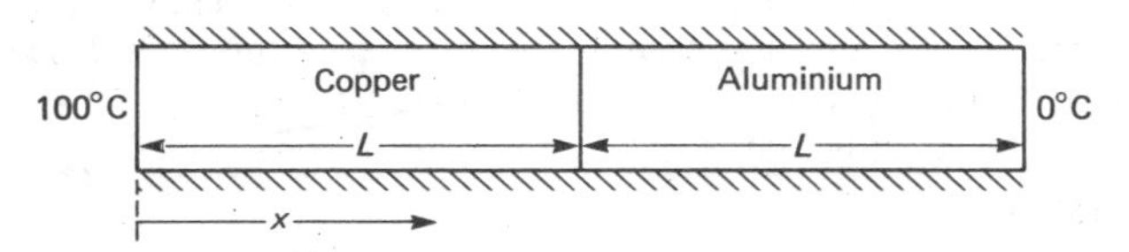

Which one of the following graphs represents the variation of temperature t with distance x along the rod in the steady state?

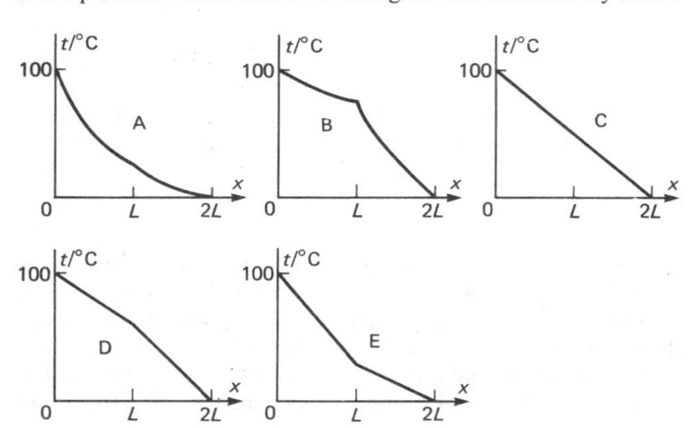

(**Cambridge**: *and all other Boards except SEB H and SEB SYS*)

There are two key points to consider:
(i) With no heat loss from the sides, is the temperature gradient constant or not? See p. 90 and F6.7 and F6.8 if necessary.
(ii) Bearing in mind that copper is a better heat conductor than aluminium, is the junction temperature likely to be more than, equal to, or less than 50°C?

6.8M Which of the following statements regarding thermal radiation is NOT correct?
A A good absorber of thermal radiation is also a good emitter.
B As the temperature of a black-body emitter rises, the frequency at which the radiated energy is a maximum decreases.
C The energy emitted by an opening in the wall of an enclosure is nearly identical with that which would be emitted by a black body at the same temperature as the wall of the enclosure.
D The rate at which radiant energy is emitted from a black body is proportional to the fourth power of its Kelvin temperature.
E When a body is in equilibrium with its surroundings, the rate of emission of thermal radiation is equal to the rate of absorption.

(**London**: *Cambridge and Oxford**)

Alternatives A, B, D and E are discussed on p. 91. A hole in an enclosure is the usual form of a practical black-body radiator. (NB. Pay particular attention to what is plotted on the *x*-axis of F6.10)

6.9M A mass of an ideal gas undergoes a reversible isothermal compression. Its molecules will then have, compared with the initial state, the same:
1 root mean square velocity.
2 mean momentum.
3 mean kinetic energy.
Answer:
A if **1, 2, 3** correct. B if **1, 2,** correct.
C if **2, 3** correct. D if **1** only. E if **3** only.

(**London**: *all other Boards except O and C Nuffield, SEB SYS*)

What does isothermal mean? See p. 92 if necessary. For 1 and 3, see p. 92. For 2, remember that momentum is a vector (ie, its direction is important) and that we are dealing with a very large number of molecules moving with random motion. What will the average momentum be?

6.10M Which one of the graphs below best illustrates the relationship between the internal energy U of an ideal gas and the temperature T of the gas in K?

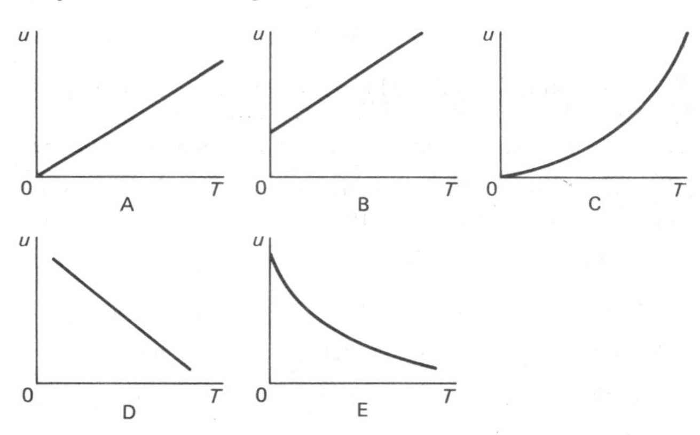

(**AEB** June 81: *and all other Boards except SUJB*)

For an ideal gas the internal energy is the total kinetic energy of random motion of the molecules. See p. 92 if necessary.

6.11M The kinetic theory of gases gives the formula $PV = \frac{1}{3} Nm\overline{v^2}$, for the pressure P exerted by a gas enclosed in a volume V. The term Nm represents:
A the mass of a mole of the gas.
B the mass of the gas present in the volume V.
C the average mass of one molecule of the gas.
D the total number of molecules present in volume V.
E the total number of molecules in a mole of the gas.

(**SEB H**: *and all other Boards*)

See E6.14 if necessary.

6.12M Two molecules of a gas have speeds of 1 km s^{-1} and 9 km s^{-1}, respectively. What is the root mean square speed of these two molecules?
A 2 km s^{-1}. B $\sqrt{3}$ km s^{-1}. C 4 km s^{-1}.
D $\sqrt{41}$ km s^{-1} E $\sqrt{82}$ km s^{-1}.

(**SEB H**: *and all other Boards*)

See p. 92 if necessary for the formula for rms speed in terms of individual speeds.

6.13M An ideal gas is heated at constant volume until its pressure doubles. Which one of the following statements is correct?
A The mean speed of the molecules doubles.
B The number of molecules doubles.
C The mean square speed of the molecules doubles.
D The number of molecules per unit volume doubles.
E The rms speed of the molecules doubles.

(—: *all other Boards except O and C Nuffield*)

Assume the mass of the gas is unchanged. Use the kinetic theory equation E6.14 to relate the various factors above to the pressure change at constant volume.

6.14M Which of the following pressure (P)–volume (V) diagrams best describe the behaviour of a gas which is returned to its original state after first being expanded adiabatically, then heated at constant volume and finally compressed isothermally?

(**AEB**: *and London, WJEC, Oxford, Cambridge and O and C*)

P V A P V B P V C

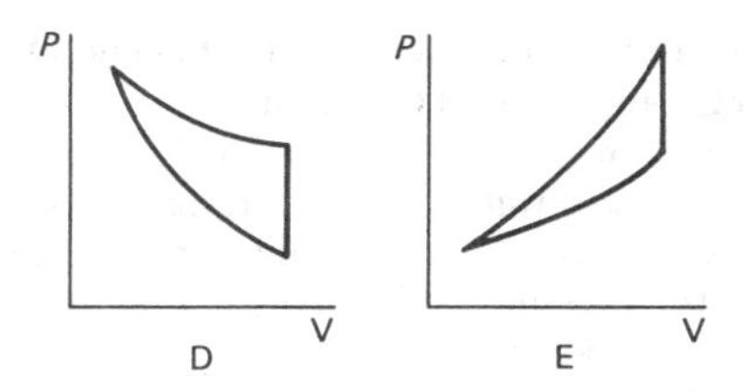

For this problem you need to be able to distinguish between the $P-V$ curves for adiabatic and isothermal changes. Remember that for an ideal gas, the curve for an adiabatic change is steeper than for an isothermal change. Take care to put the sequence of events in correct order.

6.15M Which of the following statements about the distribution of quanta of energy amongst the oscillators of an Einstein solid is/are correct?

1 In equilibrium, if the number of oscillators of a given distribution which have 1 quantum each is $0.8 \times$ the number of oscillators of that distribution with zero quanta, then the number of oscillators with 2 quanta will be $0.64 \times$ the number of oscillators with zero quanta in that distribution.
2 The hotter the solid, the steeper is the distribution.
3 Equilibrium is attained when each oscillator has the same number of quanta.

A **1** only. B **1** and **2**. C **1**, **2** and **3**. D **2** only.
E **2** and **3**.

(—: *O and C Nuffield*)

You must consider the distribution curve of F6.18 and how the steepness relates to the temperature. See p. 94 if necessary.

6.16S A clock has a brass pendulum that may be considered as a simple pendulum and which has a period of exactly 2 s when the temperature is 15°C. How many seconds will the clock lose or gain if it is maintained at a temperature of 25°C for one day? (Linear expansivity of brass $= 18 \times 10^{-6}\,\mathrm{K^{-1}}$.)

(**AEB** *only*: June 81)

Use the formula for linear expansion given on p. 90. Write down an expression for the length at 25°C in terms of the length at 0°C, and a similar expression for the length at 15°C in terms of the length at 0°C. Then obtain an expression for the ratio of the length at 25°C to that at 15°C without the length at 0°C in the expression. Next, use the standard formula for the time period of a simple pendulum (E1.7) to obtain an expression for the ratio of the time period at 25°C to that at 15°C. Hence calculate the time period at 25°C in s.

The 'time lost' (or gained?) can then be calculated from the number of oscillations in 1 day, and remembering that each oscillation moves the clock dial forward by 2 s at any temperature.

6.17S

(a) In an air-conditioned building the air is changed continuously at a rate that is equivalent to two complete changes per hour. In cold weather fresh air drawn in must be warmed up to a comfortable room temperature. (i) Use the following data to calculate the rate at which heat must be supplied to warm the air for one room in the building:

Dimensions of the room	$= 5.0\,\mathrm{m} \times 4.0\,\mathrm{m} \times 3.0\,\mathrm{m}$
Temperature of the fresh air drawn in	$= 5.0°\mathrm{C}$
Temperature of the room	$= 20.0°\mathrm{C}$
Average density of the air	$= 1.3\,\mathrm{kg\,m^{-3}}$
Specific heat capacity of air	$= 1.01 \times 10^3\,\mathrm{J\,kg^{-1}\,K^{-1}}$

(ii) State whether you would expect the actual rate of heating to be greater or less than the value you have calculated and give a reason to justify your statement.

(b) Give any two reasons why it is necessary to refer to the average density of the air in **(a)** (i).

(c) Outline an experiment you would carry out to measure the density of air.

(**SEB H**: *and all other Boards except SEB SYS*)

(a) (i) Firstly, calculate the mass of air in the room. Given that there are two complete changes per hour, calculate the mass of air to be heated up in 1 s. Then use E6.3 to calculate the heat supplied per second. (ii) You must consider heat losses due to conduction and radiation.

(b) Density of air depends upon pressure and temperature. There would almost certainly be a temperature difference between the top and the bottom of the room.

(c) See your textbook.

6.18S In this question you are expected to make sensible estimates for various quantities and then combine them in calculating the required answer. Give your answer to a justifiable number of significant figures and show the units of your estimated quantities and of your final answer.

What would be a typical mean rise in temperature of the brake-blocks when the brakes are applied to bring a bicycle to rest?
(i) List the quantities that you have estimated and need to enable you to calculate the answer.
(ii) Carry out the calculation of the estimate, using the quantities that you have listed in (i).
(iii) A practical test showed a higher temperature rise than you have calculated, give two possible reasons for this.

(**O and C Nuffield**: *NISEC and all other Boards*)

In any estimation it is first necessary to think about the physics of the problem. Where does the energy that heats up the brake blocks come from? Is it possible to calculate the energy stored by a moving bicycle (and its rider)?

For part **(i)** you are going to need to estimate the mass of the bicycle and rider, the speed of the bicycle, the mass of the four brake blocks, the specific heat capacity of rubber.

For part **(ii)** you must equate the kinetic energy lost to heat energy gained by the rubber. Don't forget to put in all the units and quote the final answer either to the nearest power of ten (order of magnitude) or to ONE significant figure if you consider that your answer is unlikely to be wrong by a factor of '10'. To estimate the one difficult quantity (the specific heat capacity of rubber) use about $\frac{1}{3}$ of the specific heat capacity of water (the specific heat capacity of water is known to be an unusually high value compared with other materials).

To answer part **(iii)** examine your calculation to part **(ii)** and decide which quantities affect the temperature change and the ways in which two of them could be altered to give a higher temperature rise. You should also ask yourself whether all the energy put into brakes is from stopping the bicycle, or is there another energy source? In our answer it has been assumed that the bicycle only has translational kinetic energy, but perhaps it also has rotational ke (see 1.8)? What if the bicycle were going downhill as it was brought to a halt?

6.19S

(a) (i) A steady current, I, flows through a conductor of resistance, R, when the steady potential difference between its ends is V. Write down an equation for the current. (ii) The ends of a perfectly lagged bar of length, L, and cross-sectional area, A, are maintained at steady temperatures, θ_1 and θ_2 where $\theta_1 > \theta_2$. If the thermal conductivity of the material of the bar is k, write down an expression for the rate of flow of heat along the bar in the steady state. (iii) A property of the bar called its 'thermal resistance', R_θ, can be defined in a similar way to the resistance of the conductor. By comparing your equations stated in (i) and (ii), deduce an equation for R_θ.

(b) The base of an aluminium kettle is 2.4 mm thick and has a cross-sectional area of $0.020\,\mathrm{m^2}$. The inside surface is coated with a uniform layer of scale 0.50 mm thick. When water is boiling inside the kettle at a steady temperature of 100°C, heat flows normally through the base at a rate of 2.0 kW. By considering the thermal resistance of the arrangement, or otherwise, calculate the temperature of the underside of the kettle base.

Thermal conductivities of aluminium and the scale are $240\,\mathrm{W\,m^{-1}\,K^{-1}}$ and $1.0\,\mathrm{W\,m^{-1}\,K^{-1}}$, respectively.

(**JMB**: *and all other Boards except SEB H and SEB SYS*)

(a) (i) See p. 102 if necessary. (ii) See E6.11 if necessary. (iii) Thermal resistance is discussed on p. 91.

(b) Thermal resistance method. Calculate the thermal resistance of the aluminium, then of the scale. Since all the heat flow through the scale also passes through the aluminium, then the scale and the aluminium are 'in series'; hence, add their individual thermal resistances to give the total thermal resistance. Then, multiply the total thermal resistance by the 'heat current' (ie, the rate of flow of heat) to give the temperature difference across the scale and aluminium. If the kettle is heated internally, then the underside temperature will be lower than 100°C, but if heated externally (via the base), then the underside temperature will be higher than 100°C; since the form of heating is not specified, then either answer is acceptable. See p. 91 for further details of this method.

Otherwise: Let the interface temperature be X. Assume internal heating. The temperature difference across the scale is $(100 - X)$, so given the rate of heat flow through the scale is 2 kW, use E6.11 to calculate X. Then the temperature across the aluminium can be written as $(X - Y)$ where Y is the underside temperature (to be determined). Use E6.11 applied to the aluminium to calculate Y.

6.20S Find the rms speed of the air molecules in a room under typical conditions (pressure 10^5Pa, density of air 1.3 kg m^{-3}.) When ammonia is released, it takes several seconds for the smell to be noticed by a person in another part of the laboratory. Explain why there is no real contradiction between this experimental fact and the answer to the first part of the question.

(**WJEC**: *and all other Boards except O and C Nuffield*)

First part: use E6.14. Assume $R = 8.3$ J mol^{-1} K^{-1}.

Second part: you should explain the process of diffusion in molecular terms. Use your textbook to find out why gas molecules moving at high speeds (of the same order as the speed of sound in air) only spread out slowly when released into another gas. The key lies in collisions between gas molecules.

6.21S At a temperature of 100°C and a pressure of 1.01×10^5 Pa, 1.00 kg of steam occupies 1.67 m^3, but the same mass of water occupies only 1.04×10^{-3} m^3. The specific latent heat of vaporisation of water at 100°C is 2.26×10^6 J kg^{-1}. For a system consisting of 1.00 kg of water changing to steam at 100°C and 1.01×10^5 Pa, find:
(a) the heat supplied to the system.
(b) the work done by the system.
(c) the increase in internal energy of the system.

(**Cambridge**: *and London, AEB, Oxford, JMB, NISEC*)

(a) Use E6.4.

(b) For work done by a gas, see p. 92. Calculate the volume change from the volume of steam and the volume of water (for mass = 1.00 kg in both cases).

(c) Use E6.15.

6.22S State the relation between pressure and volume at constant temperature for **(a)** an ideal gas, **(b)** a saturated vapour.

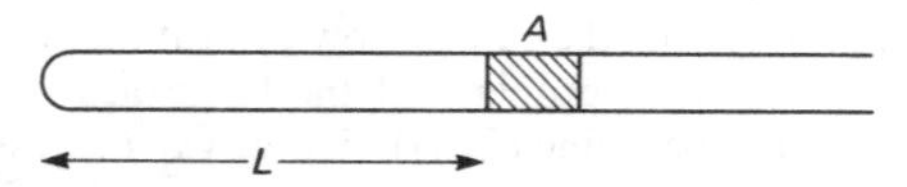

A long uniform horizontal capillary tube, sealed at one end, and open to the air at the other, contains air trapped behind a short column of water A. The length L of the trapped air column at temperatures 300 K and 360 K is 10 cm and 30 cm, respectively. Given that the vapour pressures of water at the same temperatures are 4 kPa and 62 kPa, respectively, calculate the atmospheric pressure.

(**SUJB**: *and AEB and NISEC*)

(a) For an ideal gas, see p. 91.

(b) For a saturated vapour, see p. 93.

To calculate atmospheric pressure (X), you need to assume that the 'trapped' column L contains air (assumed to be an ideal gas) saturated with water vapour. Thus the total pressure inside the tube is equal to the 'dry air' pressure + svp of water vapour. Next, you must consider the relationship between the pressures on either side of the water column A; on the open side, the pressure is atmospheric. When equilibrium is attained at each temperature, the pressure on the closed side (dry air pressure P + svp of water vapour) must balance the atmospheric pressure. Hence, write down the dry-air pressure at each temperature in terms of atmospheric pressure X and the svp at that temperature. Since the volume of the dry air is in proportion to the length L, then with dry air pressure given in terms of X, with volume and temperature known, you can use the combined gas law (see p. 91 if necessary) to write down an equation for X and so determine X.

6.23L Explain what is meant by a **scale of temperature**, discussing how a scale would be established.
The following table gives an example of a thermometer, the associated physical property and the measurement taken. Make a list giving the entries for the blank spaces (a)–(d) in the table.

Instrument	Physical property	Measurement taken
Constant-volume gas thermometer	Pressure change of gas	Difference in manometer levels supported by the pressure of a fixed volume of nitrogen.
Liquid-in-glass thermometer	(a)	(b)
Thermocouple	(c)	(d)

Why is it **not** convenient to use a constant-volume gas thermometer in most practical situations? To what is it actually put to use?
The masses of hydrogen and oxygen atoms are 1.66×10^{-27} kg and 2.66×10^{-26} kg, respectively. What is the ratio of the 'average' speed of hydrogen and oxygen atoms at the same temperature? What is usually meant by the term 'average' speed in this case?
For the thermocouple give, with reasons, a situation for which it would be particularly suitable.

(**London**: *and all other Boards except O and C, Nuffield and SEB SYS*)

To explain what is meant by a 'scale of temperature', your answer should explain thermometric property, fixed points and the defining equation (E6.1).

For (a) to (d), see F6.1.

The advantages and disadvantages of the constant volume gas thermometer are discussed in F6.1.

For the ratio of 'average' speeds, etc, assume that 'average' used here means rms speed. From E6.13 and E6.14, combine the two equations to eliminate PV so that you can establish the link between rms speed and molecular mass at constant temperature. Given the mass ratio, use the link to determine the ratio of rms speeds. The meaning of rms speed is given on p. 92.

For the advantages of a thermocouple, see F6.1.

6.24L (a) Describe, with the aid of a labelled diagram how you would use a thermocouple and a suitable potentiometer circuit to measure an unknown temperature between ice point and steam point. Include details of how you would calibrate the apparatus at 0°C and 100°C and how you would work out the temperature from the measurements. It is not necessary to give details of apparatus to attain ice point and steam point.
(b) A digital millivoltmeter accurate to 0.1 mV could be used to measure the e.m.f. of the thermocouple. Such a meter has a high input resistance. Suppose this type of meter is used with a thermocouple that produces an e.m.f. of 4.0 mV at 100°C. Is this a more precise method of measuring temperature than using the potentiometer method?
(c) Outline a practical situation in which the thermocouple is (i) more suitable, (ii) less suitable than a mercury-in-glass thermometer for measuring temperature.

(d) (i) The resistance of a platinum coil may be used to determine temperature. Outline how the resistance of such a coil can be measured using a Wheatstone bridge.
(ii) Such a coil has a resistance of 75.6 Ω at ice point, 78.0 Ω at steam point and 76.1 Ω at an unknown temperature θ. Work out the value of θ on the centigrade scale of the resistance thermometer and explain why a correction is necessary to give θ on the celsius scale.

(—: *all Boards except O and C Nuffield, London, SEB H and SEB SYS*)

(a) See p. 104 for the measurement of thermal e.m.f.s
(b) The meter has a high input resistance so negligible current passes through it. The potentiometer circuit at balance is the same in this respect. However, if the balance length at steam point is of the order of 1000 mm, each mm of balance length corresponds to a temperature interval of 0.1°C. In practice, the balance point could probably be located to within 2 or 3 mm. Use this to work out how precisely temperature can be determined and compare this with the precision of the millivoltmeter method.
(c) and (d) See p. 105 if necessary.

6.25L
(a) Describe a continuous-flow calorimeter experiment for the determination of the specific heat capacity of a liquid. State two particular advantages of this method. Explain how a correction may be made to allow for heat losses.

(b) Explain why, in dealing with gases, it is necessary to take into account the conditions of pressure and volume when considering heat capacities. Why is this not generally necessary in the case of solids and liquids?

(c) A beam of electrons, in an evacuated tube, impinges upon a silver target of mass 0.60 g, initially at 15°C. The speed of the electrons is $1.5 \times 10^7\,\mathrm{m\,s^{-1}}$ and 1.0×10^{16} strike the target each second. Assuming that all their energy is converted to heat, and that there are no heat losses, how much time will elapse before all the silver melts?
(Mass of electron $= 9.0 \times 10^{-31}$ kg; specific heat capacity of silver $= 240\,\mathrm{J\,kg^{-1}\,K^{-1}}$; specific latent heat of fusion of silver $= 1.1 \times 10^5\,\mathrm{J\,kg^{-1}}$; melting point of silver = 960°C.)

(**AEB** Nov 81: *(c) all boards except NISEC, SEB H and SYS*)

(a) A clearly labelled diagram is required. It should include the electric circuit and any device used to ensure a constant rate of flow of liquid. Your description of the method should stress the precautions taken to minimize sources of inaccuracy. In particular, you should mention that a reasonably large temperature difference is desirable and that the values of current and voltage should be read at regular intervals and subsequently averaged (maintain constant current using rheostat and record voltage readings regularly).

The method is discussed on p. 89. Consider the following points: (i) do the temperatures to be recorded change with time? (ii) is the heat capacity of the container required? (iii) is a cooling curve correction procedure (see textbook) necessary to deal with heat losses?

(b) See E6.15 and the subsequent comments. Remember that for liquids and solids the change of volume V can be normally considered negligible.

(c) If the time elapsed for complete melting to take place is t seconds then the energy delivered by the electron beam is given by (ke of each electron $\times$ the number striking the target per second $\times$ time t). By the conservation of energy this is equal to the sum of heat energy required to raise the temperature of the silver to its melting point (see E6.3) and the energy subsequently required to melt it (see E6.4).

6.26L
(a) (i) Describe how you would determine the coefficient of thermal conductivity of copper. Your account should include details of how each measurement is taken. (ii) State one way in which the behaviour of the apparatus used differs from that assumed in calculating the thermal conductivity. What is the cause of the difference and how does it affect the result?

(b) The cabin of a light aircraft can be regarded as an approximately rectangular box, sides 1.5 m by 1.5 m by 2 m. The windows are made of perspex, 10 mm thick, and have a total area of 3 m^2; the remainder of the cabin wall is made of thin aluminium-alloy sheet lined with insulating material, 20 mm thick. The cabin heater is able to maintain a temperature of 20°C in the cabin when the outside temperature is −10°C.
Assuming that the temperature difference across the aluminium alloy may be neglected, calculate: (i) the power of the heater; (ii) the percentage of the total energy which is lost through the perspex.
Explain why, in practice, the power of the heater needed would be much less than that calculated.
(Take the thermal conductivity of perspex to be $0.2\,\mathrm{W\,m^{-1}\,K^{-1}}$, and that of the insulating material to be $3.5 \times 10^{-2}\,\mathrm{W\,m^{-1}\,K^{-1}}$.)
(**Oxford**: *and AEB, London, O & C NISEC, and Cambridge)*

(a) (i) Your textbook ought to have a suitable method. (ii) When you calculate thermal conductivity (k), you need to know the heat flow per second, the area of cross-section and the temperature gradient. The first quantity (heat flow per second) assumes that the measured value of heat flow per second is the actual heat flow per second all the way along the sample; in other words, heat loss per second from the sides is assumed negligible. You should discuss whether or not heat loss causes the calculated value of k to be greater or smaller than the value you would obtain if there was no heat loss.

(b) (i) The heater power equals the heat flow per second through the perspex + the heat flow per second through the insulation; use E6.11 to calculate the heat flow per second through each of the two materials in turn. (ii) Use the values calculated in (b) (i).

To explain why the heater power in practice would be much less than your calculated value, consider the effect of still air either side of the cabin walls. Outside the cabin, the air behind (and possibly the underside of) the cabin would be relatively undisturbed and would provide another layer of insulation between the cabin wall and the cold air further away. Inside the cabin, even the presence of a very thin layer of still air against the walls will act as an effective layer of further insulation. Don't forget that people act as heaters!

6.27L State the condition necessary for (a) heat, and (b) electrical charge, to flow along a solid.
Write down equations which give the rate of flow in each case, and from them point out analogies between relevant thermal and electrical quantities.
Explain on the microscopic scale why metals are good conductors of both heat and electricity.
The external wall of a brick house is of area 16 m^2 and thickness 0.3 m. The indoor and outdoor temperatures are 20°C and 0°C, respectively. Find the rate at which heat is lost through the wall. What is the rate of loss when the internal surface of the wall is covered with expanded polystyrene tiles of thickness 20 mm, and what is the temperature at the brick-tile interface?
(Thermal conductivity of brick $= 0.5\,\mathrm{W\,m^{-1}\,K^{-1}}$; thermal conductivity of expanded polystyrene $= 0.03\,\mathrm{W\,m^{-1}\,K^{-1}}$.)

(**WJEC**: *and all other Boards except SEB H and SEB SYS*)

See p. 91 for analogies between heat flow and flow of electric charge. The equations required are E6.11 and E7.7.

Conduction of heat in metals is explained microscopically on p. 91.

To calculate the rate of loss of heat, use E6.11; the temperature difference is the same in °C as in Kelvins so there is little point here in converting temperatures into K.

With the polystyrene as above, start by drawing a cross-section of part of the wall, marking the temperature on the outside of the brick as 0°C, on the outside of the polystyrene as 20°C and at the interface as X°C.

To determine the heat loss per second, the straightforward method is to use E6.11 to obtain an expression for the heat loss/second through the brick in terms of X, and then repeat the procedure to obtain another expression for the heat loss/second through the polystyrene in terms of X also. Since the heat flow/second through the brick is the same as that through the polystyrene, then equate the two expressions to find X. Then use the value of X in either expression for heat flow to complete the problem.

An alternative method for the heat loss/second with polystyrene is to follow the 'thermal resistance' method on p. 98. Calculate the thermal resistance of the brick, and then of the polystyrene; then add the two values to give the total thermal resistance (since the brick and the polystyrene are conductors in series). To obtain the heat flow/second, divide the temperature difference across the two conductors by the total thermal resistance (similar to dividing potential difference by resistance to give current) to obtain the 'heat current' (ie, the heat flow/second).

6.28L (*a*) Explain why gases may be used to define a standard scale of temperature.

The boiling point of liquid nitrogen (N_2) is said to be 77 K. Suggest briefly how you would use gas thermometry to verify this statement, given a suitable supply of the liquid.

(*b*) In this part of the question, you should assume that the atmosphere of the Earth is at a constant temperature of 293 K.

A 'scale height' H for gases in the atmosphere is defined as the height at which the gravitational potential energy above sea level equals the mean translational kinetic energy of the gas molecules.

(i) Calculate H for N_2 gas (mass 0.028 kg mol $^{-1}$) and for Oxygen$_2$ gas (mass 0.032 kg mol^{-1}). State your assumptions.

(ii) The scale height of a gas also determines the variation of pressure p of the gas with height h. The theoretical expression is $p = p_0 e^{-3h/2H}$, where p_0 is the pressure at sea level. Use this expression to derive the variation with height of the number density of the gas (i.e. the number of gas molecules per unit volume).

Assuming that the atmosphere at sea level contains 20% O_2 and 80% N_2, estimate the percentage composition of O_2 at a height of 30 km.

(**O and C**: *and all other Boards except O and C Nuffield*)

(a) See p. 88 if necessary.

(b) (i) In each case, calculate H by equating the average KE of 1 mole ($=3RT/2$) to the PE of 1 mole ($=MgH$). Assume R = 8.31 J mol⁻ K⁻.

(ii) Since $p = Nmc^2/3V$. then p is proportional to N/V (ie, the number density). Thus N/V decreases with increasing height in the same way as p. For the % composition of O_2 at h = 30 km, use the formula to work out the pressure of each gas in terms of its sea level pressure which is (0.20 × atmospheric pressure) for O_2 and (0.80 × atmospheric pressure) for N_2. The pressure ratio is equal to the composition ratio.

6.29L Explain carefully why the pressure exerted by the molecules of a gas on the walls of its container is proportional to the 'mean square speed' of the molecules, and not to their 'mean speed'.

Define **(a)** an 'adiabatic change', and **(b)** an 'isothermal change', Describe how you would attempt to compress a gas **(a)** adiabatically, and **(b)** isothermally. Discuss the factors that determine how closely the compression of the gas in your experiments approximates to the definition you have given.

State the relationship between the pressure P and the volume V of an ideal gas for a reversible adiabatic change. From this, and the equation of state for an ideal gas, deduce the corresponding relationship between the temperature T and the volume.

Assuming that the compression stroke of a diesel engine is the reversible adiabatic compression of an ideal diatomic gas initially at $P = 100$ kPa and $T = 27$°C, calculate the maximum temperature and pressure reached by the gas when its volume is reduced from $1.5 \times 10^{-3}\,m^3$ to $1.0 \times 10^{-4}\,m^3$.
(γ for an ideal diatomic gas = 1.4)

(**O and C**: *JMB*, AEB, Oxford, WJEC*)

You must firstly be quite clear on the difference between the mean speed and the mean square speed; for N molecules of speeds $c_1, c_2, c_3, \cdots c_N$, the mean speed is given by $(c_1 + c_2 + c_3 + \cdots + c_N)/N$, whereas the mean square speed is given by $(c_1^2 + c_2^2 + c_3^2 + \cdots + c_N^2)/N$. You should study the proof of the kinetic theory equation (E6.14) in your textbook to see why the mean square speed is involved and not the mean speed.

For adiabatic changes, see p. 92. An isothermal change is one that takes place without change of temperature.

In describing the compression of a gas, remember that work is done on the gas when it is compressed. If the gas is perfectly insulated, its internal energy and hence its temperature will rise; if there is no insulation, heat will flow from the gas and its temperature will be unchanged. The speed of the change is another factor; a fast change will not give sufficient time for heat to enter or leave the gas.

The adiabatic equation E6.17 is required. This equation should be combined with E6.13 to eliminate P.

To calculate the maximum pressure, given initial pressure and volume, and final volume, use E6.17. Leave P in the units of kPa since the equation is a 'relative' equation.

To calculate the maximum temperature, either use the expression derived (relating V and T for adiabatic changes) or use E6.13 in the form PV/T = constant. Remember that T must be converted into Kelvins.

6.30L

(a) The equation $P = \frac{1}{3}\rho \overline{c^2}$, relating the pressure P and density ρ of an ideal gas, may be derived by making certain assumptions about the behaviour of the molecules of the gas. (i) List four of the assumptions that are essential to this derivation. (ii) Explain what is meant by the symbol $\overline{c^2}$. (iii) Using the above equation and the ideal gas equation $PV = RT$, find an expression for the total translational kinetic energy of one mole of ideal gas molecules in terms of the temperature T.

An athlete of mass 70 kg can run 100 m in 10 s. Find the ratio of the athlete's average kinetic energy to the total translational kinetic energy of one mole of ideal gas molecules at 300 K. (Take the molar gas constant as $R = 8\,J\,mol^{-1}K^{-1}$).

(b) (i) How do the properties of the molecules of a 'real' gas differ from those which are attributed to the molecules of an ideal gas? (ii) Draw labelled graphs of pressure against volume for a fixed mass of a 'real' gas at a number of temperatures, including the critical temperature. On these curves, label the critical isotherm, and an isotherm along which liquefaction could 'not' be achieved. Indicate also on an appropriate isotherm the part of the curve for which the substance is entirely liquid.

(c) When fuel from a cylinder of liquefied butane (such as is used for camping stoves) is released through an opening which is not itself a pressure regulator, the temperature of the cylinder being kept constant, the pressure in the cylinder remains steady until the cylinder is almost empty, when the pressure falls continuously. Account for this in terms of a molecular model.

(**NISEC**: *and all other Boards for (a)*)

(a) (i) See p. 91 if necessary. (ii) See p. 92 and E6.14 if necessary. (iii) See p. 92 (kinetic energy and temperature).

(b) (i) See p. 93 for an explanation of real gas behaviour. (ii) See p. 93 and F6.13 in particular.

(c) The liquefied butane in a cylinder is in equilibrium with its saturated vapour. A molecular explanation of the fact that svp is independent of volume is given on p. 93; however, take care since the given explanation is for a **reduction** of the volume of a saturated vapour, and allowing vapour to escape is equivalent to an **increase** of volume. The explanation needs only minor changes to meet the conditions of the question.

Unit 7 Electricity and electronics

To obtain an appreciation of the far-reaching consequences of modern electronics, an understanding and a working knowledge of basic electrical principles is essential. The Unit starts by looking at the essential principles of electricity, and attempts to show how these principles are linked together. Then, these principles are applied to deal with a range of topics from alternating circuits to modern digital electronics. The subject of electronics is developing at a rapid pace, and today's new devices soon become part of 'yesteryear' as research scientists and engineers find better techniques. For this reason, the emphasis on electronics in A-level physics courses involves understanding of general principles rather than rote-learning of details of individual devices.

The Unit is closely linked to Unit 2 through the application of the ideas of electric potential to circuit theory, and of other field ideas to capacitors and inductors in the study of dc and ac circuits. Conduction in metals and semiconductors described briefly in this Unit links back to the section on structure of solids in Unit 5. The section on measuring ac can best be tackled by a look back at Unit 3 dealing with general ideas of frequency, amplitude, etc; finally, the study of oscilloscopes can be linked to the principles of thermionic emission and electron deflection in Unit 8.

7.1 CURRENT AND CHARGE

An electric current is a flow of charge. In a metal, the charge is carried by 'conduction' electrons which are not attached to any given fixed ion of the metal. The unit of electric current is **the ampere**, defined on p. 37. All other electrical units are defined in terms which relate back to the ampere.

1 **coulomb** (C) of electric charge is the charge which passes a given point when 1A of steady current flows for 1 second. Thus a steady current of 3A for a time of 5 minutes means that a charge of $3 \times 5 \times 60 = 900$C has been passed in total. The link can be expressed in terms of the following equations:

E7.1 $Q = It$ for steady direct currents

E7.2 $I = \dfrac{dQ}{dt}$ for changing currents (dc and ac)

where Q = charge (C), I = current (A), t = time (s).

Note that the prefixes milli- (m $= 10^{-3}$) and micro- ($\mu = 10^{-6}$) are commonly used.

Conduction in metals The reason for the presence of conduction electrons (sometimes called free electrons) in metals is that the outer shell electrons of metal atoms are easily removed. In the solid state, the metal ions form a lattice structure, as in F7.1, and the free electrons move about at high speeds. When a potential difference is applied across the metal, the free electrons are made to 'drift' towards the + ve terminal, thus giving a flow of charge through the metal. In an insulator, the vast majority of the electrons remain firmly attached to atoms.

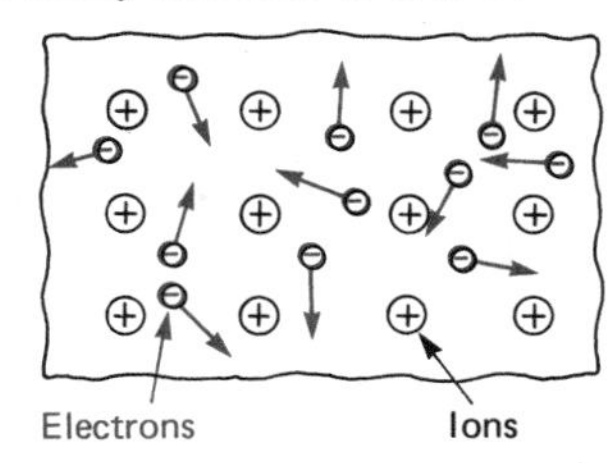

F7.1 Free electrons in a metal

Consider a metal wire of uniform cross-sectional area (A) along which a steady current (I) passes. Conduction electrons carry the current by 'drifting' along the wire towards the + ve terminal, as in F7.2. Let u be the average drift speed of a conduction electron. Referring to the diagram, in 1 second a conduction electron will move a distance u from point X to point Y. So all the free electrons between X and Y will pass Y in one second.

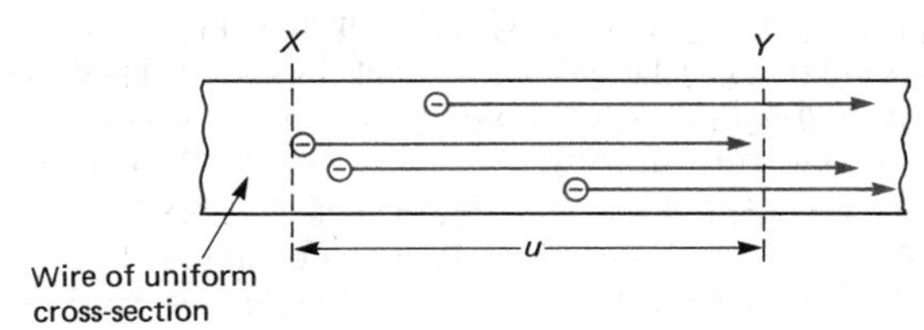

F7.2 Conduction by electrons

The volume of the wire section XY is uA, and so between X and Y there are nuA free electrons, where n is the number of free electrons per unit volume. Hence, the charge flow per second (I) is $nuAe$, where e is the charge carried by an electron.

E7.3 $I = nuAe$

where I = current (A), n = number of free electrons per unit volume (m^3), A = area of cross-section (m^2), u = drift velocity (ms^{-1}), e = electron charge (C).

7.2 POTENTIAL DIFFERENCE

To use potential difference (pd) effectively in circuit theory, its basic nature must first be understood. Remember that electric potential (see Unit 2) is essentially the potential energy of a unit + charge. If unit + ve charge is allowed to move from a point of high potential (V_1) to a point of low potential (V_2) then the energy given up by the charge is $(V_1 - V_2)$ J.

1 **volt** (V) of potential difference exists between two points of a circuit when 1J of work is done by 1C of charge in moving from one point to the other. Note that 1V is identical with 1J/C. It is important to remember that pd can only exist between **two** points (eg. across a resistor's terminals). To illustrate the nature of pd. consider the network shown in F7.3. The pd from A to C is 12V.

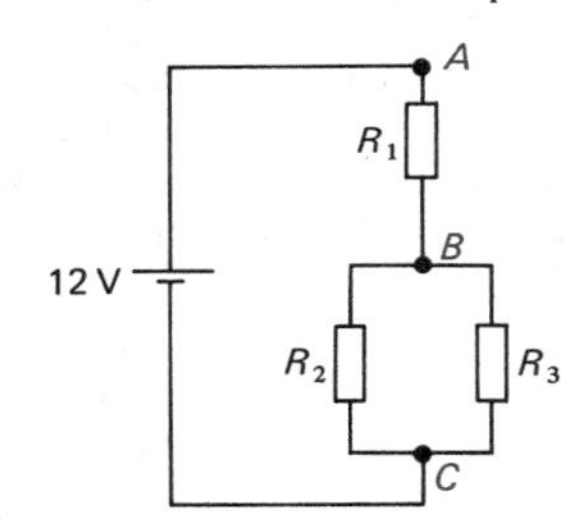

F7.3 Nature of potential difference

so that when 1 C of charge passes from A to C, either via R_2 or via R_3, then it loses 12 J of energy. An individual charge passing from A to C uses one of two possible routes, which are (i) through R_1 and R_2, (ii) through R_1 and R_3. Now suppose the pd across R_1 is 4 V; then, unit charge will use up 4 J on passing through R_1, and will give up its remaining 8 J in either R_2 or R_3 according to its route. Thus, the pd across R_2 = pd across R_3 = 8 J/C. This simple example illustrates two key points in connection with pd: (i) The pd across parallel components (R_2 and R_3 above) is **always** the same, (ii) The pd across AC = pd across AB + pd across BC.

Power taken by an electrical component is given by the following equation:

E7.4 $W = IV$

where W = power taken (watts), I = current (A), V = pd across device terminals (V).

The equation follows from the fact that a pd of V V means that each coulomb of charge which passes through the device gives up V J of energy. Since current I means that I C of charge pass through the device each second, then IV J of energy are delivered each second.

Electromotive force (emf) of a cell or battery of cells is the energy converted into electrical energy per coulomb of charge inside the cell. In the case of a solar cell the energy is converted from light energy, whilst in the case of a dynamo, the energy is converted from mechanical energy. Thus, an emf of 12 V means that each coulomb from the cell will deliver 12 J of energy. If the cell has **internal resistance** then some of the electrical energy will be converted to heat energy inside the cell when current is drawn from the cell; consequently, the amount of energy available for the external components will be reduced. For a cell of emf, E and internal resistance, r, shown in F7.4, connected up to an external 'load', then the pd across the cell terminals when current

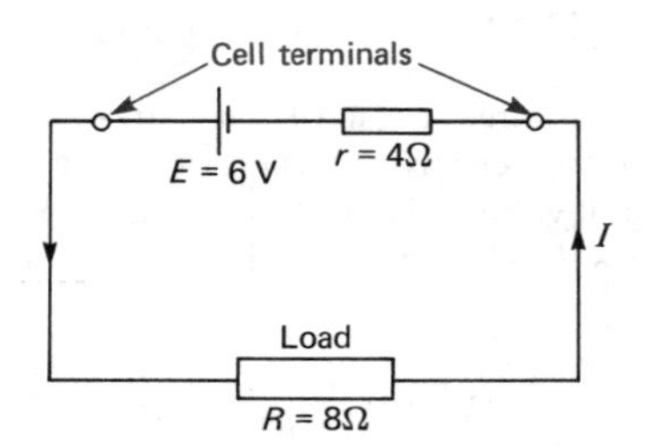

F7.4 Internal resistance

(I) flows will not be E but will be $E - (Ir)$. The reason is the loss of energy of the charge inside the cell as it tries to flow out through the cell's internal resistance. The 'lost voltage' (Ir) represents this loss of electrical energy inside the cell. As an example, consider the circuit of F7.4 in which a 6 V cell (ie, $E = 6$ V) with internal resistance 4 Ω is connected to an external 8Ω resistor. Since the total circuit resistance is 12 Ω, then the current from the cell will be 0.5 A. Hence, the pd across the external resistor will be $0.5 \times 8 = 4.0$ V, and the 'lost voltage', at $0.5 \times 4 = 2.0$ V, will make up the difference between the cell emf (6 V) and the pd (4 V) across its terminals. In energy terms, each coulomb of charge which passes through the cell will be given 6 J of energy but will use 2 J in moving through the interior of the cell; thus, only 4 J of energy will be delivered to the external resistor. Note that the more current which a cell with internal resistance delivers then the greater is the 'lost voltage' so the pd across the cell terminals falls below the emf value by an increasing amount when the current is increased. It follows that the accurate measurement of the emf of a cell requires that the pd across its terminals be determined when no current is taken from the cell (since there is no 'lost voltage' when no current is taken).

E7.5 and E7.6 summarize the ideas above:

E7.5 $V = E - (Ir)$

where V = pd across cell terminals (V), E = cell emf (V), r = internal resistance, (Ω), I = current drawn (A).

If the external load has a total resistance R, then since the external load is connected across the cell terminals so $V = IR$ then the above equation may be written as $IR = E - (Ir)$ or $E = (IR) + (Ir)$. By multiplying each term in this last equation by I, the power distribution becomes clear:

E7.6 $IE = I^2R + I^2r$

where IE = power generated by the cell (W), I^2R = power dissipated by R (W), I^2r = power 'wasted' inside the cell (W).

7.3 THE LIMITS AND USES OF OHM'S LAW

Resistance is defined by the following equation:

E7.7 $R = V/I$

where R = resistance (Ω), V = pd (V), I = current (A).

Note that 1 kilohm (kΩ) = 10^3 ohms, and 1 megohm (MΩ) = 10^6 ohms.

***I–V* Curves** The graphs of F7.5 show the current–voltage relationship for several different circuit components. They are not to the same scale. Since resistance $R = V/I$, it should be clear from the graphs that only the wire has a resistance which is independent of current; in other words, only the I-V curve for the

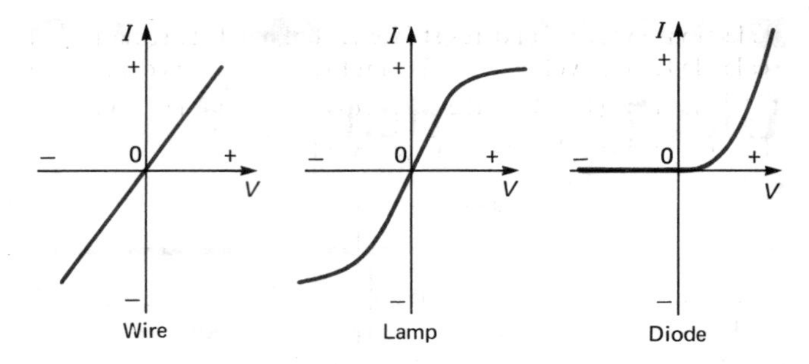

F7.5 I–V curves

wire is a straight line. For the filament lamp, the graph shows that the ratio V/I increases as the current increases; in other words, the resistance of the filament lamp increases with increased current. The reason is that the filament lamp becomes hotter at increased current, and the resistance of metals increases with increased temperature (ie, + ve temperature coefficient of resistance, see p. 90). The silicon diode has a resistance that is very large in the 'reverse' direction; in the forward direction, the resistance is large until the pd exceeds approximately 0.5 V after which the resistance falls. Of the three graphs, only the metal wire gives a straight line (ie, has constant resistance for all current values).

Ohm's law: for a metallic conductor, the current is proportional to the applied pd for constant physical conditions. It follows that the resistance of an 'ohmic' conductor is constant, and does not change when the current changes. Note that the equation $R = V/I$ is not an expression of Ohm's law; it simply defines R and does not express the fact that V/I (= R) is independent of I for an 'ohmic' conductor.

Conductivity is defined as 1/resistivity, and **resistivity** is defined by the following equation:

E7.8 $\rho = RA/L$

where R = resistance (Ω), ρ = resistivity (Ωm), L = length of specimen (m), A = area of cross-section of specimen (m^2).

Resistor combination rules

1 Resistors in series: the pd across a series combination is equal to the sum of the individual pds. Therefore, for two resistors R_1 and R_2 in series, the combined resistance R is based upon $IR = IR_1 + IR_2$ since the current I is the same in each resistor. Hence, $R = R_1 + R_2$ gives the total (ie combined) resistance.

2 Resistors in parallel: the total current is equal to the sum of the individual currents through each resistor. Therefore, for two resistors in parallel (R_1 and R_2), the combined resistance is based upon $V/R = V/R_1 + V/R_2$, since the pd V is the same across each resistor. Hence, the combined resistance R is given by the rule $1/R = 1/R_1 + 1/R_2$. The combined resistance for parallel resistors is always less than the smallest individual resistance value.

Meter conversion An ammeter or a voltmeter can have its range extended as follows:

1 Ammeters: by connecting a suitable resistance (called a 'shunt') in parallel with the ammeter. For an ammeter of resistance r with a full-scale deflection current of i, then the value of the shunt resistance which will extend its range to I will be $\frac{ir}{(I-i)}$, since the pd across the shunt will be ir, and the current through the shunt will be $I - i$. F7.6 shows the basic arrangement.

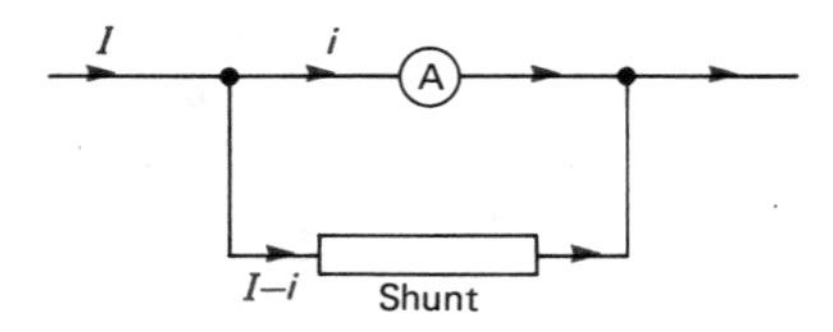

F7.6 Using a shunt

2 Voltmeters: by connecting a suitable resistance (called a 'multiplier') in series with the voltmeter. For a voltmeter of full-scale deflection voltage v and resistance r, to extend its scale to full scale deflection voltage V, the multiplier resistance must be $\frac{V-v}{v}r$, since $(V-v)$ is the pd across the multiplier, and v/r is the current through the multiplier. See F7.7.

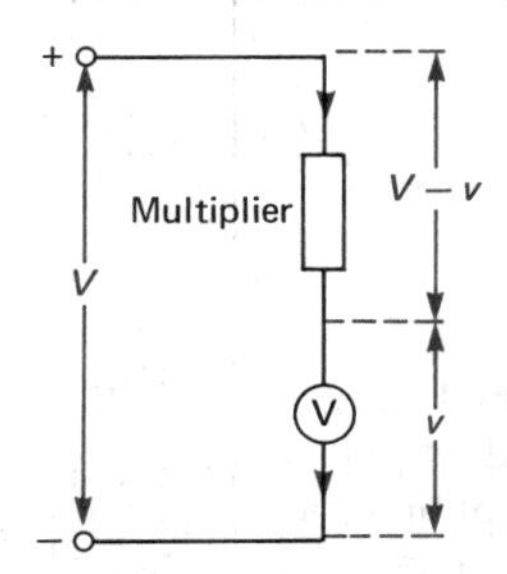

F7.7 Using a multiplier

Because moving-coil voltmeters require current for their basic operation, then the use of a moving-coil voltmeter to measure pd in a circuit will affect the current flow in the circuit being measured. For example, consider the simple circuit shown in F7.8(a). Clearly

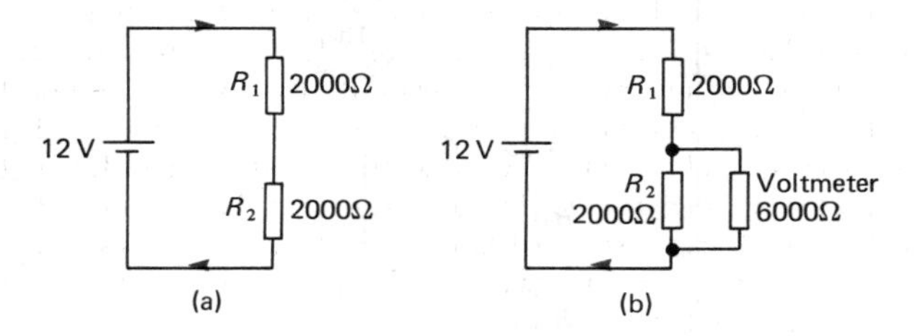

F7.8 Effect of a 'low' impedance voltmeter

the pd across the lower 2000 Ω resistor (R_2) is 6 V. However, if a voltmeter of resistance 6000 Ω is connected across R_2 to measure the pd across R_2, then the circuit has been changed to that in F7.8(b). The total circuit resistance will now be 2000 Ω+1500 Ω= 3500 Ω (the combined resistance of R_2 and the voltmeter in parallel is 1500 Ω, then add on the resistance of R_1), so that the current from the 12 V battery (negligible internal resistance) will be 12/3500 A =0.0034 A. This means that the pd across R_1 is 2000×0.0034= 6.8 V, which leaves a pd across the parallel combination of 12−6.8 =5.2 V. An alternative way of calculating the pd across the parallel combination is to multiply the current taken by the combination (0.0034 A) by the combined resistance (1500 Ω) to give 5.2 V (5.1 V actually on account of rounding off 0.00342 A too early!) The voltmeter therefore reads 5.2 V, which is less than the pd across R_2 without the voltmeter in circuit.

Because a moving-coil voltmeter takes current, its use to measure the emf of a cell will cause an error if the cell has internal resistance. Consider the arrangement shown in F7.9 where the cell has emf E and internal resistance r, and the voltmeter has resistance R. As shown, the circuit resistance will be $(R+r)$ and the current I will be $E/(R+r)$. Thus, the pd across the voltmeter will be $IR = ER/(R+r)$, and so the pd must always be less than the emf E. As a numerical example, suppose $E = 12$ V, $r = 50$ Ω and a voltmeter of resistance $R = 5000$Ω is used as in F7.9. Then, the

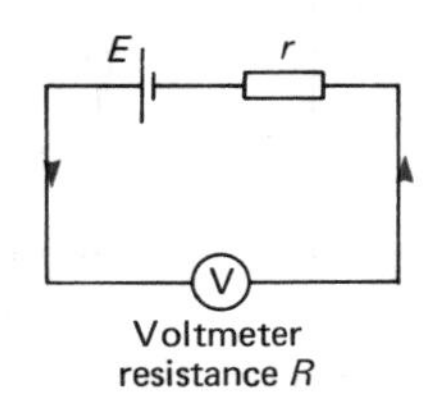

F7.9 emf measurement with a voltmeter

current will be 12/5050 A giving a pd across the voltmeter of 12 × 5000/5050 = 11.88 V on account of 'lost voltage' of 0.12 V. The bigger the voltmeter resistance R is, (compared with the cell's internal resistance r) then the smaller will be the error.

7.4 POTENTIAL DIVIDERS AND POTENTIOMETERS

Potential dividers are used to supply required levels of pd to a circuit from sources of fixed emf. The circuit in F7.10(a) shows the simplest form of the arrangement with two resistors R_1 and R_2 connected in series. With a battery supplying fixed pd V across the ends as shown, the current taken by the two resistors will be $V/(R_1+R_2)$; thus, the pd across R_2 will be $V.R_2/(R_1+R_2)$. By choosing suitable values of R_1 and R_2, the pd across R_2 can be made equal to any value from 0 to V.

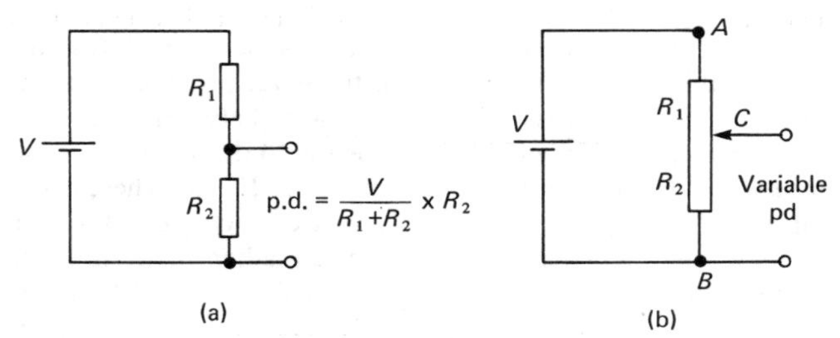

F7.10 Potential dividers

A more useful form of the potential divider is shown in F7.10(b). This time, R_1 and R_2 are adjacent sections of the same resistor R which has a 'tapping off' point at C. Contact C is a sliding contact, so that the resistance of section R_2 can be made to vary from 0 to R; in this way, the pd across R_2 (ie, between points B and C) can be made to change from 0 to V by sliding the contact C from end B to end A. In this way, the pd between B and C can be set at any specified value between 0 and V.

The principle of a **potentiometer** is based upon the variable potential divider as in F7.10(b). A potentiometer is used to measure pd without taking any current (unlike a moving coil voltmeter), and it does this by balancing up its pd (from a variable potential divider) against the pd to be measured, as in F7.11. Balance is achieved when no current is detected on the meter M in the circuit diagram.

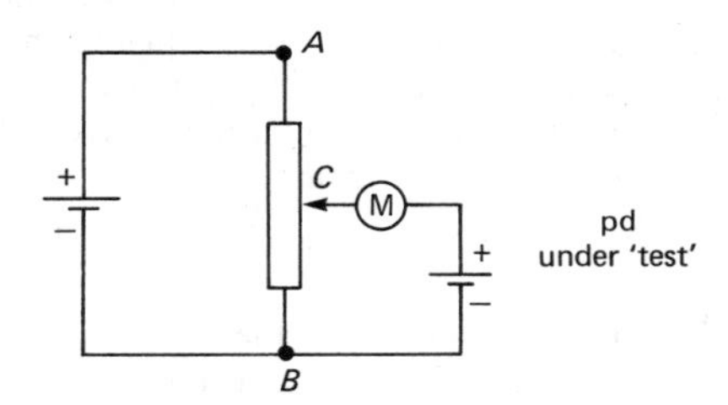

F7.11 Potentiometer principle

Useful insight into the operation of a potentiometer can be obtained by considering the simple circuits of F7.12. In circuit (a), battery A has a greater emf than battery B so current flows from the + ve terminal of A into the + ve terminal of B; in circuit (b), battery A has a smaller emf than battery C so current flows from the + ve terminal of C into the +ve terminal of A. In the third circuit (c), battery A has the same emf as battery D so no current flows either to or from A; A 'balances' D exactly.

With the potential divider arrangement of circuit F7.11, the position of the sliding contact C will determine the size and direction of the current through the meter M. When C is at end A, then current will flow through M from left to right; when C is at end B, then current will flow through M from right to left. Clearly, there will be one single position for C at which no current will flow through M. At this position, the pd to be measured will have been balanced out exactly by the pd between C and B. The position of C at balance can then be used to determine the 'unknown' pd.

Comparison of cell emfs The emfs of the two cells X and Y may be compared using the 'slide wire' form of the potentiometer shown in F7.13(a). This form of the potentiometer consists of a length

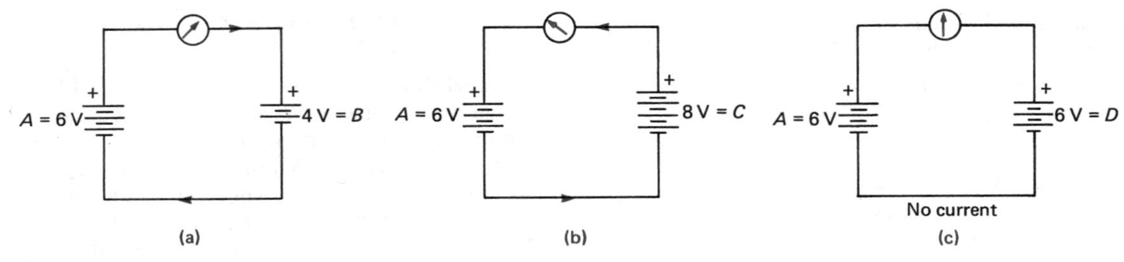

F7.12 Current flow in circuits with many cells

of resistance wire across which a fixed pd is maintained. Contact C is capable of being moved along the wire; provided that the wire has a uniform cross-section, then the pd between the contact C and one end of the wire is in proportion to the length of resistance wire from that end to the contact (ie, $V = kl$ where k is a constant). Each of the cells to be compared is connected in turn into the potentiometer circuit. A centre-reading meter must be connected in series with the 'test' cell, as shown by F7.13(a). By

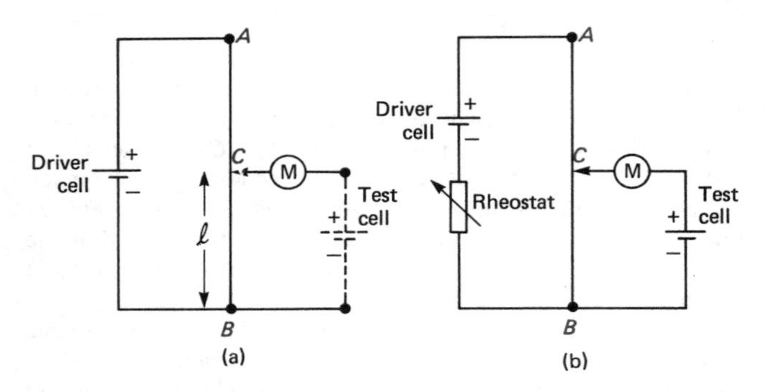

F7.13 Comparison of emfs

moving contact C along the wire, the point on the wire is found at which the meter current is zero (ie, a 'null' reading). At this point, called the balance point, the pd across the balance length CB ($= l$ in the diagram) must be exactly equal to the emf of the test cell. Thus the test cell emf $E = kl$. By measuring the balance length for cell X and then for cell Y in turn (then recheck for cell X) the emf ratio can be calculated from:

$$\text{E7.9} \quad \frac{E_x}{E_y} = \frac{l_x}{l_y}$$

If Y is a standard cell of known emf, then the emf of cell X can be calculated. The key point is that no current is drawn from the test cell when the balance point has been located; hence, there is no loss of pd inside the test cell due to internal resistance. The method is called a **'null' method** because the measurement is made at balance when the meter reads zero. The meter is required only to detect current, not to measure it.

Some common errors in setting up potentiometer circuits are:

1 Wiring the test cell with incorrect polarity when connecting it into the circuit. The test cell polarity must always be such as to oppose the pd along the wire due to the driver cell.

2 Allowing the driver cell emf to 'run down' during the experiment so that readings with the same test cell at the end will differ from readings taken at the start (with that test cell). A rheostat is sometimes included in series with the driver cell to limit the driver cell current, as in F7.13(b). Once set, the rheostat must not be adjusted otherwise the pd across the wire will change.

3 Insufficient pd across the wire to balance up a test cell. The rheostat in (2) must be set in the first place (if used) to enable the largest test emf to be balanced near the far end of the wire (end A in F7.13(b)).

Measurement of a small emf A thermocouple emf is a good example, and since the value will be no more than 10 mV, the ordinary potentiometer method is inaccurate because the balance length will be so small. For example, if the pd per unit length (k) is $1.0\,\mathrm{V\,m^{-1}}$, then a 10 mV thermocouple emf would give a balance length of only 0.01 m.

The method, as outlined above, is modified for small emfs by connecting a pair of high-value resistance boxes in series with the driver cell, as shown in F7.14. In this way, the pd across the

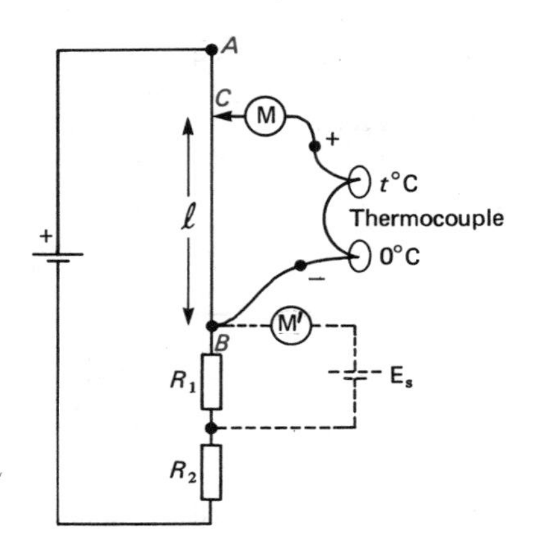

F7.14 Measuring small emfs

potentiometer wire is considerably reduced; the values of R_1 and R_2 are chosen so that the balance length of the thermocouple (l_{BC}) is a significant portion of the wire AB. The thermocouple emf is given by $E = kl_{BC}$. The value of k is determined by connecting a standard cell (ie, known emf) in series with a centre-reading meter, and then connecting the two components across R_1. Then, keeping the value of $R_1 + R_2$ in total the same, the value of R_1 is adjusted until a null (ie, zero) deflection is obtained on the meter (M'). With this value of R_1, the current flow through R_1 must be E_s/R_1, so the pd per unit length of the wire must equal $\frac{(E_s/R_1)r}{L}$ where r is the wire resistance and L is the wire length AB.

Comparison of two resistances. This can be undertaken using a potentiometer as in F7.15. The two resistors to be compared,

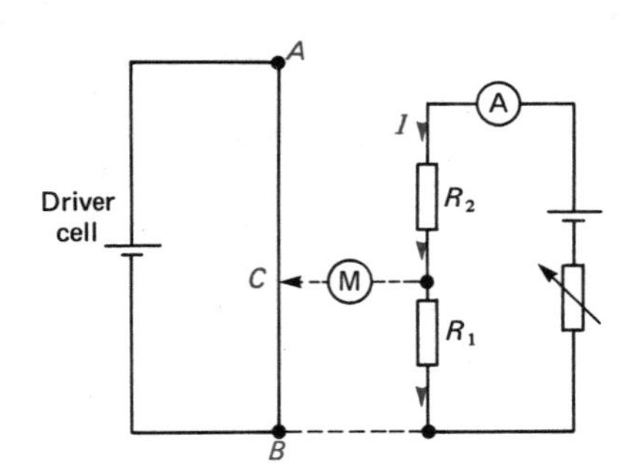

F7.15 Comparison of resistances

resistances R_1 and R_2, are connected in series and a steady current is passed through. The potentiometer is used to measure the pd across R_1, as in the diagram, and then across R_2. The same current must flow through the two resistors at balance in each case.

E7.10 $\frac{R_1}{R_2}=\frac{l_1}{l_2}$

where l_1 = balance length (m) for R_1, and l_2 = balance length (m) for R_2.

The method is useful for low resistances.

7.5 THE WHEATSTONE BRIDGE

The Wheatstone bridge network shown in F7.16(a) will be balanced (ie, zero current through M) when the resistance values satisfy the equation:

E7.11 $\frac{P}{Q}=\frac{R}{S}$

In the metre bridge arrangement, shown in F7.16(b) resistors P and Q together form a metre length of uniform resistance

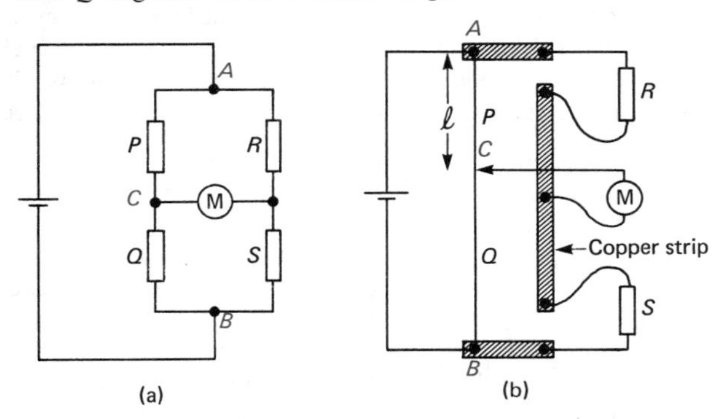

F7.16 Wheatstone Bridge

wire AB, resistor S is usually a standard resistor, and R is the 'unknown' resistance. Contact C is moved along the wire until the galvanometer current is zero. Then, the length of AB that corresponds to resistor P (length l in the diagram) is measured. R is given by:

E7.12 $R=\frac{l}{(L-l)}S$

where S = standard resistance (Ω), R = unknown resistance (Ω), l = balance length (from same end as R) (m), L = total length of the wire (m).

Some understanding of the balanced Wheatstone bridge can be achieved by considering F7.16(b) again. Consider wire AB and sliding contact C as a supplier of variable pd between B and C, and also consider R and S as supplying a fixed pd across S. When at balance, the variable pd between B and C must exactly equal the fixed pd across S.

The Wheatstone network in its metre bridge form is used to measure resistance of resistance thermometers (see p. 88) and to determine resistivities of wires, etc. To measure resistivity of a given uniform specimen, the resistance R of the specimen would need to be measured, its area of cross-section A and its length would also need to be determined. Then, resistivity could be calculated from resistance × area of cross-section/length.

7.6 CAPACITORS IN DC CIRCUITS

Capacitance (C) is defined as the charge stored per unit pd:

E7.13 $C=\frac{Q}{V}$

where Q = charge stored (C), V = applied pd (V), C = capacitance (Farads, F).

Combination rules

1 In series: the combined capacitance (C) of two capacitors in series is given by:

E7.14 $\frac{1}{C}=\frac{1}{C_1}+\frac{1}{C_2}$

Two capacitors in series, as in F7.17 each store the same charge (Q) so that the pd across C_1 is Q/C_1 and the pd across C_2 is Q/C_2. Thus, the pd across the series combination is $Q/C_1+Q/C_2$.

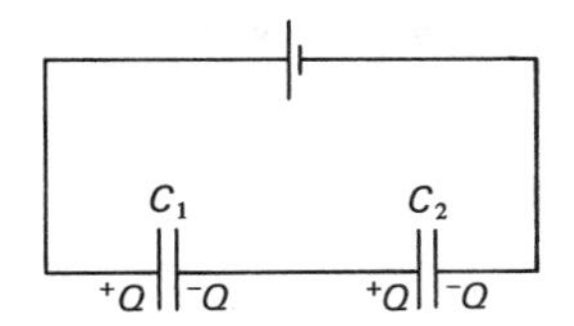

F7.17 Capacitors in series

Since the combined capacitance C is given by C = total pd/charge stored, then E7.14 follows. Note that the total charge stored is still Q.

2 In parallel; the combined capacitance C is given by:

E7.15 $C=C_1+C_2$.

Capacitors in parallel have the same pd across their terminals (V), so that the charge stored by each capacitor is in proportion to its capacitance (ie, C_1 stores charge C_1V and C_2 stores charge C_2V). The total charge stored is thus C_1V+C_2V.

When a charged capacitor shares its charge with another capacitor, as in F7.18 when switch S is disconnected from A and

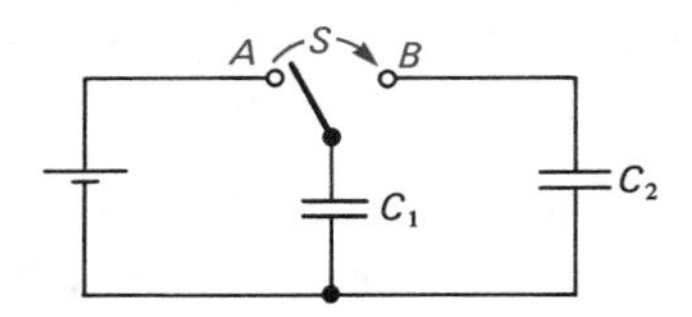

F7.18 Capacitors in parallel

reconnected to B, the final sharing of charge can be determined using two basic rules:

(i) the total final charge = the total initial charge.

(ii) the final charge is distributed in proportion to the capacitance (because the two capacitors have the same pd after sharing has taken place).

Discharging a capacitor through a resistor Initially the discharge current is large because the pd across the resistor (= pd across the capacitor) is at its greatest. Subsequently, the discharge current falls because the capacitor pd falls. A discharge circuit is shown in F7.19, together with graphs for (i) y = charge, x = time and (ii) y = current, x = time.

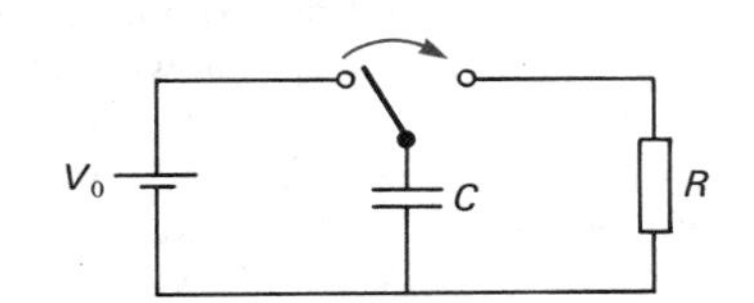

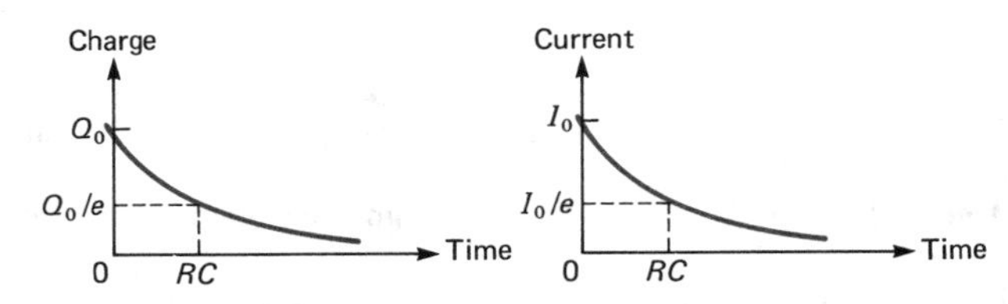

F7.19 Discharging a capacitor

Since the current is given by $I=V/R$, and the charge on the plates is given by $Q=CV$ then it follows that $I=Q/CR$. Because current I is equal to the rate of flow of charge **off** the plates (dQ/dt), then the differential equation $dQ/dt=-Q/RC$ can be used to give a formula for the variation of charge with time:

E7.16 $Q=Q_0\,e^{-t/RC}$

where Q = charge at time t (in seconds) (C), Q_0 = initial charge (C), RC = circuit 'time constant'.

The circuit **time constant *RC*** is a measure of the rate of discharge. Use of E7.16 shows that the capacitor discharges to 37% of its initial charge in a time RC. Note that the discharge current follows the same exponential decay law as the charge; in other words, $I=I_0e^{-t/RC}$ where I_0 = initial current = Q_0/RC.

Charging a capacitor through a resistor Initially, a capacitor will charge up with a high rate of flow of charge (ie, current) onto the plates, but as the charge on the plates builds up, then the charging current becomes less and less. F7.20 shows a charging circuit,

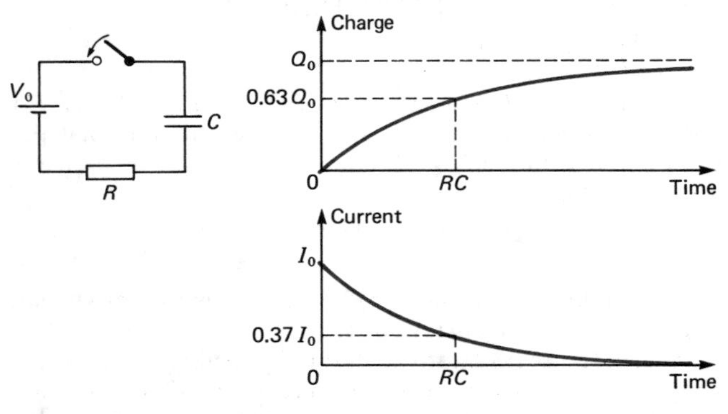

F7.20 Charging a capacitor

and the graphs show how the charge on the plates, and the current through the resistor vary with time. Note that the gradient of the charge curve gives the current curve (since $I = dQ/dt$). Also, since the pd across the resistor is equal to IR, and the pd across the capacitor is equal to Q/C, then the pd across each component varies as in F7.21. The sum of the pds is equal to the battery pd.

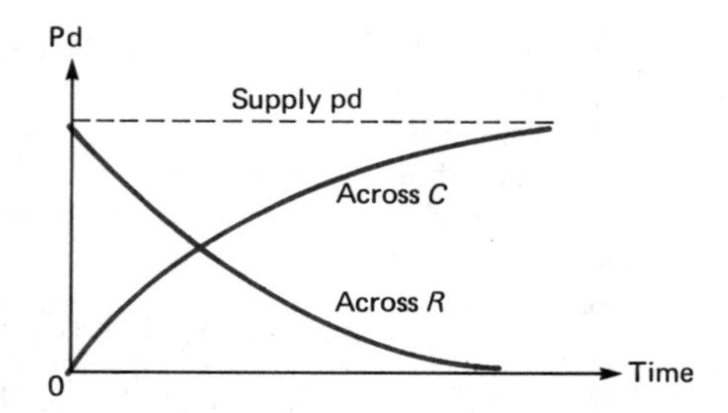

F7.21 Capacitor and resistor pds on charging

The charge curve of F7.20 is a 'build-up' exponential (because it 'builds up' to a final level), and it can be shown that the time constant RC represents the speed of charging up in a similar way to the discharge curve where it represents the speed of discharge. In fact, RC for the charging circuit is the time taken for the charge to build up to 63% (ie, 100 − 37%) of its final value.

7.7 MEASURING AC

Frequency is the number of complete cycles per second. The unit is the Hertz (Hz).

Amplitude (or peak value) is the maximum value of an alternating signal.

Root mean square (rms) value of an alternating current (or pd) is the value of the direct current (or pd) which would give the same power dissipation as the alternating current (or pd), in a given resistor. The term 'alternating' can be taken to include not only sine wave signals but any other regular waveform such as a square wave signal or a triangular wave form, as in F3.1.

E7.17 rms value $= \dfrac{\text{Peak value}}{\sqrt{2}}$ **for sine waves only**

Representing **sine wave** signals can either be by:

1 graphs of y = instant value, x = time, as in F7.22(b), or

2 an equation of the form:

E7.18 $I = I_o \sin(2\pi f t)$

where I = current at time t (A), I_o = peak current (A), f = frequency (Hz), or

3 rotating vectors sometimes called **phasors**, as in F7.22(a). The vector length is scaled to the peak value, and the vector is con-

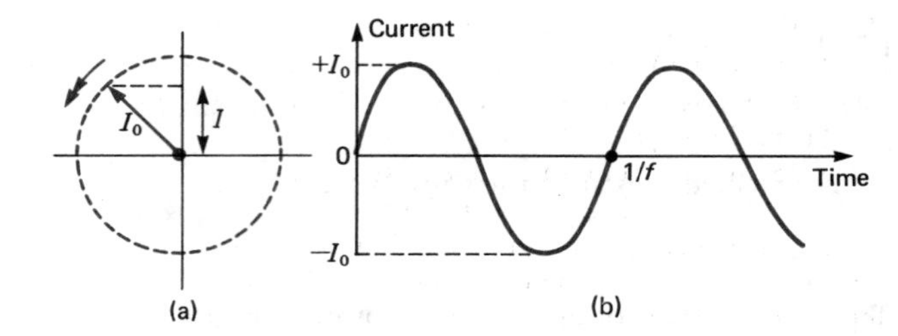

F7.22 Representing a sinusoidal change

sidered to rotate at frequency f; thus, the projection of the vector onto a straight line represents the instant value.

Rectifiers convert ac into dc. F7.23 shows a bridge rectifier with which sine wave ac may be converted into full-wave dc as illustrated by the graphs in the diagram. On one half of each cycle, diodes D_1 and D_3 conduct, whereas on the other half of the cycle diodes D_2 and D_4 conduct. A smoothing capacitor is usually included, as shown on the diagram, so as to give a steadier direct current.

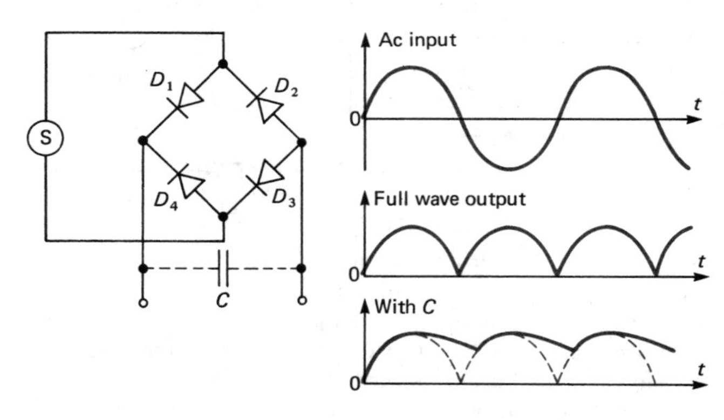

F7.23 Full-wave rectification and smoothing

Oscilloscopes are used to display and measure alternating voltage wave forms. With a calibrated time base, the time period of an alternating signal may be measured; with a calibrated Y-scale, the peak value may be measured. Note that the vertical deflection of the spot (or trace) is in proportion to the pd between the Y-input terminals. F7.24 shows an example of use of the oscilloscope to measure peak voltage and frequency. If the time base is set to a rate of 5 ms cm^{-1}, then the time period must be

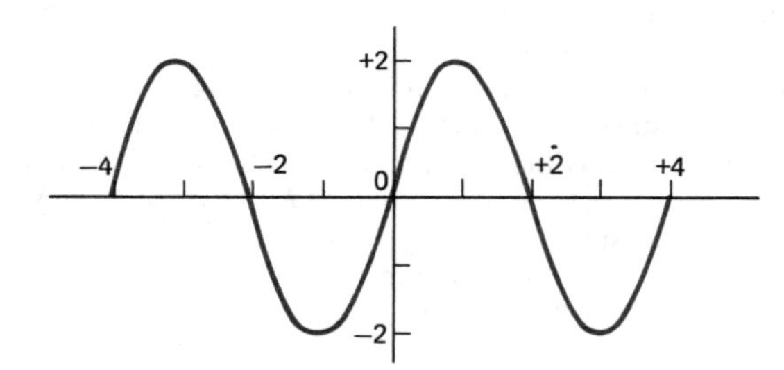

F7.24 Oscilloscope trace of sine wave

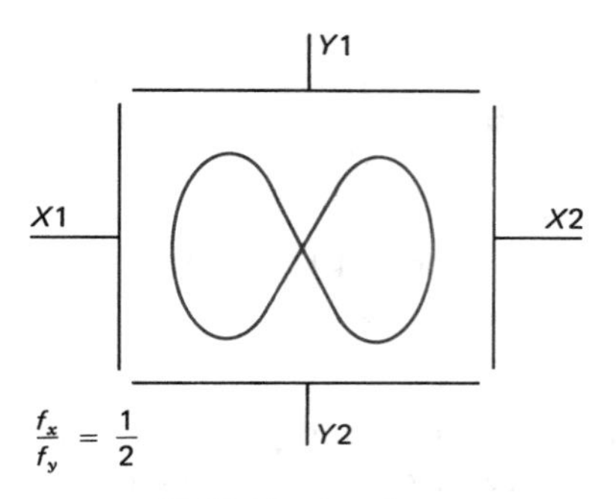

F7.25 Lissajous figure

20 ms since 1 full cycle takes 4 cm along the *X*-scale; if the *Y*-sensitivity is 0.5 V cm^{-1}, then the peak value (ie, from centre to top) will be 1.0 V since the trace is 4 cm from top to bottom.

An oscilloscope can be used to display **Lissajous'** figures, as in F7.25, by disconnecting the time base circuit from the *X*-input plates and connecting an alternating voltage of frequency f instead. Then, a second alternating voltage of the same amplitude and of frequency nf (where n is an integer) is applied to the *Y*-input terminals. In the diagram, the spot has moved up and down in the same time as it has moved across just once; ie, $f_Y = 2f_X$.

7.8 AC CIRCUITS

The effect of alternating pds upon individual components must first be understood before dealing with combinations of components. For a **sine wave** pd applied across a single component (ie, a resistor or a capacitor or an inductor), the current can be calculated from the pd if the following two quantities are known:

1 The resistance R if a resistor, or the reactance X if a capacitor or an inductor.

2 The phase difference between current and pd (ie, the fraction of a cycle between peak pd and peak current). Remember that 1 full cycle is 2π radians.

Resistance only Current is in phase with the applied pd, so that at every instant the current is given by $I = (V_0/R)\sin(2\pi ft)$ for an applied pd $V_0 \sin(2\pi ft)$.

Capacitance only Current is ahead of the applied pd by $\frac{1}{4}$ cycle. The reason is that with an applied pd $V = V_0 \sin(2\pi ft)$, the charge Q at any instant on the plates is given by:

$Q = CV = CV_0 \sin(2\pi ft)$

Since the current $I = \mathrm{d}Q/\mathrm{d}t$, then differentiation gives:

$I = (2\pi fC)\, V_0 \cos(2\pi ft)$

The peak current I_0 is when $\cos(2\pi ft) = 1$, so giving $I_0 = (2\pi fC)\, V_0$.

Since the reactance of a capacitor is defined as $\dfrac{\text{peak pd}}{\text{peak current}}$ then reactance is given in terms of capacitance by E7.19.

E7.19 $X_c = \dfrac{1}{2\pi fC}$

where X_c = capacitor reactance (Ω), f = frequency (Hz), C = capacitance (F).

F7.26 illustrates the circuit involved.

Inductance only Current is behind the applied pd by $\frac{1}{4}$ cycle. The explanation here lies in the fact that the applied pd causes current change, which causes an induced emf to match the applied pd. If $V_0 \sin(2\pi ft)$ is the applied pd, then the induced emf $[= L(\mathrm{d}I/\mathrm{d}t)]$ must equal $V_0 \sin(2\pi ft)$. By integration:

$$I = \frac{-V_0}{(2\pi fL)} \cos(2\pi ft)$$

The peak current I_0 is when $\cos(2\pi ft) = 1$, so $I_0 = V_0/(2\pi fL)$. The variation of applied pd, and of current, with time is shown in F7.27.

E7.20 $X_L = 2\pi fL$

where X_L = reactance of the inductor (Ω), f = frequency (Hz), L = self-inductance (H).

A useful aid for memorizing phase differences is **'CIVIL'** (I ahead of V for C; I after V for L).

Filter circuits A capacitor will tend to block the passage of low-frequency currents because its reactance is high at low frequency; an inductor will allow low frequency currents to pass with relative ease, but will tend to block passage of high-frequency currents since its reactance increases with increased frequency. A simple tuning circuit can be made by connecting a capacitor in parallel with an inductor, as in F7.28. The aerial will pick up many

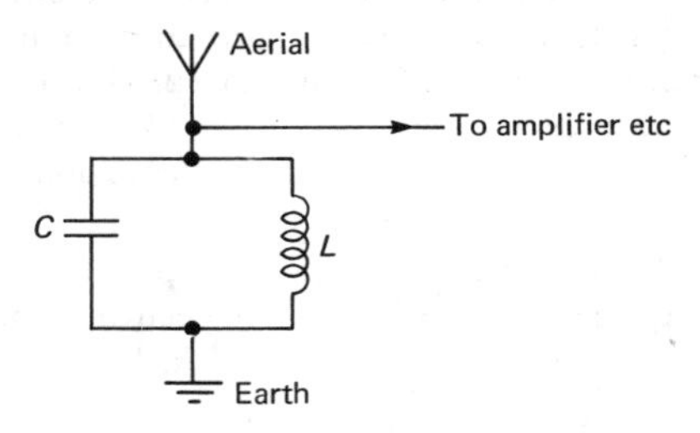

F7.28 Tuned circuit

frequencies, but the high-frequency signals will pass to earth via the capacitor whereas the low-frequency signals will pass to earth via the inductor. Frequencies of an intermediate value will not pass easily through either the inductor or the capacitor, so will pass into the amplifier, etc; if the signal to be detected has a frequency near this intermediate value, then it will pass into the amplifier. The intermediate frequency is then given by equal reactance of the capacitor and inductor, (See E7.24).

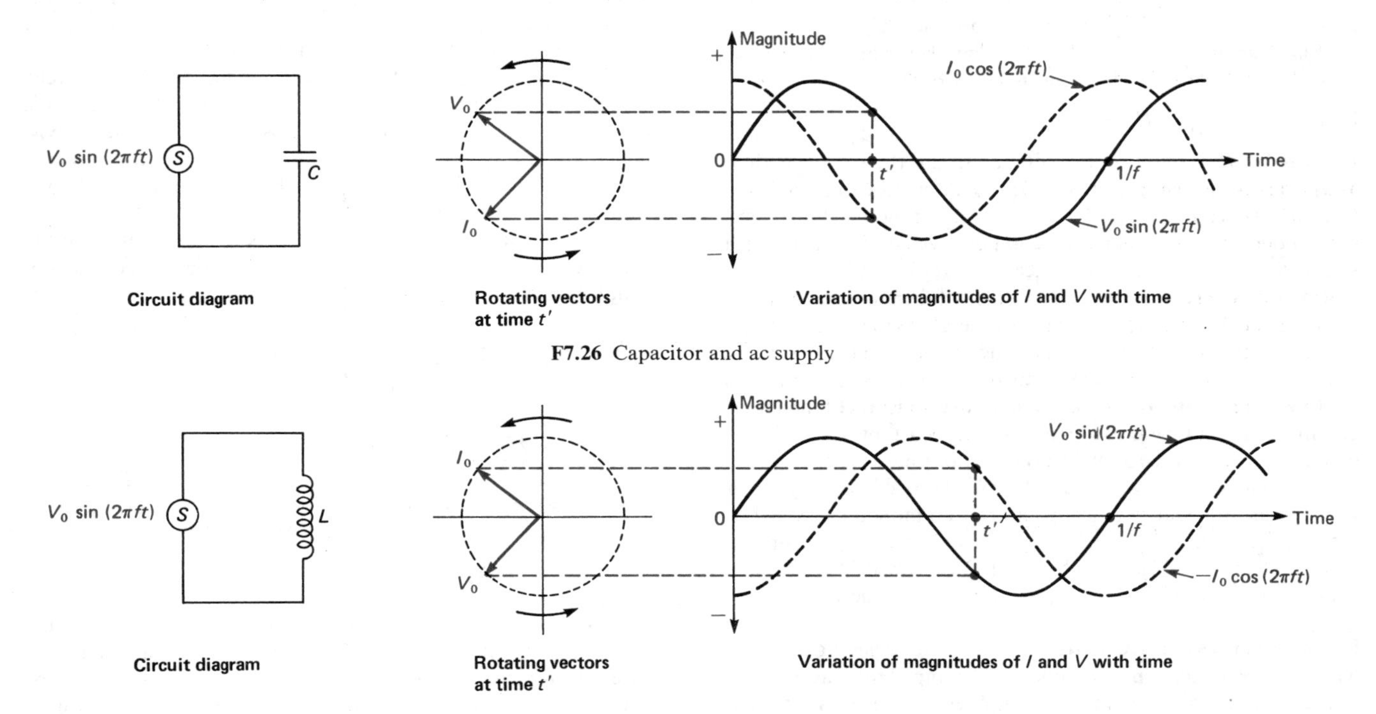

F7.26 Capacitor and ac supply

F7.27 Inductor and ac supply

Series circuits and sine wave pds Consider a series LCR circuit connected to a sine wave pd, as in F7.29. There are two key points to use in working out currents and pds:

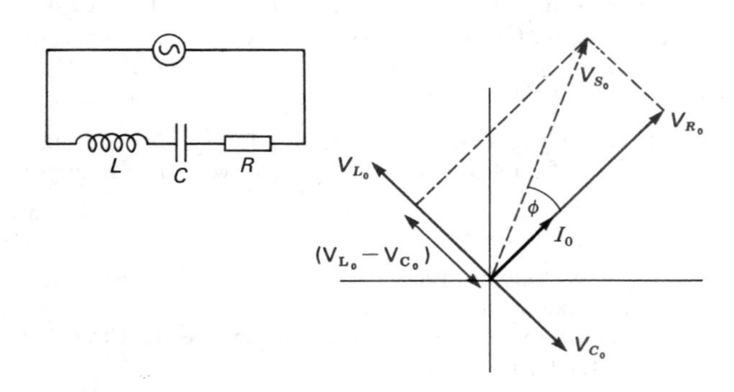

F7.29 *LCR* series circuit

1 the current at any instant is the same in each component (because they are in series).
2 the sum of the pds at any instant is equal to the source pd at that instant. In this example, currents and pds can best be dealt with by using the rotating vector (phasor) method. Since the current is the same for each component, then the individual pd vectors can be drawn relative to the current vector, as in F7.29. The sum of the three pd vectors, V_R, V_C and V_L is equal to V_S, the supply pd vector, so giving the following equation:
E7.21 $V_{S_0}^2 = V_{R_0}^2 + (V_{L_0} - V_{C_0})^2$
where V_{S_0} is the peak value of the supply pd etc.

Then since individual peak pds are related to peak current (I_0) by the following equations: $V_{R_0} = I_0 R$, $V_{L_0} = I_0(2\pi fL)$, and $V_{C_0} = I_0(\frac{1}{2\pi fC})$ then E7.21 may be rewritten as $V_{S_0}^2 = I_0^2[R^2 + (2\pi fL - \frac{1}{2\pi fC})^2]$.

Impedance (Z) of a combination of components is defined as $\frac{\text{peak applied pd}(V_0)}{\text{peak current }(I_0)}$ so that the impedance of a series LCR circuit is given by:

E7.22 $Z = \sqrt{[R^2 + (2\pi fL - \frac{1}{2\pi fC})^2]}$

where R = resistance (Ω), C = capacitance (F), L = self-inductance (H), f = frequency (Hz), Z = impedance (Ω).

Phase angle φ between the supply pd vector and the current vector for the series LCR circuit is given by:

E7.23 $\tan\varphi = \frac{1}{R}(2\pi fL - \frac{1}{2\pi fC})$

For a series RC circuit or a series RL circuit, the vector method as above applies in a similar way; the final results can be deduced from the above E7.22 and E7.23 by omitting the reactance term corresponding to the excluded component. For instance, for a series RL circuit, the impedance $Z = \sqrt{[R^2 + (2\pi fL)^2]}$. It is worth noting that the term 'coil' is usually taken to mean a device with inductance **and** resistance, so unless you are given that a particular coil has negligible resistance, you must treat it as an inductance in series with a resistance.

Power dissipation only occurs in the resistances of an ac circuit. The reason is that the pd and current for either a capacitor or an inductor are $\frac{1}{4}$ cycle out of phase, so that power (= pd × current) averages out at zero over a complete cycle for L or C. The impedance of a capacitor or an inductor is called its **reactance**, a term used to signify that the average power dissipated is zero. The power dissipated by a resistance R which passes ac of peak value I_0 is $\frac{1}{2}I_0^2R$ which equals I_{rms}^2R since the rms current $I_{rms} = I_0/\sqrt{2}$.

Resonance of a series LCR circuit Consider a series LCR circuit with a variable-frequency supply pd. The impedance, as given by E7.22, will be a minimum when the applied frequency is such that the capacitor reactance ($1/2\pi fC$) is equal to the inductor reactance ($2\pi fL$). This frequency is known as the 'resonant' frequency, and corresponds to equal values of capacitor pd and inductor pd so the two pds cancel one another out (the vectors V_C and V_L of F7.29 will be of equal length.). At resonance, the current will be a maximum (= V_{S_0}/R) and will be in phase with the supply pd:

E7.24 $f_0 = \frac{1}{2\pi\sqrt{LC}}$

where C = capacitance (F), L = self-inductance (H), f_0 = resonant frequency (Hz).

F7.30 shows a graph of y = impedance (Z), x = frequency of supply (f).

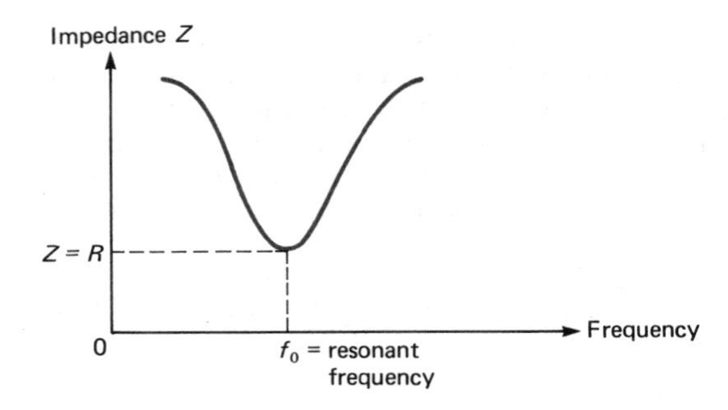

F7.30 Series resonance

7.9 TRANSISTORS

Semiconductors are the basic materials used in the manufacture of transistors and integrated circuit 'chips'. The element silicon (Si) is a widely used semiconductor. Without any atoms of different elements added (ie, without doping), the semiconductor is known as an **'intrinsic'** semiconductor; at absolute zero, the atoms have loosely attached electrons in their outer shells. At room temperature, these electrons become detached (ie, become conduction electrons), and so respond to applied pds. The resistance of intrinsic semiconductors falls as the temperature rises because more electrons become detached from 'parent' atoms. In other words, intrinsic semiconductors have a −ve temperature coefficient of resistance (compared with a +ve coefficient for metals).

Extrinsic semiconductors are made by adding controlled amounts of a different element to an intrinsic semiconductor. When the added atoms have one more outer-shell electron each than is required for bonding, the surplus electron per added atom then becomes a conduction electron; the semiconductor is then known as an **n-type** (extrinsic) semiconductor. Alternatively, by doping an intrinsic semiconductor with atoms which each have one less outer-shell electron than a 'host' atom has, then **p-type** semiconductor is produced. In p-type material, the electron vacancies are called 'holes', and these are responsible for conduction. Being produced by electron vacancies, the holes are +ve charge carriers, a fact which can be demonstrated by the Hall effect (see p. 37). F7.31 illustrates the idea of conduction by holes.

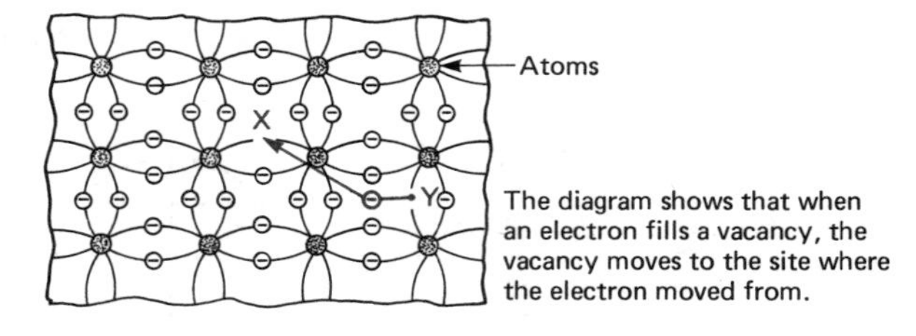

F7.31 Movement of a vacancy (hole)

Transistor action The most common form of transistor is the silicon npn junction transistor in which two regions of n-type material are separated from one another by a layer of p-type material. The arrangement and circuit symbol are shown in F7.32. To understand transistors in action, you should first

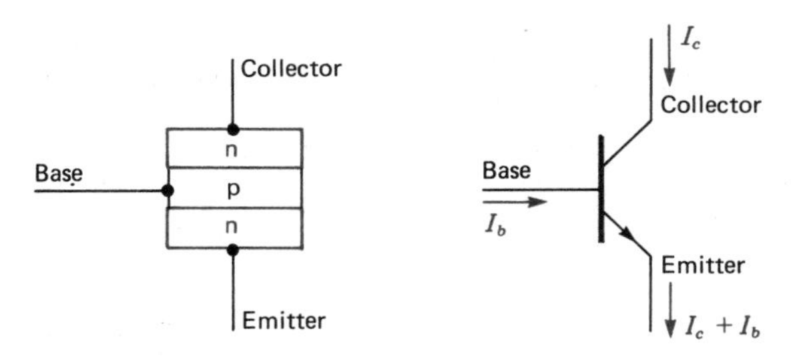

F7.32 n-p-n transistor

appreciate that current flow into the collector and out through the emitter (I_c) is controlled by a much smaller current flow into the base (I_b) (and out at the emitter). In normal operation, the collector current is always determined by, and in proportion to, the current taken by the base.

E7.25 Current gain (β) $= \dfrac{\text{change of } I_c}{\text{change of } I_b}$

where I_c = collector current, I_b = base current.

The characteristics of a junction transistor are represented by three graphs:

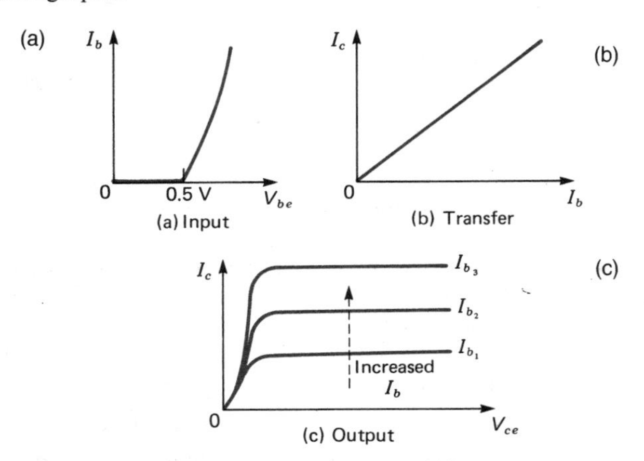

F7.33 Transistor characteristics

1 Input characteristic, as in F7.33(a) which shows how the base current varies with base-emitter pd. The graph shows that the base-emitter resistance is not constant, and that the base-emitter pd must exceed approximately 0.5 V before the base will take any current (ie, before the transistor is 'turned on').

2 Transfer characteristic, as in F7.33(b) which shows how the collector current varies with base current. The slope is equal to the current gain (β).

3 Output characteristic, as in F7.33(c), shows how the collector current and collecter-emitter pd vary in relation to one another, for fixed values of base current.

Simple ac voltage amplifier The circuit diagram F7.34 shows a simple ac voltage amplifier; it is called a 'common emitter'

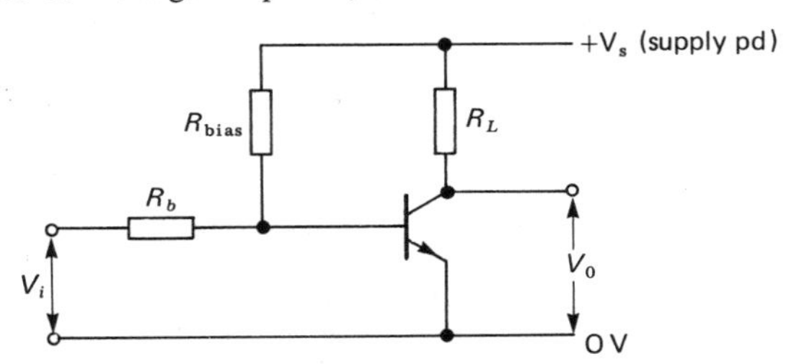

F7.34 Transistor voltage amplifier

amplifier because the input and output pd share the emitter as a common terminal. Its mode of operation is the following sequence:

1 change of input pd gives change of base current,
2 change of base current causes change of collector current,
3 change of collector current causes change of pd across the load resistor,
4 since the sum of the output pd and the pd across the load resistor must always add up to the supply pd (which is constant), then change of pd across the load resistor causes change of the output pd.

With no input pd, the bias resistor R_{bias} allows a steady current to flow into the base, so giving a steady collector current and thus a steady pd across the load resistor. The value of R_{bias} is chosen so that, with no input pd, the pd across the load is $0.5 \times$ supply pd; then, the output pd $= 0.5 \times$ supply pd.

With an input pd (eg, +ve), the base current will change from its 'bias' value (eg, rise if input pd is +ve), so causing the output pd to change from $0.5 \times$ supply pd (eg, fall if input is +ve). Thus, a +ve input pd will make the output pd fall; a −ve input pd will make the output pd rise. An alternating pd at the input stage will produce an 'inverted' alternating output pd.

The **voltage gain** is given by the following equation:

E7.26 Voltage gain $\dfrac{\Delta V_o}{\Delta V_i} = -\beta \dfrac{R_L}{R_b}$

where ΔV_o = change of output pd (V), ΔV_i = change of input pd (V), β = current gain, R_L = load resistance (Ω), R_b = base resistance (Ω).

The value of R_L must be such that the maximum collector current (= supply pd/R_L) is insufficient to overheat the transistor. The value of the base resistor R_b must be sufficiently large so as to make the base-emitter resistance insignificant, so making distortion of the input signal much smaller. The bias resistor is chosen to make the output pd $= 0.5 \times$ supply pd when there is no input.

Saturation is reached when the output pd reaches the limits set by the supply pd; therefore, input pds beyond a certain range will have no effect upon the output pd. For example, with a supply pd of 10 V and a voltage gain of × 8, an input pd can drive the output pd from 5 V (corresponding to no input) down to 0 V or up to 10 V. With a voltage gain of × 8, an input pd of +5/8 V will thus drive the output pd down to 0 V; input pds beyond 5/8 V will not alter the output from 0 V.

When a sine wave input pd is applied, and its amplitude is increased beyond the range for output changes, then the output pd will be a clipped sine wave as shown in F7.35.

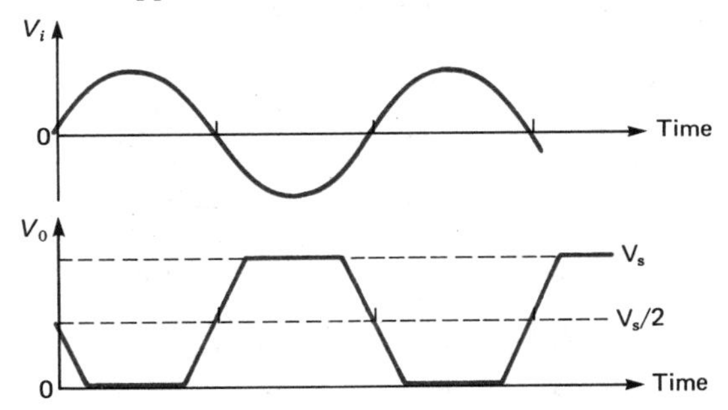

F7.35 Input/output behaviour of transistor voltage amplifier

7.10 OPERATIONAL AMPLIFIERS

Op-amps in integrated form are widely used in **analogue** circuits where output pds can take any value between the limits of the supply pd. The essential features of an op-amp are (i) a very high gain (typically 10^5), (ii) a very high input resistance (typically 10^{12} Ω in the latest type of op-amps). The circuit symbol is shown in F7.36, and represents an 'open loop' amplifier. Note

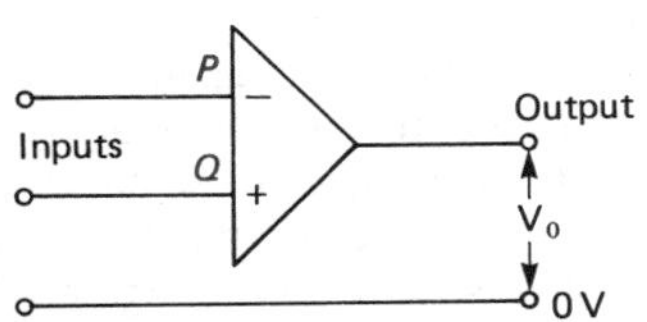

F7.36 Operational amplifier: basic connections

that there are two inputs; input P is the inverting input, and input Q is the non-inverting input. The output pd is proportional to the pd between Q and P provided it does not 'saturate' (ie, reach the limits set by the supply pd):

E7.27 $V_o = A(V_Q - V_P)$

where V_o = output pd (V), V_Q = pd at the non-inverting input (V), V_P = pd at the inverting input (V), A = open loop gain (typically 10^5).

Open-loop voltage comparator If the input to the non-inverting terminal Q is set at fixed pd, and a sine wave input is applied to the inverting input P, as in F7.37, then the output pd will be $-V_s$

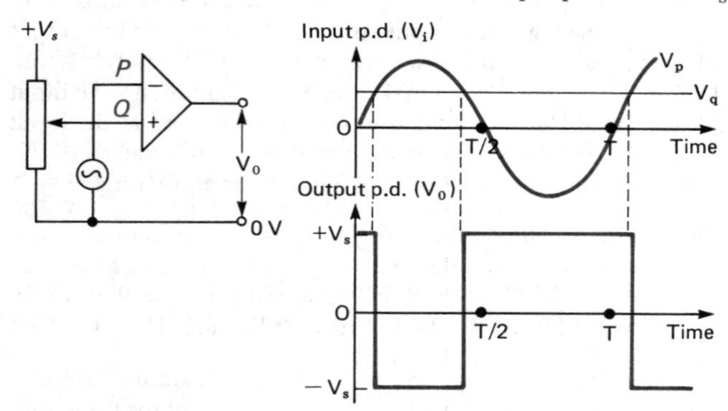

F7.37 Operational amplifier: comparator

whenever the sine wave pd exceeds the fixed pd at Q. When the sine wave pd is less than the fixed pd, then the output pd will be $+V_s$. The diagram also shows the variation with time of the inputs and the output pd. The high gain on open loop (A) forces the output to saturation whenever the pd between the inputs exceeds $\pm V_s/A$ (V_s = supply pd) so for a supply pd of $\pm$ 15 V with $A = 10^5$, the input pd only needs to exceed $\pm$ 150 μV for saturation to occur.

Inverting amplifier To amplify input pds greater than approximately 150 μV without saturation, a feedback resistor and an input resistor must be added. F7.38 shows the arrangement for an inverting amplifier (a closed-loop amplifier).

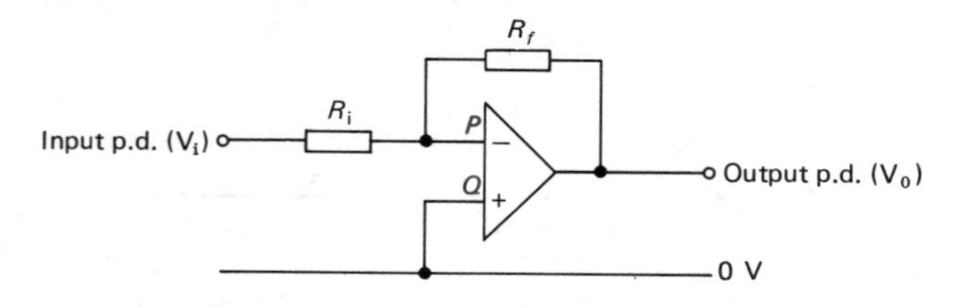

F7.38 Operational amplifier: inverting voltage amplifier

The voltage gain is given by:

E7.28 Voltage gain $\dfrac{V_o}{V_i} = -\dfrac{R_f}{R_i}$

where R_f = feedback resistance (Ω), R_i = input resistance (Ω).

This circuit provides an example of 'negative feedback', in which a fraction of the output signal is fed back so as to reduce the gain from the open-loop value. Note that the voltage gain here depends only upon the external resistors. Also in the circuit, the non-inverting input is earthed. Provided the output does not saturate, then the inverting input will be at 'virtual earth' potential. This is because $V_Q = 0$, and with $A = 10^5$ and V_o < supply pd (say 15 V), then V_P will be less than 150 μV (ie $\simeq$ 0 V).

Black boxes The op-amp provides a good example of an electronic 'black box' in that only the input/output characteristics need be known; detailed knowledge of the internal circuit is not essential to use the device. For example, consider a black box with input/output characteristic as shown in F7.39. If the sine wave pd of peak value 2 V is applied between the input terminals, the output pd can be deduced from the characteristics as follows: (i) $V_{in} \leqslant 0.5$ V, so $V_o = -10$ V; (ii) $V_{in} \geqslant 1.5$ V, so $V_o = +10$ V; (iii) $0.5 < V_{in} < 1.5$ V—in this range V_{in} is

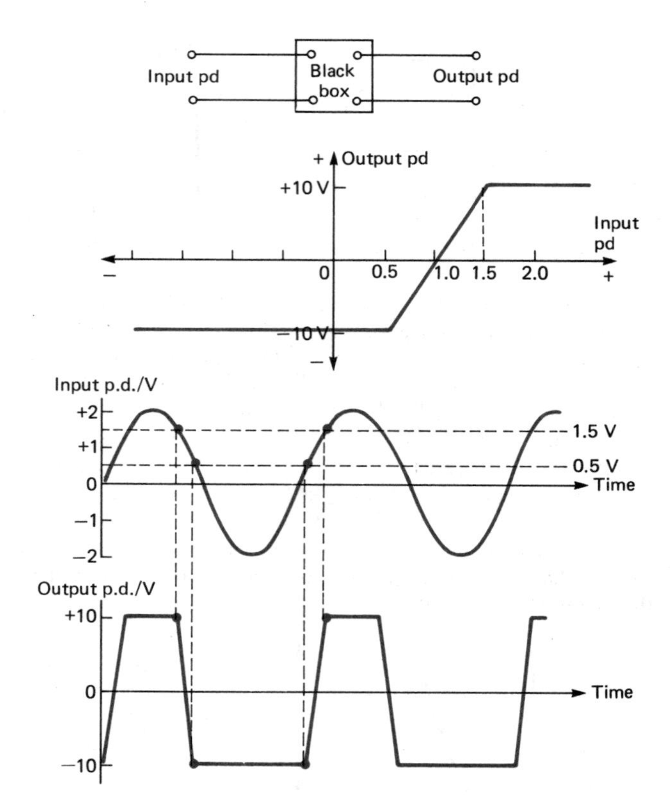

F7.39 Operational amplifier as a 'black box'

amplified, and since the voltage gain (from the gradient of the graph) is $\times 20$, then $V_o = 20(V_{in} - 1)$ in this example.

7.11 LOGIC CIRCUITS

Logic circuits give a simple introduction to **digital** electronics in which circuit units can have only two states, so that inputs and outputs are either high (ie, +ve saturation, termed a '1') or low (ie, −ve saturation, termed a '0').

The circuit in F7.40, will operate as a **logic switch** if the base resistance is sufficiently small. When the input pd is high, then the output pd is low; when the input is low, then the output will be high. The circuit is a NOT unit in logic terms; its symbol and truth table relating output to input is shown in F7.40.

The table of F7.41 shows several other simple logic units based upon transistors. The key to each is the 'truth table'.

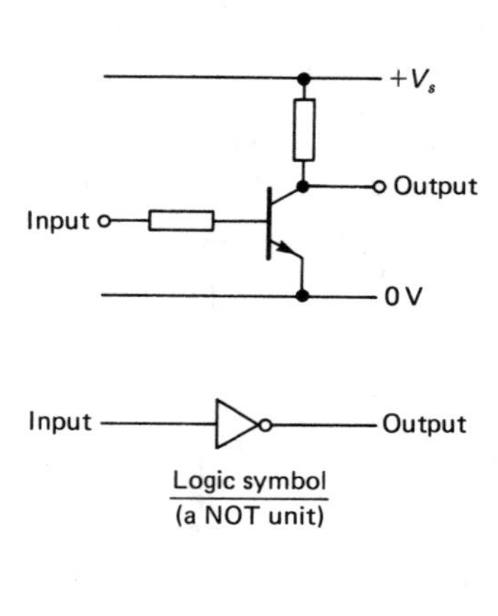

Truth table

Input	Output
0	1
1	0

F7.40 Transistor as a logic block

Name		NOR	OR	NAND	AND	AND constructed from 2 NOTs & 1 NOR
Logic symbol		A, B → V_0	A, B → V_0	A, B → V_0	A, B → V_0	A, B → V_0
Truth table	A	0 0 1 1	0 0 1 1	0 0 1 1	0 0 1 1	
	B	0 1 0 1	0 1 0 1	0 1 0 1	0 1 0 1	
	V_0	1 0 0 0	0 1 1 1	1 1 1 0	0 0 0 1	

F7.41 Logic gates

You should be able to work out the truth table of any simple combination of the above logic units. For example, consider the combination shown in F7.41 of two NOT units and a NOR unit; the output is 1 only when the two inputs are both 1, so its truth table is the same as that of an AND unit.

A **multivibrator** consists of two logic switches that are coupled together via either resistors or capacitors or both. With capacitor coupling, as in F7.42, the circuit is known as an **astable** multivibrator; its outputs change back and forth automatically between the two logic states such that when one output is high, the other is low. The diagram shows the output variation with time for one of its outputs.

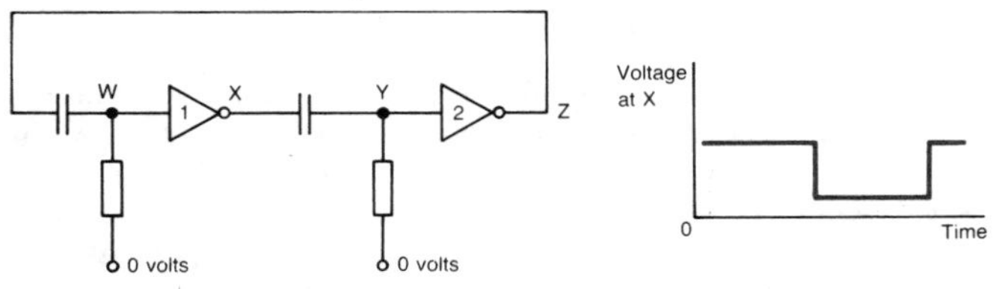

F7.42 The astable multivibrator

When the output of gate 1, X, goes high, the voltage at Y goes high temporarily. Thus Z and W go low while Y is high, keeping X high. However, as the capacitor between X and Y charges up, the voltage drops at Y. When the voltage at Y decreases below a certain level, gate 2 switches and Z goes high, taking W high temporarily – thus switching X off until the voltage at W decreases below a certain level. The sequence repeats itself automatically.

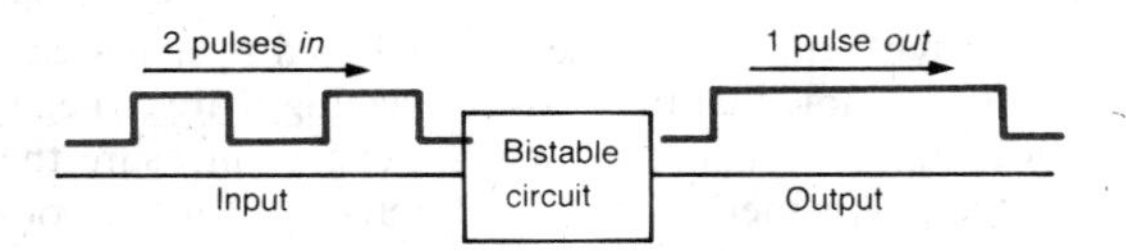

F7.43 The bistable multivibrator

A **bistable** multivibrator is a combination of logic gates that has two stable states; it can only be in one of these states at any time. An input pulse makes it change from one state to the other.

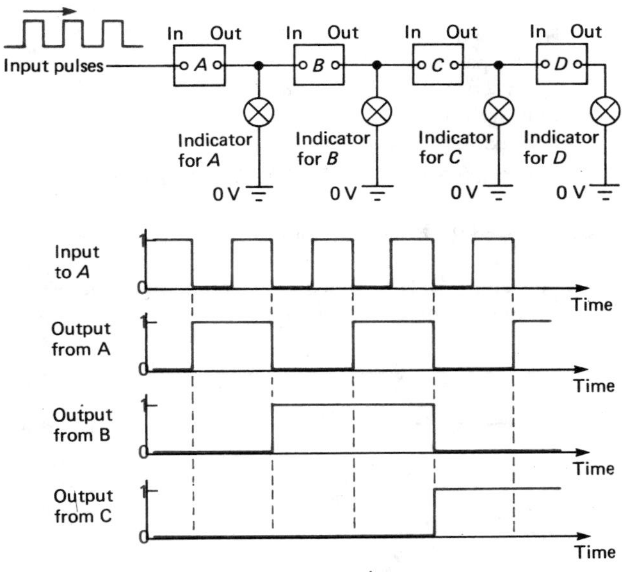

F7.44 Binary counting system

The bistable unit is the basis of the binary pulse counter, as in F7.44, in which a series of bistable units are connected together with indicator lamps (or light emitting diodes) to demonstrate the state of each bistable unit.

Fall of the input at A causes the input at B to change state, and when B's input changes from a 1 to a 0, then C changes state, etc; starting with all outputs at 0, the diagram shows the state of the outputs after n input pulses. The output states represent the binary number for n in each case; for example, after 4 input pulses (and before the fifth), output C is a 1 and all other outputs are 0s. The outputs in order $DCBA$ are therefore 0100 which is the binary number for 4.

Digital transmission Information may be transferred as a sequence of pulses, each pulse caused by switching electrical or electromagnetic carrier waves on and off.

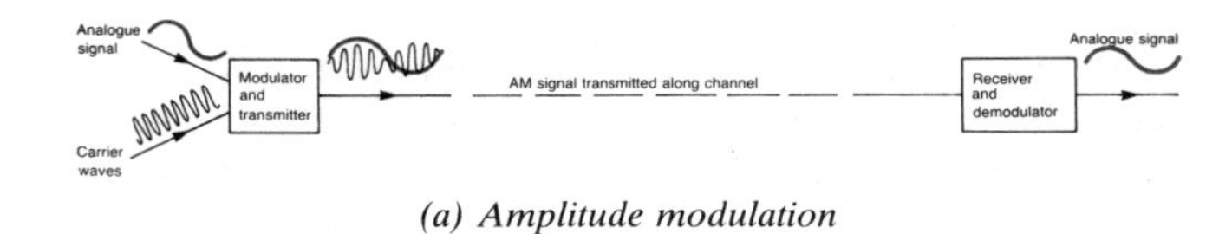

(a) Amplitude modulation

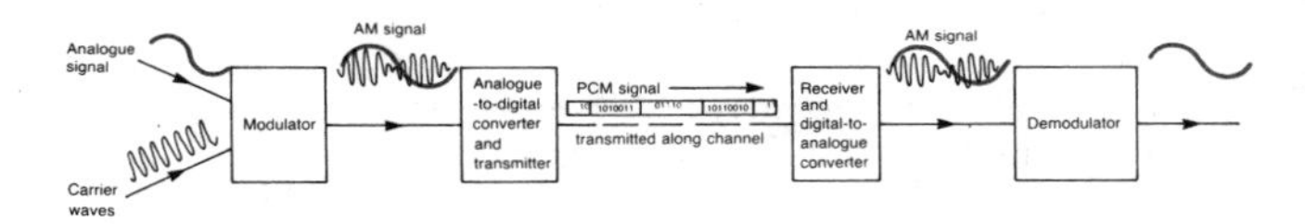

(b) PCM transmission

F7.45

Pulse code modulation (PCM) involves converting an analogue signal into a string of 1's or 0's representing the signal amplitude in binary form. The pulses may need to be **amplified** and **regenerated** at intervals along the transmission path to remove unwanted signals due to 'noise'. At the receiver, the pulses are used to recreate the original analogue signal. In comparison, **amplitude modulation (AM)** is where the analogue signal varies the amplitude of the carrier waves. Noise cannot be removed and amplifiers boost the noise as well as the signal.

UNIT 7 QUESTIONS

7.1M The diagram shows a piece of pure semiconductor, S, in series with a variable resistor, R, and a source of constant voltage, V. S is heated and the current is kept constant by adjustment of R. Which of the following factors will decrease during this process?

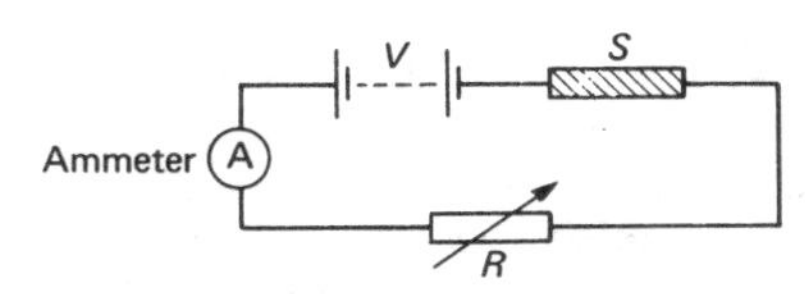

1 The drift velocity of the conduction electrons in S.
2 The dc resistance of S.
3 The number of conduction electrons in S.
Answer: A if **1**, **2**, **3** correct.

B if **1**, **2** correct.
C if **2**, **3** correct.
D if **1** only.
E if **3** only.

(**London**: *NISEC and all other Boards except Oxford, SEB H and SEB SYS*)

See p. 108 for the behaviour of semiconductors. Note that the drift velocity in semiconductors is affected by an increase in temperature in the same way as in metals (ie, reduced, due to an increase in the number of collisions with the lattice ions, which vibrate with greater amplitude at a higher temperature.)

7.2M The current in a copper wire is increased by increasing the potential difference between its ends. Which one of the following statements regarding n, the number of charge carriers per unit volume in the wire, and $\bar{v}$, the drift velocity of the charge carriers, is correct?

A n is unaltered but $\bar{v}$ is decreased.
B n is unaltered but $\bar{v}$ is increased.
C n is increased but $\bar{v}$ is decreased.
D n is increased but $\bar{v}$ is unaltered.
E Both n and $\bar{v}$ are increased.

(**Cambridge**: *and all other Boards except Oxford, SEB H and SEB SYS*)

See p. 101. The situation is not unlike that of water flow from a hosepipe; increased pressure makes the water flow faster but does not change the water density.

7.3M X and Y are resistors of value 6 ohms each. C is a cell of emf 12 V and internal resistance 3 ohms. A is an ammeter of

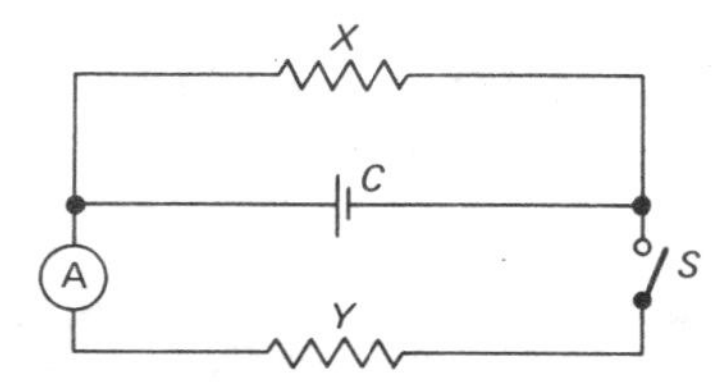

negligible resistance. When switch S is closed, the reading of A, in amperes, is:

A $\frac{1}{2}$. B 1. C 2. D 4. E 8.

(**AEB** June 81: *all other Boards*)

X and Y form a parallel combination and since their values are equal, the resistance of the combination is half the individual value. The total resistance in series with the cell is the sum of the resistance of the parallel combination and the internal resistance of the cell. Use E7.7 to calculate the cell current, and thus the current through the meter.

7.4M The circuit shown in the diagram contains a battery, a rheostat and two identical lamps. What will happen to the

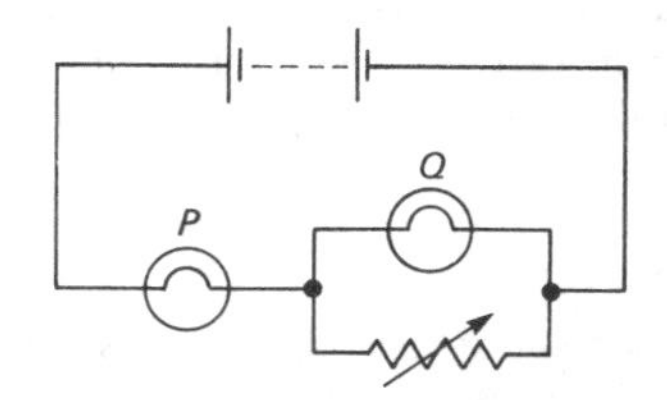

brightness of the lamps if the resistance of the rheostat is increased?

	Lamp P	**Lamp Q**
A	Less bright	Brighter
B	Less bright	Less bright
C	Brighter	Less bright
D	Brighter	Brighter
E	No change	Brighter

(**AEB** Nov 80: *and all other Boards*)

It is useful to consider two extreme cases: (i) When the resistance of the rheostat is zero the current through Q is zero since Q is short-circuited. The circuit is then essentially a battery in series with lamp P. (ii) When the resistance of the rheostat is very large almost no current flows through it so the currents through P and Q are almost equal. The circuit is essentially a battery in series with lamps P and Q.

7.5M In the circuit diagram below, two resistors X and Y are connected to a 12 V battery of negligible internal resistance. The resistance of X is known to be 6000 ohms. A voltmeter of internal

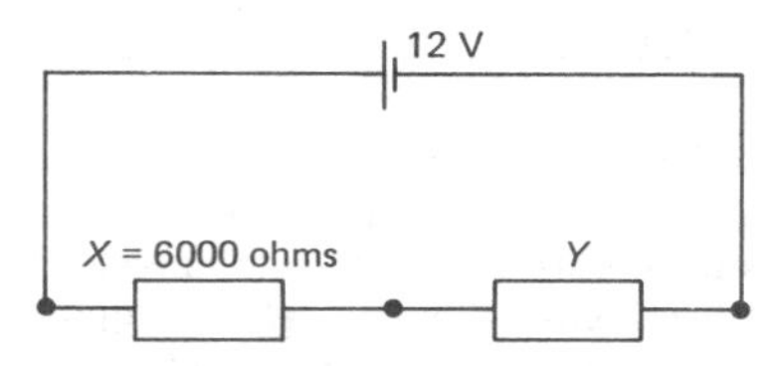

resistance 6000 ohms is then connected across X, and its reading is found to be 9.0 V. What must be the resistance of Y in ohms?

A 7500. B 6000. C 5000. D 2500. E 1000

(—: *all Boards*)

With 9 V across X when the meter is connected in circuit, the remainder of the battery pd is dropped across Y. Also, the current through Y = current through X + current through the meter.

7.6M A voltage supply has an emf E and an internal resistance r. A load of variable resistance R is connected across the supply with leads of negligible resistance. Which of the following statements is/are correct?

1 The current in the circuit is a maximum when the load R is equal to the internal resistance r.
2 The potential difference across the load R is less than the potential difference across the terminals of the power supply whilst the load is connected.
3 The power delivered to the load is a maximum when the load R is equal to the internal resistance r.

A **1** only. B **2** only. C **3** only. D **1** and **3** only.
E **2** and **3** only.

(**NISEC**: *and AEB, O and C*, SEB H, and O and C Nuffield*)

Make a quick sketch of the circuit diagram. For **1**, consider how the current will vary as the external load is increased from zero to r. For **2**, note that the leads connecting R are of negligible resistance. Imagine connecting a voltmeter to measure the pds between the points mentioned. For **3**, refer to your textbook for the maximum power theorem.

7.7M A galvanometer gives a full-scale deflection when a current of 2 mA flows through it and the potential difference across its terminals is 4 mV. Which of the following resistors would be most suitable to convert it to give a full-scale deflection for a current of 1 A?

A 0.004 Ω in series.
B 0.004 Ω in parallel.
C 0.50 Ω in series.
D 500 Ω in series.
E 500 Ω in parallel.

(**London**: *and all other Boards except O and C, SEB H, SEB SYS, NISEC*)

See F7.6 and the discussion on p. 102.

7.8M In the circuit, XY is a potentiometer wire 100 cm long. The circuit is connected up as shown. With switches S_2 and S_3 open

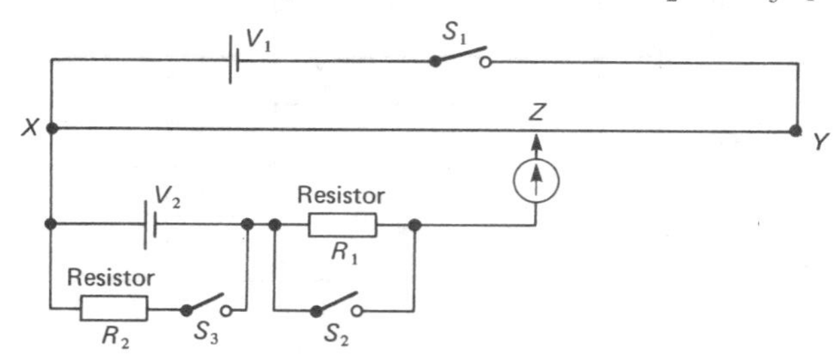

a balance point is found at Z. After switch S_1 has remained closed for some time, it is found that the contact at Z must be moved towards Y to maintain a balance. Which of the following is the most likely reason for this?
A The cell V_1 is running down.
B The cell V_2 is running down.
C The wire XZ is getting warm and its resistance is increasing.
D The resistor R_1 is getting warm and increasing in value.
E Polarization is affecting the emf of V_2.
(**London**: *and all other Boards except O and C Nuffield, SEB H, SEB SYS*)

Ignore S_3 and R_2 since they relate to the following question. Remember that a longer balance length corresponds to either V_1 being smaller or V_2 being larger than at the start. Which is the more likely?

7.9M The potentiometer in 7.8M is balanced as before with S_3 open. S_3 is then closed and the balance point again found. What happens to the balance length XZ when S_3 is closed if the internal resistance of the cell V_2 is NOT negligible?
A no change occurs.
B XZ increases, the increase being greatest for large values of R_2.
C XZ increases, the increase being greatest for small values of R_2.
D XZ decreases, the decrease being greatest for large values of R_2.
E XZ decreases, the decrease being greatest for small values of R_2.
(**London**: *and all other Boards except O and C Nuffield, SEB H, SEB SYS, and WJEC*)

With S_3 closed and the balance point located, V_2 will deliver current through R_2 and S_3 only. Since V_2 delivers current, then there will be a loss of pd across its internal resistance (see p. 103 if necessary). Will this loss change the pd (from the emf value) across the terminals of V_2? If the pd across the terminals of V_2 changes, then the balance length must change in proportion (compared with the value when S_3 was open).

7.10M In the circuit shown below the e.m.f. of a thermocouple is to be measured using a potentiometer wire PQ in series with a resistor R and a driver cell.

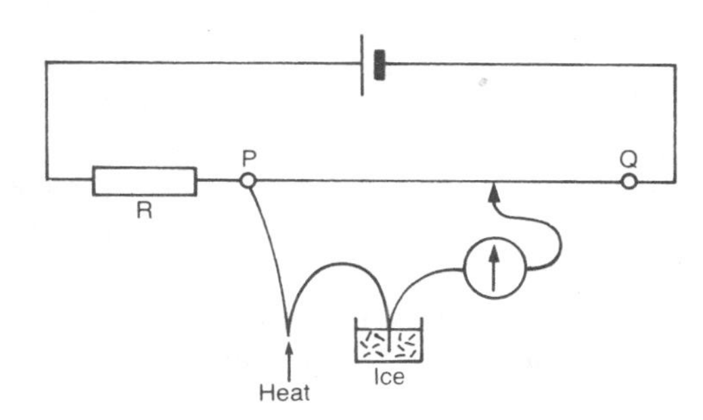

It is found impossible to obtain a balance point anywhere along the wire PQ. Which of the following may be possible causes?
1 The resistance of R may be too large
2 Connections of the thermocouple to P and to the galvanometer are the wrong way round
3 The resistance of the galvanometer is too low

A 1, 2 only correct B 2, 3 only correct
C 1 only correct D 3 only correct
(**AEB Nov '89**: *and all other Boards except O and C Nuffield, SEB H and SEB SYS*)

See p. 104 if necessary. Note that the galvanometer resistance does not affect the position of the balance point.

7.11M The balance point of a slide wire Wheatstone bridge will **not** be changed by increasing:
1 the emf of the driver cell.
2 the resistance of the slide wire.
3 the galvanometer resistance.
A **1, 2, 3** correct. B **1** and **2** correct.
C **1** and **3** correct. D **1** only correct. E **3** only correct.
(*—:JMB*, Oxford, O and C, SEB H*)

What is the equation which represents the balance condition? See E7.11 if necessary. Does that equation contain any of the above factors? For **2**, remember that it is the **ratio** of the resistances of the two sections of the slide wire that is important.

7.12M A box is known to contain three **identical** capacitors wired together in a circuit containing no other components. Two wires lead from this circuit to the outside of the box, and the measured capacitance between these wires is 30 μF. Which one of the following could be the correct capacitance of each capacitor?
A 15 μF. B 20 μF. C 40 μF. D 60 μF. E 100 μF.
(**London**: all other Boards except SEB H and SYS)

Combination rules for capacitors are given on p. 105. There are four possible arrangements:
1 All in parallel: will any of the alternatives give 30 μF in total?
2 All in series: the combined capacitance will be $\frac{1}{3}$ of the individual capacitance for three identical capacitors in series. Will any alternative give 30 μF in total?
3 Two in parallel wired with one in series: Let C be the individual capacitance, so write down the total capacitance in terms of C. Can you now choose a value for C from the alternatives that gives 30 μF in total?
4 Two in series wired with the third in parallel with the series pair: Again, let C be the individual capacitance, and use the same approach as in 3.

7.13M The combined capacitance of the arrangement shown below, in μF, is:
A 1. B $\frac{18}{11}$. C $\frac{30}{11}$. D 4. E 11.

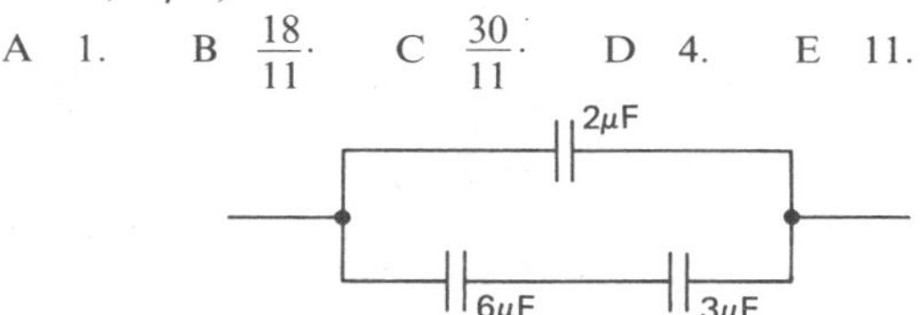

(—: *all other Boards except SEB H and SYS*)

Determine the capacitance of the pair in series firstly (see E7.14) and then consider the parallel combination (see E7.15, if necessary).

7.14M The question below consists of an ASSERTION (statement) in the left-hand column and a REASON in the right-hand column.
Select the answer:
A if both assertion and reason are true statements and the reason is a correct explanation of the assertion.
B if both assertion and reason are true statements but the reason is NOT a correct explanation of the assertion.
C if the assertion is true but the reason is a false statement.
D if the assertion is false but the reason is a true statement.
E if both assertion and reason are false statements.

ASSERTION		**REASON**
When several capacitors are connected in parallel the total capacitance is equal to the sum of the individual capacitances	because	If several uncharged capacitors are connected in parallel across a battery all must acquire the same charge.

(**NISEC**: *all other Boards except SEB H and SYS*)

Make sure you understand the instructions for the question. The assertion and reason must first be considered **separately**. Only if both stand up as true statements do you need to think about whether the reason is a correct explanation of the assertion. For the assertion, see p. 105 if necessary. For the reason as a separate statement, remember that components in parallel have the same pd.

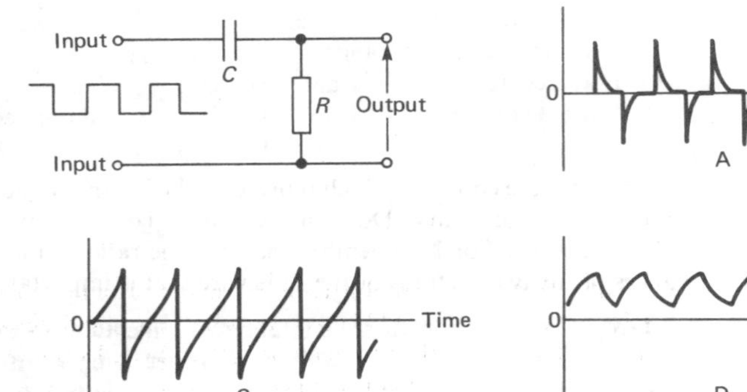

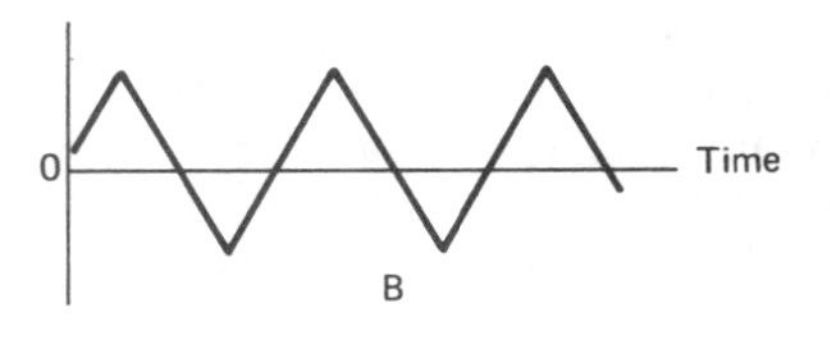

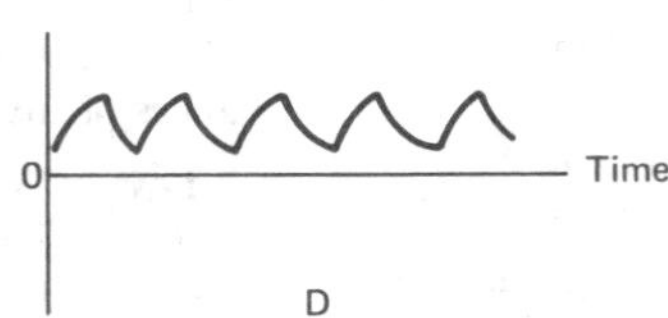

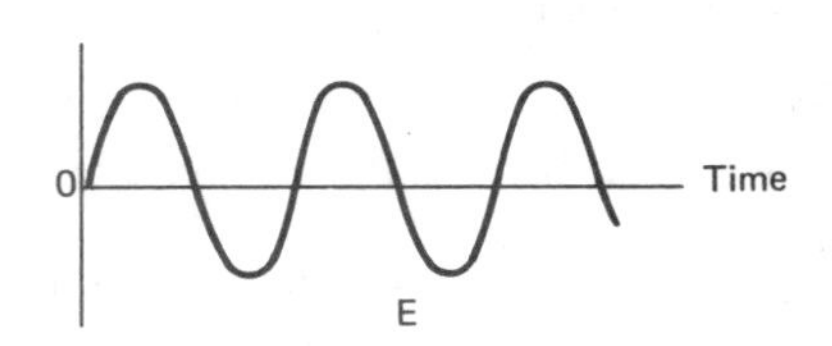

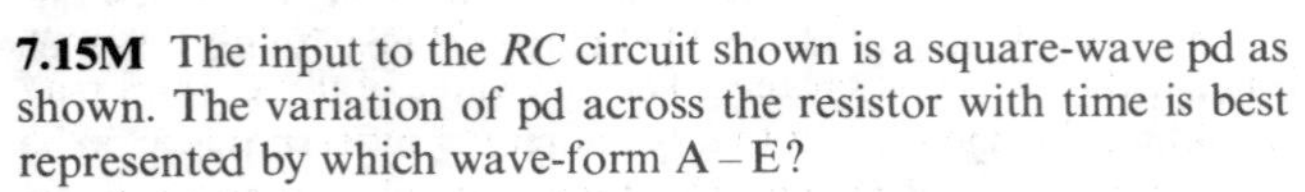

7.15M The input to the *RC* circuit shown is a square-wave pd as shown. The variation of pd across the resistor with time is best represented by which wave-form A–E?

(—: *Oxford*, O and C* and all other Boards*)

Remember that the pd across a resistor is proportional to the current. Each time that the input changes from low to high, *C* charges up thus taking a current that drops to zero as *C* charges. When the input changes from high to low, *C* discharges so a discharge current flows through *R* (in the reverse direction to the charging current).

7.16M An alternating current of 1.5 mA rms and angular frequency $\omega = 100$ rad s^{-1} flows through a 10 kΩ resistor and a 0.50 μF capacitor in series. The rms pd across the capacitor is:
A 4.8 V. B 15 V. C 30 V. D 34 V. E 190 V.

(**Cambridge**: *and Oxford*, all other Boards except SEB H*)

The circuit is a series *RC* circuit; calculate the capacitor reactance using E7.19 and referring to the comments on p. 107. Then calculate the rms pd (= rms current × capacitor reactance).

7.17M In the circuit, if the rms voltage of the ac supply remains

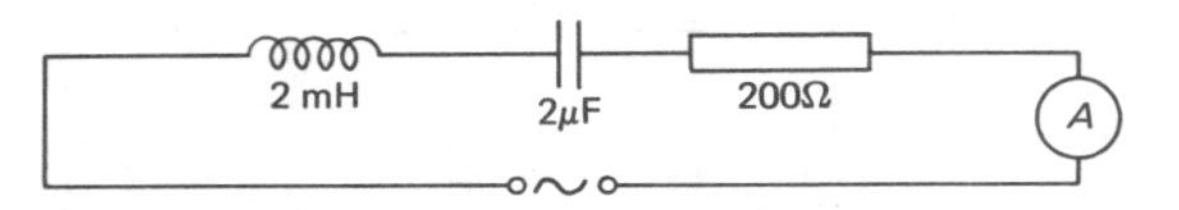

constant when the frequency is varied between 50 Hz and 50 kHz, which of the graphs below best illustrates the variation in current through the ammeter with frequency?

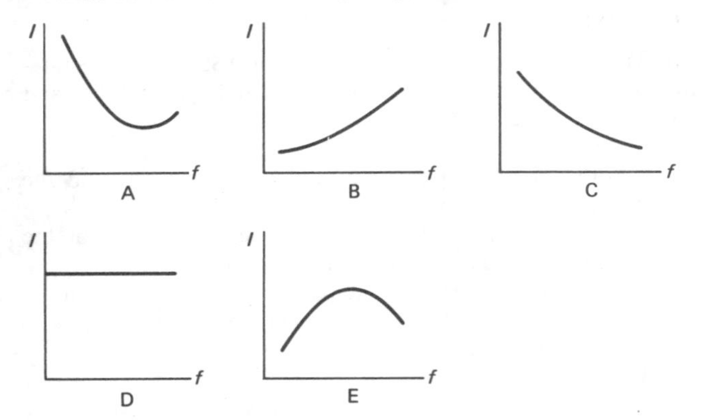

(**AEB** Nov 81: *and O and C*, all other Boards*)

For a series *LCR* circuit the resonant frequency is that at which the current is a maximum for a given supply voltage. The value of this frequency may be calculated using E7.24.

7.18M In a circuit containing a capacitor, an inductor and a resistor in series, V_C, V_L and V_R represent the potential differences across those three components and *I* represents the current through them. Which of the following statements is (are) true?
1 V_C and *I* are 180° out of phase.
2 V_R and *I* are 90° out of phase.
3 V_L and V_C are 180° out of phase.

Answer:
A if **1**, **2**, **3** are correct. B if **1**, **2** correct.
C if **2**, **3** correct. D if **1** only E if **3** only.

(**London**: *all other Boards except SEB H, SEB SYS*)

See p. 108 if necessary.

7.19M A moving-coil ammeter is to be adapted to detect small alternating currents. Which of the diagrams shows how a diode could be connected in order to make the conversion?

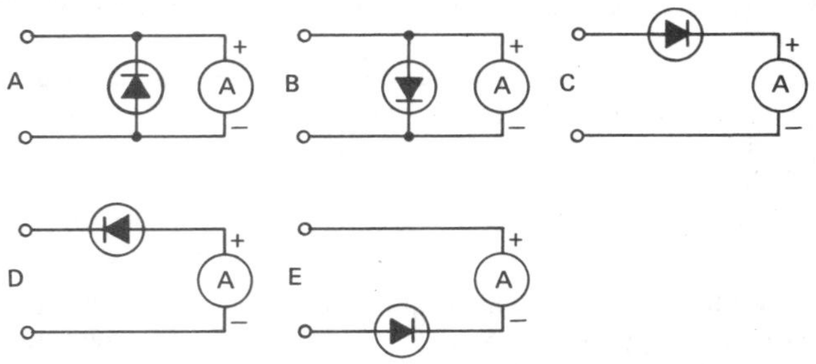

(**London**: *all other Boards except SEB H, SEB SYS*)

The current must only pass through the meter in the direction + to −. The diodes in A and B will not stop the reverse current, so can be eliminated. Remember that a diode conducts in the direction in which the triangle of the symbol is pointing.

7.20M In the circuit shown, a capacitor of 10 μF capacitance is connected in parallel with a coil of inductance 1 mH and negligible resistance. The combination is then connected in series with an ac supply of variable frequency and an ac meter.

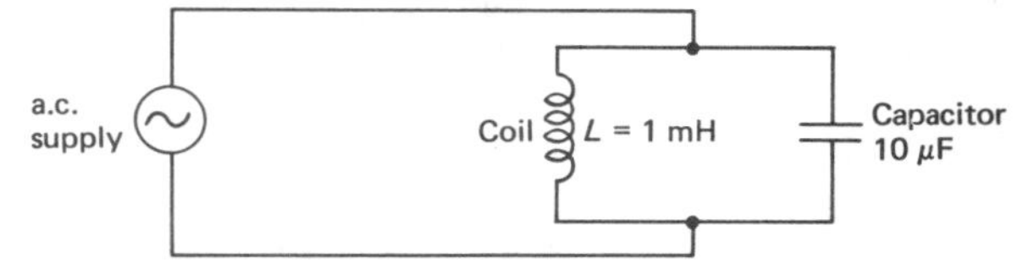

The frequency of the supply is then adjusted to 1600 Hz and the output is adjusted until the meter reads 10 mA. Without changing the output pd of the supply, the meter reading will increase when:
1 the frequency is changed to 2000 Hz.
2 the frequency is decreased to 1200 Hz.
3 the capacitance is decreased to 5 μF.
A **1** only. B **2** only. C **1** and **3**. D **2** and **3**.
E **1**, **2** and **3**.

(—: *O and C Nuffield*)

The current from the supply is at a minimum at the resonant frequency. Hence the current will drop if any of the changes 1, 2, 3 result in moving closer to resonance. Conversely the current will increase only if the suggested change results in moving further away from the resonance position. E7.24 can be used to determine the resonance frequency, this frequency is the same for 1 and 2 but changes for 3.

7.21M The resistance of an intrinsic semiconductor falls when it is heated because:
A the atoms vibrate more.

B more electrons are produced.
C the number of electron vacancies falls.
D the solid expands.
E more free electrons become available.

(—: *NISEC* and all other Boards except SEB H, and SEB SYS*)

See p. 108 if necessary.

7.22M The circuit diagram shows a logic 'combination' with the states of outputs X, Y and Z given for inputs P, Q, R and S all at

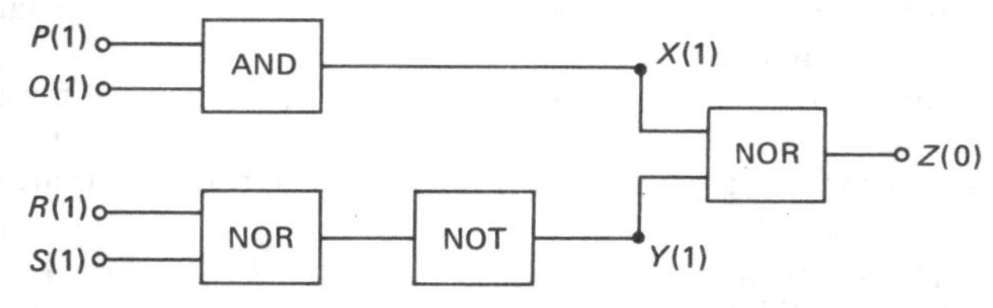

state 1 (ie, high). When inputs P and R change to state 0 (ie, low), with inputs Q and S still at 1, the condition of outputs X, Y and Z changes to:

	A	B	C	D	E
X	1	1	0	0	0
Y	0	1	1	0	1
Z	0	1	0	1	1

(—: *see page 13*)

See F7.41 for truth tables of AND, NOR and NOT units.

7.23S The overhead cables used in a 132 kV grid system consist of 7 strands of steel wire and 30 strands of aluminium wire. The 7 strands of steel wire have a combined resistance of 3.0 ohms per kilometre and the 30 strands of aluminium wire have a combined resistance of 0.17 ohms per kilometre.
(i) Show that the resistance of the cable is 0.16 ohms per kilometre of cable.
(ii) A typical current in the cable is 400 A. Calculate the power loss per kilometre.

(**SEB H**: *all other Boards*)

(i) Each kilometre of cable is equivalent to two resistors of 3 ohm and 0.17 ohm in parallel.
(ii) Power loss occurs in the cable since its resistance will cause heating. Since electrical power is given by current × pd, then the heat produced per second = I^2R.

7.24S
(a) Explain what is meant by the 'electromotive force' and the 'terminal potential difference' of a battery.
(b) A bulb is used in a torch which is powered by two identical cells in series each of emf 1.5 V. The bulb then dissipates power at the rate of 625 mW and the pd across the bulb is 2.5 V. Calculate (i) the internal resistance of each cell and (ii) the energy dissipated in each cell in one minute.

(**JMB**: *and all other Boards*)

(a) See p. 102.
(b) Draw the circuit consisting of the two cells and the torch bulb in series with one another. You should show the internal resistance of each cell clearly. Calculate the current taken by the bulb from the power and pd values given. See E7.4 if necessary. Then, calculate the pd across each internal resistance, using the fact that the total emf is 3.0 V, but only 2.5 V is dropped across the external load (ie, the bulb). Once you have determined the current in the circuit and the pd across each internal resistance (the lost voltage: see p. 102 if necessary), then calculate a value for internal resistance. For (ii), use E7.4 to calculate the power dissipated in each internal resistance (remember to use the 'lost voltage' not the cell emf for power dissipated). Then, since power = energy per second, calculate the energy dissipated in 1 minute in each cell.

7.25S A moving-coil ammeter at 0°C has a resistance of 5 ohms and gives a full-scale deflection for a current of 15 mA. What value of shunt resistance is required so that it reads 1.50 A at 0°C? What circuit current will give a full-scale deflection of the instrument if the temperature of the shunt rises to 25°C while the instrument remains at 0°C? (Temperature coefficient of resistance of material of shunt = $4.0 \times 10^{-3}\,K^{-1}$, ie, the resistance of the shunt increases by 0.4% of its resistance at 0°C for every degree rise in its temperature.)

(**AEB** June 81: *Oxford, O and C and London only*)

Using the same notation as in F7.6 we have $I = 1500$ mA, $i = 15$ mA and $I - i = 1485$ mA. By equating the pd across the shunt to that across the ammeter at full-scale deflection a value for the shunt resistance at 0°C may be found.

The resistance of the shunt at 25°C can be found using the formula given on p. 90. The change in shunt resistance causes an alteration in shunt current. The new value of this current may be obtained using the fact that the potential difference across the shunt remains unchanged. The total current is then given by adding the new shunt current to i.

7.26S (a) Fig. 1 shows an operational amplifer used with two resistors to amplify small a.c. sine-wave signals.

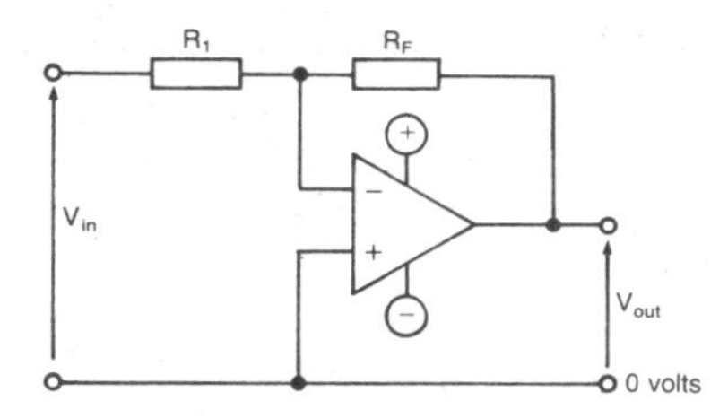

(i) What value of R_1 is necessary to give a voltage gain of −10 if R_F is 1.0 MΩ?
(ii) What is the function of the pair of resistors?
(iii) Sketch the input and output voltage waveforms when a sine wave of frequency 50 Hz and peak voltage 0.50V is applied at the input terminals.
(b) In Fig. 2, VR is a variable resistor in series with a fixed resistor R_1 and T is a thermistor in series with a fixed resistor R_2.

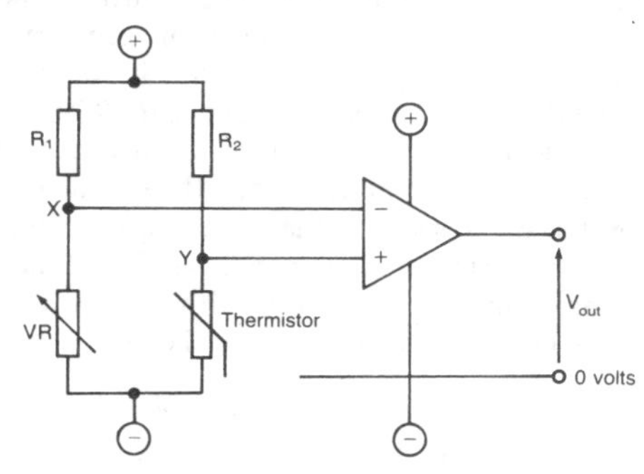

(i) How does the voltage at Y vary as the thermistor temperature is increased?
(ii) With the thermistor at constant temperature, VR is gradually increased until the output voltage switches from positive to negative saturation. Explain why the output voltage switches at a certain point as VR is increased.
(iii) The thermistor is now cooled without changing VR. Describe and explain what happens now to the output voltage.

(—: *all Boards*)

(a) (i) See E7.28 and the inverting amplifier.
(ii) See 'negative feedback' comments immediately after E7.28.
(iii) Sketch input and output waveforms on the same axes with the same time period and with the peak output voltage ten times the peak input voltage. Remember to show the two waveforms out of phase by 180°.
(b) (i) VR and R_1 form a potential divider and the thermistor and R_2 form a second potential divider. Increasing the lower resistance in either potential divider raises the voltage supplied by the potential divider. Remember the thermistor's resistance

increases as it cools.
(ii) and (iii) Each potential divider in Fig. 2 supplies voltage to the open-loop op-amp. Work out if the output is at + or − saturation according to whether or not the voltage at Y is more than or less than the voltage at X.

7.27S A student wanted to light a lamp labelled 3 V 0.2 A but only had available a 12 V battery of negligible internal resistance. In order to reduce the battery voltage he connected up the circuit shown in Fig. 1. He included the voltmeter – using it rather stupidly – so that he could check the voltage before connecting the lamp between *A* and *B*. The maximum value of the resistance of the rheostat *CD* was 1000 Ω.

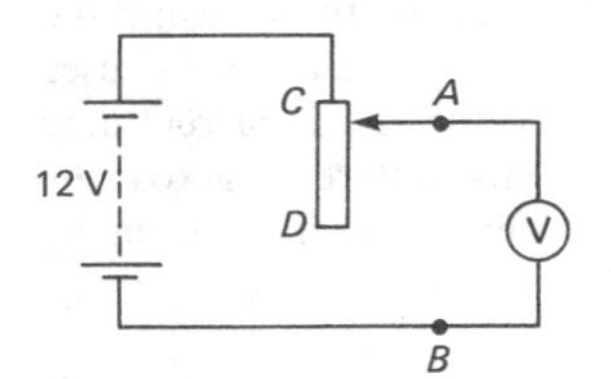

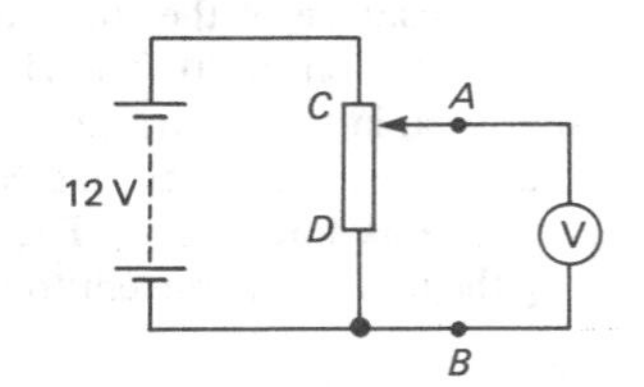

(a) He found that, when the sliding contact of the rheostat was moved down from *C* to *D*, the voltmeter reading dropped from 12 V to 11 V. What was the resistance of the voltmeter?

(b) He modified his circuit as shown in Fig. 2, using the rheostat as a potentiometer, and was now able to adjust the rheostat to give a meter reading of 3 V. What current would now flow through the voltmeter?

(c) Assuming that this current is negligible compared with the current through the rheostat, how far down from *C* would the sliding contact have been moved?

(d) The student then removed the voltmeter and connected the lamp in its place, but it did not light. How would you explain this? (The lamp itself was not defective.)

(**O and C Nuffield**: *and all other Boards*)

(a) The rheostat and voltmeter are in series here so they have the same current. Hence $(V/R)_{\text{meter}} = (V/R)_{\text{rheostat}}$, etc.

(b) Use your value for voltmeter resistance from **(a)** and the given reading of 3 V to calculate the current through the meter.

(c) See F7.10(b) and related comments if necessary.

(d) When the lamp replaces the meter, it effectively short-circuits the section of the rheostat from *A* to *B* because of the lamp's comparatively low resistance (check the statement using the lamp's rating as given). Thus the circuit resistance is provided by the other section (the upper part) of the rheostat only. How much resistance is in the upper section of the rheostat? What is the maximum battery current with this resistance, and is that amount of current sufficient to light the lamp?

7.28S Two resistance wires *A* and *B*, made of different materials, are connected into a circuit with identical resistors R_1 and R_2 ($R_1 = R_2$), a sensitive high resistance galvanometer *G*, a cell *C* and a switch *S*, as shown in the diagram.

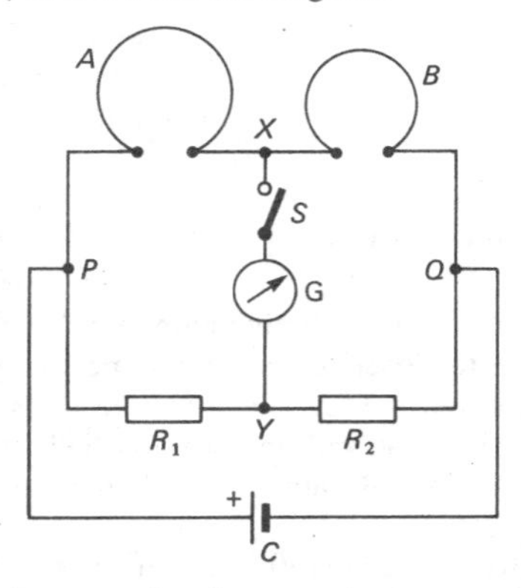

(a) If *A* and *B* have equal resistance, no current will flow through *G* when the switch *S* is closed even though the cell is still delivering a current. Explain why this is so.

(b) The diameter of *A* is twice that of *B* and the resistivity of the material of which *B* is made is 6×10^{-6} Ωm. It is found that for zero current through *G* the length of *A* has to be three times that of *B*. Calculate, showing your working, the resistivity of the material of which *A* is made.

(c) If the length of wire *B* is now reduced by a small amount so that the current through *G* is no longer zero, say which way the current will flow through *G* and explain why.

(**O and C Nuffield**: *and JMB*, Oxford, O and C, SEB H*)

(a) See p. 105 for the Wheatstone bridge circuit which is what the above circuit is.

(b) Remember that R_1 equals R_2 so the resistance of *A* must be equal to the resistance of *B*. See E7.8 for resistivity; equate $\rho L/A$ for wire *A* to $\rho L/A$ for wire *B*, and remember that the area ratio = (diameter ratio)^2.

(c) At balance, the pd from *X* to *Q* is equal to that from *Y* to *Q*. Suppose *S* is now opened, and **then** *B* is shortened; will the change of *B*'s resistance increase or decrease the pd from *X* to *Q*, and will *X* be more positive than *Y* as a result... or less positive? Once you have decided, you can then state which way current passes between *X* and *Y* when *S* is closed. The above reasoning should give you the basis of your explanation.

7.29S

(a) (i) What is the internal arrangement of a capacitor? (ii) A capacitor has an associated quantity called 'capacitance'. Define capacitance. (iii) List three factors which determine the capacitance of a parallel-plate capacitor. In each case state the relationship between the factor quoted and the capacitance. (iv) The sketch shows the plates of a charged capacitor. Indicate the shape and direction of the electric field lines between the plates. (v) Calculate the energy stored in a capacitor of capacitance 2μF when the potential difference between the plates is 100 V.

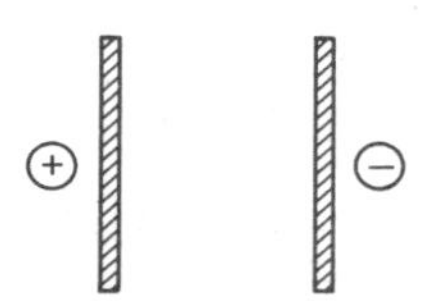

(b) A 2μF capacitor is charged to 100 V. It is then disconnected from the supply and connected in parallel with a 3 μF capacitor. (i) Calculate the capacitance of the combination. (ii) Find the voltage now occurring between the terminals of the capacitors. (iii) If the capacitors are then discharged and re-connected in series, calculate the capacitance of the series combination.

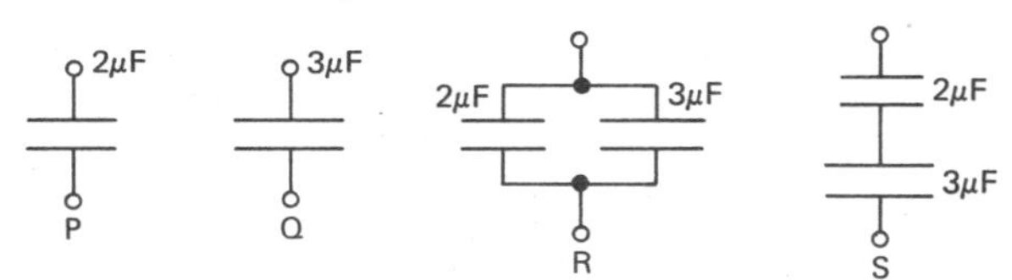

(c) (i) Each of the capacitor combinations *P, Q, R, S* is charged by connecting it to a 100 V supply. If each combination is then discharged through the same resistance the initial rate of fall of pd will be different in each case. For which of the combinations would this rate be greatest? (ii) On the axes provided sketch a graph to represent the variation with time *t* of the potential difference V_c across one of the combinations during a discharge.

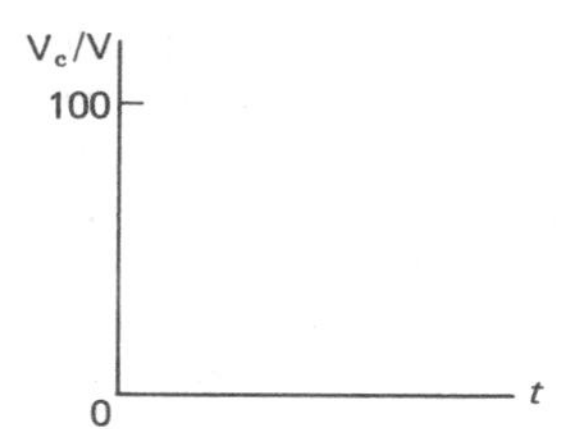

(**NISEC**: *all other Boards except SEB H and SYS*)

(a) (i) See your textbook. (ii) See p. 105. (iii) See E2.12.

(iv) See F2.1. (v) Use equation E2.14. Note that to obtain the energy in joules the capacitance must be expressed in farads.

(b) (i) Use E7.15. (ii) Use E7.13 to calculate the initial charge on the 2 μF capacitor. Now apply the same equation to the combination to find the new value of V (the total charge is unchanged, the total capacitance is that found in (b)(i)). (iii) Use E7.14.

(c) (i) The rate of fall of pd depends on the time constant RC. See p. 105. (ii) See F7.19.

7.30S In the circuit shown, the parallel-plate capacitors are identical except that the distances apart of the plates, d, are as

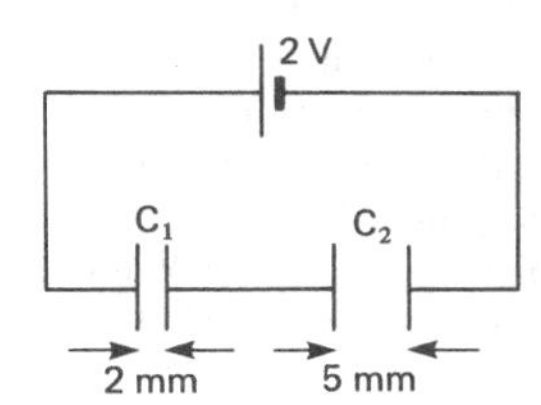

shown. Find the potential difference across **each** capacitor and the electric field intensity between **each** pair of plates.

(**WJEC**: *all other Boards except SEB H and SYS*)

There are two possible methods here:

1 The formal method: calculate the combined capacitance (see E7.14 if necessary). Then calculate the charge stored by the combination (see E7.13). Since each capacitor stores that amount of charge (because the two are in series), then use E7.13 again to calculate the pd across each capacitor. Once you have calculated the pd across each capacitor, then use E2.9 to calculate E across each gap.

2 The 'intuitive' approach: the two capacitors have the same charge since they are in series, so the pd ratio is the inverse of the capacitor ratio (because $V = Q/C$); since the capacitance is inversely proportional to the spacing, then the pd ratio is therefore proportional to the spacing ratio (ie, 2/5). Since the pds add up to 2 V, then you can calculate the pd across each. Then proceed as in **1** for E.

Yet another approach is to use the fact that the electric field strength depends only upon charge and area. Since the charge and area of the two capacitors are the same, then the electric fields must be the same, so the situation is identical (as regards E) to having a single 7 mm gap with 2 V across. So E can be calculated directly, and then the pd across the 2 mm gap can be determined from the value for E.

7.31S Explain what is meant by the root mean square value of an alternating current. Why is this value used when making calculations involving the power dissipated in a resistor carrying an alternating current?

(**AEB** Nov 81: *all other Boards except SEB H*)

Begin your answer by defining the rms value of an ac as the square root of the mean value of the square of the current taken over a complete cycle. Give a diagram to help explain what is meant by the 'mean value', etc.

The second part of the question is best answered by reference to p. 106.

7.32S In the circuit shown, the source has negligible internal impedance. Find **(a)** the rms current in the circuit, **(b)** the mean rate of heat production.

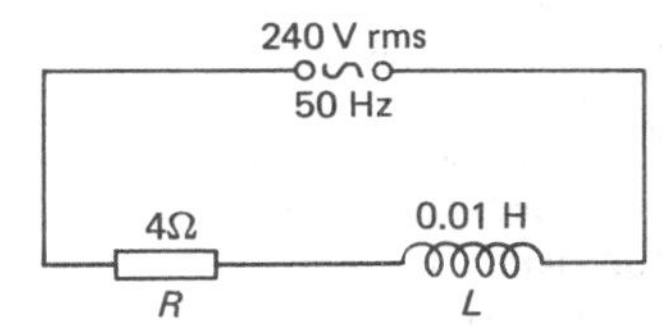

(**WJEC**: *and Oxford*, O and C*, and all other Boards except SEB H*)

(a) Use an appropriately modified form of E7.22 to calculate the circuit impedance. To modify E7.22, leave out the capacitance term as discussed on p. 108 after the equation. The rms current can then be calculated from the impedance value and the rms value of the supply pd.

(b) Heat is only produced in the resistor of the circuit. See the comments on 'power dissipation' on p. 108.

7.33S Fig. 1 shows how, for a self-contained electronic unit enclosed in a box, the output voltage V_{out} changes when the input voltage V_{in} is varied over a range from -4 V to $+4$ V.

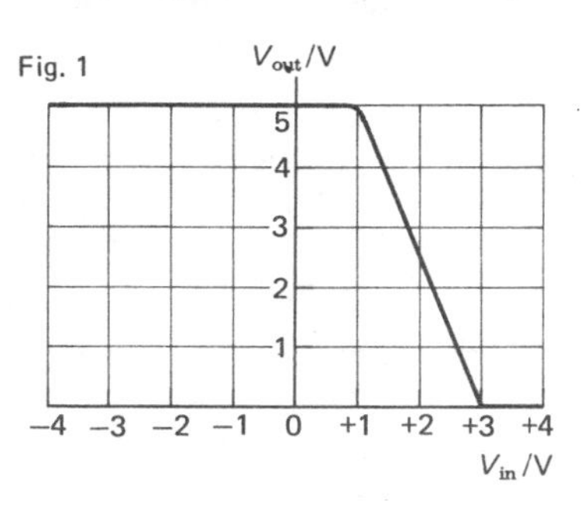

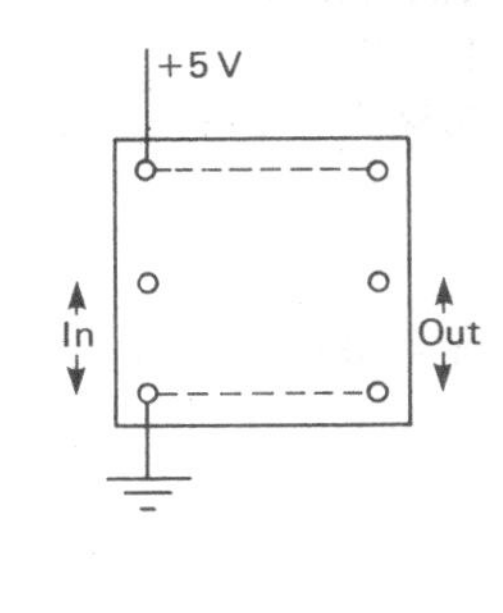

(a) The box, with its input and output terminals marked, is shown in Fig. 1. Add to the drawing of the box input and output circuits which could have been used to obtain the values of V_{in} and V_{out} used in plotting Fig. 1. Label your added circuit components.

(b) Two different inputs to the box are represented on the upper pair of axes as in Fig. 2. On the axes already provided under each input, carefully draw graphs representing the corresponding outputs from the box.

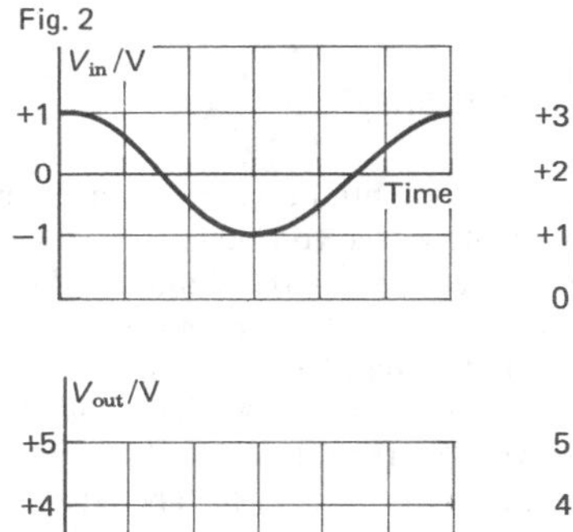

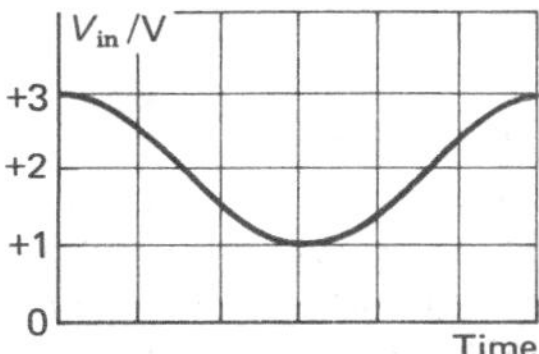

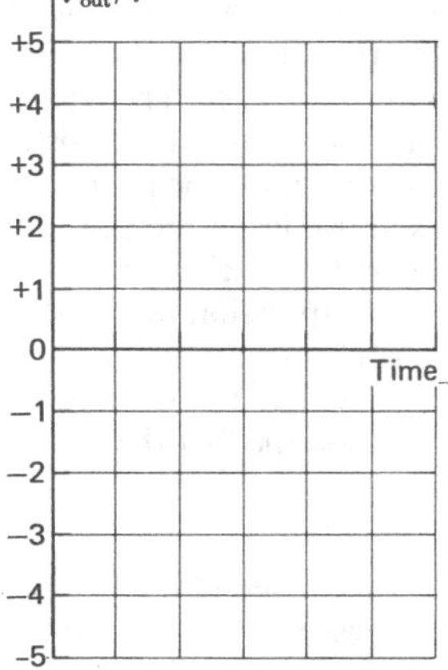

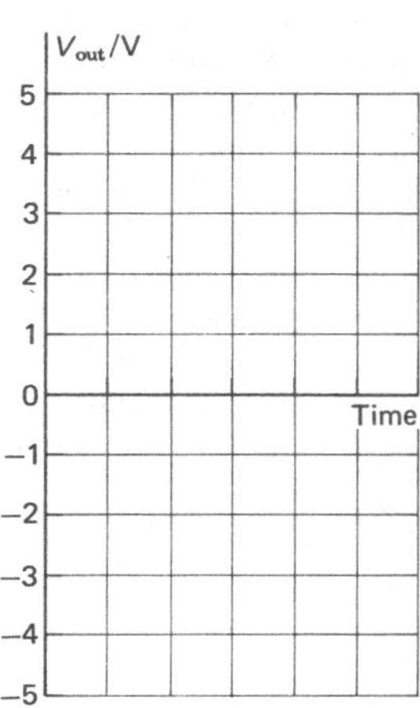

(**O and C Nuffield**: *and JMB, Cambridge, O and C, WJEC, SEB H**)

(a) To give a variable-input pd, you need to add a potential divider with a battery from -4 V to $+4$ V across its ends so that any value of pd from -4 to $+4$ V can be supplied to the input terminals. Show voltmeters, with ranges, to measure input and output pds.

(b) The first input has a mean value of 0 and a peak value of 1 V. Thus it never exceeds $+1$ V. Use the input/output graph to determine the output in this case. The second input varies from $+3$ to $+1$ V and back. The simplest way to proceed is to mark the input graph each quarter cycle and read off the value of input pd after each quarter cycle; then, use the input/output characteristic to determine the corresponding output value each quarter cycle, and plot the value on the output/time axes. Then sketch the wave form over a full cycle, using the plotted points to guide you. Note that since the input varies from +1 to +3 V which is exactly the range of the 'sloped section' of the input/output graph, there will not be any 'saturation'; the output should be inverted compared with the input because the slope is negative. You might find it useful to determine the voltage gain from the gradient of the slope (see p. 110 if necessary), and that should tell you how many times 'bigger' the amplitude of the output wave is compared with the input wave.

7.34L State Ohm's law.

For the circuit shown in the diagram, derive an expression for the potential difference V_1 in terms of R_1, R_2 and V, where the symbols have their customary meanings and the cell has negligible internal resistance.

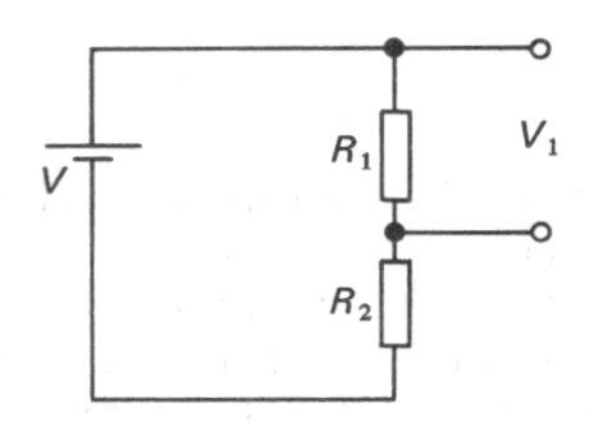

Deduce expressions for the potential differences indicated by a moving-coil voltmeter of resistance R when it is connected (a) across R_1, (b) across R_2.

Calculate the readings on the meter if

$$V = 2\,V,\ R_1 = R_2 = 1 \times 10^3\Omega \text{ and } R = 500\Omega.$$

Comment on the fact that your calculated values do not add to 2 V. Hence, discuss the factors that affect the choice of voltmeters for practical purposes.

(**Cambridge**: *and all other Boards*)

For Ohm's law, see p. 102.

The circuit here is essentially a potential-divider arrangement which is discussed on p. 103. (a) When the voltmeter is connected across R_1, the circuit is no longer that of the diagram above. Sketch the new circuit. Then, derive an expression for the combined resistance of R_1 and R in parallel, and add on R_2 to give the total circuit resistance (assume the cell has negligible internal resistance). The current from the cell is then V/R_T where R_T is the expression for the total circuit resistance. The pd across the voltmeter is then given by (cell current × combined resistance of R_1 and R in parallel). (b) Your expression for the pd across R_2 can be derived by the same steps as above using the new circuit with R in parallel with R_2 this time. Since interchange of R_1 and R_2 takes you from (a) to (b), your final expression for (b) should be as for (a) with R_1 and R_2 interchanged.

Use your derived expressions to calculate the readings.

Your comments should be based upon the fact that moving-coil voltmeters take current. See p. 103.

7.35L

(a) The diagrams below show two circuits commonly used to determine the value of an unknown resistance. Both ammeter and voltmeter are moving-coil instruments. State which circuit you would use if the unknown resistance had a value similar to the resistance of the voltmeter. Justify your choice, commenting on the position of the ammeter in each case, and the errors likely to result if the other circuit were used.

Fig. 1

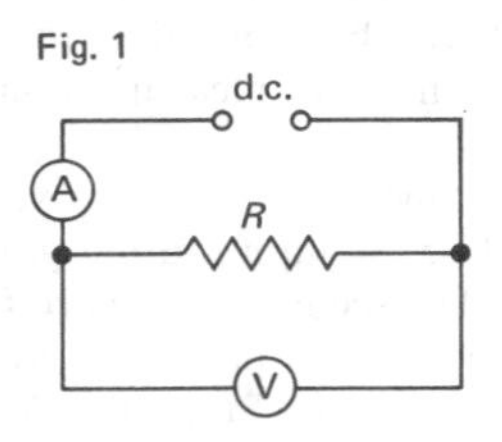

Fig. 2

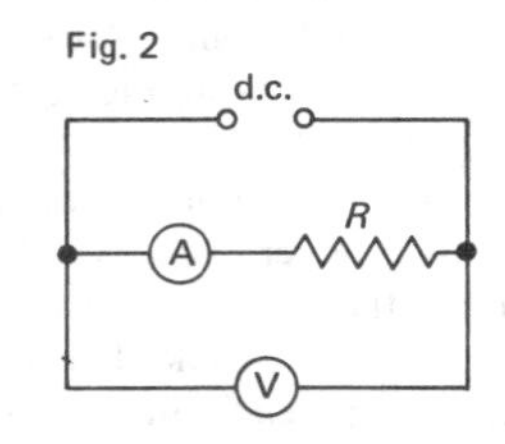

(b) A galvanometer of resistance 10 ohms gives its maximum deflection for a current of 50 mA. (i) What is the maximum pd the galvanometer can measure if it is used as a voltmeter? (ii) How would you convert the voltmeter so that it could read values of pd up to 10 V? A numerical answer is required. (iii) After conversion, find the current in the meter if the scale reading is 2 V.

(**SEB H**: *and all other Boards except O and C and NISEC*)

(a) In which circuit does the ammeter correctly record the current through R? Remember that R and the voltmeter have similar resistances. The key to your answer lies in the point that in one circuit the ammeter measures only the resistor current, and since the ammeter has low resistance, then the voltmeter reading is the same in either case. Use of the other circuit will give the same voltmeter reading but the ammeter will not record the resistor current only.

(b) (i) See E7.7 if necessary. (ii) See p. 103 if necessary. A quick method is to calculate the total resistance which will draw 50 mA from a 10 V supply. Then subtract the meter resistance to give the multiplier resistance. (iii) 10 V gives 50 mA through the meter, so 2 V gives ? mA.

7.36L

(a) Describe an experiment to determine how (i) the electrical resistance of a lamp filament, and (ii) the power converted by the filament, vary with the potential difference applied to it. Indicate with a clearly labelled circuit diagram.

(b) In such an experiment on a 12 V, 24 W lamp, the graph showing how the resistance R of the lamp filament varied with the applied pd was obtained. Use the graph to (i) calculate the

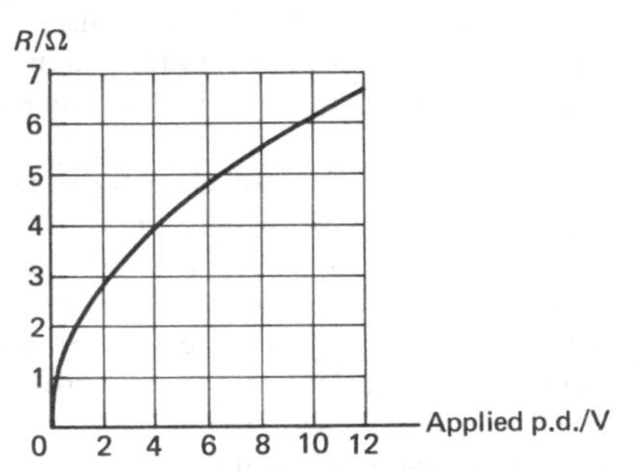

power of the lamp under normal working conditions, ie, when the applied pd is 12 V, (ii) estimate the temperature of the lamp filament under normal working conditions, assuming that the resistance of the filament is proportional to its temperature in Kelvin and that room temperature is 300 K, (iii) calculate the resistance of a fixed resistor which, when placed in series with the lamp and a 20 V dc supply, allows the lamp to work normally.

(c) When such a lamp is switched on, it may take about 0.5 s to achieve full brightness, at which time the current through the lamp is about 2 A. Explain why the current is initially greater than this and calculate its maximum value.

(**London**: *and all other Boards*)

(a) As $R = V/I$ and Power $= IV$, you need a circuit that will enable you to measure the pd across the bulb and the current through the bulb for different applied pds (ie, a rheostat must be included). A voltmeter across the bulb will cause error unless it has a very high resistance (which should be stated) – a potentiometer would be better.

(b) (i) Use Power $= V^2/R$, finding the value of R at 12 V from the graph. (ii) From the graph find R at room temperature (ie, 0 V) and under working conditions (ie, 12 V) and then use simple proportions. (iii) Find the **current** needed to operate the bulb normally (using R from the graph, **not** the nominal value of 24 W). This current will also be the current in the resistor. If the supply is 20 V and the bulb needs 12 V, the difference must be dropped across the resistor, hence its value can be found from $R = V/I$.

(c) When the lamp is switched on, the filament will initially be at room temperature (and of resistance found in (b)(ii) above).

7.37L

(a) Describe qualitatively how the magnitude of the drift velocity of free electrons in a current-carrying metal conductor depends on (i) the temperature of the conductor, (ii) the potential difference between the ends of the conductor. A copper wire of cross-sectional area 1.0 mm^2 and length 1.5 m carries a current of 2.0 A. Each copper atom contributes one free electron to conduction. Calculate how long it takes for a free electron to drift from one end of the wire to the other. (Electron charge $e = -1.6 \times 10^{-19}$ C; density of copper $= 8.9 \times 10^3$ kg m^{-3}, 0.064 kg of copper contains 6.0×10^{23} atoms.)

(b) (i) A resistance network is formed by connecting twelve equal lengths of wire, each of resistance 6Ω as the edges of a cube. A current of 3 A enters at *S* and flows symmetrically through the network, leaving it by the diagonally opposite corner *Z*. The directions of all the currents and the magnitudes of some are indicated in the diagram. Find the magnitude of the current in

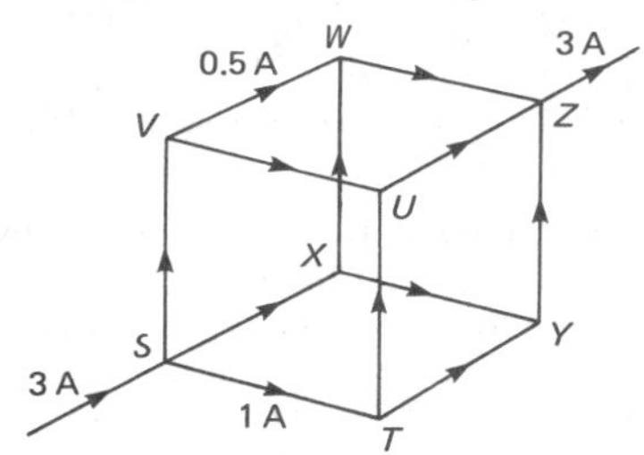

each wire. The potential difference between the points *S* and *Z* may be calculated as the algebraic sum of the potential differences along any path taken between the points *S* and *Z*. Hence show that the resistance of the network between *S* and *Z* is 5Ω. (ii) If a solid cube of edge 20 mm were made from metal of resistivity 5×10^{-8}Ωm, what would be the resistance between opposite faces of the cube?

(c) Comment briefly on the possibility of using a potentiometer to compare the values of the resistance of the network in (b)(i) and of the cube in (b)(ii).

(**NISEC**: *and all other Boards except Oxford, SEB H, SEB SYS*)

(a) (i) See p. 101. With increased temperature, the vibrations of the lattice ions increases, and this causes an increase in the frequency of collisions between free electrons and lattice ions. What effect will this increase have on the mean drift velocity of the electrons as they try to move through the lattice? (ii) A greater pd across the ends of the conductor makes the electrons accelerate more between collisions, and it is this accelerated motion superimposed on the random motion of the electrons which gives the electrons their drift velocity. With greater acceleration, will the drift velocity be more?... less? etc. Use E7.3 to calculate the drift velocity; you will need to do a preliminary calculation of the number of copper atoms (and hence electrons) in the wire, so use its dimensions, density and the given value of Avogadro's number to do this. Remember to put the area into m^2, and that n in E7.3 is the number of free electrons per unit volume. Lastly, calculate time taken from the value of drift velocity and the length.

(b) (i) The key lies in the fact that the resistances of each edge are equal, and the current flows symmetrically through the network. Hence the currents in *ST*, *SX* and *SV* will be equal and will also equal the currents in *WZ*, *UZ* and *YZ*. At *V*, the total current arriving must equal the total current leaving, hence obtain the current in *VU*. The currents in the remaining conductors may be determined by symmetry (or by applying Kirchhoff's first law to each remaining junction – ie, total current in = total current out). For the resistance from *S* to *Z*, choose a short path from *S* to *Z* (eg, *SV*, *VW*, *WZ*). Apply E7.7 to each conductor in turn, and then add the separate pds to find the total pd between *S* and *Z*. Then divide the total pd by the total current taken to give the resistance. (ii) Rearrange the resistivity equation to give $R = \rho L/A$. Use this equation, but take care with units.

(c) See p. 104, in particular the section on 'comparison of two resistances'. The circuit is given in F7.15. Only a brief answer is expected; you should not give details of the experimental method. The main point concerns the relative orders of magnitude of the two resistances to be compared.

7.38L

(a) Describe, with a circuit diagram, a potentiometer circuit arranged (i) to compare the emf of a cell with that of a standard cell, and (ii) to measure accurately a steady direct current of approximately 1 A. What factors determine the accuracy of the current measurement?

(b) A 12 V, 24 W tungsten filament bulb is supplied with current from n cells connected in series. Each cell has an emf of 1.5 V and internal resistance 0.25 Ω. What is the value of n in order that the bulb runs at its rated power? An additional resistance R is introduced into the circuit so that the potential difference across the bulb is 6 V. Why is the power dissipated in R not 6 W? Is it greater or less than 6 W?

(**O and C**: *and AEB, Oxford, Cambridge, (b) only for JMB, NISEC, O and C Nuffield*)

(a) (i) See p. 103 for the comparison of cell emfs. (ii) See your textbook for a suitable method. Most methods involve passing the current to be measured through a standard resistance, and then using the potentiometer to measure the pd across the standard resistance. In addition to the potentiometer circuit, draw the circuit which you would use to supply a steady current through a standard resistor. Explain how you would calculate the current from your measurements. Since the driver cell will give a pd across the potentiometer wire of about 2 V, then the resistance of the standard resistor must not be any more than 2 Ω (otherwise 1 A through it will give a pd that cannot be balanced by the fall of pd along the potentiometer wire). Contact resistance at the terminals of the standard resistor must therefore be minimized; see your textbook if necessary.

(b) To determine the number of cells necessary to run the bulb at its rated power, first calculate the current taken by the bulb when it runs at that power. With the cells in series, each cell will pass that amount of current, so you can calculate the 'lost voltage' due to internal resistance of each cell. The pd across the terminals of each cell can then be calculated by subtracting the 'lost voltage' from the emf of each cell; then you can state how many cells will give a total pd of 12 V. See p. 102 and E7.5 if necessary. Sketch the new circuit with the same number of cells as before, and the additional resistor R in series with the bulb. Next, consider the bulb resistance; show that it is 6 Ω at its rated power, and consider if it is more or less than 6 Ω when run at low power (see p. 102 and F7.5). Thus, show that the bulb current must be more than 1 A but less than 2 A when its pd is 6 V. The next stage is to show that the pd across R must be between 6 V and 9 V; then, since power = current × pd, show that the power dissipated in R must exceed 6 W.

7.39L This question is about the properties of *RC* circuits fed with square pulses.

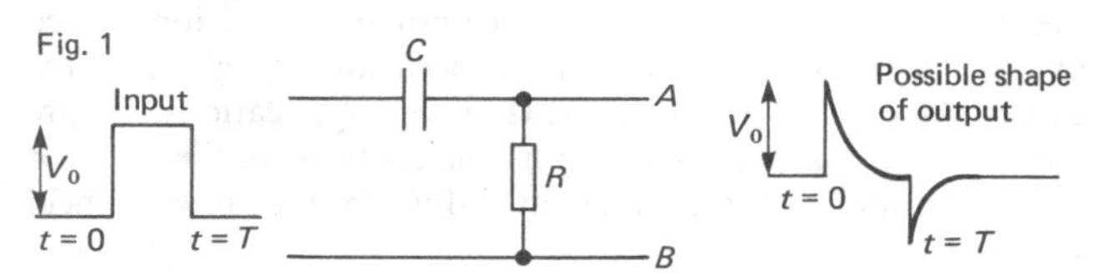

Below are four sets of ideas (a) to (d) about the effects of *RC* circuits on square pulses. **(i)** Explain fully why each statement (a) to (d) is correct. **(ii)** Say briefly how you would demonstrate each statement experimentally, giving typical magnitudes of components you would need to use. **(iii)** Describe one practical application of each of the two circuits shown.

(a) When a square pulse is first applied (at $t = 0$) to the circuit in Fig. 1, the potential difference first appears in full across *AB*, and then decays exponentially. An inverse effect is found when the pulse is switched off at $t = T$.

(b) The variation with time of the potential difference observed across AB depends on how RC compares with the length of the pulse T. Very different effects are seen for $T \gg RC$, and $T \ll RC$.

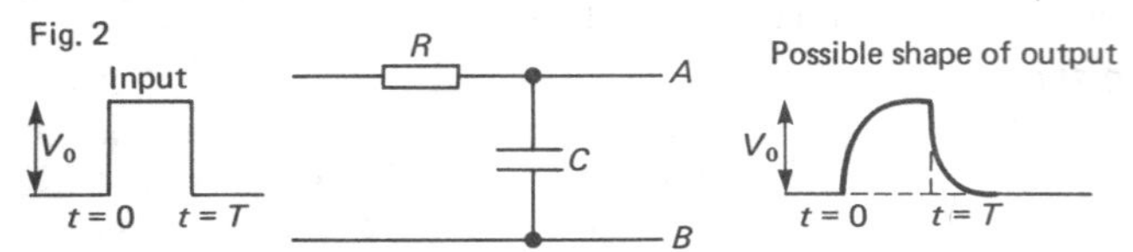

(c) A re-arrangement of the circuit (Fig. 2) leads to quite different effects across AB. These effects again depend on whether $T \gg RC$, or $T \ll RC$. This new type of circuit is called an integrating circuit.

(d) The results in (c) can be related to those for the circuit of (a), by remembering that $V_{input} = V_R + V_C$ at all times.
(**O and C Nuffield**: *and Oxford*, O and C*, and all other Boards except SEB SYS*)

(i) These circuits are simply understood by assuming that the input voltage varies from a level of 0 to V_0 volts. When the input is at 0 V then the circuits behave as a standard capacitor discharge circuit; when the input is at V_0 then the circuits behave as a capacitor-charging circuit. If the output voltage is measured across the resistor then it is indicating the charge/discharge current (as the voltage across the resistor is proportional to the current in the circuit); this happens in part (a). If the output voltage is measured across the capacitor (as in part (c)), then it is monitoring the pd across the capacitor (or the charge stored, as $Q = CV$). See p. 105 for charge/discharge information; remember that when a capacitor 'empties', the current flows in the reverse direction to charging and hence is drawn on the graph as a negative value. The output for both circuits shows that the time 'allowed' by the input (T) is easily enough to completely charge or discharge the capacitor. If T is considerably increased or decreased the output shape changes (see Nuffield Student Book Unit 6). In the circuit of part (c) if $T \ll CR$ this shape is a 'triangle' wave form (of very small amplitude), which can be shown to be the integral (area under the graph) of the input. Part (d) is simply stating for these circuits the standard electrical law that the sum of the pds around a circuit ($V_C + V_R$) must equal the emf (V_{input}).

(ii) Choose the voltmeter that you connect between A and B with care, or else it may affect the resistance within the circuit. Values of CR must be chosen to be about ¼ of T to produce the outputs shown. The frequency f $(1/T)$ of the signal generator can then be varied to show $T \gg CR$ and $T \ll CR$. Quote frequencies needed to do this. Part (d) can be demonstrated in many ways, but in all of them remember that when the capacitor discharges the pd across the resistor is in an opposite direction to that across the capacitor (ie, it has a negative value – a centre-zero meter may be helpful).

(iii) The circuit of (a) can be used as a differentiating device or as a filter that removes low-frequency signals (as the bigger T becomes, the less is the output). Conversely circuit (c) can integrate or remove high-frequency signals (as the smaller T becomes, the less is its output). Much of the explanation needed to describe these effects may have been already given in the earlier answers; do not repeat it! However 'applications' require more information; talk about hi-fi tone controls for the filters or analog computing for the integrator/differentiator (see Nuffield Unit 6 Books).

7.40L Explain what is meant by the 'peak value' and the 'root mean square value (rms)' of an alternating current. State the relation between these two quantities. Why is the rms value the most convenient measure of an alternating current?
Describe the construction and principle of operation of an instrument for measuring alternating current.
A filament lamp, which is designed for operating with a potential difference of 100 V, is connected in series with an ammeter and a pure inductor across 200 V, 50 Hz supply mains. It is observed that the current in the circuit is 0.5 A rms, and that the lamp is operating at normal brightness. What is the power rating of the lamp, and what is the power taken from the mains? (Assume the lamp is a pure resistor). What is the inductance of the inductor?
(**SUJB**: *Cambridge*, O and C*, SEB H*, and all other Boards except Oxford*)

For an explanation of the meaning of peak value and rms value of an alternating current, see p. 106. The relationship for sine wave alternating current is given by E7.17. The reason why the rms value is the most convenient measure lies in the link with direct current and power.

For details of an instrument to measure alternating current, see your textbook.

To tackle the circuit problem, firstly sketch the circuit; it consists of the supply pd, the lamp and the inductor in series with one another. Since the lamp operates at normal brightness, then the rms potential difference across its terminals must be 100 V, and since it takes 0.5 A rms, then calculate its power rating. See p. 108 if necessary. Remember that a pure inductor does not take any power, so the power taken from the mains is by the lamp only.

To calculate the inductance L, first calculate the circuit impedance from the supply pd (assume 200 V is its rms value) and the current. See p. 108 if necessary. Then calculate the lamp resistance, given its pd is 100 V rms and its current is 0.5 A rms. Then use E7.22 to calculate L; E7.22 is for the series LCR circuit, so to modify it to the series LR circuit, just leave out the capacitor term $(1/2\pi fC)$.

7.41L A series circuit consists of a 100 Ω resistor, a 20 μF capacitor and a 0.2 H inductor driven by an alternating source of frequency f. The potential difference across the inductor, measured with a high impedance voltmeter, is 50 V rms; and that across the capacitor is 200 V rms. Find f, and the rms current in the circuit.
Draw a vector diagram showing in magnitude and phase the peak potential differences across each of the circuit components (R, L and C) and the peak emf of the source. Find the phase difference between the emf and the current.
Find the resonant frequency for this circuit, and point out with the aid of a sketch how the above vector diagram becomes altered at resonance.
How would you detect resonance experimentally in such a circuit by means of a cathode ray oscilloscope?
(**WJEC**: *and Oxford*, O and C*, NISEC, and all other Boards*)

The series LCR circuit is discussed on p. 108. Start your answer by making a circuit diagram and label the components. Since the components are in series, the ratio of (capacitor pd/inductor pd) is the same as the ratio of (capacitor reactance/inductor reactance). Using the reactance formulae (see E7.19 and E7.20) and the given values of L and C, you ought to be able to calculate f. Then use the calculated value of f to determine the reactance of either L or C, and so calculate the rms current

$$\left(= \frac{\text{rms potential difference}}{\text{reactance}}\right)$$

For the vector diagram, see F7.29 if necessary. Before drawing the diagram to scale, you must first calculate the peak pd across the resistor (= peak current × resistance) and the peak pds across L and R from the rms values. See E7.17 for the link between peak and rms values. The phase angle can either be measured off your diagram or you can calculate the phase angle. See E7.23 for the appropriate formula.

Resonance of a series LCR circuit is discussed on p. 108. To calculate the resonant frequency, use E7.24.

At resonance, the circuit impedance is a minimum, so for a fixed supply pd, the current will be a maximum. Since cros measure pd, then the pd across the resistor will give a trace in proportion to the current. You should redraw the circuit showing where you would connect the cro to determine the current and then to check the supply pd remains unaltered. Refer to your diagram in describing the procedure you would follow, and state what trace measurement(s) you would make and how you would detect resonance from these measurements.

7.42L In a laboratory exercise, a student is given a sealed box containing an ac source and either an inductor or a resistor. The source and the other electrical component are connected in series to the two output terminals of the box. See the diagram.

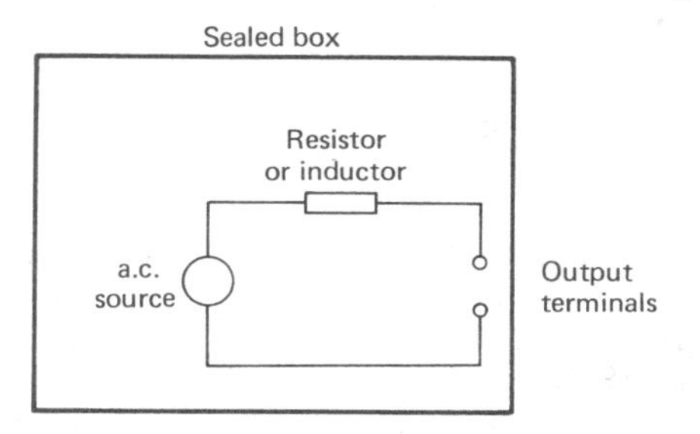

The student is asked to determine the resistance R or the inductance L of the unknown component.

(a) He first measures the voltage across the terminals with a high-resistance voltmeter and then measures the current when he connects a resistor of resistance 1000 ohms between the terminals.

Results

Voltage across the terminals on open circuit = 10.0 V
Current in 1000 Ω resistor = 7.0 mA

Using this data, and assuming that the unknown component is a resistor, what values are obtained for the emf of the source and for R?

(b) He replaces the resistor by a 1.0 mH inductor between the terminals and finds that the voltage across it is 5.0 V and the current in the inductor is 5.0 mA. (i) Explain whether the measurements are consistent with his previous results and assumption. (ii) Then the student assumes that the unknown component is an inductor. What values are obtained for the inductance and for the frequency of the ac source?

(c) As a check, he replaces the inductor by a $1.0 \times 10^{-3}\ \mu F$ capacitor and is pleased to find that the current is too large to be measured. Explain why.

(d) As a final check he connects another $1.0 \times 10^{-3}\ \mu F$ capacitor in parallel with the first between the terminals. What value is obtained for the current in the capacitors?

(**Scottish SYS**: *Oxford*, O and C*, N Ireland and all other Boards except Scottish H*)

(a) Even though the circuit is ac, the essential ideas are discussed on p. 102. Remember that the emf = pd on open circuit. Use E7.5 to calculate R.

(b) (i) You must now consider the circuit as a series LR circuit, as on p. 108. Remember that the pds are related to one another by the Pythagoras rule (ie, $V_S^2 = V_L^2 + V_R^2$ where V_S is the source emf). Hence calculate V_R given that $V_L = 5.0$ V. Then, knowing that the current through R is the same as that taken by inductor L, calculate the value of R again. Is it the same as in (a)? (ii) If the unknown component is an inductor, then with the 1.0 mH inductor across the terminals, half the source emf (ie, 5.0 V) will be dropped across the unknown component since the pd across the 1.0 mH inductor will be in phase with the unknown inductor's pd. Thus the two inductors have the same pd and the same current values, etc. For the frequency f, use the information given in the equation for reactance of an inductor (E7.20).

(c) The circuit is obviously at or near resonance so the capacitor 'cancels' the internal (no longer unknown) component. Use the L and C values to calculate the resonant frequency (see E7.24) to check if it is actually at resonance.

(d) Use $I = E/Z$ where Z is given by E7.22 with $R = 0$ and $C = 2 \times 10^{-3}\ \mu F$.

7.43L This question is about the design of an experiment to investigate the frequency response of an amplifier. In the experiment the gain of an amplifier is to be investigated over a range of frequencies.

The diagram shows the circuit of the amplifier to be investigated. The operational amplifier may be assumed to be ideal. It is to be operated using a − 15 V — 0 — + 15 V supply which is not shown.

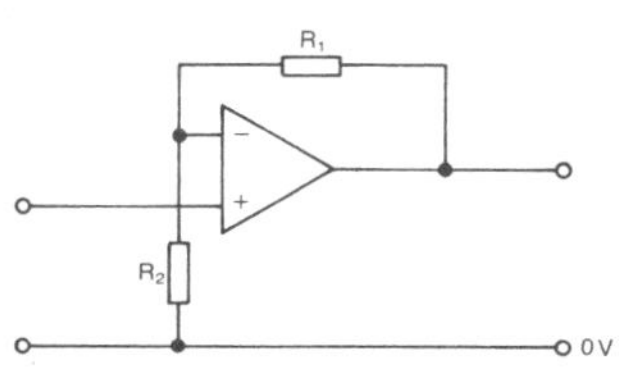

(a) (i) Select from the following list the two resistors you would use for R_1 and R_2 so that the gain of the amplifier at low frequencies would be as near to 30 as is possible.

1 kΩ, 2.7 kΩ, 4.7 kΩ, 10 kΩ, 39 kΩ, 150 kΩ

(ii) Calculate the expected low frequency gain of the amplifier using the resistors you have chosen.

(b) A sinusoidal input signal is to be provided by an uncalibrated oscillator which has a variable frequency with a range 1 Hz to 1 MHz. The output has a peak value which is constant at 2 V for all frequencies.

(i) Explain why this voltage is too large for investigating the frequency response of the amplifier.

(ii) Draw a diagram to show how you would reduce this voltage to a suitable magnitude using components from the list in (a) (i) and calculate the new peak output voltage.

(iii) Explain how you would proceed to measure the d.c. gain of the amplifier given that a 2 V d.c. supply is available. Indicate clearly the instrument(s) you would use, where the instruments would be connected and the measurements you would make.

(iv) Describe how you would use an oscilloscope to calibrate the oscillator and use the calibrated oscillator to investigate the frequency response of the amplifier.

(v) Draw a graph indicating the shape of the frequency response graph you would expect.

(**AEB June '89**: *and JMB*, O and C*, and all other Boards*)

(a) The circuit is a non-inverting amplifier. Its voltage gain = $(R_1 + R_2)/R_2$. Choose values from the list to give a voltage gain as near as 30 as possible. Then use the chosen values to work out the voltage gain exactly.

(b) (i) See comments on p. 110 re. saturation.

(ii) A potential divider is needed. See p. 103.

(iii) Give a circuit diagram showing a potential divider to supply a variable input voltage and showing voltmeters to measure the input and output voltages.

(iv) and (v) See your textbook if necessary.

7.44L (a) What do you understand by a *logic gate?*

(b) Fig. 1 illustrates a circuit which has inputs I_1 and I_2 and output S.

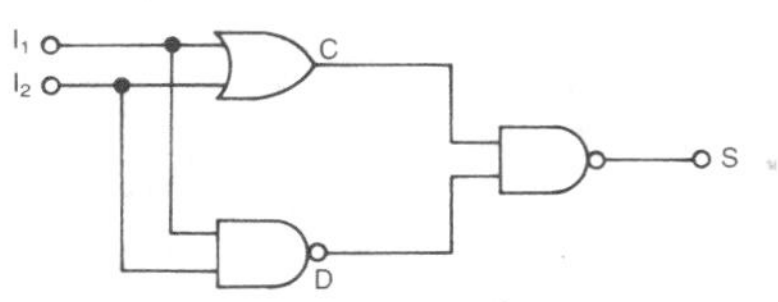

(i) Identify the logic gates shown in Fig 1 and write out their respective truth tables

(ii) Copy out and complete the truth table shown below for the circuit.

I_1	I_2	C	D	S
0	0			
0	1			
1	0			
1	1			

(iii) Describe in words the logic function of the circuit.

(c) An electric motor is to be controlled by three switches P, Q and R. The motor is to be running (logic state 1) when switches P and Q are in the same state. Whenever P and Q are in different states, the motor is to be controlled by switch R, such that the motor is running when R is in logic state 1.
(i) Write out a truth table for the control circuit.
(ii) Hence, using the circuit of Fig. 1 or otherwise, design a circuit which could be used to control the motor.
(d) In the operational amplifier circuit of Fig. 2, a sinusoidal e.m.f. of 2.0 V (r.m.s.) and frequency 50 Hz is applied to the non-inverting input. The inverting input is at earth potential.

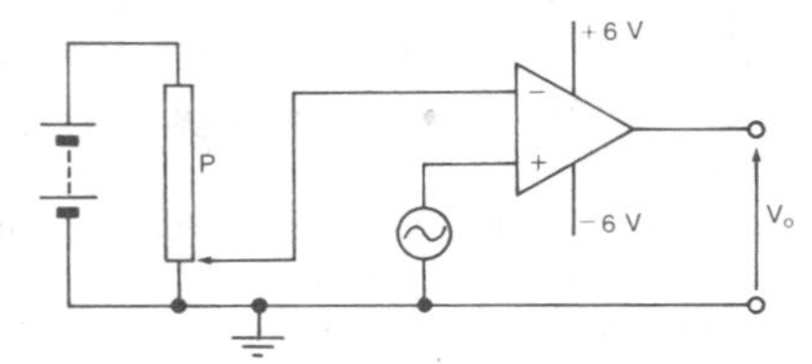

Draw sketch graphs on the same axes to show the variation with time of
(i) the potential at the non-inverting input,
(ii) the output potential V_o.

The potential at the inverting input may be made positive with respect to earth by adjustment of the potentiometer P. Draw a sketch graph to show, in detail, how V_o varies with time t when the inverting input is held at +2.0 V.

(**Cambridge**: *and all other Boards except London, O and C, and SEB SYS*)

(a) and (b) See pp. 110 and 111. In this circuit, S = 1 whenever the input states are the same. The circuit is therefore a comparator.
(c) (i) With three switches P, Q and R, there are 8 possible input conditions. The output to the motor must be 1 when P is the same as Q or when R = 1. Use this statement to write out the truth table.
(ii) The statement above should tell you what to add to the circuit in (b) to make the control circuit.
(d) The circuit is an open-loop op-amp with the output at + or − saturation according to whether the non-inverting input voltage is more than or less than the inverting input voltage. When P is set to give non-zero voltage, the output is an 'uneven' square wave. See Fig. 7.37 if necessary.

7.45L
(a) The operational amplifier shown in the diagram can be used as a voltage amplifier. Describe how you would determine experimentally its dc input/output characteristic for positive and negative input voltages. Include a labelled circuit diagram showing how the amplifier is connected to a suitable power supply and explain how you would obtain different input voltages using a potential divider. Suggest suitable ranges of voltmeters you would use to measure V_1 and V_o given that the resistance of $R_1 = 10\,\mathrm{k}\Omega$ and the resistance of $R_2 = 100\,\mathrm{k}\Omega$.

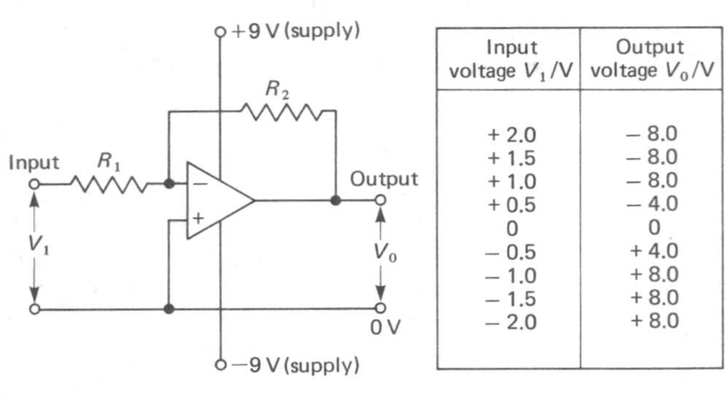

Input voltage V_1/V	Output voltage V_o/V
+ 2.0	− 8.0
+ 1.5	− 8.0
+ 1.0	− 8.0
+ 0.5	− 4.0
0	0
− 0.5	+ 4.0
− 1.0	+ 8.0
− 1.5	+ 8.0
− 2.0	+ 8.0

(b) The table shows typical results for a voltage amplifier similar to that shown above. Draw the input/output characteristic and, by reference to it, explain what is meant by (i) voltage gain, (ii) saturation and (iii) inversion. State the range of input voltages for which the amplifier has a linear response and calculate the voltage gain within this range.
(c) A sinusoidal voltage of frequency 50 Hz is applied to the input terminals of the amplifier described in (b). Sketch graphs on one set of axes showing how the output voltage varies with time when the input voltage is (i) 0.5 V rms, (ii) 1.0 V rms. In each case indicate the peak value of the output voltage and comment on the wave form.

(**JMB**: *London*, O and C* and all other Boards except AEB*)

(a) The circuit given should be redrawn, and a potential divider added (see F7.10(b)) to give a variable input pd. The voltage supply for the potential divider should be such that the potential divider output can be varied over the full range of input voltages (ie, from + to 0 to −). Show clearly on your diagram voltmeters to measure the input and output pds.

(b) Plot y = output pd, x = input pd. See p. 109 for the meaning of the terms in (i), (ii) and (iii). You should explain these terms by reference to your graph. A linear response is where the output is in proportion to the input; use your graph to decide on the input range which gives a linear response.

(c) (i) Calculate the peak input pd; see E7.17 if necessary. Then make a sketch of the input pd (for one complete cycle) against time. Then draw the output pd against time by 'multiplying the input wave by the voltage gain value'; in other words, if the input is a sine wave of peak value V_i, then the output will be a sine wave of peak value $V_i \times$ voltage gain. If the voltage gain is −ve, the output is inverted compared with the input. Then draw the limits of the output pd on your graph. If the output wave exceeds the limit, then you must 'clip' the wave; see p. 109 and F7.39. (ii) As (i).

Unit 8 Structure of the atom

Despite the endeavours of modern science, an individual atom still remains too small to be seen, even with the aid of the most powerful microscopes. Yet such is the importance of the atom that, despite this invisibility, atomic structure and behaviour is remarkably well understood. This Unit examines the theories, experiments and effects that have enabled this understanding to be gained. Two distinct areas of study exist; that of the atomic electrons where typical experimental energies are normally of the order of electronvolts (eV), and that of the atomic nucleus where the energies involved are a million times larger (MeV).

Knowledge from other Units is used extensively in this Unit. Key ideas of Unit 1 about force and energy are used in analysing the motion of charged particles, and in understanding the production of nuclear energy. The study of the motion of charged particles also draws extensively from Field concepts in Unit 2, so a look back at 2.8 and 2.10 would be valuable at this stage. The sections on Energy levels and on Spectra within this unit rely heavily on the Photon theory, so re-reading the relevant parts of 1.10 and 3.7 would also be helpful.

8.1 PROPERTIES OF THE ELECTRON

General information Electrons were originally discovered during experiments involving the passage of electric currents through gases. The direction of electron travel (cathode to anode) coupled with information from the deflection by electric and magnetic fields proved that electrons must carry a negative charge.

The charge of an electron: Millikan's experiment A charged oil drop is kept in static equilibrium. The downward force upon the drop (its weight $= mg$) is balanced by an equal and opposite upward force due to the uniform electric field (force $= qV/d$, see E2.9). Equating the forces gives:

E.8.1 $q = \dfrac{mg}{V/d}$

where q = total charge upon oil drop (C), m = mass of the oil drop (kg), g = acceleration due to gravity (m s^{-2}), V = pd between parallel plates (V), d = separation of the parallel plates (m).

The mass of the drop is obtained by measuring its terminal velocity when falling through air in the absence of the electric field. Remember that q is the charge upon the oil drop and not the charge on an electron. However this charge q is found to be an integral (whole number) multiple of a value e (eg, $3e$, $1e$, $6e$, etc), but never a fractional multiple of e (eg, $1.5e$, $7.8e$, etc). Electric charge is hence quantized (see 1.10) in equal size quanta of size e (-1.6×10^{-19}C). So the smallest amount of charge available is e; this is assumed to be the charge on one electron.

The charge/mass ratio of an electron (the specific charge of an electron) There are two methods commonly used to obtain a value for e/m (the charge/mass ratio of an electron). The first method is normally carried out using a **'fine beam tube'** and involves bending an electron beam around in a circle under the influence of the uniform magnetic field provided by a pair of Helmholtz coils. The force Bev on each electron (see E2.16) always acts at right angles to the direction of travel of the electrons. If this force can be kept constant (uniform magnetic field and electrons travelling at constant speed) then the electrons must travel in a circle (see 1.7). The centripetal force required (mv^2/r see E1.12) is supplied by the magnetic field effect (Bev) so:

E8.2 $\dfrac{e}{m_e} = \dfrac{v}{Br}$

where e/m_e = charge / mass ratio of an electron (C kg^{-1}), v = speed of the electrons (m s^{-1}) (see E8.4), B = strength of the magnetic field (T), r = radius of the circle in which electrons move (m).

Beware of putting a voltage value into E8.2 in mistake for the speed of the electrons v. You will often be given the accelerating voltage V_A to use in E8.4 to calculate the electron speed v. It is often necessary to combine E8.4 and E8.2 to obtain the charge–mass ratio. F8.1 illustrates the experiment; use the left-hand rule

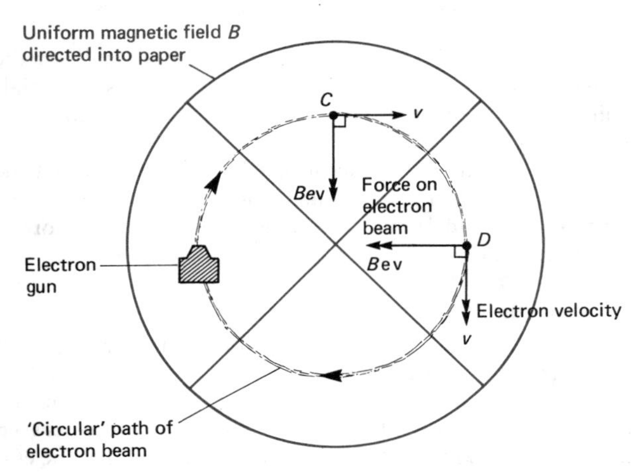

F8.1 Electron path in a uniform magnetic field

to check on the direction in which the electron beam 'bends' at C and D. Remember that you do not see the electrons in the experiment! The light is emitted by atoms that have been excited (see 8.4) by the electron beam passing through them.

The second method of measuring e/m_e is a 'null' method involving the cancelling of two equal and opposite effects (other examples of **'null' methods** include using potentiometers to measure voltages and Millikan's method as already described). The experiment is carried out using a **'deflection tube'** and requires the deflection in one direction by an electric field to be cancelled by the use of a uniform magnetic field that will deflect the beam in the opposite direction. F8.2 shows the electron beam

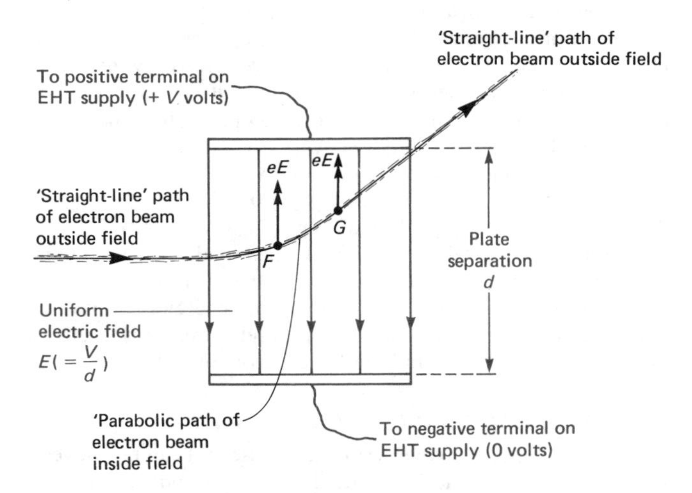

F8.2 Electron path in a uniform electric field

deflected upwards by an electric field; the left-hand rule shows that the magnetic field must be directed into the paper to deflect the beam back down. For an undeflected beam the force due to the magnetic field (Bev) must equal the electric field force (eV/d) so:

E8.3 $v = \dfrac{V}{Bd} = \dfrac{E}{B}$

where v = speed of the electron (m s^{-1}), V = pd used to deflect the electrons (V), B = strength of the magnetic field (T), d = separation of the deflecting plates (m), $E = V/d$ = strength of the electric field (V m^{-1} or N C^{-1}).

e/m_e is calculated by combining the above formula with E8.4. Often the same pd (V) is used both to accelerate the electrons and deflect them. In both experiments for e/m_e the electron beam is often found to 'spread' considerably as the electrons acquire a range of speeds owing to collisions with gas atoms inside the tubes. The formulae are only correct for the fastest moving electrons (ie, those that have not lost kinetic energy in collisions); this can be allowed for by using the largest radius in the first method and the largest magnetic field that still leaves part of the beam undeflected in the second method.

Mass, speed and energy of electrons: velocity selector and electron-volts The result of Millikan's experiment (e) can be combined with a value for the charge–mass ratio of the electron (e/m_e) to enable a value for the mass of the electron to be obtained ($m_e = 9.1 \times 10^{-31}$kg).

If an electron with no kinetic energy is accelerated by a pd (V_A) the work done on the electron ($V_A e$) supplies ke ($\frac{1}{2}m_e v^2$) hence:

E8.4 $v = \sqrt{\dfrac{e}{m_e} \cdot 2V_A}$

where v = speed of the electron (m s^{-1}), e/m_e = charge–mass ratio of the electron (C kg^{-1}), V_A = pd used to accelerate the electron (V).

It is also possible to measure the speed of electrons (or any beam of charged particles) using the experiment summarized by E8.3. E8.3 has far wider applications as it demonstrates that the charged particles left undeflected by a combination of electric and magnetic fields will all have the same velocity REGARDLESS of the mass or charge on any of the particles. Hence a beam of charged particles all travelling at the same velocity can be obtained from a beam of assorted particles with many different velocities; the chosen velocity is selected by the value of E/B and the device is known as a **velocity selector**. Such a device is very useful in mass spectrometers.

A word of caution must be sounded in using the formulae E8.3 and E8.4 to calculate speeds of electrons. The formulae are based on classical physics and cannot be applied to electrons travelling near the speed of light (see 1.10), when relativistic effects become important.

In atomic physics the situation keeps occurring where electrons are being either accelerated or decelerated by the use of pds. The change in energy of an electron as it moves through a pd of 1V is 1.6×10^{-19}J (energy = charge × pd). This amount of energy is called **one electronvolt** (1eV).
The following conversions are important:
1 To convert from J to eV: divide by 1.6×10^{-19}.
2 To convert from eV to J: multiply by 1.6×10^{-19}.
Remember that the value in J will be a much smaller quantity than the same energy expressed in eV. Note that any energy can be expressed in eV: this method of measuring energies is not only applied to electrons; eg, alpha-particle and gamma-ray energies are usually measured in millions of electronvolts (MeV).

8.2 FREE ELECTRONS IN METALS

Thermionic emission **Free electrons** are those that can be 'moved' by electric or magnetic fields, ie, those not bound into atoms. There are large numbers of free electrons present in all metals (about one free electron per atom on average). These free electrons are the charge carriers of electric current in metals, they can also be referred to as the **'conduction' electrons**. It is possible to remove these free electrons from the surface of the metal; the energy that must be given to a free electron to make this happen is called the **work function** (φ). Removing electrons from a metal in this way is very similar to the process of evaporation in liquids. A liquid molecule needs energy to enable it to escape from the surface of the liquid to become a vapour molecule. Molecules in a liquid can collide with one another and as a result have a wide range of energies; similarly free electrons in a metal acquire a wide range of energies as a result of collisions with each other and the metal atoms. In a liquid if the molecules with high energies reach the surface they can escape; this process is known as evaporation. In a metal at normal temperatures the free electrons virtually never acquire enough energy to be 'emitted' in a similar fashion; however if the metal is heated to a sufficiently high temperature the atoms and electrons have far more ke and free electrons can escape from the metal surface. This process is known as **thermionic emission**. Whenever a beam of electrons is to be produced (eg, in a cathode ray tube, electronic valve, or X-ray tube) it is usually generated by thermionic emission, the number of electrons per second (tube current) being controlled by the temperature of the thermionic emitter, which is heated electrically.

Photoelectric effect Another method of supplying free electrons near the surface of a metal with sufficient energy for them to escape from the metal is by transferring energy from a photon (of electromagnetic radiation) to the free electron. This is known as the **photoelectric effect**. It is essential that each photon has at least enough energy to enable an electron to escape, this minimum energy is of course the work function (φ); since the energy of a photon (hf) depends upon its frequency (f) then there must be a minimum frequency of radiation needed to cause the photoelectric effect. This frequency is known as the **threshold frequency** (f_o) and no beam of radiation at a lower frequency than this, however intense, can cause photoelectric emission to occur.

E8.5 $hf_o = \varphi$

where f_o = threshold frequency of the metal (Hz), h = Planck's constant (J s), φ = work function of the metal (J).

The number of electrons emitted per second determines the **photoelectric current** and clearly is proportional to the number of photons per second hitting the metal surface (assuming that each photon is above the threshold frequency). The photoelectric effect provides important evidence of the need for a quantum theory in physics, as the photon model of electromagnetic radiation is essential to explain photoelectricity. Occasionally a photon may deliver all its energy to a single electron; the conservation of energy requires that the photon energy should equal the kinetic energy of the emerging electron added to the energy needed to extract the electron from the metal giving **Einstein's equation** for the photoelectric effect (E8.6). This equation makes it possible to calculate the maximum ke with which an electron may be emitted from the **photocathode**. This maximum energy can also be measured experimentally. Normally the electrons pass through the photocell from the negative photocathode to the positive anode. However if the supply voltage is reversed, then the electric field will actually repel the electrons and cause a decrease in the photoelectric current. As this reverse voltage is increased the photoelectric current will reduce to the point where it becomes zero. At this point the potential difference across the cell will just be sufficient to stop the fastest electrons (ie, those with maximum kinetic energy); this potential difference is called the **stopping voltage** (V_s). The work done by these 'fastest' electrons against the pd they have travelled through is eV_s and must equal the kinetic energy they have lost giving:

E8.6 Max. ke of electrons $= eV_s$
$= hf - \varphi$
$= hf - hf_o$

where V_s = stopping voltage (V), e = charge on electron (C), h = Planck's constant (J s), φ = work function (J), f = photon frequency (Hz), f_o = threshold frequency (Hz).

From E8.6 it can be seen that a graph of the stopping voltage plotted against the frequency of the photons incident upon the photoelectric material will be as shown in F8.3. In F8.3: **slope**, h/e; **area**, no physical significance; **intercept**, the threshold frequency (f_o).

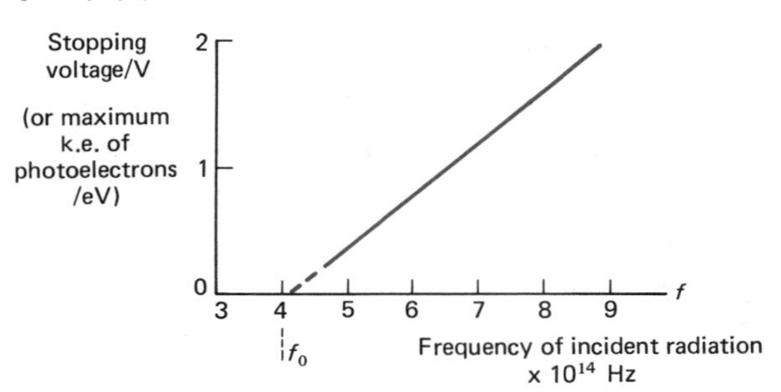

F8.3 Stopping voltage plotted against frequency

Few materials exist with a threshold frequency that lies in the visible region of the electromagnetic spectrum; most materials require higher frequencies (ultraviolet radiation is particularly useful for photoelectric work).

It is possible for the reader to use F8.3 to confirm that the slope of the graph is equal to h/e and that the threshold frequency of the chosen photoelectric material is unusually low (it corresponds to red light of wavelength 750 nm).

8.3 ELECTRONS WITHIN ATOMS

Energy levels, ionization and excitation In metals there are some 'free electrons', but the majority of electrons in metals and all materials are 'bound' to individual atoms. These electrons are 'held in the atom' by the force of attraction that exists between their negative charge and the positive charge on the nucleus. Hence each 'bound' electron must have electrical potential energy (see 2.2) and the work that must be done to remove an individual electron from the atom (ie, to 'infinity') is called the **ionization energy** of that electron. The energy needed to ionize atoms can be supplied in many ways; eg, by radioactivity, heat, electron collisions, X-rays, etc. Experiments (notably by Franck and Hertz) showed that each atom exhibits its own unique pattern of ionization energies. Hence atoms can be identified from their ionization energies and electrons in any atom can only have specific values of pe, ie, electron energies are quantized (see 1.10). These specific values are referred to as **energy levels** and

F8.4 shows four of the energy levels (−19, −12, −7, and −3 eV) for an imaginary atom. There are many more energy levels available in the shaded region but they have been left out of the diagram.

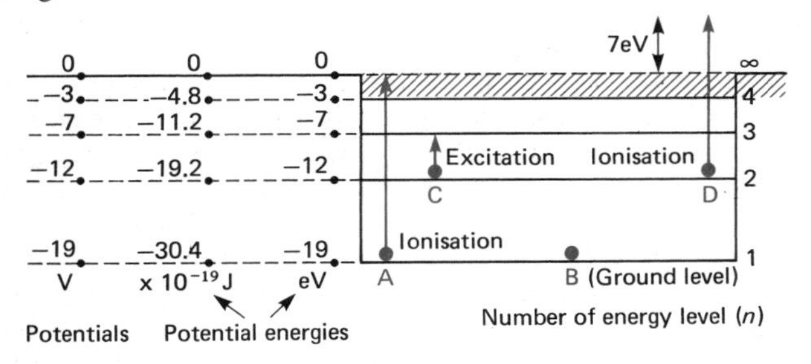

F8.4 Electron energy levels in an imaginary atom

The pd needed to give a free electron sufficient energy to enable it to ionize an atom is called the **ionization potential**. For electron A in F8.4 this is 19 V, which corresponds to an ionization energy of 19 eV (or 30.4×10^{-19} J). If on the other hand 19 eV of energy were given to electron *D* it would have 7 eV more energy than it needed for ionization (12 eV), hence it would emerge from the atom with 7 eV of ke. It is possible to give an electron extra energy that is not enough to remove it from the atom; in this case the electron must be given precisely enough energy to move it to a new level within the atom. This process is called **excitation**. It is possible to excite electron *C* as shown with precisely an extra 5 eV of energy to move it up to the next level. However the electron would not accept 4 or 6 eV of extra energy, yet it would accept 9 eV of extra energy which would move it up to level 4. It can be seen from the diagram that the energy levels are numbered (n) starting from the lowest potential energy level ($n = 1$), this lowest level is often referred to as the **ground level**.

Energy levels in hydrogen An equation for calculating the pe levels (in J) of the hydrogen atom was developed by Balmer:

E8.7 $$E_n = -\frac{21.8 \times 10^{-19}}{n^2}$$

where E_n = pe of nth level in hydrogen atom (J), n = number of level – see F8.4.

As hydrogen only has one electron in its atom, which will usually be found in the ground level ($n = 1$) it should be possible for the reader to predict that the ionization potential of an unexcited hydrogen atom is 13.6 V (using E8.7).

Pauli exclusion principle It is a general rule in science that all systems will try to adopt their position of minimum pe. If this were to happen to electrons within atoms then we should find all the electrons on the $n = 1$ level (in the ground level); however this does not happen! There is a maximum number of electrons allowed on each level, as shown in F8.5.

Number of level (n):	1	2	3	4	etc.
Maximum number of electrons allowed on the level:	2	8	18	32	etc.
Letter code for level:	K	L	M	N	etc.

F8.5 Electron shells

The set of rules that predicts F8.5 is collectively referred to as the Pauli exclusion principle. Basically the total energy of each electron in an atom is defined by a set of **quantum numbers**; our level number, n is one of the quantum numbers. The exclusion principle states that no two electrons in an atom may have the same set of quantum numbers, hence a limited number of electrons can be found on each level (where the quantum number n stays fixed but other quantum numbers may alter). Energy levels are sometimes referred to as **shells**, and the shells are given code letters as shown in F8.5. Hence a *K* shell electron will be found in the ground level ($n = 1$) of an atom etc.

8.4 ENERGY EMITTED BY ELECTRONS

General information Electrons can lose energy in two distinct ways:

1 The energy is transferred to other particles or bodies (eg, electrons striking a target will heat it up as electron ke is converted into molecular ke – or heat energy – in the target).

2 The energy is converted into a photon of electromagnetic radiation (eg, an electron changing to a lower energy level in an atom emits a photon to get rid of the energy it must lose).

Electrons changing energy levels F8.4 shows the four electrons in an imaginary atom occupying the lowest levels that they can according to the exclusion principle (see F8.5 – only two electrons are allowed in the ground level). This atom is said to be in the **ground state** or **unexcited**. F8.6 shows the same atom in an **excited** state (ie, it is possible for one or more of the electrons to move to lower energy levels).

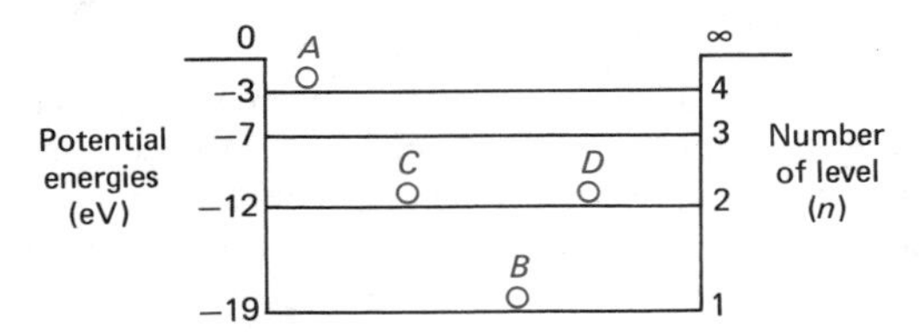

F8.6 Imaginary atom in an excited state

It can be seen in F8.6 that electron *A* has been excited to the $n = 4$ level (*N* shell) by receiving 16 eV of energy. This excited atom will spontaneously (ie, of its own accord) return to the ground state (its position of minimum allowable potential energy). However for F8.6 there are four different methods by which the electrons can move levels to achieve this as shown in F8.7.

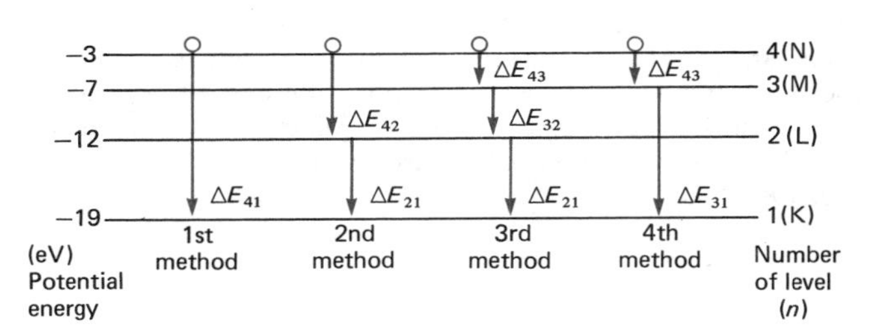

F8.7 Transitions from $n = 4$ to $n = 1$ level

Any electron on a level may be involved in a transition; for example in the 2nd method in F8.7 after the ΔE_{42} transition electrons *A*, *C* and *D* will all be on level $n = 2$ and any one of them could then carry out the ΔE_{21} transition.

An electron involved in a downward transition must lose energy, eg, a transition from level $n = 3$ to level $n = 2$ involves an electron energy loss (ΔE_{32}) of 5 eV. This energy is emitted as a photon of electromagnetic radiation giving:

E8.8 $$\Delta E = hf = hc/\lambda$$

where ΔE = energy difference between the levels (J), h = Planck's constant (J s), f = frequency of photon emitted (Hz), c = speed of light (m s^{-1}), λ = wavelength of photon emitted (m).

The wavelengths of photons emitted from atoms can vary from infrared right through the entire visible spectrum and ultraviolet to X-rays; it should be pointed out that the energy needed for X-rays is so large (of the order of keV) that only transitions between the lower levels (small n and large ΔE) in larger atoms can produce X-ray emissions. The light from excited atoms is used in neon and sodium lamps and the ultraviolet emissions from excited mercury vapour provide the energy needed for fluorescence in fluorescent lights. An **emission spectrum** of the radiation given out by excited atoms can be plotted. Part of the emission spectrum from our imaginary atom is plotted in F8.8.

F8.8 is called a **line spectrum** (compared with a **continuous spectrum**, eg, F8.9) because only certain frequencies are emitted

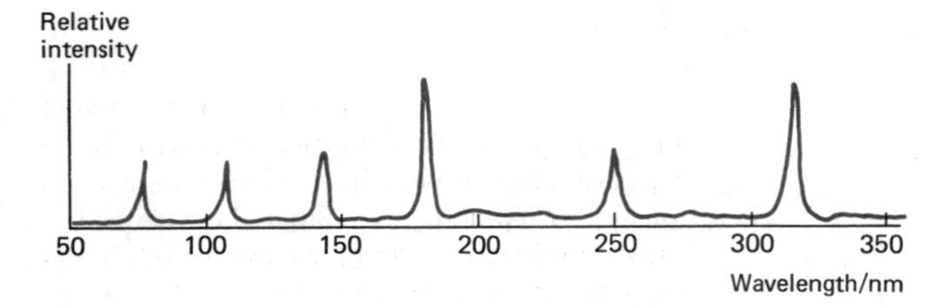

F8.8 Part of emission spectrum from an imaginary atom

as defined by E8.8, which gives rise to a spectrum of 'vertical lines'. The wavelengths most frequently emitted give 'higher' lines. The six wavelengths of emission in F8.8 correspond to the six different transitions (ΔE_{41}, ΔE_{42}, ΔE_{21}, ΔE_{43}, ΔE_{32}, ΔE_{31}) of F8.7. The reader can use E8.8 to check that this is the case. Emission spectra can be used to identify atoms as each atom emits a unique line spectrum; obviously the spectra can also be used to deduce information about the energy levels of the electrons. It is also possible to study atoms by making them absorb energy (ie, by exciting the atoms) and studying the associated **absorption spectrum**; atomic absorption will also produce a line spectrum.

Photons created by very high energy electrons: X-rays X-rays are the photons produced (energy 1 keV or greater) when very high energy electrons suddenly lose energy. They are produced in an X-ray tube when electrons of energy 10 keV or more are 'fired' into a dense target (frequently tungsten). Well over 90% of the ke lost by the electrons as they hit the target is converted into heat energy (ke of the target's molecules). However some of the electron energy is converted into high-energy photons or X-rays. This can happen in two distinct methods.

1 The electrons lose energy in inelastic collisions with nuclei of the target atoms; the lost ke is emitted as X-ray photons; the electrons are slowed down in this process and hence the X-rays are also called **Bremsstrahlen** – this is German for 'braking radiation' and refers to the way in which they are made; electrons can lose any fraction of their energy in a collision with a nucleus – this means photons of a continuous range of wavelengths are produced resulting in a **continuous spectrum** for Bremsstrahlen, see F8.9.

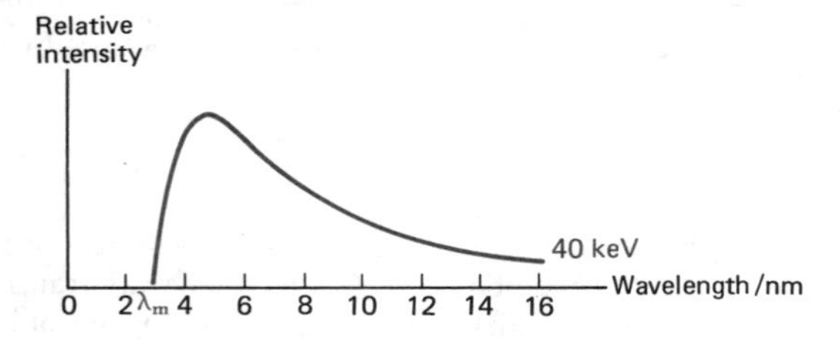

F8.9 40 KeV continuous X-ray spectrum

2 The electrons lose energy in collisions with electrons in the target atoms; the target atoms become excited and on returning to the ground state emit photons of well defined energies producing **X-ray line spectra**.

The shape of the continuous Bremsstrahlen spectrum of F8.9 depends only upon the energy of the electrons and not upon the nature of the target. There will always be a clearly defined minimum wavelength of emission (λ_m) which corresponds to the largest energy photon that can be emitted; this must refer to a collision in which a maximum energy electron converts all its energy into a photon (40 keV in F8.9). It is possible for the reader to use E8.8 to confirm that λ_m in F8.9 is equivalent to a photon energy of 40 keV.

If the X-ray spectrum for 40 keV electrons is examined it will show both types of spectra superimposed on top of each other. F8.10 shows the full X-ray spectrum for 40 keV and 80 keV electrons 'fired' at the same target element.

The wavelengths defined by the line spectrum in F8.10 depend only on the nature of the target; ie, the X-ray line spectrum will identify the element from which the target is made. The extra

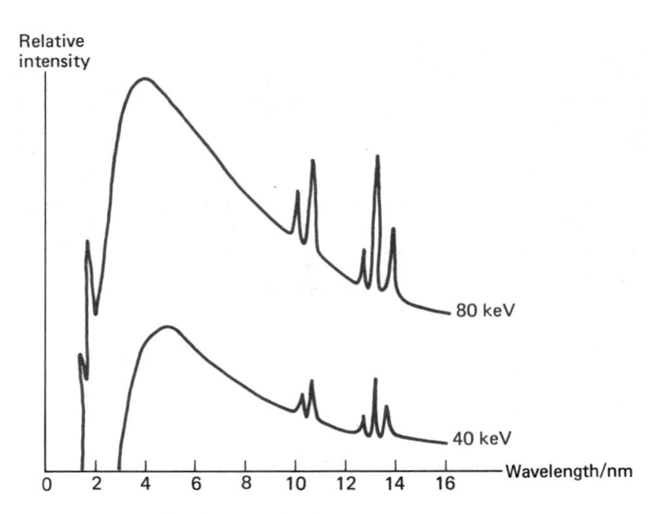

F8.10 40/80 KeV X-ray spectra (same target element)

lines appear in the 80 keV spectrum only because 40 keV electrons do not have enough energy to excite the electrons that are causing these lines.

8.5 PROPERTIES OF THE NUCLEUS

The nuclear model: neutrons, protons and isotopes Under the direction of Lord Rutherford, alpha-particle scattering experiments that proved the existence of a nucleus within the atom were carried out by Geiger and Marsden in 1909. In a typical medium-sized atom (a sphere of diameter about 10^{-9} m) the nucleus (a sphere of diameter about 10^{-14} m) occupies a minute fraction of the total volume of the atom (which is mostly 'empty space'). Yet as virtually all the mass of the atom is concentrated in the nucleus, the nuclear density has a phenomenally high value (about 10^{15} kg m^{-3}). The nucleus consists of **nucleons** (neutrons and protons) bound together by a very strong (but very short range) 'nuclear field force' that exists between two or more nucleons. The basic properties of the nucleons are summarized in F8.11 where an important comparison is made with the properties of electrons.

Nucleon	charge		mass	
PROTON	$+1.6 \times 10^{-19}$ C	+1(e)	$M_p = 1.673 \times 10^{-27}$ kg	about $1800m_e$
NEUTRON	Zero	0(e)	$M_n = 1.675 \times 10^{-27}$ kg	about $1800m_e$

F8.11 Properties of nucleons

A nucleus is identified as being of a particular element by its number of protons: the number of protons in a nucleus is called the **atomic number** (Z). For example a nucleus containing 1 proton ($Z=1$) is hydrogen, a nucleus containing 92 protons ($Z=92$) is uranium. The atomic number hence defines the positive charge carried by a nucleus. The total number of nucleons (neutrons + protons) in a nucleus is referred to as the **mass number** (A).

It can be seen that the **neutron number** (N) is defined by:

E8.9 $N = A - Z$

where N = neutron number, A = mass number = total number of nucleons, Z = atomic number = total number of protons.

It was discovered using mass spectrometers that nuclei of the same element can have different masses as they can contain different numbers of neutrons; each different nucleus is said to be an **isotope** of that element. Two important isotopes of uranium are uranium-235 and uranium-238; both isotopes contain 92 protons but one has 143 neutrons and the other 146 neutrons. A graph of N plotted against Z (an $N-Z$ plot, see F8.14) has several important interpretations in nuclear physics.

Nuclear equations In nuclear physics a particle is identified by its mass number (ie, how many nucleons it contains) and its charge compared with the charge on a proton (for any nucleus this is also its atomic number). The particle itself can be labelled by its chemical symbol (eg, He for helium nuclei), by a capital letter (eg, *X*, *D*, *E*, etc.), or by a 'code symbol' (eg, *e* for electron, *a* for alpha-particle, etc.). Some examples are given in F8.12.

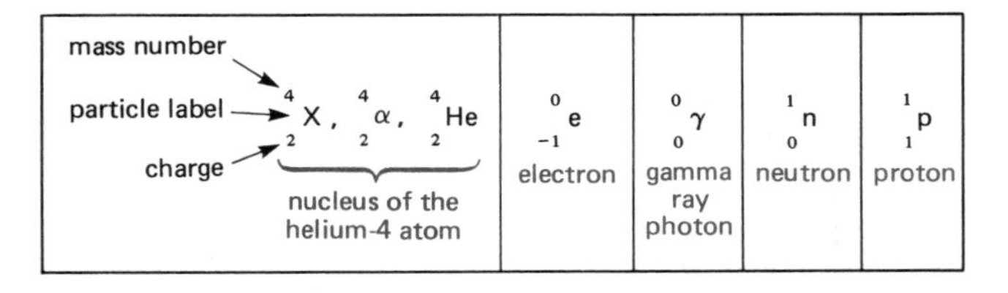

F8.12 Notation of particles

In any nuclear physics reaction both the total number of nucleons and the total charge are conserved. Consider the following nuclear interaction of an alpha-particle colliding with the nucleus of a nitrogen atom:

E8.10 $^4_2a + ^{14}_7N \rightarrow ^{17}_8O + ^?_?X$

On the left-hand side of the equation the total mass number is 18; hence the mass number of *X* is 1 to ensure that there are also 18 nucleons on the right-hand side of the equation. On the left-hand side of the equation the total charge is + 9; hence *X* must carry a charge of + 1 (no sign on the charge means that it is positive). Hence as *X* is one nucleon ($A = 1$) carrying the charge of a proton ($Z = 1$) it must be a proton (^1_1p)! This is the nuclear interaction observed by Rutherford when he discovered the **proton** in 1919.

In nuclear physics it is important to realize that although charge and the total number of nucleons will be conserved in a reaction, neither mass nor energy need be conserved as one can be converted into the other (see 1.10 and 8.7).

8.6 SPONTANEOUS EMISSION OF NUCLEAR ENERGY: RADIOACTIVITY

General information about radioactivity If an isotope is **stable** it will never spontaneously (of its own accord) emit energy. However there are many unstable isotopes which will eventually emit energy spontaneously; such nuclei are said to be **radioactive**. The three best-known radioactive energy emissions (alpha a, beta β, and gamma γ) are summarized in F8.13.

Radioactive emissions occur as a result of the basic principle that any system will adopt a state of lower total potential energy if possible. The pe of the **parent** nucleus must be higher than that of the **daughter** nucleus that will be formed after the radioactive decay. The energy difference must be sufficient to permit the radioactive emission to take place. For large radioactive nuclei ($Z >$ about 80), a or β emissions are possible, but a-particles are not emitted from smaller nuclei (ie, with $Z < 60$).

In an **alpha-particle emission** two protons and two neutrons are ejected as one particle (the alpha-particle – a helium-4 nucleus); so if *D* is the parent nucleus and *E* the daughter:

E8.11 $^A_ZD \rightarrow ^{A-4}_{Z-2}E + ^4_2a$ alpha-particle emission

eg, $^{226}_{88}Ra \rightarrow ^{222}_{86}Rn + ^4_2a$

The arrow indicates that the process is 'one-way', ie, not reversible.

In a **beta particle emission** a neutron in the parent nucleus (*D*) is turned into a proton and an electron (the beta-particle) so producing a new nucleus, the daughter (*E*):

E8.12 $^1_0n \rightarrow ^1_1p + ^0_{-1}\beta$ beta-particle emission

or $^A_ZD \rightarrow {}_{Z+1}^{A}E + ^0_{-1}\beta$

eg $^{90}_{38}Sr \rightarrow ^{90}_{39}Y + ^0_{-1}\beta$

All beta-particle emissions are accompanied by emission of extra energy which is carried away by a particle of no mass and no charge called a **neutrino** ($^0_0\upsilon$). Beta-particles are electrons and are no different to any other free electrons of the same energy; the term 'beta-particle' refers to the way in which the electron was

Type	a	β	γ
Nature	Helium-4 nucleus (2 neutrons and 2 protons)	Electron	Photon of electromagnetic radiation
Charge, e	+ 2	− 1	0
Mass, kg	6.7×10^{-27}	9.1×10^{-31}	0
Mass, u	4	1/1836	0
Energy, MeV	0.5–10	0.01 – 10	0.01 – 10
Ionizing power	Very good ionizer	About $\frac{1}{10}$th of a	About $\frac{1}{10\,000}$ of a
Speed	Up to 0.01 c	(0.01 − 0.9) c	c
Absorbed by (range)	Thin paper	mm of aluminium	Partially absorbed by cm of lead
Detected by	Thin-window Geiger tube, Solid-state detector, Cloud/bubble chambers, Ionization chamber, Photographic film, Scintillation counter, Spark counter,	Geiger tube, Solid-state detector, Cloud/bubble chambers, Photographic film, Scintillation counter,	Geiger tube, Solid-state detector, Photographic film, Scintillation counter,
Uses/applications	Artificial disintegration of nuclei, Luminous paint	Radioactive dating Thickness gauges, Medical tracing	Gamma-ray photography through metals, Thickness gauges, Medical tracing

F8.13 Properties of radioactivity

created, ie, by radioactive emission. Although electrons can be ejected from nuclei (by neutrons 'turning into' protons) it is impossible to find an electron within a nucleus. The wave properties of such an electron (see 1.10 for particle/wave equivalence) would result in energy of the order of GeV (10^9eV) for the electron; this energy would shatter any nucleus!

A key feature of all alpha- and beta-decays is the fact that the daughter is a different element to the parent; radioactivity hence enables one element to spontaneously turn into another! In E8.11 radium (Ra) turns into radon (Rn) as a result of an alpha-decay, and in E8.12 strontium (Sr) turns into yttrium (Y) as a result of a beta-decay. Most daughter nuclei created by radioactive decay are radioactive themselves and undergo further decays. It can be possible to trace a whole sequence of radioactive decays from a single parent; a complete sequence of possible decays is called a **radioactive series**.

All unstable nuclei will eventually undergo radioactive decay. It is possible to predict likely radioactive decays by examining a plot of the stable nuclei; F8.14 shows such a plot. The $N-Z$ **plot** (as it is called) has the neutron number (N) plotted on the y-axis and the atomic number (Z) plotted on the x-axis. It is not necessary for A level to know the position of any points that represent the stable nuclei, so instead the 'locus' of these points is shown (shaded area on graph).

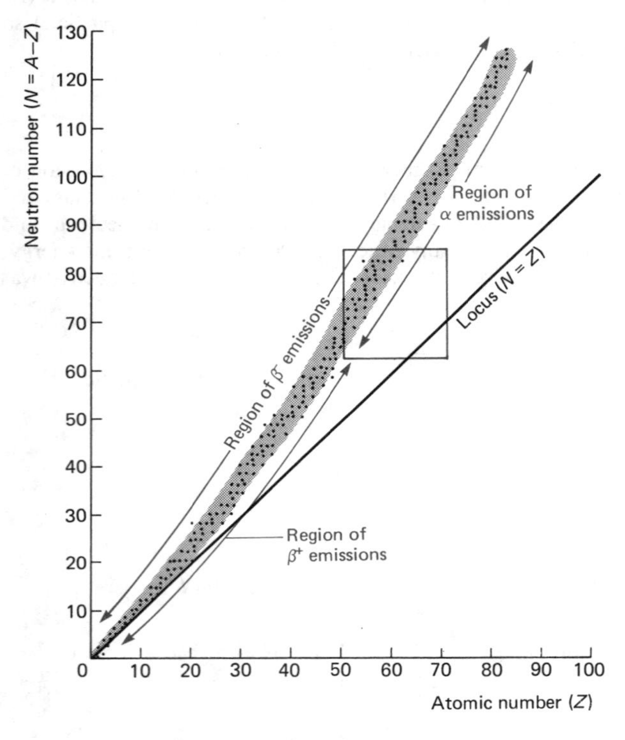

F8.14 N – Z plot for stable isotopes

If a radioactive isotope (ie, an unstable nucleus) is plotted on F8.14 it is possible to make a good prediction of its likely radioactive decay from its position on the plot in relation to the stable nuclei. To illustrate this more clearly, F8.15 shows an enlargement of a portion of the plot as defined by the coloured rectangle on F8.14

In F8.15 the stable nuclei are shown as black dots, the radioactive nuclei (X, Y, Z and E) are plotted in colour. Arrows show the effect of three basic radioactive decays on E. The position of the daughter nucleus on the plot in relation to the parent nucleus can be summarized as follows: for an alpha-decay (N and Z reduced by two) the nucleus moves at 45° down to the left, for a beta-minus decay (Z up one, N down one) the nucleus moves at 45° down to the right, for a beta-plus decay (N up one, Z down one) the nucleus moves at 45° up to the left. In a positive beta-decay a proton is turned into a neutron by emission of a positive electron (**positron**) or by the nucleus capturing an electron; an

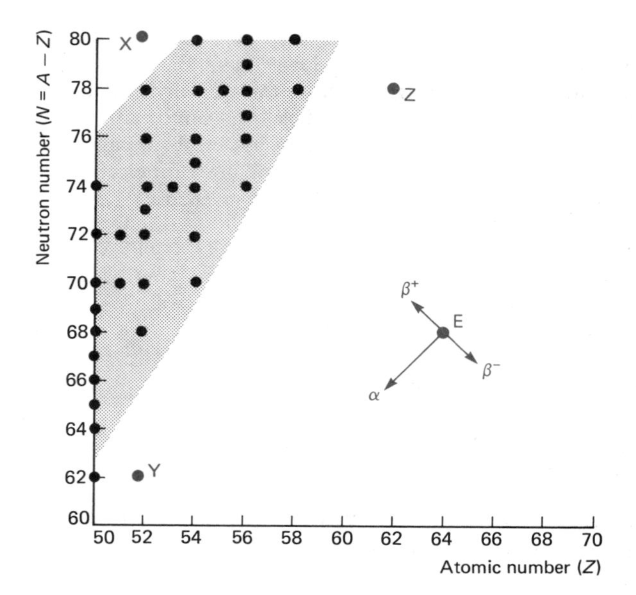

F8.15 Detail from the $N-Z$ plot

understanding of these decays is not required for most A level courses. For X, Y and Z it can be seen that the nuclei can be moved closer to the zone of stability in the following ways; X down to the right (β^-), Y up to the left (β^+), and Z down to the left (α). Clear examples have been chosen, but sometimes the appropriate beta-decay or an alpha-decay will seem to move a nucleus nearer to stability; indeed some nuclei are found to undergo either type of radioactive decay! An important point about the $N-Z$ plot is that for 'small' and 'medium' sized nuclei the plot follows the line defined by $N = Z$ (see F8.14). An alpha-decay will move a nucleus parallel to this line and hence no nearer stability. This means that alpha decay occurs only for large nuclei ($Z >$ about 60) in the region marked on the plot, smaller nuclei below the region of stability tend to undergo β^+-emissions. β^--emitters can be found anywhere above the region of stability.

In the same way that excited atoms (see 8.4) lose energy as electrons drop into lower energy levels, an **excited nucleus** can lose energy as the nucleons carry out a similar process. In each case the energy is emitted as photons of electromagnetic radiation. However the photon energies of nuclear emissions (MeV) are much higher than those from the atomic electrons (eV to keV). Photons of electromagnetic radiation emitted from excited nuclei are called **gamma-rays**. On the completion of an alpha- or beta-decay most nuclei are left in an excited state; this means that most alpha- and beta-decays are also accompanied by gamma-emissions. Excited nuclei are indicated with an asterisk (*) hence:

E8.13 ${}^{A}_{Z}X^* \rightarrow {}^{A}_{Z}X + {}^{0}_{0}\gamma$ Gamma-ray emission

It is possible to get X-rays with the same wavelength as gamma-rays, they are of course identical and are only given different names as they have been produced in different ways.

All three basic types of radioactivity have sufficient energy to ionize atoms (see F8.13) and it is their ionizing properties that enable them to be detected. Dectectors that work on this principle include cloud and bubble chambers, Geiger – Müller tubes, spark counters, ionization chambers, solid-state detectors and scintillation counters. In scintillation devices the electrons produced by ionization (due to the radioactivity) then excite other atoms which proceed to emit light. Photographic emulsions that use the energy of the radioactivity to produce a chemical effect are one of the few detectors not to rely on ionization. It is possible to distinguish between the three types of radiation by their different ionizing powers (ie, thickness of tracks in cloud and bubble chambers) or by using absorbers to classify them (see F8.13). As gamma-rays undergo negligible absorption and travel in straight lines in air, they will obey an inverse square law as in E8.14:

E8.14 $I_r \propto \frac{1}{r^2}$

where I_r = intensity of gamma-ray source a distance r away (Watts or photons per second; per square metre), r = distance of detector from source (m).

As both alpha- and beta-particles undergo significant absorption in air, their intensity drops off at an even greater rate than that predicted by E8.14 for gamma-rays. It is important to remember that the two most vital concepts in radioactive safety are time of exposure and distance from the source; a surface 10 cm away from a source will be exposed to at least 100 times the dosage that would be received 1 m away over the same period of time.

Mathematical model of radioactivity: decay laws and half-life If it were possible to watch a group of identical radioactive nuclei, the time that would pass before any particular nucleus carried out its radioactive emission would appear to be random. Such is the random nature of radioactivity. However if the average time were taken for a large number of the same nuclei to decay, this time would have a consistent value. Large numbers of objects that individually behave randomly can follow very predictable patterns; eg, the toss of a coin is random (head or tail), but the result of a large number of tosses (half heads, half tails) can be reliably predicted, see also 6.10. The number of radioactive particles emitted per second from a particular type of nucleus, the **activity**, is the same quantity as the rate of change of the number of that nucleus actually present (dN/dt); furthermore the number of radioactive emissions per second must be proportional to the number of radioactive nuclei actually present, hence:

E8.15 $\frac{dN}{dt} = -\lambda N$

where dN/dt = number of radioactive emissions per second (s^{-1}), λ = radioactive decay constant (s^{-1}), N = number of radioactive nuclei present (no units).

This equation defines the **radioactive decay constant** (λ), the negative sign indicates that the number of radioactive nuclei present is being reduced (not increased) and hence as time progresses the decay rate drops. E8.15 can be represented graphically by plotting the number of radioactive nuclei present (N) against time (t) as shown in F8.16. In F8.16: **slope**, decay rate (dN/dt) (s^{-1}); **area**, no physical significance; **intercepts**, etc., N_0 is the number of radioactive nuclei initially present – after sufficient time the graph asymptotically approaches the t axis. The graph has a constant half-life ($T_{\frac{1}{2}}$).

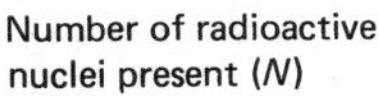

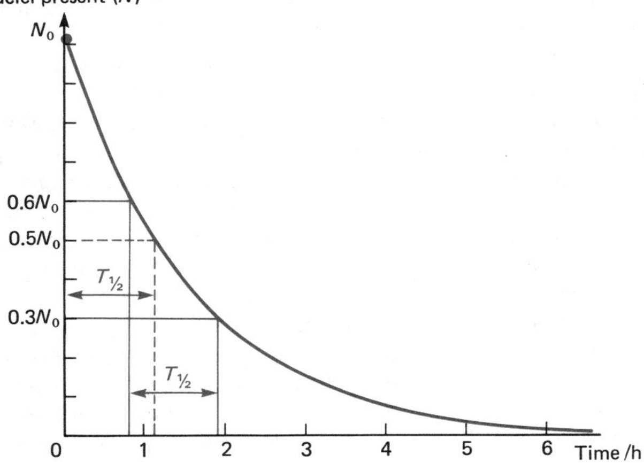

F8.16 Theoretical curve of radioactive decay ($T_{\frac{1}{2}} = 1.1$h)

Like all exponential graphs F8.16 has a constant ratio property. In radioactivity this property is defined by measuring the time taken for the number of radioactive nuclei present to be halved; this time is called the **half-life** ($T_{\frac{1}{2}}$) of the radioactive isotope. On the graph two half-lives have been measured; one from $N = N_0$ to when $N = \frac{1}{2}N_0$, the other from when $N = 0.6\,N_0$ to when $N = 0.3N_0$. It is possible for the reader to measure some more half-lives from the graph; like the two shown they will all be found to be 1.1 hours. Half-lives for radioactive isotopes can vary from millions of years to millionths of seconds. The half-life and decay constant are all linked as in E8.16:

E8.16 $T_{\frac{1}{2}} = (\log_e 2)/\lambda$

where λ = decay constant (s^{-1}), $T_{\frac{1}{2}}$ = half life (s).

The numerical value of $\log_e 2$ is 0.693. It is possible to use E8.15 and E8.16 to help find the number of radioactive nuclei present after a time (t) has elapsed when N_0 radioactive nuclei were originally present. The equation needed (E8.17) is obtained by integrating E8.15:

E8.17 $N_t = N_0 e^{-\lambda t}$

where N_t = number of radioactive nuclei present at time = t (no units), N_0 = number of radioactive nuclei originally present, ie, at time $t = 0$ (no units), λ = decay constant (s^{-1}), t = time that has elapsed from when $N = N_0$(s).

The numerical value of e is 2.718. It is possible to use E8.15 and E8.17 in several different forms by replacing N, N_0 and N_t with quantities that are proportional to them; eg, R, R_0 and R_t, ratemeter readings (count rates), or I, I_0 and I_t, ionization chamber currents, etc. Similarly the vertical axis of F8.16 may be relabelled with these quantities, though an actual experiment will produce the graph shown as F8.17 rather than the theoretical result of F8.16. In F8.17: **slope**, rate of change of count rate (s^{-1}); **area**, total number of radioactive particles detected; **intercepts**, etc., C_0 is the count rate at the start of the experiment, C_B is the count rate due to the background radiation.

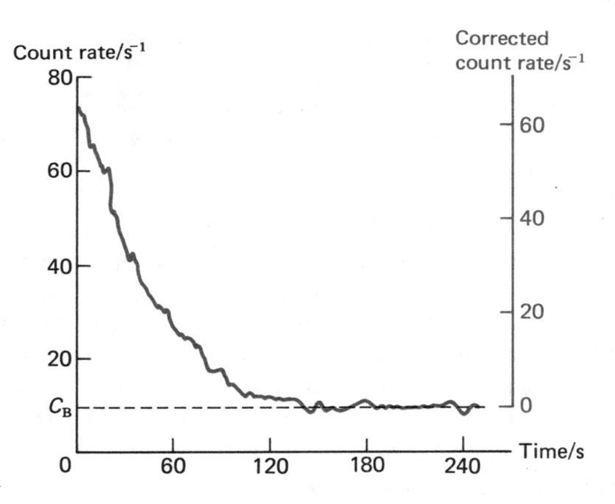

F8.17 Practical decay curve

The graph can be seen to show random fluctuations typical of the radioactive process but it still adopts the theoretical shape defined by F8.16. An important further difference is that when the radioactive decay has ceased there is still a 'steady' count rate (C_B) due to the ever present **background radiation** that is found on the Earth. Before the radioactive decay can be analysed it is necessary to subtract the background count from all the readings; this can simply be done by rescaling the vertical axis as shown by the numbers in colour on the right-hand side of the graph. The reader can use this scale to check the half-life to be 30 s.

8.7 INDUCED EMISSION OF NUCLEAR ENERGY: FISSION AND FUSION

Measuring the energy of nuclear interactions Any unstable nucleus will eventually emit energy (radioactivity); it is also possible to induce a stable nucleus to emit energy. This is usually done by causing an interaction to occur between two nuclear particles; E8.10 is an example of an induced release of nuclear energy (the alpha-particle induces a stable nitrogen-14 nucleus to form an oxygen-17 nucleus and the energy released is carried away as ke of the proton and oxygen-17 nucleus).

The nucleons in a nucleus are 'held together' by the short-range attractive nuclear force (which is far stronger than the electric field repulsive force between the protons). Thus to remove a nucleon from the nucleus requires work to be done against the attractive force so the pe of the nucleon is increased; this is a similar concept to the work done against the attractive force of gravity as the pe of a rocket is increased as it leaves the surface of the Earth. The total energy needed to remove a nucleon from a nucleus is called the **binding energy of the nucleon**, ie, the energy with which it was originally bound into the nucleus. The total energy needed to separate all the nucleons from a nucleus so that they are all 'free' is called the **binding energy of the nucleus**. The lower the potential energy of a nucleon the greater will be its binding energy; in much the same way the lower the potential energy of a rocket (ie, the nearer it is to the surface of the Earth), then the greater will be the energy needed to free it from the Earth's gravitational field.

Nuclear interactions do not obey the normal classical mechanics laws of conservation of energy and of mass (see 1.10) as it is possible for it is possible for energy to be converted into mass and vice-versa (see E1.17 for the conversion formula). So as mass and energy are 'equivalent' in nuclear physics it is the total of 'mass + energy' that is conserved. This leads to a simple method of analysing nuclear interactions; consider an interaction in which particles A and B interact to create particles C and D:

A	$+B$	$\longrightarrow$	C	$+D$
If Mass +	Mass	is greater	Mass +	Mass
of A	of B	than	of C	of D

then the missing mass on the right hand side will have been turned into energy and the interaction will release this amount of energy. But:

If Mass +	Mass	is less	Mass +	Mass
of A	of B	than	of C	of D

then to make the equation balance the missing mass must be supplied to the left-hand side of the equation in the form of energy, ie, this interaction needs an energy supply if it is to happen.

The 'missing mass' in a nuclear interaction measures either the energy released in the interaction or the energy needed to make it happen, it is a vital property of any nuclear interaction.

The measurement of mass is of such importance in nuclear physics that a new unit is defined for greater convenience; the **atomic mass unit** (abbreviation amu) is a unit mass equal to 1/12th of the mass of an atom of carbon-12. The units of amu (u) can be converted into mass (in kg) or energy (in MeV or J):

E8.18 1 atomic mass unit $= 1\,u$
$= 1.66 \times 10^{-27}$ kg
$\equiv 931$ MeV $\equiv 1.49 \times 10^{-10}$ J

It can be seen that each nucleon corresponds to nearly $1\,u$ of mass. The following example illustrates how to calculate the energy involved in an interaction:

E8.19 ${}^2_1H + {}^2_1H \rightarrow {}^3_2He + {}^1_0n$

Total mass of $({}^2_1H + {}^2_1H) = (2.015 + 2.015) = 4.030\,u$

Total mass of $({}^3_2He + {}^1_0n) = (3.017 + 1.009) = 4.026\,u$

Hence $0.004\,u$ of mass 'goes missing' and is converted into the 3.7 MeV of energy released in this process.

The reader can calculate the energy released when an alpha-particle is formed from the following information:

Mass of proton $= 1.0076\,u$

Mass of neutron $= 1.0090\,u$

Mass of alpha-particle $= 4.0028\,u$

The energy released should be 28.3 MeV, which is therefore the binding energy of the alpha-particle. It is possible to calculate the binding energy of any other nucleus in a similar fashion; the 'missing mass' difference between the mass of the nucleus and the mass of the 'free' nucleons from which it is made up is called the

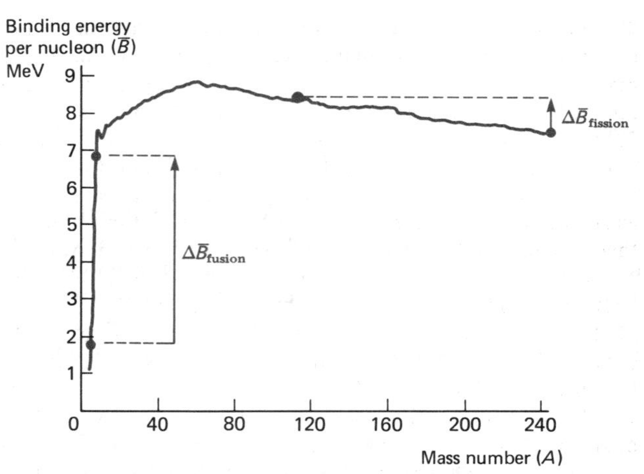

F8.18 Average binding energy per nucleon against mass number

mass defect. If we divide the binding energy of a nucleus by the number of nucleons it contains we have measured the average **binding energy per nucleon** ($\bar{B}$). A graph of this quantity plotted against mass number (F8.18) is one of the most important graphs in nuclear physics.

Fission and fusion Such a phenomenal amount of energy can be generated from mass (10^{17} J kg^{-1}, see E1.17) that extracting energy from the nucleus in this fashion has been the subject of over 40 years of concentrated research. Two features of nuclear interactions are of key importance in designing nuclear energy systems (bombs or reactors):

1 Only a small proportion of the mass is converted into energy in most nuclear reactions.
2 It is very difficult to create the experimental conditions that will cause nuclear reactions to take place sufficiently often to generate significant quantities of energy.

The amount of energy released in a single nuclear reaction is equal to the increase in binding energy (reduction in mass) that occurs. Hence the amount of energy released per unit mass of 'fuel' is equal to the increase in binding energy per unit mass; this can be expressed as the increase in binding energy per nucleon ($\bar{B}$) which can be determined from F8.18. The nuclear reactions which will be most efficient at generating energy will be those that result in the biggest increase in $\bar{B}$, two basic reactions are suggested by the graph.

Fusion: very small nuclei ($A \sim 2$, $\bar{B} \sim 2$ MeV) are fused (joined together) to form a larger nucleus ($A \sim 4$, $\bar{B} \sim 7$ MeV) creating an increase in binding energy per nucleon (or energy release per nucleon) of $\Delta\bar{B}_{fusion} \sim 5$ MeV.

Fission: very large nuclei ($A \sim 240$, $\bar{B} \sim 7.5$ MeV) undergo fission (splitting) creating two smaller nuclei ($A \sim 115$, $\bar{B} \sim 8.5$ MeV) resulting in a release of energy per nucleon of $\Delta\bar{B}_{fission} \sim 1$ MeV.

A typical fusion reaction has already been analysed (see E8.19). Fuel for fusion is plentiful (there is an almost limitless supply of hydrogen on the Earth in the form of water). However fusion reactions only take place at extraordinarily high temperatures (typically 10^6 K) from which their alternative name of **thermonuclear reactions** is derived. It is possible to produce such temperatures for uncontrolled reactions (hydrogen bombs), but scientists have yet to build a successful thermonuclear reactor for power generation.

Despite the facts that fission generates less energy (per kg of fuel) than fusion and that fission fuel (eg, uranium and plutonium) is far less plentiful and more difficult to process, fission reactors remain the only viable means of producing controlled nuclear power at the moment. However atomic bombs (fission) have in the main been superseded by the more powerful thermonuclear (fusion) weapons. Fission processes do not need the same high temperatures as fusion; a typical fission reaction (E8.20) illustrates that the process can be self-sustaining:

E8.20 ${}^{235}_{92}U + {}^1_0n \rightarrow {}^{148}_{57}La + {}^{85}_{35}Br + 3{}^1_0n$

The energy released by a nucleon when it changes from being a member of a U-235 nucleus to being a member of a nucleus of mass $\cong \frac{1}{2} \times$ U-235 mass is about 0.8 MeV. Hence the energy released by the fission of E8.20 is about 185 MeV ($\cong 0.8 \times 235$). The fission is **induced** by a neutron colliding with the parent nucleus of U-235. The fission creates several further free neutrons (in E8.20, 3 further free neutrons are released); if at least one of these neutrons goes on to create a further fission, then a **chain reaction** of continual fissions will be underway.

In the 'thermal' type of **nuclear fission reactor**, fission neutrons (produced with energies $\cong$ MeV) must be slowed down to 'thermal' energies (of the order of eV) to produce further fission of U-235: otherwise, the fast neutrons released after fission will be absorbed by U-238 atoms (since the fuel is mostly U-238) without further fission. The slowing-down of fast neutrons to become thermal neutrons is achieved by setting the fuel pins into a **moderator** block, as in F8.19. Here fast neutrons

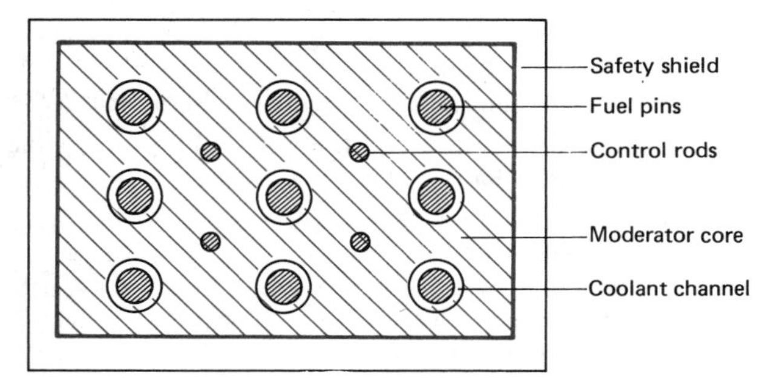

F8.19 Core of a fission reactor

leave the fuel pins after fission and enter the moderator where collisions with moderator atoms (usually light atoms, such as carbon for maximum energy transfer) cause the neutrons from the fuel pins to lose energy quickly. Eventually the slow (by now) neutrons in the moderator re-enter the fuel pins so producing further fission. In this way, the fission energy is transferred to heat energy of the moderator which is then removed by a **coolant** fluid. To prevent the chain reaction from proceeding too rapidly, **control rods** of high neutron-absorbing power can be lowered into the 'core' to maintain a constant neutron flux in the core. In this way, a steady chain reaction is maintained.

A rough estimate of the energy per unit mass of fuel for a thermal reactor can be made from the following figures:
Assume fuel is 1% U-235, and 200 MeV of energy are released from every U-235 fission.
For complete fission of U-235 in 1 kg of fuel, 10 g of U-235 containing $(10/235) \times N_a$ atoms will be fissioned (N_a = Avogadro's number). Thus, the energy released from 1 kg of fuel will be $(10/235) \times 6 \times 10^{23} \times 200 \times 1.6 \times 10^{-13}$ J, which is approximately 10^{12} J kg^{-1}.

It is possible to release nuclear energy with the type of reaction illustrated in E8.10. However the energy released per unit mass in these reactions tends to be far less than in fission or fusion. Further, these reactions are very rare and cannot be induced frequently enough to be viable energy producers.

UNIT 8 QUESTIONS

8.1M The diagram shows a charged oil drop between two horizontal plates connected to a high voltage source V. Which of the following statements is/are correct?

1 If the drop is stationary, then it must carry a +ve charge.
2 If the drop is falling at 'terminal velocity', then a resultant force must be acting on it.
3 If the drop is stationary and then it suddenly starts to rise, then it must have gained extra electrons.

A **1** only. B **2** only. C **3** only. D **1** and **2**.
E **2** and **3**.

(—: *JMB*, and all other Boards except SEB H*)

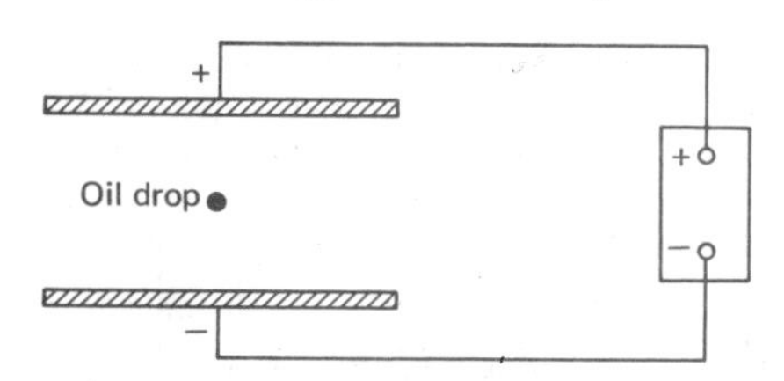

For **1**, see p. 123 if necessary.
For **2**, remember Newton's first law: see p. 17 if necessary.
For **3**, a sudden rise means that the drop's attraction to the top plate must have suddenly become greater.

8.2M An electron is moving with velocity v. It enters a region of uniform magnetic flux density B, and uniform electric field, E. E and B are mutually perpendicular. The velocity v will not remain constant unless v is:

A perpendicular to E and parallel to B.
B perpendicular to B.
C parallel to E.
D perpendicular to E and to B and of magnitude B/E.
E perpendicular to E and to B and of magnitude E/B.

(**London**: *all other Boards except SEB H*)

See p. 123 and apply E8.3. Also, think of how you arrange your fingers for Fleming's left-hand rule!

8.3M Electrons travelling at a velocity of 2.4×10^6 m s^{-1} enter a region of 'crossed' electric and magnetic fields, as shown in the diagram. If the electric field is 3.0×10^6 V m^{-1}, and the flux density of the magnetic field is 1.5 T, the electrons, upon entering the region of the crossed fields, will:

A continue to travel undeflected in their original direction.
B be deflected upwards out of the plane of the diagram.
C be deflected downwards into the plane of the diagram.
D be deflected upwards in the plane of the diagram.
E be deflected downwards in the plane of the diagram.

B - into the plane of paper

(**AEB** Nov 81: *all other Boards except SEB H*)

It is important to understand that all forces and deflections occur in the plane of the paper and that at a certain velocity the deflection will be zero. This velocity may be calculated using E8.3. For velocities other than this, the deflection is determined by whether the magnetic force is greater or less than the electric force.

8.4M A beam of alpha-particles and protons of the same ke E enters a uniform magnetic field at right angles to the field lines, so that the particles bend on circular paths. The ratio $\frac{\text{path radius for alpha particles}}{\text{path radius for protons}}$ is equal to:

A $\frac{1}{2}$. B 1. C $\sqrt{2}$. D 2. E 4.

(—: *all Boards except SEB H*)

The centripetal force on each particle is provided by the magnetic field, so that $mu^2/R = Bqu$ (see E8.2 if necessary). Since $E = \frac{1}{2}mu^2$, combine the two equations to obtain an expression for R in terms of E, B, q and m. Finally use your knowledge of q and m values in atomic units for the two types of particles to obtain a numerical value for the ratio.

8.5M An electron of mass m travelling with a speed u collides with an atom and its speed is reduced to v. The speed of the atom is unaltered, but one of its electrons is excited to a higher energy level and then returns to its original state, emitting a photon of radiation. If h is the Planck constant, the frequency of the radiation is:

A $\frac{m(u^2 - v^2)}{2h}$. B $\frac{m(v^2 - u^2)}{2h}$. C $\frac{m(v^2 + u^2)}{2h}$.

D $\frac{mu^2}{2h}$ E $\frac{mv^2}{2h}$ (**London**: *all other Boards*)

Apply E8.8 where ΔE will be the change in ke of the electron.

8.6M Emission of electrons from a metal plate illuminated with monochromatic electromagnetic radiation will always take place provided:

A the radiation is sufficiently intense.
B the work function of the plate is less than the energy of a single photon and the plate is uncharged.
C the wavelength of the radiation exceeds a minimum value.
D the plate is **always** negatively charged.
E the plate is freshly cleaned.

(—: *all Boards except SEB SYS*)

The key to the question is in the second sentence '...will always take place provided...'. You may well know all the relevant facts (see p. 124 if you do not), but if you interpret the question wrongly you will arrive at the wrong choice! For example, a freshly cleaned plate will only exhibit photoelectric emission if the radiation frequency is greater than a certain minimum value, so if the radiation frequency is less than the minimum value (the threshold frequency), then emission does not occur – so E is out.

8.7M An ultraviolet light source causes the emission of photo electrons from a zinc plate. A more intense source of the same wavelength would give

	Maximum energy/electron	No. of electrons/second
A	More	The same
B	The same	More
C	The same	The same
D	More	More
E	Less	More

(**SEB H**: *all other Boards except SEB SYS*)

Intensity of incident light is the energy per second per unit area of the beam. Increased intensity of the same wavelength means more photons of the same energy. What effect will this have on the emission of electrons? See p. 124 if necessary.

8.8M In a series of photoelectric emission experiments a number of different metals were illuminated with monochromatic light of a number of different photon energies and of different intensities. The variables in the experiments were thus:

1 work function of the metals.
2 photon energy.
3 light intensity.

It was found that, for each experiment, the emitted electrons emerged with a spread of kes up to a certain maximum value. This maximum ke depends on:

A **1** only. B **1** and **2** only. C **1** and **3** only.
D **2** and **3** only. E **1**, **2** and **3**.

(**NISEC**: *all other Boards except SEB SYS*)

Note that an increase of light intensity causes more electrons to be emitted, but their individual kinetic energies are not increased by the increase of light intensity. See p. 124 if necessary.

8.9M The energy levels of an atom of an element X are shown in the diagram, with values given in electronvolts. Which electron transition A – E will produce photons of wavelength = 620 nm?

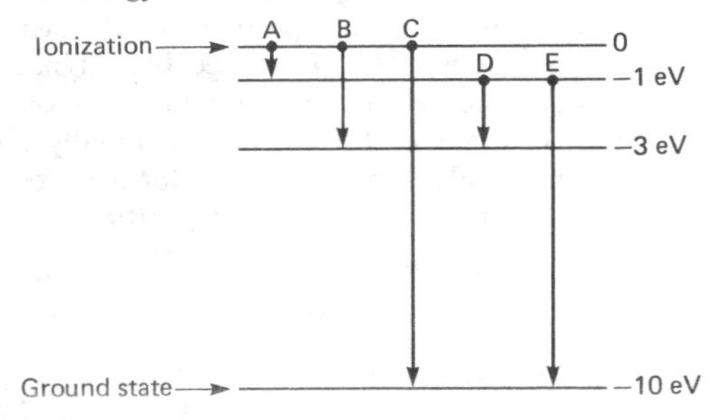

(—: *all Boards except SEB SYS*)

Use E8.8 to calculate the photon energy in J. Then convert to eV, remembering that 1 eV = 1.6×10^{-19} J. Assume that the speed of light, $c = 3 \times 10^8$ m s^{-1}, and that Planck's constant, $h = 6.6 \times 10^{-34}$ J s.

8.10M In an atom containing its normal complement of electrons:

1 the ground state of the atom is the state in which all its electrons are in their lowest possible energy levels.
2 the gaps between excited energy levels are evenly spaced as the atom approaches ionization.
3 an electron while in an excited energy level is continuously radiating energy in multiples of h, where h is the Planck constant:

Answer:
A if **1**, **2**, **3** correct. B if **1**, **2** only. C if **2**, **3** only.
D if **1** only. E if **3** only.

(**London:** *all other Boards except SEB SYS*)

For 1, see p. 125. For 2, see F8.4. For 3, see E8.8 (noting the use of the word transition!).

8.11M In an X-ray tube the penetrating power of the X-ray beam emitted can be increased by:

1 increasing the pd between the anode and cathode.
2 narrowing the beam by using lead collimating plates.
3 decreasing the cathode to anode distance while maintaining the same pd between them.

A **1**, **2**, **3** all correct. B **1**, **2** only correct.
C **2**, **3** only correct. D **1** only correct.
E **3** only correct.

(—:*JMB, WJEC and NISEC*)

The penetrating power of X-rays is increased if their frequency is increased. The minimum frequency from an X-ray tube is given by equating the energy of a single photon (hf) to the work done on a single electron accelerated through the tube pd (Ve). See p. 126.

8.12M Both X-rays and γ-rays:

1 might have wavelengths of the same order.
2 can be detected by a Geiger counter.
3 originate in the nucleus of an atom.

Answer:
A if **1**, **2**, **3** correct. B if **1**, **2** only. C if **2**, **3** only.
D if **1** only. E if **3** only.

(**London**: *all other Boards except O and C, SEB H and SEB SYS*)

See p. 128, E8.13 and also F8.13.

8.13M The half-life of a certain radioactive element is such that 7/8 of a given quantity decays in 12 days. What fraction remains undecayed after 24 days?

A 0. B $\frac{1}{128}$. C $\frac{1}{64}$. D $\frac{1}{32}$. E $\frac{1}{16}$.

(**London**: *and all other Boards except SEB SYS*)

If 7/8 decays in 12 days, then 1/8 remains undecayed after 12 days. Hence after another 12 days (24 days in total) then...

8.14M Two radioactive isotopes P and Q have half-lives of 10 minutes and 15 minutes, respectively. Freshly prepared samples of each isotope initially contain the same number of atoms as one another. After 30 minutes, the ratio

$\frac{\text{the number of atoms of } P}{\text{the number of atoms of } Q}$ will be:

A 0.5. B 2.0. C 1.0. D 3.0. E 0.25.

(—: *all Boards except SEB SYS*)

30 minutes is 3 half-lives for P and 2 half-lives for Q.

8.15M A radioactive source contains two materials. One decays by the emission of an α-particle with a half-life of 4 days while

the other emits β-particles with a half-life of 3 days.
The initial corrected count rate is $176\,s^{-1}$ but this becomes $80\,s^{-1}$ when a piece of tissue paper is placed between the source and the detector. What will be the corrected count rate after 12 days without the tissue paper present?
A $5\,s^{-1}$. B $11\,s^{-1}$. C $12\,s^{-1}$. D $17\,s^{-1}$.
E $22\,s^{-1}$. (**SEB H**: *and all other Boards except SEB SYS*)

Which component of the radiation, α or β, does the tissue paper absorb? See p. 127 if necessary. Hence decide what the contribution to the initial corrected count rate of $176\,s^{-1}$ is made by α-particles. Then determine what the contribution will be from the α-particles after 12 days.

Now consider the contribution to the initial corrected count rate made by β-particles. What contribution will β-particles make after 12 days? Then add the two contributions (of each component after 12 days) to give the final answer.

It is worth noting and using the fact that 12 days is a whole number of half-lives in both cases; after one half-life, the contribution drops to 50%; after two half-lives, the contribution drops to 25% of the initial value, etc.

8.16M A radioactive atom X emits a beta-particle to produce an atom Y which then emits an alpha-particle to give an atom Z:
1 The atomic number of X is less than that of Z.
2 The atomic number of Y is less than that of Z.
3 The mass number of X is the same as that of Y.
A **1, 2, 3** correct. B **1, 2** correct. C **2, 3** correct.
D **1** correct. E **3** correct.

(**NISEC**: *and all other Boards except SEB SYS*)

It is probably easier to consider the last two statements before the first one since they both involve comparison between an atom and **its** daughter product. Can you make your choice by considering 2 and 3 only? See p. 127 for the basic ideas, if necessary!

8.17M The initial activity of a certain radioactive isotope was measured as 16 000 counts per minute. Given that the only activity measured was due to this isotope, and that its activity after 12 hours was 2100 counts per minute, its half-life, in hours, is nearest to:
A 9.0. B 6.0. C 4.0. D 3.0. E 2.0.

(—: *all Boards except SEB SYS*)

EITHER divide 16 000 by half, then by half again, etc, until you have reached approximately 2100. How many half-lives have passed? Hence choose your answer.

OR, use E8.17 in the form $A = A_0 e^{-\lambda t}$ to calculate the decay constant λ (A is the activity that equals dN/dt), and then calculate the half-life from E8.16.

8.18M As a result of a number of successive decay processes in a radioactive series the nucleon number decreases by 4 while the proton number is unchanged. The particles emitted are:
A 1 alpha- and 1 beta-particle.
B 1 alpha- and 2 beta-particles.
C 2 alpha- and 1 beta-particle.
D 2 alpha- and 2 beta-particles.
E 4 alpha-particles.

(**London**: *all other Boards except SEB SYS*)

Remember that the nucleon number is the number of neutrons **and** protons, and that the nucleon number is unchanged by beta-decay. See E8.11 and E 8.12 if necessary.

8.19M The deviation of α-particles by thin metal foils through angles that range from 0° to 180° can be explained by:
A scattering from free electrons.
B scattering from bound electrons.
C diffuse reflection from the metal surface.
D scattering from small but heavy regions of positive charge.
E diffraction from the crystal lattice.

(**Cambridge**: *all other Boards except Oxford, SEB SYS and O and C*)

See p. 126 if necessary.

8.20M Alpha-particles and protons of the same initial velocity are fired in turn at nuclei of carbon-12. Which of the following statements is/are correct?
1 The alpha-particles have a greater initial ke than the protons.
2 The alpha-particles are able to approach the carbon-12 nuclei closer than the protons can (assuming 'head-on' collision in both cases).
3 Each incident particle which is deflected by a carbon-12 nucleus has a smaller speed after deflection than before.
A **1** only. B **2** only. C **1** and **2**. D **1** and **3**.
E **2** and **3**.

(—: *all Boards except SEB SYS and SEB H*)

For **2**, the closest distance (d) is given by equating the initial ke of the 'projectile' to the pe at the closest distance of approach. E2.3 gives the electrostatic pe.

8.21M A helium atom, a hydrogen atom and a neutron have masses of 4.003 u, 1.008 u and 1.009 u (unified atomic mass units), respectively. Assuming that hydrogen atoms and neutrons can fuse to form helium, what is the binding energy of a helium nucleus?
A 2.017 u. B 2.014 u. C 1.017 u. D 0.031 u.
E 0.0031 u.

(**London**: *JMB* and all other Boards except SEB SYS and AEB*)

See the definition of **mass defect** on p. 130. What is the structure of a helium nucleus in terms of numbers of neutrons and protons? Remember that a hydrogen nucleus is a single proton.

8.22M Given that the atomic mass of $^{14}_{7}N$ is 14.003 074 u and that the sum of the atomic masses of $^{1}_{1}H$ and $^{13}_{6}C$ is 14.011 179 u, it would be reasonable to suppose that the nuclear reaction:
$^{1}_{1}H + ^{13}_{6}C \rightarrow ^{14}_{7}N$
A can only happen if there is a net supply of energy.
B could not take place at all.
C could occur only in conditions of zero gravity, ie, of 'weightlessness'.
D must involve the emission of a further unchanged atomic particle.
E will result in the emission of energy.

(**London**: *JMB* and all other Boards except SEB SYS and AEB*)

See **nuclear reactions** on p. 130.

8.23S The plates of a horizontal parallel-plate capacitor are 1 cm apart and the potential difference is 10 kV. It is found that an oil drop of mass 1.96×10^{-13} kg remains stationary between the plates. How many electronic charges does the oil drop carry? Indicate briefly how the mass of the drop could be determined experimentally assuming its charge is not known. (Electronic charge $= 1.6 \times 10^{-19}$ C).

(**WJEC**: *JMB* and all other Boards except SEB H*)

To calculate the number of electronic charges carried by the drop, you must first calculate its charge in C. Use E8.1 to determine the drop charge. Assume $g = 9.81\,m\,s^{-2}$.

To determine the drop mass, the drop must be timed over a measured distance while it falls in zero electric field (ie, with zero pd across the plates). Because of air resistance, the drop falls at a steady velocity (terminal velocity) and by measuring the terminal velocity, the drop mass can be calculated. Use your textbook to find further details.

8.24S When a beam of electrons moving with speed $8.8 \times 10^5\,m\,s^{-1}$ was subjected to a uniform magnetic field of flux density 5 mT directed at right angles to the original beam, it was found that the electrons described a circular path of radius 1 mm. Find the charge-to-mass ratio for the electron.

Use E8.2. (**WJEC**: *all other Boards except SEB H*)

8.25S When a light of frequency 5.4×10^{14} Hz is shone on to a metal surface the maximum energy of the electrons emitted is 1.2×10^{-19} J. If the same surface is illuminated with light of frequency 6.6×10^{14} Hz the maximum energy of the electrons emitted is 2.0×10^{-19} J. Use this data to calculate a value for the Planck constant.

(**London**: *and all other Boards except SEB SYS*)

Use E8.6, substituting the appropriate values to give two equations, each with two unknowns. Subtract the two equations to eliminate the work function and then solve for h (remembering that h has units!).

8.26S Draw a sketch showing the energy levels of the electron in a hydrogen atom. Indicate on your diagram (a) the ground state of the atom, (b) the first excited state, (c) the ionization energy. How may information about the energy levels of atoms be obtained?

(**Cambridge**: *and all other Boards except SEB SYS*)

The energy levels of the hydrogen atom are given by the Balmer equation E8.7. Draw the energy levels in the form of a 'ladder' diagram, and mark which 'rungs' correspond to the ground state, the first excited state and the ionization level. Since the ground state level is 13.6 eV below the ionization level, take care to put the first excited level at the appropriate spacing, as given by the Balmer rule. On your diagram, mark the energy values of each level taking the ionization level as the zero level. See p. 124 for ionization energy.

Information about energy levels is obtained when electrons change levels. You must discuss how electron transitions give rise to photon emission, and you should explain how the photon frequency relates to the energy-level diagram. See p. 125 and F8.7 and F8.8.

8.27S The accelerating voltage across an X-ray tube is 33.0 kV. Explain why the frequency of the X-radiation cannot exceed a certain value and calculate this maximum frequency. (The Planck constant = 6.6×10^{-34} J s; the charge on an electron = 1.6×10^{-19} C.)

(**AEB**: *JMB, WJEC and NISEC*)

Your explanation should make it clear that:

(i) the amount of ke gained by a 'beam' electron in an X-ray tube is determined by the accelerating voltage.

(ii) the ke of the beam electrons is released as photons when the electrons strike the tube. A single electron may produce one or more photons; if only one photon is produced by a given electron, then all the electron ke becomes energy of a single photon.

(iii) The link between energy and frequency for a photon is given by E1.16. Hence show that there is a maximum frequency value for a given tube voltage V.

In the calculation, remember that the energy gained by an electron accelerated through a pd V volts is Ve joules.

8.28S Radon-222 is a radioactive gas for which the decay constant is $2.1 \times 10^{-6}\,s^{-1}$. If the initial decay rate is $5.6 \times 10^{10}\,s^{-1}$ calculate:

(a) the initial number of radioactive atoms present, and

(b) the time which will elapse before the activity is reduced to one-quarter of its initial value. ($\ln 2 = 0.693$).

(**AEB** June 81: *NISEC and all other Boards except SEB H, and SEB SYS*)

(a) Use E8.15.

(b) Use E8.16 to find the half-life of the radon. To calculate the time, remember that the activity halves for every half-life (ie, time equal to one half-life) which passes.

8.29S Here is some information about two radioactive isotopes which might be used as fuels in a nuclear power unit:

Name	Particle emitted	Half-life in years	Particle energy
Strontium-90	Beta	27	0.54 MeV
Plutonium-239	Alpha	2.4×10^4	5.15 MeV

(a) A sample of strontium-90 has an initial activity of 400 curie. Calculate the maximum theoretical power output of this source. Give your answer in watts and show the steps in your calculation.

(b) What will be the maximum theoretical power output of the source after 54 years?

(c) Make a rough numerical estimate of the total energy emitted by this source in the first year. Show the steps in your calculation.

(d) Safety considerations apart, give, for each isotope, one advantage it has over the other as a fuel in a nuclear power unit.

(**O and C Nuffield**: *NISEC and all other Boards except SEB H and SEB SYS*)

(a) Maximum power output must occur when the source is 'new', when it will have had no chance to decay. You will find the numerical value of the curie on the front of the exam paper; in this book it is included in the Table of values (pp. 186–89). From the number of particles emitted per second it is possible to calculate the energy emitted per second (watts); see p. 124 for conversion of eV to J, many candidates in the examination will have forgotten to convert the energy into the right units.

(b) 54 years is two half-lives. What has happened to the decay rate, and hence to the power output, after this period of time?

(c) In a year the source will have decayed little, hence you can assume (for the purposes of a 'rough estimate') that the decay rate stays constant at 400 curies. Simply multiply the answer to part (a) by the number of seconds in a year (this number was quoted on the front of the examination paper, many candidates lost time by failing to notice this).

(d) A comparison of the particle energies should make one advantage of using plutonium easy to spot. However to claim an advantage for strontium is not so easy. It is difficult to argue in favour of beta-particles rather than alphas, and at first sight the longer life of plutonium seems also to be advantageous! The material with the longer half-life will decay far more slowly; for equal masses of isotopes, this will mean a far lower count rate, but if the same power is to be generated....

8.30S A scientist wished to find the age of a sample of rock in which he knew that radioactive potassium ($^{40}_{19}$K) decays to give the stable isotope argon ($^{40}_{18}$A). He started by making the following measurements:

Decay rate of the potassium in the sample = 0.16 disintegrations/second

Mass of the potassium in the sample = 0.6×10^{-6}g

Mass of the argon in the sample = 4.2×10^{-6}g

(a) Show how he could then calculate that for the potassium (i) the decay constant (λ) was $1.8 \times 10^{-17}\,s^{-1}$, (ii) the half-life was 1.2×10^9y.

(b) Calculate the age of the rock, assuming that originally there would have been no argon in the sample. Show the steps in your calculation.

(c) Identify and explain a difficulty involved in measuring the decay rate of $0.16\,s^{-1}$ given earlier in the question.

(**O and C Nuffield**: *NISEC and all other Boards except SEB H and SEB SYS*)

(a) (i) Use E8.15. The decay rate is given in the question, but it is necessary to calculate the number of radioactive nuclei present (ie, the number of potassium-40 nuclei in 0.6×10^{-6}g). There are two methods, both of which will incur a very slight error (owing to the mass defect of the nucleus – see p. 130). The front of the exam paper will give Avogadro's number and the masses of neutrons and protons. The first method involves using Avogadro's number and the fact that a mole of potassium-40 nuclei would weigh 40 g. The second method involves calculating the mass of an individual potassium-40 nucleus from the neutron and proton masses. (ii) Use E8.16 and the answer to (a)(i).

(b) Use E8.17. However it is not necessary to calculate the number of nuclei; the masses can be used instead of N_0 and N_t as only a ratio is required. If the mass lost in decay is assumed negligible then originally there must have been 4.8 g of potassium. Time can be saved by noticing that as the potassium has decayed to an 'eighth' of the original quantity, an exact number of ½-lives must have passed.

(c) The decay rate is so small that it may be necessary to count for many hours to get a reasonable total count. However this is not the major difficulty; it will be extremely difficult to detect the potassium decays, the reason is illustrated on F8.17 as C_B.

8.31S When a deuteron of mass 2.0141 u and negligible ke is absorbed by a lithium nucleus of mass 6.0155 u, the compound nucleus disintegrates spontaneously into two alpha-particles, each of mass 4.0026 u. Calculate the energy, in joules, given to each alpha-particle. ($1\,u = 1.66 \times 10^{-27}$ kg, speed of light in a vacuum $= 3.00 \times 10^{8}\,\mathrm{m\,s^{-1}}$.)

(**London**: *JMB* and all other Boards except SEB SYS and AEB*)

Write the interaction in the form of E8.19 and evaluate the mass difference. Convert this into energy by using $E = mc^2$ (having put m into kg). By conservation of momentum, the alpha-particles must have velocities that are equal in magnitude and opposite in direction, and so they will share the energy equally between them.

8.32L

(a) Define the quantity **electrical potential difference**. What is the relationship between this quantity and electrical field strength?

(b) A small drop of oil of mass m which carries a charge Q may be held stationary in a vertical electric field of intensity E. Write down the expressions for the forces on the drop which are then in balance. Describe how this idea can be used to measure Q. Explain how the results of such an experiment indicate that all charges consist of an integral number of basic units of charge, e. In such an experiment an average of 20 doubly charged drops are produced per second. If this charge is provided by the current from the supply, what is the value of this current? Why does measuring this current not provide the basis of a practical method of estimating the value of e?

(The value of e is -1.6×10^{-19} C.)

(**London**: *JMB* and all other Boards except SEB H*)

(a) Note that potential **difference** is required. See p. 101 if necessary. For electric field strength, see p. 31, and E2.1 for the relationship.

(b) See E8.1, but remember that if the drop is in air it will also experience an upthrust. You should refer to a textbook for a full description of Millikan's experiment. The fact that $Q = ne$, where n is an integer, is discussed after E8.1.

(c) Use $I = \mathrm{d}Q/\mathrm{d}t$. The magnitude of your answer should indicate to you why this would not be a very practical method, particularly when you consider that there is air in the apparatus (which will conduct to some extent).

8.33L

(a) (i) Electric field strength can be expressed in $\mathrm{V\,m^{-1}}$ or $\mathrm{N\,C^{-1}}$. Show that these are equivalent. (ii) A charged plastic sphere of mass 3.0×10^{-15} kg is held at rest between two horizontal parallel metal plates as shown in the diagram. The distance between the plates is 5.0×10^{-3} m and a potential difference of 310 V is applied across them. Calculate the charge carried by the sphere and indicate any assumptions made in the calculation of this charge. (iii) If the polarity of the plates is suddenly reversed, calculate the initial acceleration of the sphere.

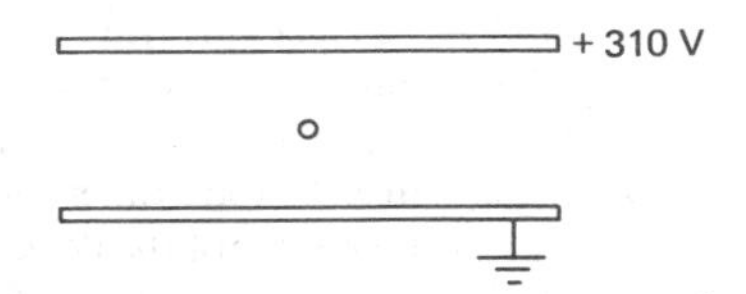

(b) Electrons are emitted from the cathode in an evacuated tube. The electrons start from rest and are accelerated through a potential difference of 1150 V. Some of the electrons pass through

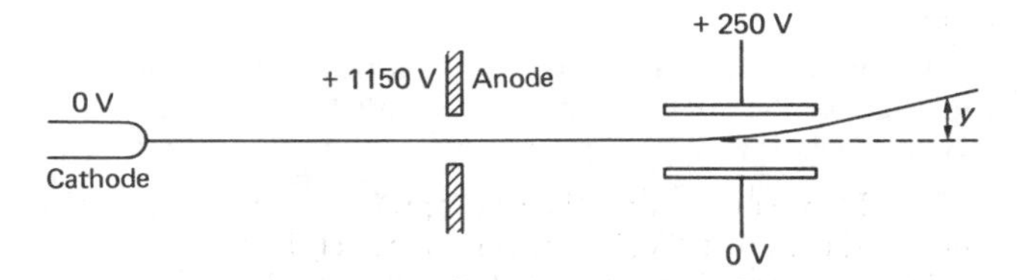

the anode into a region where the electric field may be assumed to be zero. Calculate (i) the speed of the electrons when they reach the anode; (ii) the deflection y that they undergo after passing between two parallel plates which are 2.0×10^{-2} m long and 1.0×10^{-2} m apart, and between which a potential difference of 250 V is maintained.

(**SEB SYS**: *all other Boards except SEB H*)

(a) (i) Remember volts = joules/coulomb and joules = newtons × metres. (ii) Assume the field between the plates is uniform, and so calculate its strength (see E2.9). The weight of the sphere is balanced by the force due to the electric field (qE). Hence calculate q. Since you know that the top plate is + ve, then you can state the sign of the charge on the drop. (iii) Sudden reversal means both weight and electric field force (= weight) will act downwards, giving a total downwards force on the drop of $2\,mg$. Hence initial acceleration = force/mass = etc.

(b) (i) Use E8.4. (ii) Use E1.3 applied to the vertical motion of the electrons between the plates. The initial vertical speed is zero, the vertical acceleration is the electric field force/mass (ie, qV/md), and the time taken can be calculated from the length of the plates and the electron's horizontal speed which was found in (i).

8.34L

(a) Describe and explain how to obtain a stream of fast-moving electrons. (Such a stream of electrons must also be focused if a beam is required, but you are not asked to describe the focusing system). Briefly describe one method of detecting or tracking the path of an electron beam.

(b) Describe an experiment which can be used to measure the specific charge of the electron (e/m_e) and explain how the result is obtained from the observations.

(c) An electron beam, in which the electrons are travelling at $1.0 \times 10^{7}\,\mathrm{m\,s^{-1}}$, enters a magnetic field in a direction perpendicular to the field direction. It is found that the beam can pass through without change of speed or direction if an electric field of strength $1.1 \times 10^{4}\,\mathrm{V\,m^{-1}}$ is applied in the same region at a suitable orientation. (i) Calculate the strength of the magnetic field. (ii) If the electric field were switched off, what would be the radius of curvature of the electron path?

(Electron charge $e = -1.6 \times 10^{-19}$C; electron mass $m_e = 9.1 \times 10^{-31}$ kg.)

(**NISEC**: *and Oxford* and all other Boards except SEB H*)

(a) Describe the 'electron gun' part of a cathode ray tube. See your textbook for details. To 'track' the **path**, the beam must pass over a suitably positioned sheet of fluorescent material, or a low pressure of gas must be maintained in the tube so that ionization caused by the beam makes its path visible in a darkened room. Describe **one** of these methods – see your textbook for details.

(b) See p. 123 for the principles of two methods, and use your textbook to find details of the apparatus. You should preferably choose to describe an experiment you have carried out yourself (or one you have seen demonstrated). Include a diagram of the apparatus and remember to explain how the result is worked out from the measurements.

(c) (i) Use E8.3. You may not be able to remember this equation in an examination, so make sure that you know how to derive it. Remember to give the units in which your answer is expressed. (ii) Use E8.2. You should be able to derive this equation by equating the centripetal force to the force on each electron due to the magnetic field. See p. 123 if necessary.

8.35L

(a) When atoms absorb energy by colliding with moving electrons, light or X-radiation may subsequently be emitted. For each type of radiation, state typical values of the energy per atom which must be absorbed and explain in atomic terms how each type of radiation is emitted.

(b) State **one** similarity and **two** differences between optical atomic emission spectra and X-ray emission spectra produced in this way.

(c) Electrons are accelerated from rest through a potential difference of 10 000 V in an X-ray tube. Calculate (i) the resultant energy of the electrons in eV, (ii) the wavelength of the associated electron waves and (iii) the maximum energy and the minimum wavelength of the X-radiation generated.
(Charge of electron = 1.6×10^{-19}C, mass of electron = 9.11×10^{-31} kg, Planck's constant = 6.62×10^{-34} J s, speed of electromagnetic radiation *in vacuo* = 3.00×10^{8} m s^{-1}.)

(**JMB**: *and WJEC, NISEC*)

(a) See 8.4. The key word here is in the first sentence; 'When **atoms** absorb...'. The question is about how line-emission spectra are produced for X-rays and for light; for typical values, write down typical wavelengths for each and use E8.8 to calculate typical energies per atom. If you can state the energies directly, there is no need to do the calculations. F8.7 should be useful here. In your explanation, you should mention energy levels (with diagram), electron transitions, photon emission and how photon frequency is linked to the energy-level diagram. You should discuss a transition which gives a visible photon, and one which gives an X-ray photon.

(b) You must discuss similarities and differences of the **spectra**, so think in terms of the presence of a continuous background, of lines and of the wavelengths (or energies) involved.

(c) (i) See p. 124. (ii) Use the De Broglie equation E1.18. (iii) See p. 126 and F8.9. for the maximum photon energy. Use E8.8 to calculate the minimum wavelength; remember that minimum wavelength corresponds to maximum frequency (and hence maximum energy).

8.36L

(a) What is meant by the photoelectric effect? Describe a practical application of this effect.

(b) A freshly cleaned zinc plate is connected to a gold-leaf electroscope. The plate is negatively charged and the leaves of the electroscope diverge. The plate is now irradiated with a mixture of ultraviolet and visible radiation and the leaves collapse. If a sheet of glass is held between the ultraviolet lamp and the plate, the collapse of the leaves is halted. If the plate is positively charged the leaves do not collapse when the ultraviolet radiation falls on the plate. Explain these observations.

(c) Explain what is meant by the work function of a metal. Discuss how the value of the work function influences the kinetic energy of the electrons liberated by photoelectric emission.

(d) A photoemissive metal will only emit electrons if the frequency of the incident radiation exceeds 5.0×10^{14} Hz. What is the value of the work function of the metal? What would be the maximum kinetic energy of the emitted electrons if the incident radiation were of wavelength 330 nm?
(The Planck constant = 6.6×10^{-34} J s; velocity of electromagnetic radiation = 3.0×10^{8} m s^{-1}.)

(**AEB** June 81: *all other Boards except SEB SYS*)

(a) It is sufficient to state that the photoelectric effect is the emission of... due to... etc. See p. 124 if necessary. You must include what the emission is from. For a practical application, think of some arrangement which uses a photocell (is the cell photoemissive or photoconductive or photovoltaic? See your textbook if necessary.) Give a description with the aid of a labelled diagram.

(b) The collapse of the leaves with the plate negatively charged must be due to loss of electrons from the plate. Which component of the incident radiation causes the loss of electrons? With the glass plate, consider whether or not the glass lets through all wavelengths of the incident radiation. For the third observation, beware of using the mistaken argument that a positively charged metal contains insufficient electrons for the photoelectric effect to occur. Think in terms of an electron trying to leave a positively charged metal.

(c) See p. 124 if necessary. It is important to note that the definition involves **minimum** or **least** energy.

(d) Use E8.5 and E8.6.

8.37L

(a) Explain what is meant by the photoelectric effect. Indicate how it depends on (i) the frequency of the light, (ii) the intensity of the light. Explain how your answers to (i) and (ii) are related to a theory of the nature of light.

(b) A clean zinc plate is mounted in an ionization chamber, just below a wire mesh, as shown. The chamber is connected in series with a dc supply and a sensitive current meter. The current meter amplifies any small current in the circuit by a factor of 10^4 and displays the amplified current on a microammeter. The zinc plate is illuminated by an ultraviolet lamp. Describe how you would use the apparatus to show that any small current in the circuit was due to the photoelectric effect.

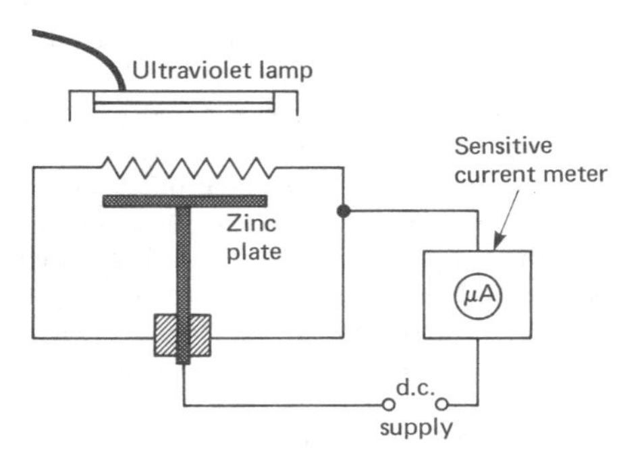

(c) The ultraviolet lamp above is replaced by a radioactive source. The distance of the source from the wire mesh is altered and for a number of source positions the amplified current indicated by the microammeter is recorded. The following results are obtained:

Height of source above wire mesh (cm)	5	4	3	2	1
Current (μA)	0.0	0.0	0.5	15	20

(i) What type of radiation is being detected? Give a reason for your answer. (ii) What would be the effect, if any, of reversing the polarity of the dc supply?

(**SEB H**: *all other Boards except SEB SYS*)

(a) (i) See p. 124. (ii) Remember that the intensity of light is the energy per second per unit area, so a more intense beam of a given frequency means more photons, each of the same energy as before. Provided that the frequency of the incident light is greater than the threshold frequency, then a more intense beam will cause more electrons to be emitted.

(b) The dc supply must be connected so that the zinc plate is − ve. Then the ultraviolet radiation will cause photoelectric emission of electrons, and these electrons will be repelled by the plate and attracted by the wire mesh so giving a current. A glass plate placed between the lamp and the mesh will cause the current to drop to zero because glass absorbs ultraviolet radiation.

(c) The radiation is of the type that, of the three types, is most easily absorbed. See p. 127 if necessary. If the dc supply polarity is reversed, the effect of the radiation will be the same as before because it is the ionization caused by the radiation which is responsible for the current.

8.38L When the stable isotope of manganese $^{55}_{25}$Mn is irradiated in a nuclear reactor the radioactive isotope $^{56}_{25}$Mn is produced. $^{56}_{25}$Mn decays to a stable isotope $^{56}_{26}$Fe. Write equations for these nuclear transformations.
$^{56}_{25}$Mn has a half-life of a few hours. Give a full account of an experiment to measure the half-life of an isotope such as $^{56}_{25}$Mn. What factors affect the precision of the value you obtain?
Calculate the difference in mass between the particles present before and after the decay of $^{56}_{25}$Mn, given that the total energy liberated in the decay of a $^{56}_{25}$Mn nucleus is 5.9×10^{-13} J.
When $^{56}_{25}$Mn decays to $^{56}_{26}$Fe the iron nucleus is left in an excited state. The iron nucleus radiates this excitation energy as a single γ-ray of wavelength 1.47×10^{-12} m. Calculate the energy carried by the γ-ray photon, and hence the maximum kinetic energy carried by the other decay products.
($h = 6.6 \times 10^{-34}$ J s, $m_e = 9.1 \times 10^{-31}$ kg.)

(**O and C**: *JMB*, and all other Boards except SEB SYS and AEB*)

There are two nuclear transformations that you must write equations for:
1 $^{55}_{25}\text{Mn} + X \rightarrow ^{56}_{25}\text{Mn}$
2 $^{56}_{25}\text{Mn} \rightarrow ^{56}_{26}\text{Fe} + Y$
Use your knowledge of nuclear structure, in terms of neutrons and protons, to work out what the particles X and Y above ought to be. See p. 126 for the meanings of atomic number and mass in terms of neutrons and protons if necessary.

For experimental determination of the half-life, see your textbook.

The formula for the scale of conversion of mass into energy (and vice versa) is given by E1.17. Assume the total energy liberated has been produced by the conversion of mass into energy.

To calculate the energy of a photon, use E1.16. You will first need to calculate the photon frequency from its wavelength value and the value of the speed of light (3×10^8 m s^{-1}). See E3.7 if necessary.

The maximum ke carried away by the other decay products can be calculated from the difference between the total energy liberated and the energy of the γ-ray photon.

8.39L
(a) Describe in terms of nuclear structure the three different stable forms of the element neon, which have nucleon numbers 20, 21 and 22. Why is it impossible to distinguish between these forms chemically? A beam of single ionized atoms is passed through a region in which there are an electric field of strength E and a magnetic field of flux density B at right angles to each other and to the path of the atoms. Ions moving at a certain speed v are found to be undeflected in traversing the field region. Show (i) that $v = E/B$, and (ii) that these ions have any mass. If the emerging beam contained ions of neon, suggest how it might be possible to show that all three forms of the element were present.

(b) Explain what is meant by the **binding energy** of the nucleus. Calculate the binding energy per nucleon for $^4_2$He and $^3_2$He. Comment on the difference in these binding energies and explain its significance in relation to the radioactive decay of heavy nuclei.
(Mass of $^1_1$H = 1.007 83 u, mass of ^1_0n = 1.008 67 u, mass of $^3_2$He = 3.016 64 u, mass of $^4_2$He = 4.003 87 u, 1 u = 931 MeV.)

(**London**: *JMB*, and all other Boards*)

(a) Isotopes are discussed on p. 126. State how many protons and neutrons there are in each of the three neon isotopes. Remember that the nucleon number is the number of neutrons **and** protons in the nucleus. Note that it is the orbital electrons which determine the chemical properties of an element. (i) and (ii): see p. 123. To separate the three isotopes, the emergent beam must be deflected in just an electric or magnetic field – the principle of the mass spectrograph, which is described in most textbooks.

(b) Binding energy is discussed on p. 130 and a similar calculation performed in E8.19. The binding energy per nucleon for $^4_2$He is much greater than that for $^3_2$He and so consideration of the stability of $^4_2$He as a unit should explain the alpha-decay of heavy nuclei.

8.40L
(a) (i) Sketch a graph showing how the number of neutrons N varies with the number of protons Z for stable nuclei. Draw the line $N = Z$ and indicate approximate numerical values of N and Z on the axes. (ii) What can be deduced from the graph concerning the numbers of neutrons and protons in stable nuclei?

(b) Describe briefly the process of α-emission and β-emission by which unstable nuclei may spontaneously decay. For each process state the relation between the masses of the initial and final atoms and the emitted particle, for the decay to be possible.

(c) (i) Calculate the energy released when $^{64}_{29}$Cu nucleus decays into $^{64}_{30}$Zn.
Atomic mass of $^{64}_{29}$Cu = 63.929 76 u
Atomic mass of $^{64}_{30}$Zn = 63.929 14 u
1 atomic mass unit, u = 931.5 MeV
(ii) If the $^{64}_{29}$Cu nucleus is at rest, will the resulting $^{64}_{30}$Zn nucleus also be at rest? Give a reason for your answer.

(**JMB***: *(b), (c) all Boards except SEB SYS and AEB*)

(a) See F8.14 for the graph of $y = N$, $x = Z$ for stable nuclei. In sketching the graph, remember that stable nuclei go up to about $N = 140$, $Z = 90$. Indicate clearly the range over which stable nuclei have equal numbers of neutrons and protons. From your graph, give the N/Z ratio for light nuclei. Then give the ratio approximately for heavy stable nuclei.

(b) In your description of the two types of emission, you should state what happens to the nucleus, in each case, in terms of neutrons and protons. See p. 127 for further details. For a spontaneous decay, the energy produced must come from conversion of mass. For α-decay, you should therefore state how the mass of the initial atom compares with the mass of the final atom and the α-particle. If you make a similar statement for β-decay, remember that the final atom is ionized (because its atomic number goes up by 1, but its number of 'bound' electrons is unchanged) so you should state that the final atom is ionized.

(c) (i) Write down the equation for the decay. Remember that the initial atom has 29 'bound' electrons so the final atom (Zn) has 29 electrons after the nuclear change. Thus the mass of the products (ie, atom of Zn with 29 electrons plus the emitted particle) is the same as the mass of an uncharged Zn atom (ie, 63.929 14 u). Since the initial mass is that of an uncharged copper atom (ie, 63.929 76 u), then use E1.17 to calculate the energy released; use the given information (that 1 u of mass converted into energy gives 931 MeV) to deal with the calculation quickly. (ii) You must consider conservation of momentum here. The Cu nucleus emits a high-energy particle, so what happens (in terms of movement) to the nucleus?

Unit 9 Data analysis

The ability to analyse scientific data is a key skill of A-level physics. Data analysis is taught continually on a conventional course through the medium of the normal subject teaching and practical work; it is rarely covered as subject in its own right. This means that its importance is often overlooked; for example there is little coverage of the subject in most textbooks. Yet A-level examinations directly test data-analysis skills, and the current trend seems to be towards setting more questions in this subject area. This Unit provides comprehensive information on the three main topics of data analysis; Units and dimensions, Graphs and Errors. Much of the material will be required by all candidates during the normal written examination papers, but the extensive information on errors and deducing equations from graphs will be of greatest value to those involved in practical work/examinations, and projects.

To tackle this unit a basic GCSE knowledge of graph plotting and use of scientific units will be needed.

9.1 UNITS AND DIMENSIONS

Base units, dimensions and derived units The system of scientific units (SI – Système International) is founded upon seven internationally agreed **base units.** All units for scientific quantities can be derived from the base units, which are hence called 'dimensionally independent' – in that no one base unit depends upon any other. The base units supply the seven **dimensions** that may be used for dimensional analysis; however, in A-level physics only the dimensions of mass [M], length [L] and time [T] (these define all the mechanical quantities) are normally required. Occasionally questions are set involving electric current [A] or temperature [θ]. F9.1 lists the 'base units'. The units of any physical

Base quantity	Abbreviation of unit	Name of the unit	Dimension
Length	m	metre	[L]
Mass	kg	kilogram	[M]
Time	s	second	[T]
Electric current	A	ampere	[A]
Thermodynamic temperature	K	kelvin	[θ]
Amount of substance	mol	mole	Not used at A level
Luminous intensity*	cd	candela*	Not used at A level

*The 'candela' does not feature on most A-level courses

F9.1 Base units of SI

quantity that does not appear in F9.1 can be derived from the base units; units derived in this way are called **derived units**. Some derived quantities and their units are shown in F9.2.

Writing units Although physical quantities could have their units expressed in terms of the base units, this leads to very lengthy sets of units for many quantities! Hence many derived units (and two of the base units – A and K) are abbreviated by being named after scientists. Such units are easily recognised as they are written with capital letters; F9.2 includes some examples.

To help reduce the length of very large or very small numbers that often have to be written in front of units a system of abbreviations has been devised. This system is summarized in F9.3.

Two 'oddities' of the SI system are worth making clear.

1 Although mass is measured in kg, fractions and multiples of kg tend to be expressed in terms of g (g = gram = 10^{-3} kg). Hence: 1 Mg = 1000 kg (1 tonne), 1 g = 10^{-3} kg, 1 mg = 10^{-6} kg, 1 μg = 10^{-9} kg, etc.

2 Although mass is measured in kg, the other base unit relating to amount of substance (the mole – *mol*) is based upon the 'mole' consisting of a fixed number (Avogadro's number) of particles. The table of values and constants (pp. 186-89) gives many more examples of quantities and their units as well as making use of the prefixes of F9.3.

Dimensional analysis (balancing units) In any scientific equation the units on each side of the equation must be the same. For example in the equation $E = mc^2$ the units of E must be the same as those of mc^2. However if this method is always to work

Derived quantity	Name of unit	Abbreviation of unit	Other common form of units	Base units
Frequency	hertz	Hz	—	s^{-1}
Force	newton	N	—	$kg\,m\,s^{-2}$
Energy	joule	J	N m	$kg\,m^2\,s^{-2}$
Power	watt	W	$J\,s^{-1}$	$kg\,m^2\,s^{-3}$
Pressure	pascal	Pa	$N\,m^{-2}$	$kg\,m^{-1}\,s^{-2}$
Electric charge	coulomb	C	—	A s
Electric pd	volt	V	$J\,C^{-1}$	$kg\,m^2\,A^{-1}\,s^{-3}$
Electric resistance	ohm	Ω	—	$kg\,m^2\,A^{-2}\,s^{-3}$
Electric capacitance	farad	F	$C\,V^{-1}$	$A^2\,s^4\,kg^{-1}\,m^{-2}$
Magnetic flux	weber	Wb	$T\,m^2$	$kg\,m^2\,A^{-1}s^{-2}$
Magnetic field strength (Magnetic flux density)	tesla	T	$Wb\,m^{-2}$ or $N\,A^{-1}\,m^{-1}$	$kg\,A^{-1}s^{-2}$
Speed/velocity	—	—	—	$m\,s^{-1}$
Temperature gradient	—	—	—	$K\,m^{-1}$

F9.2 Some derived units of SI

Factor	Name of prefix	Symbol
10^{12}	tera-	T
10^9	giga-	G
10^6	mega-	M
10^3	kilo-	k
10^{-3}	milli-	m
10^{-6}	micro-	μ
10^{-9}	nano-	n
10^{-12}	pico-	p
10^{-15}	femto-	f
10^{-18}	atto-	a

F9.3 Prefixes used with SI units

correctly then it is essential to reduce the units on each side of the equation to base units (F9.1); checking equations by balancing their base units is called the method of **dimensional analysis.** Dimensional analysis is not required by all examination Boards; some Boards only expect students to balance units in equations without being concerned over whether they are using base or derived units; however, the technique to be used is the same in each case. Most A-level 'dimensional analysis' problems are linked only to mechanical quantities (ie, dimensions of mass [M], length [L] and time [T]); however, the balancing of units often includes problems involving thermal, optical, and electrical quantities, etc.

Dimensional analysis is a powerful technique and can assist in all the following situations:

1 testing the correctness of equations.
2 assisting recall of important formulae.
3 helping to solve physical problems theoretically.
4 suggesting relationships between fundamental constants.

For example, a student without the knowledge of E3.13 considers that the velocity of transverse waves along a stretched string (v) may be related to the tension in the string (T), the mass of the string (m) and the length of the string (l). First a relationship linking the quantities is suggested (note that this must be written as a proportionality):

$v \propto T^a m^b l^c$

where a, b and c are constants to be determined.

The next step is to write the units of each term in the base units of mass (kg), length (m), and time (s); then these are converted into the dimensions of mass [M], length [L] and time [T] (note that dimensions are conventionally written in square brackets):

units of v = ms^{-1} so base units of v = m s^{-1}
so dimensions of v = [L T^{-1}]

units of T = N so base units of T = kg m s^{-2}
so dimensions of T = [M L T^{-2}]

units of m = kg so base units of m = kg
so dimensions of m = [M]

units of l = m so base units of l = m
so dimensions of l = [L]

These steps were easy except in the case of the units of T, which being a force is measured in N (newtons) – a derived unit. It is not necessary to remember the base units of all derived quantities, usually any equation involving the quantity will enable the base units to be worked out. The easiest equation involving force (Force = mass × acceleration) makes it easy to find the base units of force. Now all the terms in the equation are replaced by their dimensions giving:

$$[\mathrm{L\,T^{-1}}] \propto [\mathrm{M\,L\,T^{-2}}]^a [\mathrm{M}]^b \; [\mathrm{L}]^c$$
$$\propto [\mathrm{M}]^{a+b} \cdot [\mathrm{L}]^{a+c} \; [\mathrm{T}]^{-2a}$$

Now each dimension is considered in turn; its power on the left-hand side of the equation being compared to its power on the right-hand side of the equation:

Considering dimensions of M: LHS $0 = a + b$ RHS
Considering dimensions of L : LHS $1 = a + c$ RHS
Considering dimensions of T : LHS $-1 = -2a$ RHS

Solving these simultaneous equations gives $a = \frac{1}{2}$, $b = -\frac{1}{2}$ and $c = \frac{1}{2}$. This gives a possible relationship of $v \propto \sqrt{Tl/m}$ (see E3.13).

The following important limitations of the method of dimensional analysis should be appreciated:

1 It is only possible to 'suggest' relationships using the method of dimensional analysis. The relationship must then be checked by experiment or further theoretical methods.
2 Using only the three dimensions [M, L, T] means that only three unknowns (a, b, c) can be found.
3 It is impossible to check for dimensionless quantities (ie, those with no units, such as strain, π, etc); so any relationship suggested by dimensional analysis can only be written as a proportionality not as an equation.

Students not required to understand dimensional analysis are still encouraged to study the method, but to apply it to units and not worry about differences between base units, derived units and dimensions. However it is often necessary to 'juggle' units to balance an equation; an example follows.

It is required to show that the units of the expression $T = CR$ are consistent (ie, that they balance). In the equation T represents a time constant (s), C represents capacitance (F) and R represents resistance (Ω).

Units of LHS = Units of T = s
Units of RHS = Units of CR = FΩ

At this stage the expression must be inspected for methods of simplifying it. The LHS is already in a base unit so the units of the RHS must be reduced to base units. From $Q = CV$ it can be seen that units of capacitance (F) can be written as CV^{-1}. From $V = IR$ it can be seen that units of resistance (Ω) can be written as VA^{-1}. This gives:

Units of RHS = FΩ = CV^{-1}·VA^{-1} = CA^{-1}

Now the expression has one base unit (A) but still includes a derived unit (C). Coulombs (C) are units of charge Q so using $Q = It$ we can rewrite the units of charge as A s. This gives:

Units of RHS = CA^{-1} = A s.A^{-1} = s

Hence it has been shown that both sides of the equation have the same units.

When handling physical quantities, units and dimensions, beware of the 'confusing' terms in the Table below.

There are many more that can be added to this list. Each student has his/her own particular set of terms that cause confusion; it is a good idea to keep a list of them to help cure the problem!

9.2 GRAPHS

Plotting graphs The first decision to be made in graph plotting is the choice of axes. It is a general rule that the quantity that is 'set' (**independent variable**) is plotted on the x-axis (horizontal axis or **abscissa** – hint for mnemonics, *a*bscissa goes *a*cross) – and the quantity that is 'measured' (**dependent variable**) is plotted on the y-axis (vertical axis or **ordinate**). This rule is occasionally broken when the graph will be easier to interpret if the axes are interchanged, for example if graphs involve 'time' or 'length' these quantities are usually plotted on the x-axis. Hence an experiment could involve a 'set' force (F – the independent variable) which is used to stretch a sample of rubber to produce a 'measured' extension (Δl – the dependent variable); but the general rule (F on the x-axis, Δl on the y-axis) is broken in this case as it is easier to interpret the 'length' on the x-axis and the graph is plotted as shown in F9.4. When talking about a graph the quantity on the y-axis is mentioned first, so F9.4 is a graph of the force used to stretch a rubber sample (y-axis) plotted against the extension (x-axis). In questions on graphs this convention also helps in making the choice of axes.

Choosing a sensible scale normally enables the graph to 'cover' as much of the graph paper as possible, it may not be necessary or desirable to include the **origin** ($x = 0$, $y = 0$). It should be possible to see the points on the graph, the recommended plotting methods include 'crosses' (×, +) and 'points' with circles, triangles or squares around them (⊡ △ ⊙); some teachers have a distinct preference for 'crosses' on the basis that it is easier to define the plotted point by intersecting lines rather than dots. When plotting several different sets of data onto the same axes it is important to use a different type of point for each set, a change of colour can be used in addition to the five methods

Table of some of the confusing physical quantities, units and terms

Physical quantity 'tension' – T	[T] – dimensions of 'time'
Physical quantity 'electric current' – I	A – units of electric current
Units of length – m	[M] – dimensions of mass
Physical quantity 'Acceleration due to gravity' – g	g – units of mass (grams)
Physical quantity 'mass' – M	M – for 'Mega-' denoting 10^6
(There are more like this, G, T etc.)	
Physical quantity 'inductance' – L	[L] – dimensions of 'length'
Physical quantity 'capacitance' – C	C – units of 'electric charge' (Coulombs)
Units of temperature 'kelvin' – K	k – for 'kilo-' denoting 10^3
Physical quantity 'spring constant' – k	k – physical constant 'Boltzman's constant'

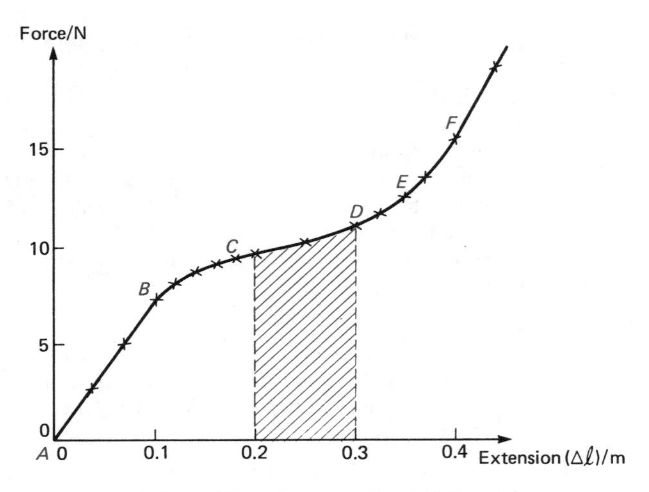

F9.4 Force (F) against extension (Δl) for rubber

already described. The scale should be marked along each axis and the axes clearly labelled with the quantity plotted followed by a stroke (/) and its associated units; eg, F/N, or Force used to stretch rubber/N.

For straight-line graphs five or six points can be sufficient; they should cover as wide a range of X and Y values as possible. When curves 'bend' considerably (eg, from B to C and D to E in F9.4) more points are needed to define the shape; it becomes vital to plot extra points when trying to locate a **maximum** or **minimum** ('peak' or 'trough'). Remember to ensure that a graph passes through the origin if this is known to be the case!

Interpreting graphs When examining graphs, apart from the general trends illustrated by the shape, the physical significance of the following quantities should be considered; slope (gradient), area under the graph, intercepts, other key points (maxima, minima, etc.). F9.5 shows the right (large 'triangle') and wrong

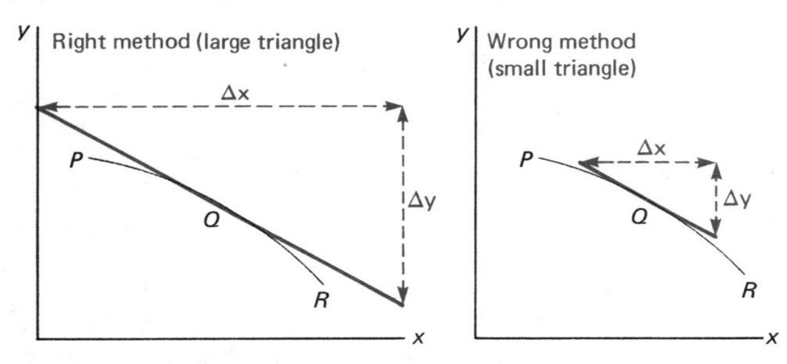

F9.5 Measuring slope at Q on curve PQR

(small 'triangle') method of measuring the **slope** at a point Q on a curve PQR. The slope ($\Delta y/\Delta x$) in F9.5 will have a negative value as an increase in y means a decrease in x (ie, the graph goes 'down'). Remember that horizontal lines have 'zero' slope, and that vertical lines have 'infinite' slope. Tangents must be drawn to points on curves to define the slope at that point (see F9.5). To decide upon the physical significance of a slope, divide the quantity on the y-axis by the quantity on the x-axis and consider the resulting units. For F9.4 the slope is force/extension measured in units of $\mathrm{N\,m^{-1}}$; over the straight line region (AB) where the rubber obeys Hooke's law, the slope represents the spring constant (k) of the rubber band (the force needed to produce an extension of 1 m).

The physical significance of the **area** under the graph can usually be found by multiplying together the quantities plotted on each axis and considering the resulting units. For F9.4 this gives force × extension measured in N m (or J); this represents the work done in stretching the rubber. Hence the blue shaded area shown on the graph is the energy (about 1 J) needed to stretch the rubber from an extension of 0.2 to 0.3 m; this energy is stored as extra potential energy in the stretched rubber. If the equation that defines a graph is known it is possible to measure the area mathematically by the process of **integration** (likewise the slope can be mathematically measured by **differentiation**).

Intercepts are the points where a graph crosses the axes. The x intercept ($x_{y=0}$ where y is zero) and the y intercept ($y_{x=0}$ where x is zero) can be read directly off the graph (see F9.6); however sometimes the axes chosen will not permit this, then the intercepts

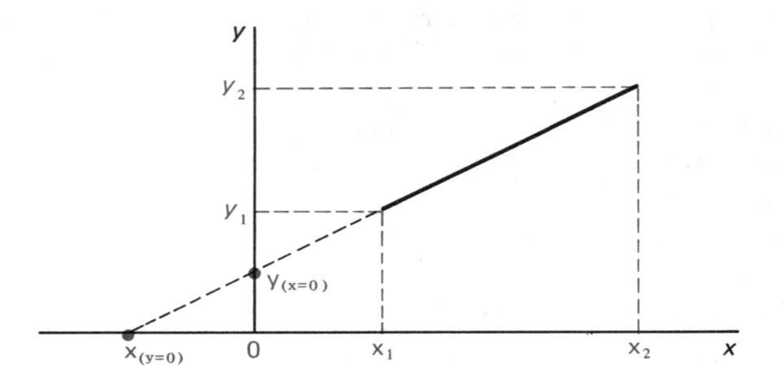

F9.6 Intercepts on a straight line graph

can be estimated in the case of a curve or calculated (from two points) in the case of a straight line using E9.1., E9.2 and E9.3:

E9.1 $y = mx + c$
E9.2 $y_{(x=0)} = c$
E9.3 $x_{(y=0)} = -c/m$

where x & y are the coordinates of a point on the straight line, m = slope of the straight line, c = value of intercept on the y-axis, $y_{x=0}$ = value of intercept on the y-axis, $x_{y=0}$ = value of intercept on the x-axis.

The coordinates of two points are sufficient to enable m and c to be found from substitution in E9.1, or the slope m can be measured conventionally (see F9.5) and c found by substitution. When calculating intercepts remember the negative sign in E9.3. Important intercepts in A level physics include: F1.2, x intercept defines range; F2.5(a), x intercept gives equilibrium separation; F8.16, y intercept gives the original number of radioactive nuclei; etc. Occasionally a graph fails to pass through the origin as expected and the resulting intercepts can be helpful in assessing the cause; an example to follow up in a textbook is the effect of 'stray capacitance' in the reed switch experiment to measure capacitance.

Other key points can also be found on graphs. Examples of important **maxima** and **minima** include F1.10, F2.5(b), F3.2, F4.9, F5.2 – point B, F6.10, F7.30, etc. **Asymptotic** behaviour occurs when a graph gradually approaches a fixed value on one axis without theoretically ever reaching that value! Examples of important asymptotes can be found on F2.5 and F2.6, F7.19 and F7.20, F8.17, etc. **Discontinuities** occur when a graph 'unexpectedly' changes shape. Important examples of discontinuities include F5.2 – y_1, etc., F6.13, F7.33(a), F8.10 – the line spectra, etc.

The ability to 'read' graphs is becoming an increasingly more important part of all A-level physics examinations. This is reflected by the treatment of the graphs featured in this book. A final point of caution should be made; there is not always some sensible physical significance of the area and gradient (etc.) of every graph!

Deducing equations from graphs Although the need actually to deduce equations for graphs is limited mainly to those involved with project work, the recognition of graph shapes and the physical situations to which they apply is a skill demanded of all A-level students. There are six basic graph shapes that frequently occur in A-level physics, these are shown in F9.7. Once a 'shape' has been recognised it may be necessary to try each of the equations (E9.1, E9.4 to E9.9) that might describe the shape. There are two 'unknowns' in each equation which can be found by substituting the coordinates of two points from the graph, the equation can then be 'checked' by making sure that other points from the graph 'fit' it. When 'spotting' graph shapes it is im-

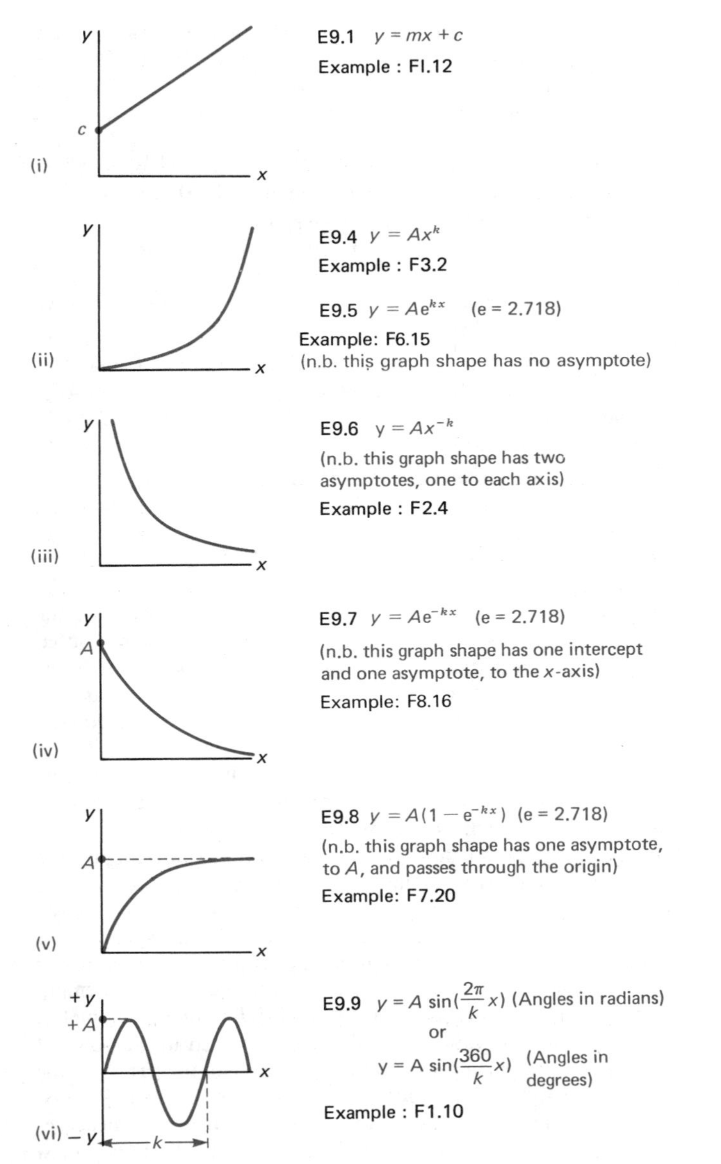

F9.7 Basic graph shapes and equations

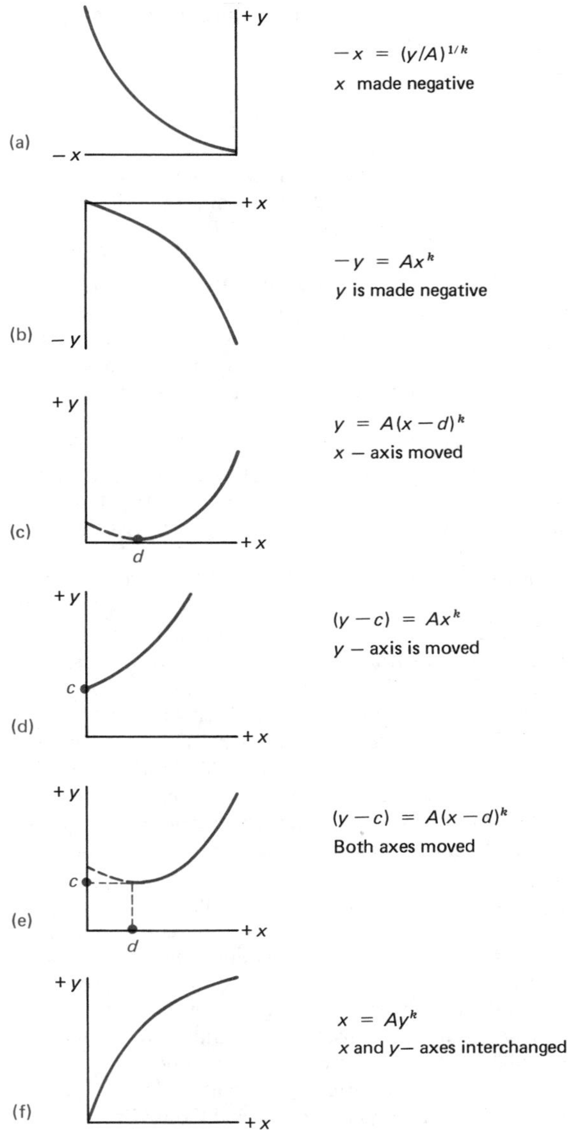

F9.8 Variations on shape of graph $y = Ax^k$ (F9.7b)

portant to look for the effect of moving the origin, making x or y negative, or interchanging x and y. The effects of this are shown in F9.8 for the graph whose equation is $y = Ax^k$ (see F9.7 and E9.4). The 'guessing' method of deducing equations from graphs is poor if the points exhibit any scatter (ie, errors) from their true position, as is frequently the case when plotting experimental data; finding an equation that will be a 'good fit' to scattered points is made much easier if the graph can be converted to a straight line. There are two important techniques that enable this to be done. It is possible to convert all graphs of the form $y = Ax^k$ (ie, E9.4 and E9.6 where k is negative) to a straight line by taking **logarithms** of both sides. For simplicity logarithms to the base 10 (logs) shall be used:

$$y = +Ax^k.$$

so $\log(y) = \log(Ax^k)$

E9.10 $\log(y) = k\log(x) + \log(A)$

Hence if a graph of $\log(y)$ (y-axis) is plotted against $\log(x)$ (x-axis) a straight line should be obtained of slope k and y intercept $\log A$.

It is possible to convert all graphs of the form $y = Ae^{kx}$ (ie, E9.5 and E9.7 where k is negative) to a straight line by taking logarithms of both sides once more. In this exponential (ie, involving e) case for simplicity a logarithm to the base e (ln) is used:

[NB. $\ln(y) = \log_e(y)$]

$$y = +Ae^{kx}$$

so $\ln(y) = \ln(Ae^{kx})$

E9.11 $\ln(y) = kx + \ln(A)$

Hence if a graph of $\ln(y)$ (y-axis) is plotted against x (x-axis) a straight line should be obtained of slope k and y intercept $\ln(A)$.

The reader can try out these two techniques. If E9.10 is used to plot a new version of F2.4(a), a slight modification is needed; treat the field strength values as positive (you cannot take logs of negative numbers!). k should be found to be -2 and $\log(A) = -11.0$. It is possible to analyse F8.17 using E9.11; take care to use the 'corrected count rate' for y. In this case $k = 0.012$ and $\ln(A) = +4.1$. It is left to the reader to construct the two equations and (most importantly) to ascertain the units of k and A in each case. Some research in 2.5 and 8.6 will enable the answers to be checked!

9.3 ERRORS

Types of error There are two methods of quoting the error in a value; they are to state the **absolute error** and to state the **percentage error**. To illustrate the two methods consider the case of a voltmeter observed to read 5.0 V when the error is assessed to be

such that it is considered that the reading might reasonably represent a value between 4.9 V and 5.1 V. The reading can be quoted in two ways:
as (5.0 ± 0.1) V – Absolute error quoted
or as 5.0 V ± 2% – Percentage error quoted
When a measurement is made it is possible to describe two fundamental types of error that may occur. **Random error** is said to occur when repeated measurements of the same quantity can give rise to different values. **Systematic error** refers to an effect that influences all measurements of a particular quantity equally.

Random errors can be demonstrated using the example of the voltmeter where they can arise from two obvious causes:
1 Friction in the bearings of the system (analogue meter) and deficiencies in its mechanics can cause the pointer to settle in slightly different positions, even though the meter might be repeatedly used to measure the same pd.
2 The interpretation of the actual position of the pointer is a skill that is subject to random error; this error is less if the observer is aided by a finely marked scale and a mirror to reduce parallax error! Skilled observation can help minimize (but never eliminate!) this type of random error.
Systematic errors of both an absolute and percentage nature might arise when using a voltmeter:
1 If the voltmeter has a 'zero error' then a systematic absolute error will occur. For example if a 'disconnected' voltmeter reads + 0.2 V, then readings will be 0.2 V 'too large'; ie, an absolute systematic error of + 0.2 V will occur in all measurements.
2 If the voltmeter is incorrectly 'calibrated' then a systematic percentage error will occur. For example if 'full scale deflection' (fsd) of the voltmeter is actually 10 V when the meter scale indicates 9 V then all readings taken with the meter on that range will be 10% too small; ie, they will have a percentage systematic error of −10%.

A key difference exists when dealing with results that contain random or systematic errors. If a reading is thought to be subject to random error, then repeating the measurement several times and taking an average (mean) can improve the 'confidence' in the reading (ie, reduce the likely random error in it). However this technique is useless for dealing with systematic error which will affect all the observations equally.

Systematic errors are notoriously difficult to detect (because they affect all results equally) and can usually only be found by checking the instrument in which the error is suspected against a known reliable instrument. For our voltmeter example, the zero error is easily found by comparing with a known value of 'zero volts' by disconnecting the meter, and to show up a faulty calibration requires comparison with another voltmeter of known accuracy.

Calculations involving errors To assess the total error in a calculation it is necessary to evaluate the **likely** error in all the values involved in that calculation. Then the calculation can proceed using the following three rules:
1 When two quantities are to be **added or subtracted**, then **add together their absolute errors** to obtain the absolute error in the answer.
2 When two quantities are to be **multiplied or divided** then **add together their percentage errors** to obtain the percentage error in the answer.
3 When a quantity is to be **raised to the power** n then **multiply the percentage error by** n to obtain the percentage error in the answer.
It should be pointed out that a full treatment of error calculation goes far beyond this simple scheme and far beyond the scope of any A-level course. An example can illustrate the three simple A-level rules for calculating errors.

A value for the mass of a large ball is required. An experiment to measure the density of the ball yields the result (300 ± 8) kg m^{-3}. The ball is placed upon a metre rule in order to assess its diameter. An observer notes that one end of a diameter is opposite the 35 cm mark and the other end is at 78 cm; the likely error in each of these measurements is assessed to be 1 cm:

D = diameter of ball = (78 ± 1) − (35 ± 1) cm = (43 ± 2) cm (using rule 1)
D = (0.43 ± 0.02) m (absolute error)
D = 0.43 m ± 4.7% (percentage error)
V = volume of ball $= \frac{4}{3}\pi\frac{D^3}{8} = 0.0416\ \text{m}^3 \pm 14.1\%$ (using rule 3, we need to treble the percentage error in D)
V = 0.0416 m^3 ± 14.1% (percentage error)
ρ = density of ball = (300 ± 8) kg m^3 (absolute error)
ρ = 300 kg m^{-3} ± 2.7% (percentage error)
M = mass of the ball = ρV = 12.48 kg ± 16.8% (using rule 2 we need to add the percentage errors in ρ and V).

Finally it is important to remember never to quote an answer to any more significant figures than the error assessment can justify, so the final value for the mass of the ball is quoted as: 12 kg ± 17% or in absolute terms as (12 ± 2) kg.

Experiment design and errors In an experiment to measure the mass per unit length of a metal rod a student measured the length with vernier callipers (accuracy within 1%), and the mass using a spring balance (accuracy about 10%). This is a classic example of bad experiment design, because the 11% error in the final answer was almost entirely due to the error in one of the two measurements (10% in the mass). If the student had measured the mass using a chemical balance (accuracy within 1%) the result would have been within 2%, and the experiment would have been well designed with each measurement contributing equally to the error in the calculation.

It is a general rule when experimenting that all measurements should contribute equally to the error that will be obtained in the final answer. This normally means that equipment should be chosen to enable all measurements to be made to roughly the same percentage error; an important exception to this suggestion occurs when a measurement is to be raised to a power n in subsequent calculation. The following examples will clarify this.

In the sample error calculation the total error in the answer for the mass of the ball was 17%. The error in each measurement was similar, 2.7% for the density, and 4.7% for the diameter. Unfortunately the value of the diameter is raised to the power 3 (cubed) in the calculation, which means that the error in the volume is 14.1% (three times the error in the diameter). So most of the error in the answer for the mass arises from the measurement of the diameter; however if the diameter had been measured to about 1% then the error in the answer would have been about 6%, with 3% contribution from each measurement; this would be much more satisfactory. On the other hand, it can be seen that a measurement that is to be 'square rooted' will have its percentage error halved in the calculation, and hence its value need not be known as accurately as others used in the calculation.

Good experiment design involves realizing that the required accuracy of measurements is governed by how the results are to be processed in further calculations.

Assessing errors A systematic error can be removed by modifying all the observations to remove the error. The difficulty is suspecting that systematic errors might exist! They are detected as a consequence of analysis of the observations against 'expected' results and discovering a consistent error (percentage or absolute). Checking can be done sometimes by repeating the suspect measurement using a standard instrument. Often systematic errors are found by graphs behaving in an unexpected fashion; two good examples from A-level physics are 'stray capacitance' in the reed switch experiment (see your textbook) and the background count in radioactivity (see F8.17). The systematic errors are the 'stray capacitance' and the 'background count', respectively.

Random errors can be assessed by repeating the same measurement (ideally using different observers) and examining the fluctuations in the results. The following example will illustrate this.
Several observers were asked to time 10 oscillations of the same simple pendulum and the following results were obtained: 20.0 s,

20.2 s, 20.1 s, 19.8 s, 20.0 s, 20.0 s, 20.0 s, 18.1 First, the decision should be made to ignore the last observation as it is obvious from the consistency of the other results that the observer has only timed 9 oscillations. (Results can only be ignored if it can be clearly shown that the observations have been 'wrongly' made because their error is far greater than it ought to be for the particular observations!) The average of the rest of the observations is 20.017 s. The largest deviation from this average is 0.2 s (20.2 and 19.8 seconds); however, to quote this as the 'likely error' is rather exaggerated, as most of the results are within $\pm$ 0.1 s. (Students of statistics may choose to quote the standard deviation as 'likely error', especially as modern scientific calculators have statistical facilities). Quoting the answer to the correct number of significant figures gives a final value of (20.0 ± 0.1) s. Note that one of the most common Examiners' complaints is that students quote answers to calculations to ridiculous numbers of significant figures, so don't copy down a mass of numbers from your calculator but quote just enough significant figures to ensure that the error only affects the last figure!

The above estimate of random error assumes that the error is mainly due to the observer rather than inaccuracy in the instrument itself. This is true in many cases, but an awareness of the manufacturer's specification for the instrument can be important. For example most dc electrometers have a random error of about 10% in the system, far greater than the error incurred by reading the meter. A sensible balance between the two effects must always be maintained when assessing random errors. This becomes very necessary when it is impracticable to take an average of several readings to help assess the random error. A useful 'rule of thumb' for estimating random error effects is to evaluate the biggest possible error and then quote half of it as the likely error. One final example will illustrate this.

A manufacturer claims that a voltmeter will give readings accurate to within 1% of its fsd. A meter on fsd of 10 V will hence be subject to a random error of 0.1 V (1% of 10 V). The observer estimates that the meter can be read to an accuracy of $\pm$ 0.1 V. Hence if both errors add together a maximum error of 0.2 V will occur in any reading. However, it is unreasonable to consider that the errors will always 'add up' (sometimes they will 'cancel each other'). A sensible quote of 'likely error' is half the maximum error; ie, $\pm$ 0.1 V.

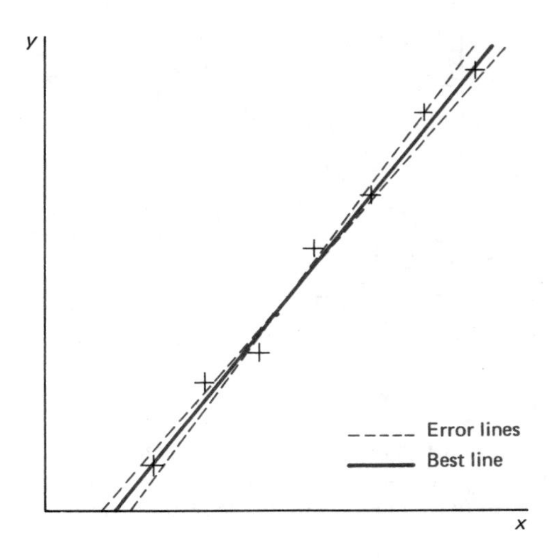

F9.9 Drawing error lines on a straight-line graph

If a straight line graph is to be plotted, a good estimate of the likely random error can be obtained from the 'scatter' of the points. A common technique is to draw the 'best' line (solid coloured line in F9.9) and the two acceptable straight lines that deviate most from the best line (dotted coloured lines in F9.9). The error in the slopes or intercepts (etc) given by these error lines is used as a basis for the quote of 'likely' error in measurements involving the 'best' line.

UNIT 9 QUESTIONS

9.1M If p is the momentum of an object of mass m, then the expression p^2/m has the dimensions of:
A Power. B Impulse. C Force.
D Acceleration. E Energy.

(**SEB H**: *and all other Boards*)

Obtain the dimensions of momentum (1.3) and work out the dimensions of p^2/m. Then you will have to work out the dimensions of A–E in turn.

Alternatively write momentum as mv so p^2/m becomes $(mv)^2/m$ or mv^2. Putting in a constant without dimensions ($\frac{1}{2}$) gives $\frac{1}{2}mv^2$; this should make it easy to recognise the answer. Remember that 'dimensions' are the same as 'base units' (see 9.1).

The techniques can be summarized as follows:

1 Work out the dimensions of the expression and compare with the answers.
2 Leave the expression in terms of physical quantities and find an answer from A–E which has the same units.

In questions like this with physical quantities quoted, it is best to try method 2 first as it can save much time, reverting to method 1 if stuck! If the answers A–E are quoted as dimensions, then it is best to use method 1 straight away.

9.2M and **9.3M** Five graphs showing the relationship between a physical quantity y and another physical quantity x are lettered A to E. Select from these graphs the one which best represents the relationship between the following pairs of physical quantities. Each graph may be used once, more than once, or not at all.

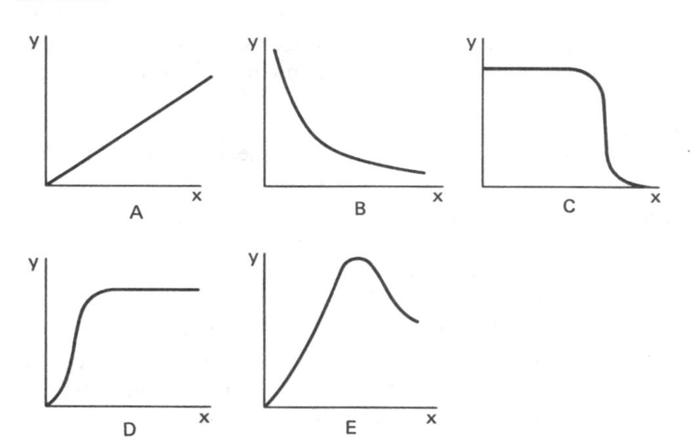

	y-axis (vertical axis)	*x*-axis (horizontal axis)
9.2M	Power transferred from a battery of finite internal resistance to a resistor connected across it.	The resistance of the load resistor.
9.3M	Flux density along the axis of a long current-carrying solenoid.	The distance from the centre of the solenoid along the axis.

(**NISEC**: *all other Boards except SEB H*)

When 'graph spotting', two basic techniques can be employed if the answer is not 'known':

1 Identify some key points on the graph to isolate the answer.
2 Predict some feature about the shape of the graph and obtain the answer from that.

Sometimes a combination of both methods is necessary.

9.2M Using method **1**, firstly consider the point $x = 0$ (ie, when the load resistance is zero). Then the pd across the load resistor must be zero and all the power supplied by the battery is dissipated in its internal resistance. Thus the graph passes through the origin, limiting the answer to A, D or E. Then consider a second point at 'large x' so that the load resistance is very large and the current will tend towards zero; once again the power dissipated in the load (IV) must tend towards zero. Only one graph fulfils both criteria.

Using method **2** is easy if you are aware that at the right finite value of load resistance, the power transferred to the load is a maximum, and only one graph behaves like this.

9.3M To solve this it is necessary to know that the field inside a solenoid is uniform giving a constant y value until on emerging from the solenoid (large x) the field rapidly drops off towards zero. See p. 37.

9.4M and 9.5M The diagram below shows axes which represent the 'logarithms' of pairs of quantities y and x listed below.

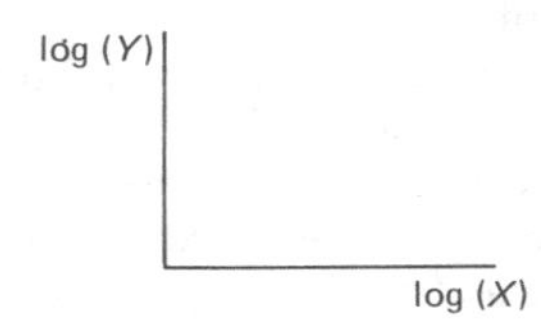

When the graphs of log (y) against log (x) are drawn, the slopes are among the numbers listed below under the letters A to E:
A -1. B $-\frac{1}{2}$. C $\frac{1}{2}$. D 1. E 2.
For each of the pairs of quantities in questions 9.4M and 9.5M, choose the slope that would be obtained. Each response may be used more than once, once, or not at all.

	y Quantity	x Quantity
9.4M	Frequency of small vertical oscillations of a loaded spring.	The mass of the load.
9.5M	Energy stored in a given capacitor.	Pd across the capacitor.

(**NISEC**: *all other Boards except SEB H*)

To answer each question, it is necessary to remember the equation that links the two physical quantities and then use the analysis of E9.10.

9.4M Use E3.5.

9.5M Use E2.14; remember that the capacitance in this question is fixed so use the $\frac{1}{2}CV^2$ version.

9.6M If P is the pressure of a gas and V is its volume, in what unit could the quantity PV be measured?
A Newton. B Watt. C Newton/metre.
D Newton second E Joule.

(**SEB H**: *all other Boards*)

Pressure is force/area and has units N m^{-2}; volume has units m^3. Hence the units of PV are N m. These are the units of force $\times$ distance, you should be able to relate them to one of the quantities above; note that with the knowledge that newton metres are the required units, answers A, C, and D are eliminated!

9.7M The dimensions of specific heat capacity are.
A $[L][T]^{-2}[\theta]^{-1}$ B $[L][T]^{-1}[\theta]^{-1}$ C $[L]^2[T]^{-2}[\theta]^{-1}$
D $[L]^2[T]^{-1}[\theta]^{-1}$ E $[M]^{-1}[L]^2[\theta]^{-1}$

(**AEB** June 81: *all other Boards except WJEC and NISEC*)

You need to work out the dimensions of specific heat capacity by remembering either the definition of the quantity (heat energy per unit mass per unit change of temperature) or by remembering an equation involving the quantity (E6.3). Remember that dimensions of length (units m) are [L]; many students write the dimensions down as [M] (following the units) without thinking, so causing confusion with the dimensions of mass, which is what [M] represents. The dimensions of energy can be derived in many ways; a popular way is to use work done (= force $\times$ distance) and find the dimensions of force from $F = ma$. Another approach is to remember that ke is given by $\frac{1}{2}mv^2$, and use that to give the dimensions. See p. 138 for converting units to dimensions.

9.8M In an experiment, the external diameter d_1 and the internal diameter d_2 of a metal tube are found to be (64 ± 2) mm and (47 ± 1) mm, respectively. The percentage error in $(d_1 - d_2)$ expected from these readings is at most:
A 0.3%. B 1%. C 5%. D 6%. E 18%.

(—: *all Boards except SEB H*)

The question clearly asks for the 'worst possible' error, which must be 3 mm in the measurement of 17 mm. It is easy now to find out the percentage error that this represents. (See p. 141 if necessary.)

9.9M and 9.10M The following are five relationships between quantities x and y, k being constant:
A $x = y + k$. B $x = ky$. C $x = k/y$.
D $x = k/y^2$. E $x = ky^2$.
Which of these relationships applies to the following?

9.9M x is the pd across the plates of Millikan's apparatus needed to hold a certain oil drop stationary, and y is the distance between the plates.

9.10M x is the speed of an electron as it moves in a vacuum from rest near a negatively charged plate towards a parallel positively charged plate, and y is the time.

(**London**: *Oxford* and all other Boards*)

First a hint of warning to mathematicians; the unknowns x and y have reversed their normal positions in standard equations for this question! Hence care must be taken (eg, in using F9.7 to help solve these equations) not to get the answer the 'wrong way up'.

9.9M Rewrite E8.1 with x for V and y for d and rearrange the equation. Always try a quick check on the answer when possible; in this case you should know that the closer together the plates (smaller y) the smaller the voltage (smaller x) that is needed to maintain the same electric field and keep the drop stationary (this eliminates C and D).

9.10M Between parallel plates the electric field is uniform and the electron will hence experience a constant force (see E2.9). A constant force means a constant acceleration; use E1.1 with x for v and y for t. Remember that $u = 0$ since the electron starts from rest. (C and D must be 'wrong' again.)

9.11S The period T of vertical oscillations of a mass M suspended by a spiral spring is given by

$$T^2 = \frac{A}{g}M + \frac{A}{3g}m$$

where A is a constant depending on the stiffness of the spring and m is the mass of the spring itself.

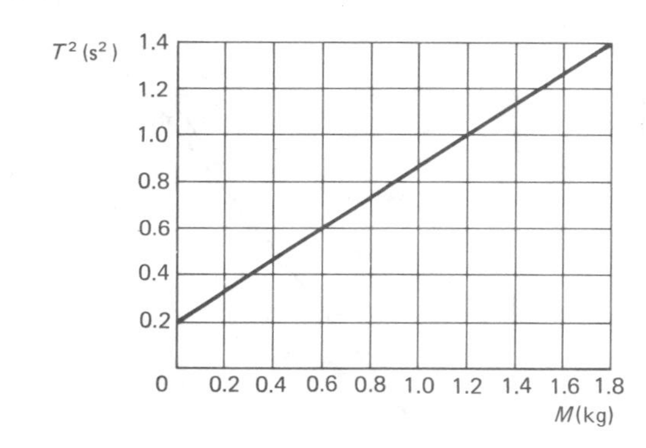

The graph shows the results of measurements of the period for various values of M. Use the graph to determine the constant A and the mass m of the spring. What are the dimensions of A?

(**SUJB**: *all other Boards*)

The graph shows a straight line whose equation is given above. Which part of the equation gives the slope? Hence calculate a value for A, assuming $g = 10$ ms^{-2}. Which part of the equation represents the y-intercept? Hence calculate m, using the previously determined value for A/g.

Each term of the equation has units of s^2, so use the dimensions of g and M to determine the dimensions of A.

9.12L (30 mins) (a) Distinguish between a 'systematic' and a 'random' error in the measurement of a physical quantity.

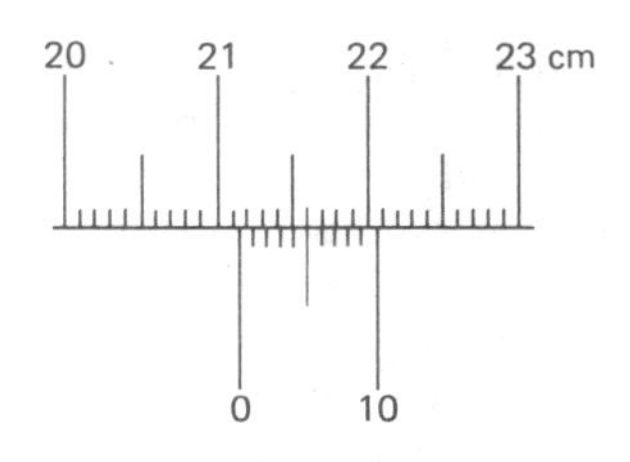

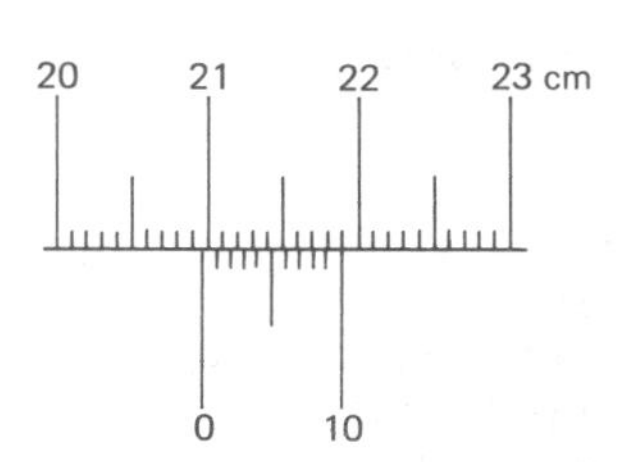

A travelling microscope fitted with a vernier scale is used to measure the internal diameter of a capillary tube. Figs 1 and 2 show the vernier when the microscope is adjusted so that the cross-wires are aligned at opposite ends of a diameter.

(i) Write down the two vernier readings.

(ii) What is the maximum uncertainty in a single reading of the vernier?

(iii) Hence find the maximum percentage uncertainty in the area of cross-section of the capillary that could arise if it were calculated from these two readings.

(iv) Explain why taking the mean of several microscope readings of the diameter tends to reduce random error.

(b) (i) How do you check a formula for its dimensional consistency? Why does this method of checking not give definite confirmation that an equation is correct?

(ii) Express the unit of force and of charge in terms of the SI base units kilogram, metre, second and ampere. Hence by reference to Coulomb's Law

$$F = \frac{1}{4\pi\varepsilon_0}\frac{Q_1 Q_2}{r^2}$$

express the unit of ε_0, the permittivity of the vacuum, in terms of these base units.

A unit for μ_0, the permeability of a vacuum, is kg m $s^{-2}A^{-2}$. Use this unit, and your unit for ε_0 to decide which one of the following relations between ε_0, μ_0 and c, the speed of light in a vacuum, is dimensionally consistent:

$$\varepsilon_0\,\mu_0 = c^2; \quad \varepsilon_0\,\mu_0 = c; \quad \varepsilon_0\,\mu_0 = c^{-1}; \quad \varepsilon_0\,\mu_0 = c^{-2}$$

(**Cambridge**: *all other Boards*)

(a) See 9.3.

(i) and (ii) Each reading can be determined to within one division either way on the lower scale, so convert this into a value in mm since the ten divisions of the lower scale 'correspond' to 1 mm.

(iii) From (ii), calculate the absolute error of the diameter's value. Then, convert this into a % error for the diameter and hence for the area. See p. 142 if necessary.

(iv) See p. 142

(b) (i) See p. 138–9.

(ii) Make ε_0 the subject of the equation, then replace the symbols for force, distance and charge with the appropriate combination of base units in each case. Hence express ε_0 in the base units.

Multiply the base units for ε_0 by those for μ_0 to give the base units of $\varepsilon_0\,\mu_0$. Hence choose by considering the base units of the right-hand side of each relationship in turn.

9.13D (45 minutes) The results in the table give corresponding values of the emf, E, mV of a thermocouple and the temperature, θ, in °C, of the hot junction when the cold junction is maintained at 0°C. The emf is related to the temperature by the expression $E = b\theta + c\theta^2$ where b and c are constants.

E/mV	θ/°C
5.5	50
10.0	100
13.5	150
16.0	200
17.5	250
18.0	300
17.5	350
16.0	400

(a) Plot a graph of E (y-axis) against θ (x-axis).

(b) Find the slope P of the curve at points $\theta = 50$, 100, 150, 200 and 250°C. Record your results in a suitable table.

(c) Plot a graph of P (y-axis) against θ (x-axis).

(d) P is known as the thermoelectric power and is related to θ by the expression $P = b + 2c\theta$. From your graph determine the value of b and the temperature at which P is zero.

(**AEB**: *all other Boards*)

(a) Provide a title for your graph and label the axes with the appropriate quantities and their units. Use a sharp pencil!

(b) Draw your tangents as accurately as possible. In measuring the x and y values of your gradient triangles make sure that you use the correct units (ie, do not simply count the graph paper divisions). When recording your values of P remember to state the units.

(c) The same comments apply here as for (a).

(d) The gradient (ie, slope) of your graph of P against θ is $2c$, and its intercept on the x-axis (ie, θ axis) is the value of θ for which $P = 0$, (ie, when $0 = b + 2c\theta$). Thus $b = -2c\theta_{(P=0)}$

Part III Examination technique

Tackling different styles of examination paper

The modern A-level student has to demonstrate a considerable degree of versatility, for it may be necessary (including a project) to tackle up to six fundamentally different examination papers. These can be listed as multiple choice, short answers, long answers, comprehension, data analysis, practical problems (Nuffield), and conventional practical examinations. This section discusses the needs and approach required for each of these papers. Students must check which papers they will sit using p. 16 or asking their teacher/Board. Do not be concerned at what may seem a large number of papers; it is this diversity of examination that has lent greater interest to A-level physics courses, and sitting several papers no longer means that a 'disaster' in one of them (eg, due to a headache) will have too much effect on your overall result. Practice at all the styles of question paper is available in Part IV (Test yourself), with the exception of a conventional practical examination!

MULTIPLE CHOICE

These papers are sometimes called multiple selection, coded answer or objective tests. Do **not** read through the paper first; you have to attempt all the questions and there are too many to make it worthwhile to have a quick scan through the entire paper. Start at the beginning and work steadily through, but leave out any questions that appear difficult at first sight. On completion go back and attempt the problems that you missed out initially. Just before time is up, make a guess at all the remaining questions; **this is vital**, as marks are not deducted for wrong answers. Above all, do not panic and spend a long time on one question. Just keep going, you are bound to find some that you can answer easily, and these will restore your confidence.

A unique feature of all multiple-choice papers is the fact that you are provided with several answers, only one of which is right. Many students fail to appreciate that instead of looking for the right answer, it can sometimes be much easier to identify the wrong ones! This approach will often enable a guess at an unknown answer to be 'one from two' instead of 'one from five'. The technique is demonstrated in some of the question practice commentaries, but one example may help.

The moment of inertia of a sphere about an axis through its centre of mass is $2mr^2/5$, where m is its mass and r its radius. When on a flat surface the sphere rolls without sliding with a horizontal velocity v, the ratio of its translational kinetic energy to its rotational kinetic energy is:

A $2r$:5. B $5r$:2. C $2m$:5. D 5:2. E 2:5.

The ratio is of two energies, so it must have no units. Hence answers A, B and C must be wrong.

SHORT ANSWERS

You are usually expected to answer all the questions, in which case there is no need to read right through the paper; this advice can be important to the many who have time troubles on short-answer papers. However you should read through each question completely; frequently the questions are in several parts that are linked. Ideas from later parts of the question may help you follow the earlier parts; you may avoid the pitfall (for example) of giving an answer in part (a) that is expected in part (b). Again move on quickly to the next question as soon as you get bogged down; leave space on your answer paper/book to enable you to return and complete the previous question later. It is amazing how the **subconscious** will unravel a tricky problem for you as soon as you have set to work on something else; this is like the experience of trying hard to recall something you know well that has suddenly slipped your memory, stop concentrating on it and it comes rushing back!

LONG ANSWERS

Most long-answer papers offer you a choice of questions. It is therefore important that you read through the whole paper before choosing which questions you intend to attempt. If the question itself is long it can often be tackled without much planning; the structure of the question will in effect generate the plan for you. However in general on long answers a plan is needed, which means that a significant amount of the total examination time (maybe up to $\frac{1}{3}$rd) is spent in thinking rather than writing an answer. This time is allowed for when setting the questions; time troubles invariably are due either to failure to stick to the terms of reference of the question, or failure to use diagrams effectively to save many words, or to describing in many extra sentences what has already been described in the first sentence! Try to identify which fault applies to you.

Once again you should read the whole question before you start so that you have an overall picture of the complete question. Calculations are often based on earlier descriptive work, so look for clues to help you tackle the problem. Information on planning long answers appears in the next section (p. 148) and in the question practice.

COMPREHENSION

Acquire a general idea of the content of the passage by reading through the text at your normal reading pace, avoiding any backtracking (ie, going back and re-reading sections). **If** you have time, it is a good idea to read all the questions before you start writing answers; the content of the questions can often add to your understanding of the passage and earlier questions. As you do a question, re-read the section of the text that is relevant before you write your answer. Always answer in your own words, only quote from the passage when asked to do so. As in many other papers the structuring of questions can help you; for example assume that the questions are laid out as follows.

Question 1.
Question 2. (a)
(b)
Question 3. (a)
(b) (i)
(ii)
(iii)

There will usually be some connection between 2(a) and (b), and 3(a) and (b); but there can be very strong links between 3(b)(i), 3(b)(ii) and 3(b)(iii).

DATA ANALYSIS

The questions are normally all compulsory but as they follow the same theme it is worthwhile to read them all before you start. The skills needed for data analysis are outlined in Unit 9 together with its question practice; however you are particularly advised to practise if you are slow at graph drawing or analysing results. Candidates without a **scientific calculator** or the ability to operate it expertly can be at their greatest disadvantage on this paper. It is foolish to wait until the examinations to buy your new calculator; would you enter a golf competition playing with a set of clubs you had only just bought? Practice is essential.

PRACTICAL PROBLEMS

This examination is unique to Nuffield, though some of the JMB short experiments in their practical examination are comparable. However, several Boards set similar questions in their short- and long-answer papers, but naturally they supply the results to the experiments rather than asking the candidates to do them.

Most of the marks are available for the 'theory' part of the questions, and Nuffield candidates should complete the experiments quickly and with an accuracy appropriate to the apparatus with which they have been supplied. If you finish a question early, read up about the next one so that when you move on to the apparatus you are already well prepared, or you can use the time to complete some earlier questions. If short of time, always try and complete the new question before finishing off the old one; apart from other reasons it is better for your morale!

PRACTICAL EXAMINATIONS

Read the whole question before you start so that you can plan what you are going to do. A few minutes spent thinking about the best way to arrange your apparatus can save considerable time when you start to take readings. The experiments usually require readings to be taken and then processed, normally with the aid of a graph.

The amount of 'writing up' you are required to do varies according to the Board; however, the question will clearly indicate what is needed. Sadly, many candidates write a full 'report' on their experiments when this is not asked for, and vice versa. Frequently a fully labelled diagram of your apparatus will be necessary, and virtually all Boards (in their 'rubric') ask for details of special precautions or experimental techniques that you may use.

Investigate the range over which observations can be made and then take readings at convenient intervals over the whole of that range. Take as many readings as you can in the time available, but always leave sufficient time to process them. Repeat readings if possible, particularly where considerable judgement has to be exercised (as in timing and focusing experiments). Leave your apparatus set up; this will enable you to check any 'strange' points after the graph has been plotted or to take further points in 'important' regions (eg, when trying to locate a maximum or when drawing a tangent to a curve to find the slope).

In all practical work, check the zero error (see p. 142) of all instruments (eg, meters, micrometers, stopwatches) and record that you have done this. Record **all** the observations that you make, in tabulated form if possible, so that it is easy to check back for any mistakes in the analysis of your results. Ensure that all results of data processing (and the initial readings) are quoted to a reasonable number of significant figures. Finally, if the results of your experiment are not those expected, comment on this and offer some sensible explanation if possible.

How to answer the questions

A most common complaint of examiner and teacher is about the student who fails to 'answer the question'. All the following misdemeanours can be classified under this heading; answer too long, answer too short, irrelevant material included, relevant material excluded, absence of diagrams/calculations/graphs that are clearly asked for, entire parts of questions missing, etc. 'Answering the question' is a skill that will only gradually be perfected as you practise throughout your course, and the question practice in this book has many commentaries that are specifically designed to help you develop the necessary technique. This section contains further hints, comments and ideas to help you organize a style of approach to questions that will assist you fully to realize your potential under examination conditions.

'A-level questions are a lot more difficult to answer... they are not so straightforward, ie, you have to work out what the question is asking before answering it'.

'Lots more depth is required in the answers'.

STATE, DEFINE, EXPLAIN, DISCUSS, COMPARE, DESCRIBE, COMMENT

Questions on all types of examination paper (except multiple choice) may use any of these key words in a question. They all have different meanings and interpreting them wrongly can be very expensive in terms of marks or time lost.

State: the briefest possible answer will suffice. It may only be one word!

Define: a full statement is needed, together with the appropriate equation if possible. Give the meaning of the symbols used in the equation and the units of the quantity to be defined. Beware of padding your precise definition with vague statements that might contradict your definition and lose marks.

Explain: you are given a piece of information that must be described in detail; normally you must use basic physical principles to justify or clarify what you are required to explain. A graph or diagram will often be invaluable.

Discuss: usually you are given something to consider that may or may not be true. You need to indicate the various possibilities, giving evidence and arguments for and against each proposition.

Compare: two or more items have to be compared with each other. You must describe similarities and differences between them. It will not be necessary to describe each item in detail unless the question asks you to 'compare and describe' or 'compare and discuss'.

Describe: most often you are asked to describe experiments. Full details of method, measurements, apparatus and processing of results are needed. Please be careful to cut back on the detail if the question asks you to 'describe briefly' or 'describe concisely'; here your ability to isolate the key features of the experiment/topic is being tested.

Comment: usually you are being asked to think about something unusual or slightly strange. You must relate the information supplied for comment to your own knowledge of the topic and draw some conclusion. Answers to these questions are rarely longer than a sentence or two.

A common difficulty amongst candidates is the judgement of the length required of an answer. Help can be provided in the form of the space supplied if you write your answer on the question paper, or if the mark scheme is supplied you can relate this to the length of the question (normally marks are related to the time needed to complete an answer). You must always use the wording of the question (eg, state, explain) to assist you determine the length of your answer. The following example question and answer may help you to understand the 'jargon'.

Question

(a) **State** one example each of an electrical insulator, semi-conductor and conductor.

(b) **Define** resistivity.

(c) **Explain** how a material can be classified as an insulator, conductor or semi-conductor.

(d) **Compare** the terms 'resistance' and 'resistivity'.
Discuss whether either of these quantities can have a negative value.

(e) **Briefly describe** how you would carry out an experiment to measure the resistivity of copper. **Comment** on the fact that a student doing such an experiment finds that as he leaves the apparatus switched on, the resistivity of copper seems gradually to increase.

Answers

(a) Insulator, glass; semi-conductor, silicon; conductor, copper.

(b) The resistivity ρ (measured in Ω m) of a material is defined as the electrical resistance per unit length of a sample of the material with unit cross-sectional area. Hence if a sample has resistance R, length L, and cross-sectional area A then $R = \rho L/A$.

(c) Classification of a material as an insulator, semi-conductor or conductor is based upon the value of its electrical resistivity at room temperature. Materials with resistivities of the order of 10^{11} Ω m or larger are called insulators, those with a value of 10^{-6} Ω m or smaller are termed conductors. If the value lies between these limits then the material is called a semi-conductor.

(d) 'Resistance' and 'resistivity' are terms both used to describe the opposition offered to the passage of electrical current. However resistance can only be defined for a specifically sized sample of a given material, whereas the resistivity is a property of the material in general and will have the same value for any size of sample (this is the same comparison as can be drawn between mass and density, or spring constant and Young's modulus). If a negative value of resistance could occur it must also infer that the resistivity of the material is negative. But if the resistance were negative then either the current would reduce as the pd across the sample were increased, or the current would flow in the opposite direction to the applied pd (the two alternatives can arise from different interpretations of the term 'resistance'); such behaviour is not encountered in A-level physics.

(e) It is left to the reader to find out how this may be done; but even in a 'brief' description the difficulty of obtaining a copper specimen of large enough resistance must be identified. You are expected to comment that the rise of resistivity of the copper suggests that its temperature is rising during the experiment, so the student is probably using a large enough current to heat up his copper specimen.

PLANNING LONG ANSWERS

Many long-answer questions are structured (ie, in several parts) and will need little or no planning, but others may require you to produce an essay style answer that will need a plan. Some of the questions of this nature in the question practice have commentaries designed to assist you with your planning. However it is worthwhile to outline the method of producing a plan and illustrate it with an example.

After reading the question make the plan as follows:

1 Jot down broad headings for the main topic areas of the question.

2 Under each of these headings list the points to be made (eg, facts, formulae, definitions, units, key experiments).

3 Arrange your information so that your answer can be written in a well organized, concise and coherent manner. Make sure that you are answering the 'actual' question, as opposed to the one you had 'hoped for'.

Some long answers involve a wide range of topics that need to be linked together. It can be a great help to put your notes in diagramatic form with such links (eg, the revision card for resistance on p. 6); similarly it is possible to plan long answers in a similar fashion as shown in the following example.

Question

Describe how you would explain electromagnetic induction to a friend studying A-level physics who had missed the teaching of this particular subject.

Plan

(First, note the topics to be explained): the laws, self and mutual induction, transformers.

(Suitable brief notes under these headings can be made as described in 2 above; alternatively, the plan can be constructed in diagram form as shown. More ideas can be added by including extra branches; it is easy to write a logically sequenced answer from such a plan.)

Remember that your examiner will have to read through many scripts. If you give a poorly planned, rambling answer, any valuable points may prove difficult for the examiner to find; a well structured answer will not present such problems. Avoid the tragic mistake of giving several pages of largely irrelevant information which will demonstrate your knowledge, waste your time, and score no marks. The sole question for entry to The Imperial Chinese Civil Service was always 'Write down all you know'; too many A-level candidates seem determined to answer this question. Sadly marks can even be lost by the candidate who spoils a good answer with extra 'padding'; try and train yourself to be relevant in answering all questions and you will not let yourself down.

Some long answers will require descriptions of experiments. Here you should plan to describe the experiment as you would perform it. A well labelled diagram of the apparatus will usually head the answer (no need to use vast volumes of words to describe the apparatus). Next you describe what measurements you take and how they are taken (eg, naming important precautions or procedural points), a table with headings can show how the results can be expressed. Finally demonstrate how the results are processed; note that sketch graphs are quick to produce and can secure valuable marks.

Some modern A-level questions provide considerable amounts of data, and the candidate has to choose which items of information are appropriate to the particular question that is being answered (83L is a good example). If the data are scattered throughout the question (and it is sometimes to be found on the front of the exam paper as well, or on a formula sheet), then try and write down a list of all values and equations you may need before you answer the question. As you do each part, consult the 'complete' list; this should enable you to answer the question with greatest accuracy and in the least time. Sometimes this technique can help you to unravel a comprehension paper as well, especially if there is much information hidden in the passage rather than being supplied in tabulated form.

NUMERICAL PROBLEMS: CALCULATIONS AND ESTIMATIONS

In calculations you are asked to find a numerical value based on data that will be supplied in the question or on the front of the exam paper; in an estimation you have to supply the values from which the estimation will be made using your own knowledge of physical constants and values. Apart from the way in which the data for the question is obtained, the method of procedure for both styles of question is the same. Estimations tend to be more difficult because in a calculation the data supplied can provide good clues as to which equations and techniques are needed to produce the answer; however in an estimation the candidate has no such data and must think of a technique to carry out the estimation without it. Once a technique has been established, the data must also be estimated. Students learning to estimate will find the table of constants and numerical values (pp. 186-89) very helpful.

In a calculation, start by summarizing the data, using the accepted symbols and converting each quantity into its correct units; sometimes a diagram of the situation will be helpful. From this written information you should be able to spot the principles and equations that relate to the situation; there are marks to be gained from simply noting the right ideas and equations, even if you don't know how to use them. See whether you need to make sensible assumptions or approximations, indicate clearly where you have done this in your answer. Each line of your answer should be a complete sentence, though it may be in the form of mathematical symbols or an equation; this makes it easy to

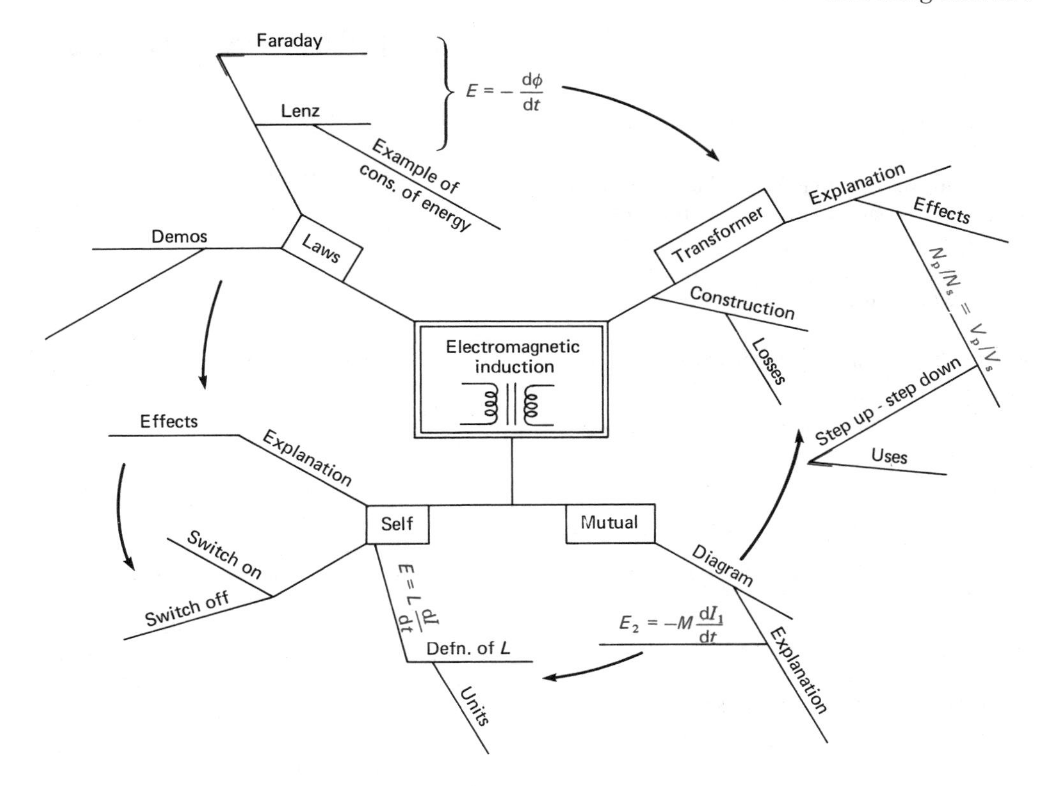

follow your line of argument and spot any mistakes (considerable credit is given for the right method). Your final answer should be quoted to the same number of significant figures as the data, together with suitable units. Check that your answer is sensible, if you have time then look back for silly mistakes; you must always comment on a wrong answer, even if you cannot identify your error.

When estimating, it is first necessary to decide upon a method of producing an answer. Physical principles, laws and equations must be considered. There are often several methods of carrying out an estimate; spotting the shortest and simplest one is skill that only comes from experience. Some students read the estimating questions early in the examination, but do not tackle them until the end. Often a good idea arises during the course of doing another question, as your subconscious comes to your rescue! Once the 'method' is found, it is necessary to estimate the quantities that will be needed to carry out the method, these quantities are normally the first material to appear in your written answer. They should be quoted to a sensible degree of accuracy (usually either one significant figure or the nearest order of magnitude). Once the quantities to be used have been written down, proceed as with a normal calculation; take particular care to quote your answer to the right degree of accuracy (usually one significant figure or the nearest power of ten). If you know what the answer to your estimate should be, then this will occasionally help you to spot mistakes or obtain values (for data quantities) of which your are not sure!

Two further points about calculating and estimating are worth making. You should work with symbols as far as possible; only put in numbers when you have an equation linking the quantity you require to those that you have available as data. Secondly, try not to leave your answer in incomplete numerical form (eg, $2\pi g$) unless you are running short of time. A sample calculation can illustrate most of the techniques needed to answer these questions.

Question

A standard resistor is made from wire having resistivity of $2.5 \times 10^{-7}\,\Omega$ m. In order not to overheat it, the current density (ie, the current per unit cross-sectional area of the wire) should not exceed 15 mA mm^{-2}. What is the maximum potential gradient that may be safely applied to the wire of the resistor?

Answer

(Summarize the information with correct units and usual symbols)

$\rho = 2.5 \times 10^{-7}\,\Omega$ m

$I/A = 15\text{mA mm}^{-2} = 15 \times 10^{-3}\text{ A mm}^{-2}$

$= 15 \times 10^{-3} \times 10^{6}\text{ A m}^{-2}$

$= 15 \times 10^{3}\text{ A m}^{-2}$ = maximum current density

(Write down any relevant equations)

$R = \rho L/A$

$R = V/I$

(Potential gradient is a term you may not recognise. However you could well imagine that as temperature gradient means temperature per unit length, potential gradient means pd/per unit length or V/L. You can certainly work this out from the information in the question, and indeed it is the right answer. So continue...)

$V/I = \rho L/A$

so $V/L = \rho I/A = 2.5 \times 10^{-7} . 15 \times 10^{3} = 3.75 \times 10^{-3}$

(Remembering the units and number of significant figures)

Maximum safe potential gradient $= 3.8 \times 10^{-3}\text{ V m}^{-1}$

AVOIDING HOWLERS

A howler is a mistake so stupendous that it makes an answer nonsensical or comic. It can result from injudicious use of a long word (describe something in simple words if you are unsure of the proper technical term), from simple spelling mistakes (eg, metre for meter), from outrageous misunderstanding of basic concepts (eg, currents across devices, or voltages passing through components), or from failure to read through work (eg, the average speed of a London Underground station is 40 mph). Use your marked work to try and analyse your mistakes and keep a checklist of your faults for reading shortly before examinations. You can also learn from the howlers of others, so some are supplied here in the hope that you will not perpetrate them!

A student was asked to design a 'bobbin' for a voltmeter that would have a pd of 990 mV across it at fsd of the associated meter. He wrote '... this robin would have to absorb 990 V...'. (So that's why they have red breasts?)

'... air cannot be compressed. In a compressed air pump...'.

'I would use a square piece of wood'. (A piece of wood may have a square cross-section but it is a 3-dimensional object.)

'The voltmeter jumped'.

'I estimate Young's modulus for wood to be about 10 N m^{-2}'. (As the Young's modulus for metal is about 10^{11} N m^{-2}, this candidate suggests that wood is ten thousand million times easier to 'stretch' then metal!)

'From the oscilloscope the wavelength is 3 cm'. (The *X*-axis of the oscilloscope was displaying time as usual, and the candidate wanted to measure the period.)

'Energy was lost'. (In principle energy is conserved, but it can be converted from one type to another; ie, it is permissible to say ke was lost.)

'Compressing the two metals is like stretching them'.

'The thermometer rises very quickly'.

'... the kinetic energy of the trampoline ...'.

'... a gymnast weighing 350 kg ...'.

'I would connect the dc electrometer to a flame thrower'.

'... and the line shows that the stress–strain curve obeys Ohm's law ...'.

A favourite method of measuring small distances is to use '... an accurate ruler ...'.

'You can use a spring balance to measure the number of neutrons'. (newtons!)

'The wavelength has to be in whole wavelengths'.

'The numbers 44°, 50° and 74° represent the time that ...'.

'The number of ions in the tank is 3.226×10^{23}, but you cannot have 0.226 of an ion so the answer is 3.0×10^{23}'.

'As temperature drops time increases'. (Presumably as temperature rises then time decreases?)

'Box B contains the capacitor'. and two lines later 'Box B cannot contain the capacitor because ...'.

'The line should be straight with a very slight curve'.

A typical example of the English that drives examiners into despair: 'The temperature doesn't uniformally also with the room temperature so high the temperature doesn't get a chance to drop as far in each interval as one would expect it'.

'Monochromatic white light'.

'Radioactivity can be used as a fertilizer'. (Sterilizer!)

'In a solid the room temperature will move really slow'.

'The equation is only valid if ω^2 is negative'. (Hence turning shm into a complex motion!)

'X-rays and thermionic emission are radioactive effects'. (Radiation effects).

'The structure is investigated by causing X-rays to be reflected off a powered piece of copper'. (Powdered!)

How to ruin a reasonable answer with a typical waffle finish: 'This theory proposes that the momentum of, for instance, a moving truck is equal to that of its new momentum plus that given to a truck it collides with; after allowing for friction and energy loss due to sound, heat, etc.'.

'Put a short across the meter with a resistance of 9900 ohms'.

'The voltage would run through the metre'.

'Copper consists of many atoms which are not very free to move, so it is a much easier job to direct X-rays at them'.

The student was asked to talk about the **disadvantages** of a certain type of ammeter: 'You cannot put a very large current into the meter or the cut-out switch will go'.

'The frequency of vibration is equal to the wavelength'.

'... so the pressure is 100 N m ...'. (This can often prompt the comments from the marker – 'units, you nit!')

'I estimate that the time taken to burn the candle completely would be 5 hours 27 mins 15.673 seconds – approximately'.

'The stopwatch is used to measure the temperature drop of the ice–water mixture accurately'.

Do's and don'ts for the examination

Most of the advice in this section concentrates on information important to you on the actual day of the examination. Some of the key points mentioned elsewhere in the text are repeated to reinforce them, but there are many new hints. You should add to the list anything that is important to your own preparation that might have been omitted.

BEFORE THE EXAMINATION

Don't enter the exam room without a plan for tackling the specific examination paper that will enable you to avoid time trouble or choosing the wrong questions.

Do familiarize yourself thoroughly with any formula sheet that you might be permitted to use.

Do ensure that your calculator batteries will not run out during the exam, that you have spare pens and sharpened pencils (in a range of colours but not red), and a ruler and rubber. Some Boards ban the use of correction fluid; check on this.

Don't enter the exam room at the last possible moment. Early candidates may be able to choose their seats, and there is much to do before the examination starts.

Do have a supply of chewing-gum (mints, etc.) if these aid your concentration and are permitted in your examination room.

Don't sit an examination if you are unwell without informing the invigilator **before** the start of the paper.

Don't drink (alcohol) before an exam, or eat too much or too little.

Do use the 'waiting' time sensibly before the 'official start'. You can write your name and examination details onto all the answer books/paper. You can study the front of the exam paper for the expected rubric and any constants and data; recall important equations and principles involving the information on the front of the paper. In a practical examination you can study the apparatus carefully.

Don't sit an examination without a convenient means of keeping track of the time that is elapsing.

DURING THE EXAMINATION

Do read all the questions. Many students ignore a question after starting to read it, only to realize that had they read it through to the end it would have proved far easier than they had imagined.

Don't spend too much time on one question. It is essential to develop techniques for doing this to avoid panic setting in.

Do call the invigilator **immediately** you have any problem, the worst that he/she can do is to refuse to help you. This is particularly vital if you are threatened with a 'call of nature'.

Don't be too fussy about the neatness of your diagrams. Make sure that they are well labelled and large (large 'freehand' diagrams are much easier to follow or correct than small ones and the Exam Board is paying for the paper!)

Do leave plenty of space to finish questions that you leave when you run into difficulties.

Don't forget your units; it is best to convert all quantities to SI units before you use them.

Do write in short sentences, avoid too many 'ands' and 'buts' in mid-sentence. It is not easy for both you and the examiner to follow an argument written in long sentences.

Don't dive into questions without planning or reading the entire question. It can frequently lead to disasters.

Don't spend time reading answers or checking for mistakes until you have completed the required number of questions.

Do comment on any answer that you suspect to be wrong.

Do try to teach yourself to enjoy the challenge of the examination. There is much skill involved in doing yourself justice; remember that you can always sit it again if things go wrong. Examinations are important, but they are not 'life and death' matters.

AFTER THE EXAMINATION

Don't fret over the many 'obvious' mistakes you will have made and should have avoided. All the other candidates will have made similar errors, and there is nothing you can do to retrieve the situation.

Do forget the paper and start to prepare for the next one. Think positively; only carry out an inquest on a paper you have just sat if similar material may appear on a paper that you are still to sit (ie, you can profit from learning about your mistakes).

Do remember that if you have found a paper hard, then in probability so have the rest of the candidates (whether they are ready to admit it or not); this could simply mean that the 'pass-mark' on this paper could be correspondingly lower.

Don't get overconfident if you found a paper easy (for the opposite reasons to the previous piece of advice).

Do take great care to make sure that your script is securely tied before you hand it in and that all your sheets/booklets bear your name and all the relevant information.

Part IV Test yourself

Multiple-choice questions

There are 50 questions, if they are to be done at once as a test then the time allowance should be $1\frac{1}{2}$ hours.

1M A constant pd is maintained across a piece of intrinsic semiconductor. When the semiconductor is heated, the current through the semiconductor increases because:

A the atoms vibrate more.
B the conduction electrons move faster.
C the number of conduction electrons increases.
D the cross-sectional area of the conductor increases.
E the electric field across the semiconductor increases.

(—: *NISEC and all other Boards except Oxford, SEB H and SEB SYS*)

Electrical conduction arises from the non-random drift of charge carriers under the influence of an electric field. Remember that atoms cannot act as charge carriers in a solid, and that an increase of cross-sectional area would cause an increase in current, but its magnitude would be several orders too small.

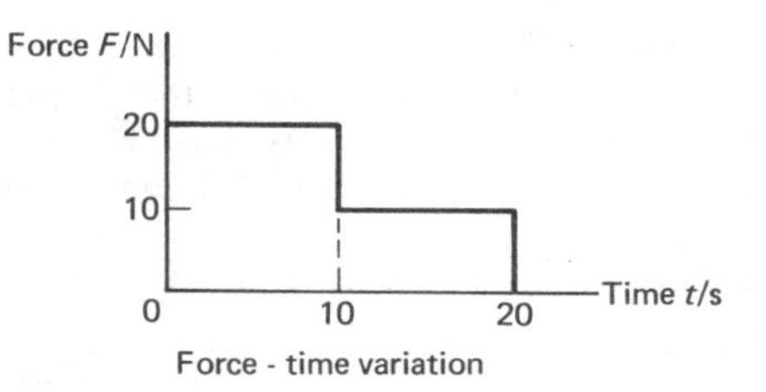

Force - time variation

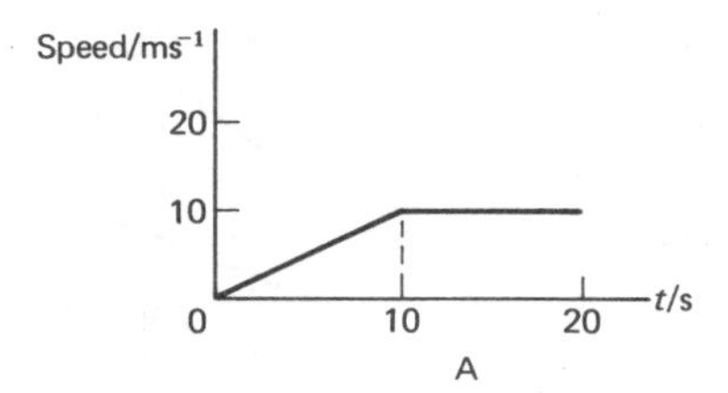

A

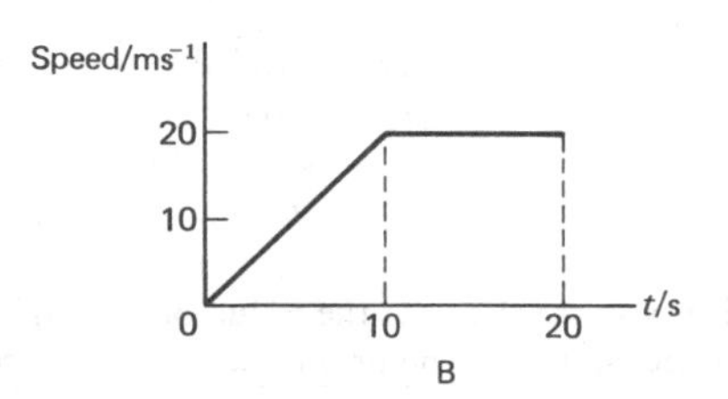

B

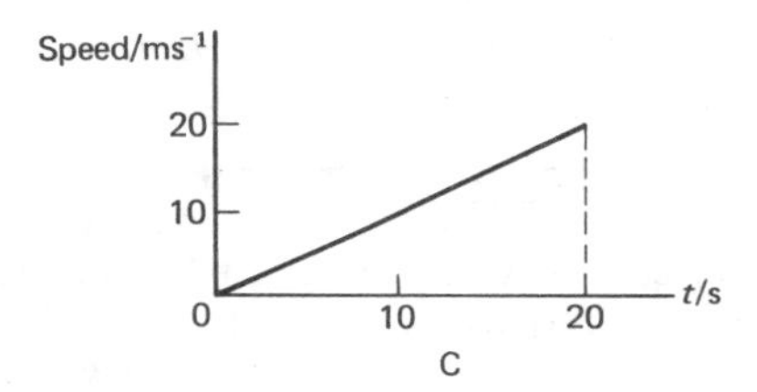

C

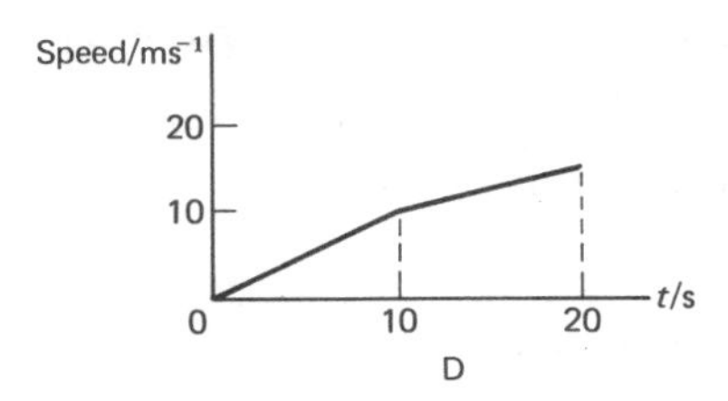

D

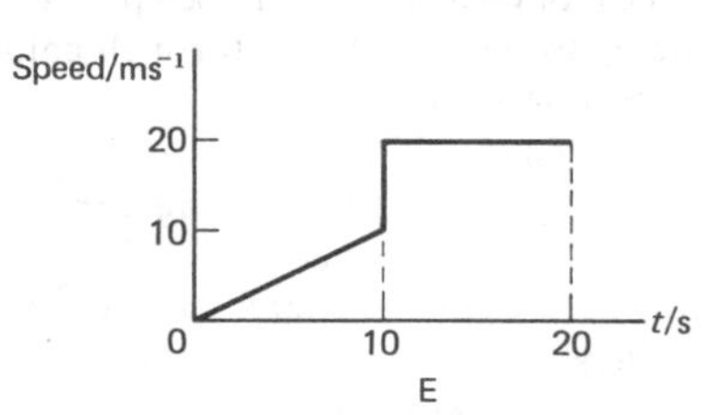

E

2M The resultant force (F) acting upon a 20 kg mass varies with time (t) as shown by the diagram above. At $t = 0$, the speed of the mass is zero. Which of the graphs A–E shows the variation of speed with time?

(—: *all boards*)

You must consider the motion in two separate parts. Use E1.5 to calculate the speed at the end of the first part of the motion. Since the speed at the end of the first part is the same as the speed at the start of the second part, you can use E1.5 again (with new force and initial speed values) to determine the speed at the end of the second part.

3M The extension x for various applied loads F is plotted for two similar wires having different values of the Young's modulus. Which of the following statements is/are true?

1 Specimen Q has a larger value of Young's modulus than specimen P.
2 For the same extension the energy stored within specimen Q is greater than in specimen P.
3 The two graphs can be made to coincide by decreasing the initial length of specimen P.

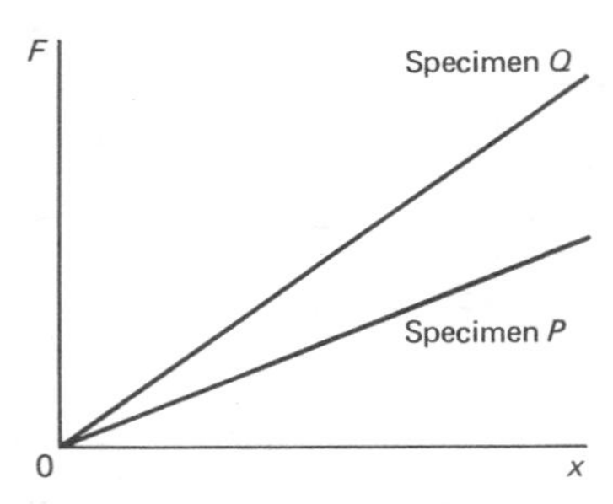

A 1, 2, 3 all correct B 1, 2 only correct
C 2, 3 only correct D 1 only correct
E 3 only correct

(**AEB** Nov 81: *NISEC and all other Boards except SEB H, SEB SYS*)

For 1 and 3, you must relate the value of Young's modulus to the gradient (ie, slope) of the graph; see E5.1 if necessary.

For 2, remember that the area under the graph gives the energy stored.

4M There is a pd of 6 V between the ends L and M of the conductor shown. Then:

L [conductor] M

1 if there is a current of 2 A between L and M, 3 J of energy will be dissipated in each second as heat in the conductor.
2 electrons in drifting from L to M lose, on average, 6 eV of energy each.
3 In order that 1 C of charge can flow between L and M, the conductor must have been supplied with 6 J of energy.

Answer: A if **1**, **2**, **3** correct.
B if **1**, **2** correct.
C if **2**, **3** correct.
D if **1** only.
E if **3** only.

(**London**: *and all other Boards except SEB SYS*)

See E7.4 and its explanation on p. 101, for 1,

For 2, how is 1 eV of energy defined? See p. 124 if necessary. Electrons lose energy to the lattice ions through collisions.

For 3, see p. 101 if necessary.

5M In Young's double-slit experiment, both slits are illuminated by the same light source to ensure that the light from the slits is:

A polarized. B coherent.
C diffracted. D of equal intensity.
E of equal frequency.

(**AEB** June 80: *and all other Boards*)

See p. 63.

6M When a two-slit arrangement was set up to produce interference fringes on a screen using a monochromatic source of green light, the fringes were found to be too close together for convenient observation. It would be possible to increase the separation of the fringes by:

A decreasing the distance between the slits and screen.
B increasing the distance between the source and slits.
C increasing the distance between the two slits.
D increasing the width of each slit.
E replacing the light source with a monochromatic source of red light.

(**London**: *and all other Boards*)

Use E4.3 and remember red light has a longer wavelength than green.

7M Consider the arrangement shown in the diagram. At a suitable ac frequency, the wire vibrates as shown (its fundamental mode). If the tension in the wire is now increased considerably,

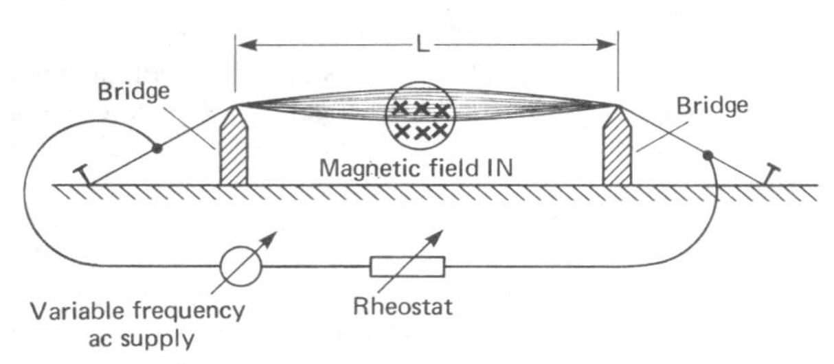

which of the following steps will bring the wire back into vibration in its fundamental mode?

1 Increase the frequency.
2 Increase the current.
3 Increase the length L of the wire between the two 'bridges'.

A 1 only. B 1 and 2. C 1 and 3. D 2 and 3.
E 3 only.

(—: *all Boards except SEB H*)

When the wire vibrates in its fundamental mode, the 'applied' frequency is equal to the natural frequency of vibration of the wire. The equation for the natural frequency can be obtained by combining E3.7 and E3.13, and by making use of the fact that the wire length L is one-half wavelength of the standing wave pattern. Thus, the natural frequency of fundamental vibrations is given by $f=\frac{1}{2L}\sqrt{\frac{T}{\mu}}$. Use this formula when you consider each statement.

8M The graph shows the distribution of kinetic energies E_k among the constituent molecules of a gas at a uniform temperature. (N is the number of molecules each having energy in a small

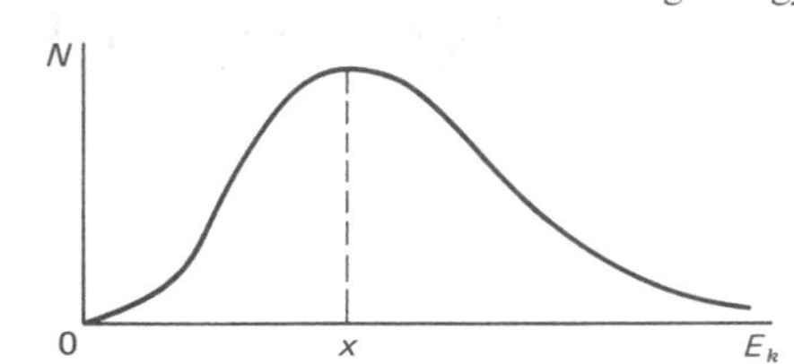

energy band around E_k). Which of the following statements is true?

A provided that the temperature does not change, the ke of each molecule is fixed.
B The commonest value of ke is also the greatest ke of any of the molecules.
C The total ke of the molecules is independent of the temperature of the gas.
D The value, x of E_k at which the peak of the curve occurs increases when the temperature rises.
E The value, x, of E_k at which the peak of the curve occurs is termed the binding energy.

(**London**: *and JMB, AEB and Cambridge only*)

For A, remember that molecules are continually in collision with one another.

For B, look at the above graph carefully.

For C, see p. 92 if necessary.

For D remember that the rms speed increases with temperature increase.

For E, see p. 130 for the term 'binding energy'.

9M In an experiment to measure the energy stored in a spring ($\frac{1}{2}kx^2$) the following measurements were made. A value for the likely error in making these measurements is also quoted: k = spring constant = $(100 \pm 5)\,\mathrm{N\,m^{-1}}$, x = extension of spring = (90 ± 3) mm. On the basis of these results the following statements are made.

1 The extension of the spring has been measured more accurately than the spring constant.
2 There is an error in the value obtained for the energy stored in the spring; the error involved in measuring the spring constant will have a greater effect on the value for the energy stored than the error involved in measuring the extension.
3 The errors quoted for k and x above are 'absolute' errors.
Which of the above statements are true:
A **1** only. B **2** only. C **1** and **3** only.
D **2** and **3** only. E All of them.

(—: *all Boards except O and C Nuffield and SEB H*)

Help in solving this can be obtained from 9.3. Remember that the 'more' accurate measurement can be evaluated by comparing the 'percentage' errors involved in making the measurements. Beware of 'squaring' the error in the extension when considering the second statement.

10M Two small objects X and Y, of mass M and $2M$, respectively, are released from rest at heights of 10 m for X and 5 m for Y above level ground. Which of the following properties is the same for X and Y? (Ignore effects of air resistance.)
A Impact speed. D Accelerating force just before impact.
B Impact ke. E Momentum just before impact.
C The time taken to fall to the ground. (—: *all Boards*)

Use $v^2 = u^2 + 2as$ to calculate impact speeds, and follow up with impact momentum (ie, just before impact) if necessary. You can use your speed values to determine ke values if necessary, or use ke gain = pe lost. Remember that all freely falling objects accelerate at the same rate in a uniform gravitational field.

11M A rigid body is in equilibrium due to forces of 2 N, 3 N and 4 N acting upon it. If the 2 N force is suddenly removed, the resultant force at the instant of removal is, in N:
A 7. B 5. C 4. D 3. E 2.

(—: *all Boards*)

Remember that the resultant of any two of the forces is balanced out by the third force when the body is in equilibrium.

12M A charge of 2.0 μC is moved through a uniform electric field, of strength 10 V m^{-1}, such that it traverses a path 3.0 m parallel to and then 4.0 m perpendicular to the field lines. The change in μJ in pe of the charge between the end points of the path is:
A 30. B 40. C 60. D 80. E 100.

(**AEB** June 81: *and all other Boards*)

Remember that the force on the charge acts along the field line direction, so when the charge is moved along the field line(s), its pe changes. However, when the charge is moved at right angles to the field line(s), then its pe remains the same. Calculate pd through which charge is moved (E2.9) and hence the pe (2.2).

13M The graph shows the typical variation of the force, F, between two molecules of a given substance as a function of their separation, r. Which of the following features of the graph is **most closely** related to the specific latent heat of sublimation (ie, direct change from solid to gas) of a solid composed of the molecules concerned?

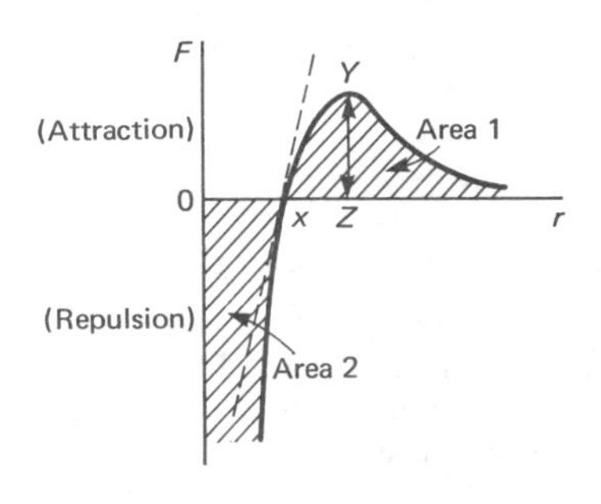

A The area 1.
B (The area 2) − (the area 1).
C The distance OX.
D The force YZ.
E The slope of the graph at X.

(**London**: *Cambridge, WJEC*)

Refer to F2.5(a) and the discussion that follows, remembering that work done is equal to the area under a force – distance curve. (Beware! The question has a non-conventional positive F for attraction).

14M A beam of monochromatic light of wavelength λ falls normally on a diffraction grating of line spacing d. The angle θ between the 'second'-order diffracted beam and the direction of the incident light is given by:
A $\sin\theta = \lambda/d$ B $\sin\theta = d/\lambda$ C $\sin\theta = 2\lambda/d$
D $\sin\theta = 2d/\lambda$ E $\sin\theta = d/2\lambda$

(**NISEC**: *other Boards*)

See F4.6 and E4.8. Note that the question refers to the 'second'-order beam.

15M The meter in the circuit shown has an uncalibrated linear scale. With the circuit as shown, the scale reading is 20. When another 2000 ohm resistor is connected across XY, the scale reading is:
A 10. B 16. C 25. D 28. E 40.

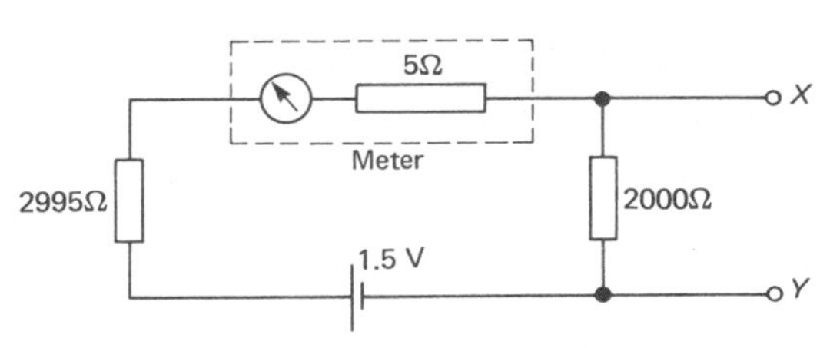

(**Cambridge**: *all other Boards*)

Without the extra resistor, calculate the total resistance of the circuit and then determine the current through the meter. Remember that this current gives a deflection of 20 units.

With the extra resistor, calculate the total resistance of the new circuit and then determine the new current, Assuming that the current is proportional to the deflection, calculate the new deflection.

16M Two pupils are asked to find the value of an unknown resistance. They each use the same equipment – a dc supply of negligible internal resistance, an ammeter and a voltmeter. Jane

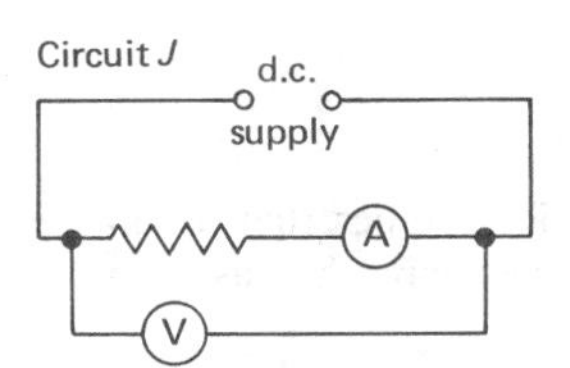

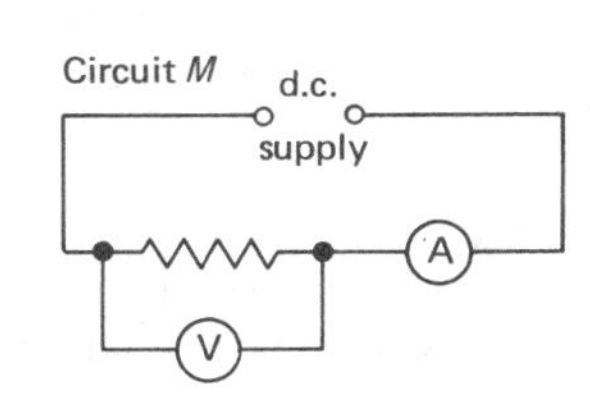

uses circuit J and Margaret uses circuit M. Jane obtained the values: reading on V = 5 volts; reading on A = 1 mA. Margaret's results would be:

	Reading on V (volts)	Reading on A (mA)
A	more than 5	more than 1
B	more than 5	less than 1
C	5	1
D	less than 5	more than 1
E	less than 5	less than 1

(**SEB H**: *and all other Boards*)

In circuit J, the ammeter records the current through R alone, whereas the voltmeter records the pd across $R + A$. In circuit M, the ammeter records the current through R and V, while the voltmeter records the pd across R only. See p. 103 if necessary.

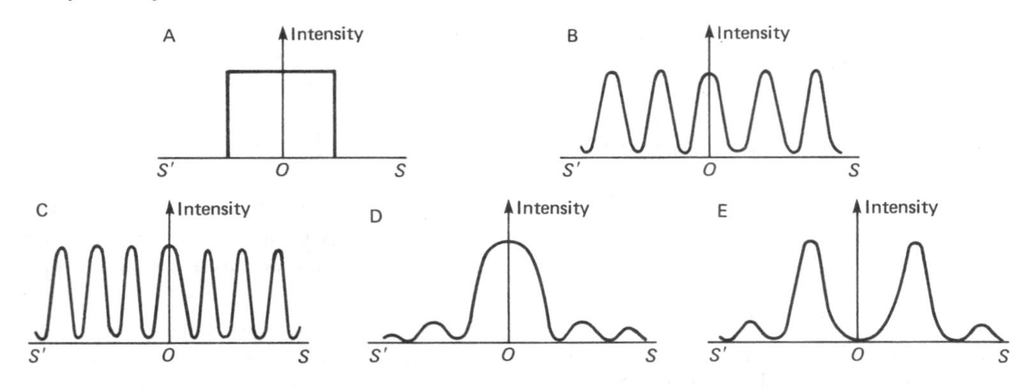

17M The diffraction pattern produced by a single slit may be demonstrated by illuminating the slit with plane waves of monochromatic light and observing the pattern on a screen SS' some

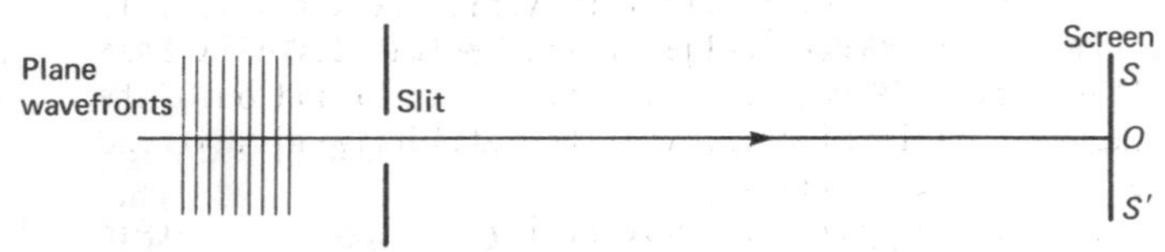

distance from the slit. If the slit is wide compared with the wavelength of the light used, which of the sketches above best represents the pattern seen on the screen?

(**NISEC**: *JMB* and all other Boards except WJEC, SEB H*)

See p. 65 and F4.5 if necessary.

18M The maximum speed of electrons emitted from a given metal surface when illuminated by suitable radiation depends on the:

1 intensity of the radiation.
2 frequency of the radiation.
3 work function of the metal.

Answer:

A if **1**, **2**, **3** correct B if **1**, **2** only. C if **2**, **3** only.
D if **1** only. E if **3** only.

(**London**: *and all other Boards except SEB SYS*)

For **1**, see the comments before E8.5.
For **2** and **3**, consider E8.6.

19M Which one of the following experimental phenomena provides evidence for discrete electron energy levels in atoms?

A the spectrum of a tungsten filament lamp.
B the spectrum of a sodium discharge lamp.
C the photoelectric effect.
D the emission of β-particles by radioactive atoms.
E the emission of γ-rays by radioactive atoms.

(**Cambridge**: *and all other Boards*)

See p. 125 and F8.7 and F8.8. Remember that a tungsten filament lamp gives a continuous spectrum whereas a sodium discharge lamp gives a line spectrum.

20M A spiral spring was hung vertically with one end attached to a fixed support. A mass of 0.20 kg was hung from the other end, and then made to oscillate vertically with an amplitude of 0.04 m. Its time period was measured and found to be 2.0 s. To increase the time period to 4.0 s with the same spring, it is only necessary to increase:

A the mass to 0.80 kg. B the mass to 0.40 kg.
C the amplitude to 0.16 m. D the amplitude to 0.08 m.
E the mass to 0.40 kg and the amplitude to 0.08 m.

See E1.8. (—: *all Boards except SEB H*)

21M A disc is rotating about an axis through its centre and perpendicular to its plane. A point P on the disc is twice as far from the axis as a point Q. At a given instant what is the value of the ratio of

$$\frac{\text{the linear velocity of } P}{\text{the linear velocity of } Q}?$$

A 4. B 2. C 1. D $\frac{1}{2}$. E $\frac{1}{4}$.

(**London**: *all other Boards except SEB H*)

A sketch will help you visualize the situation. Both P and Q will have the same **angular** velocity, ω, but different **linear** velocities. See E1.11 if necessary.

22M A wire that obeys Hooke's law is of length l_1 when it is in equilibrium under a tension F_1. Its length becomes l_2 when the tension is increased to F_2. The energy stored in the wire during this process is:

A $(F_2 - F_1)(l_2 - l_1)$. B $\frac{1}{4}(F_2 + F_1)(l_2 + l_1)$.
C $\frac{1}{4}(F_2 + F_1)(l_2 - l_1)$. D $\frac{1}{2}(F_2 + F_1)(l_2 + l_1)$.
E $\frac{1}{2}(F_2 + F_1)(l_2 - l_1)$.

(**Cambridge**: *and all other Boards except SEB H and SEB SYS*)

Consider the force–length graph for the wire. The work done in stretching the wire is given by the shaded area of

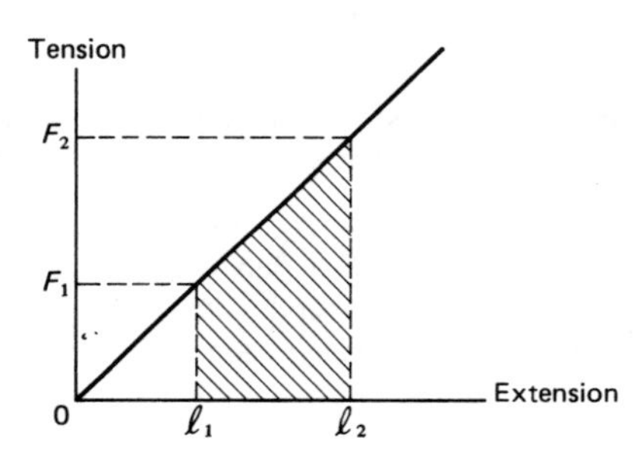

the graph. Clearly, that area depends upon the difference of the lengths, not the sum. Also, the area depends upon the average force.

23M The diagram shows a horizontal plane OXY with axes OX and OY at right angles. Which one of the following directions

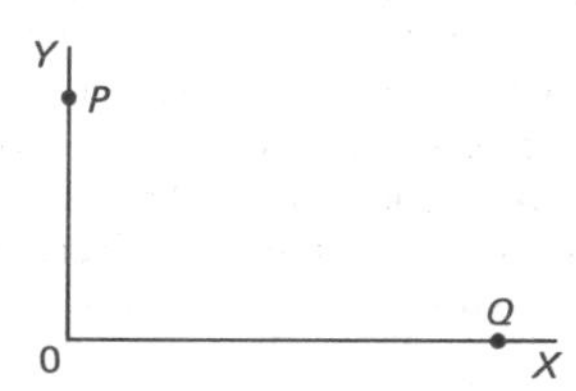

for a current in a straight conductor will produce a magnetic flux density at O in the direction $\overrightarrow{OX}$ (ie, from O towards X)?

A vertically downwards at P.
B vertically upwards at P.
C vertically downwards at Q.
D vertically upwards at Q.
E horizontally above OX in the direction $\overrightarrow{OX}$.

(**London**: *and all other Boards except SEB H*)

Try sketching the magnetic field due to a straight wire at P and Q (see p. 37 if necessary). Use the corkscrew rule to find the direction of the field in each case.

24M Using the potentiometer circuit shown, the balance (null) point is at X. A balance point to the left of X would be obtained by:
1 increasing the resistance of R.
2 increasing the resistance of S.
3 replacing cell P by a cell with a greater emf.

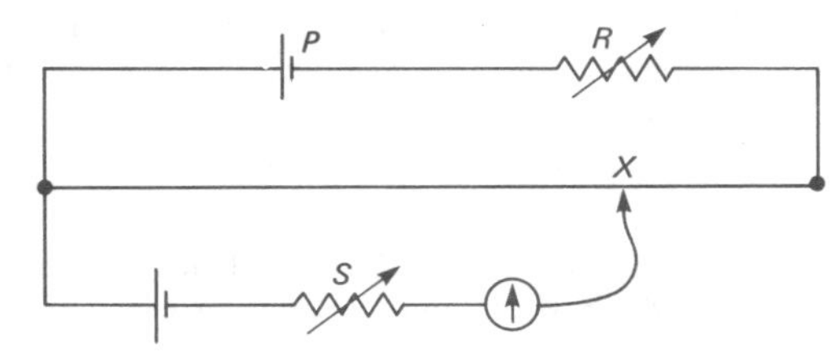

A **1, 2, 3** all correct. B **1, 2** only correct.
C **2, 3** only correct. D **1** only correct.
E **3** only correct.

(**AEB** Nov 80: *and all other Boards except O and C Nuffield, SEB H, SEB SYS*)

For **1** and **3**, consider if each change in turn increases or decreases the pd across the potentiometer wire. Since the test cell emf is unchanged, then is more or less balance length required to give the same pd across the balance length as before?

For **2**, remember that at the balance point, there is no current through the meter. See p. 103 if necessary.

25M The circuit diagram above shows how a potentiometer may be used to measure a small emf produced by a thermocouple connected between X and Y. The cell has a negligible internal resistance and the uniform resistance wire PQ, of resistance

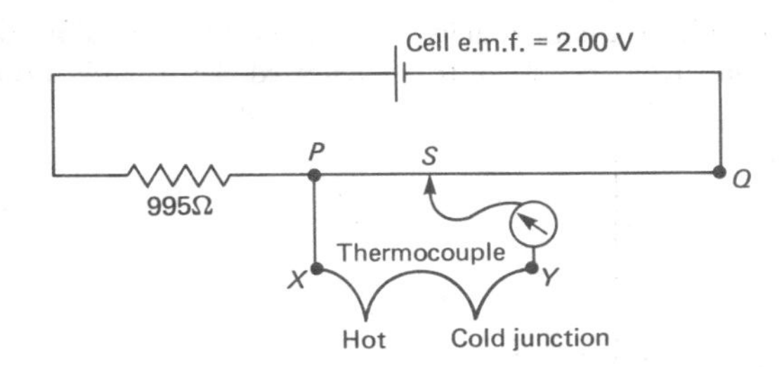

5 ohms, is 1.00 m long. The balance point S is found to be 400 mm from P. The emf, in mV, generated by the thermocouple is:

A 2. B 4. C 10. D 80. E 800.

(**NISEC**: *all other Boards except O and C Nuffield SEB H, SEB SYS and London*)

Concentrate on the main potentiometer circuit, ignoring the thermocouple. The wire PQ is in series with the 995 ohm resistor and the 2.00 V cell, so calculate the current from the cell. At balance, all the current from the cell passes through the wire so your value above is the wire current at balance. Hence calculate the pd across the wire given its resistance is 5 ohms. The pd between P and S is in proportion to the pd across the whole wire PQ, according to the lengths, and at balance the thermocouple emf is equal and 'opposite' to the pd between S and P.

26M Two capacitors of capacitance C and $2C$ are charged to pds V and $2V$, respectively. If the two positive plates are connected together and the two negative plates are connected together then this system of capacitors:
A gains charge but loses energy.
B gains energy but loses charge.
C loses both energy and charge.
D loses charge but not energy.
E loses energy but not charge.

(**London**: *and all other Boards except SEB H*)

See p. 105. As the two capacitors initially have different potentials, there will be a pd when they are connected which will cause a current to flow. This will dissipate energy.

27M A converging lens of focal length f is placed at a distance 0.3 m from an object to produce an image on a screen 0.9 m from the lens. With the object and the screen in the same positions, an image of the object could also be produced on the screen by placing:
A a converging lens of focal length $3f$ at a distance 0.3 m from the screen.
B a converging lens of focal length f at a distance 0.3 m from the screen.
C a converging lens of focal length $3f$ at a distance 0.1 m from the screen.
D a converging lens of focal length f at a distance 0.1 m from the screen.
E a converging lens of focal length $3f$ at a distance 1.1 m from the screen.

(**Cambridge**: *and London, AEB, Oxford, JMB, NISEC, SEB H*)

A methodical approach to answering the question is to calculate f from the given object and image distances, using E4.9. Then work through each alternative in turn to find out if the specified focal length agrees with the stated image distance and calculated object distance.

A quicker method is to realize that if $u = +0.3$ m, $v = +0.9$ m for focal length f then $u = +0.9$ m, $v = +0.3$ m will also give a

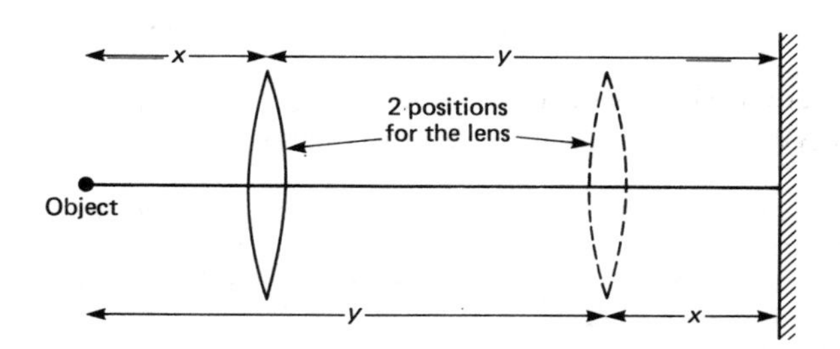

clear image with focal length f (because both sets of figures will fit E4.9). In other words, for a given distance between object and screen, if the lens gives a clear image on the screen then in general there will be one other position of the lens at which a clear image on the screen can be seen; the two lens positions are 'symmetrical' as shown in the diagram.

28M At 0°C the value of the density of a fixed mass of an ideal gas divided by its pressure is x. At 100°C this quotient is:

A $\frac{100}{273}x$ B $\frac{273}{100}x$ C $\frac{273}{373}x$ D $\frac{373}{273}x$ E x

(**NISEC**: *and all other Boards except O and C Nuffield and SEB SYS*)

The quotient is density/pressure. Since density = mass/volume, x can be expressed as $\frac{\text{mass}}{\text{volume} \times \text{pressure}}$. Mass is fixed but pressure times volume is proportional to the absolute temperature; thus the quotient is inversely proportional to the absolute temperature.

29M For the construction of a thermometer, one of the essential requirements is a thermometric substance which:
A remains liquid over the entire range of temperatures to be measured.
B has a property that varies linearly with temperature.
C has a property that varies with temperature.
D obeys Boyle's Law.
E has a constant expansivity.

(**Cambridge**: *all other Boards except SEB SYS, SEB H*)

You cannot assume that any particular type of thermometer is relevant; in other words, alternative A is essential only for liquid-in-glass thermometers but not for any other type. Choose the alternative that applies to all types.

30M Young's modulus of elasticity has dimensions of:

A $[M][L][T]^{-2}$. B $[M][L]^{-1}[T]^{-1}$.
C $[M]^{-1}[L]^{-1}[T]^{-2}$. D $[M]^{1}[L]^{-1}[T]^{-2}$.
E $[M][L]^{2}[T]^{-1}$.

(—: *all Boards except SEB H, SEB SYS*)

EITHER use E5.1 to obtain an expression for Young's modulus in terms of load, extension, etc, and then work out the dimensions of each term and combine them. OR remember that Young's modulus has the same units as stress so work from there!

31M The resistance of a semiconductor device is measured as the pd across the device is varied. The graph shown is obtained.

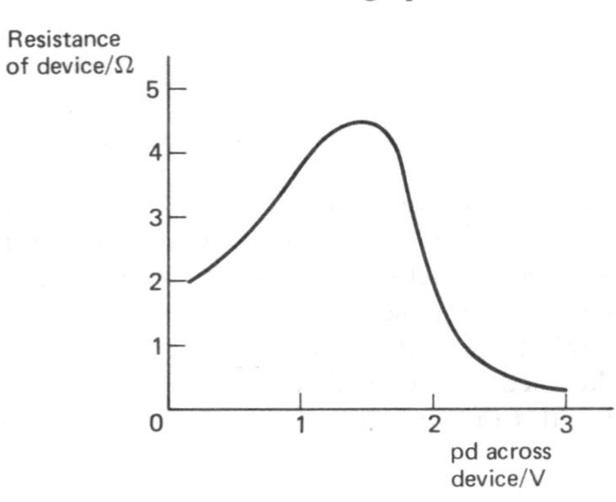

Which of the following deductions (made from the graph) is **incorrect**:

A That the intercept on the vertical axis suggests that the device has a resistance of about 2 ohms when there is zero pd across it.
B That the area underneath the graph has no physical significance.
C That the slope of the graph represents the current flowing through the device.
D That the graph asymptotically approaches the horizontal axis.
E That the change in resistance per volt change in pd is greatest when the pd across the device is between $1\frac{1}{2}$ and 2 volts.

(—: *all Boards except SEB H*)

This question is basically testing your understanding of the 'terminology' of graphs (see 9.2), though an understanding of $V = IR$ is also needed. The vertical intercept occurs when the pd is zero. The units of the area would be those of resistance multiplied by those of pd; we would say that there may be some physical significance if an equation can be recalled multiplying these quantities or if the units represent a 'standard' measurement. Remember that the slope can be interpreted from dividing the vertical axis quantity by that on the horizontal axis. Asymptotes are explained in 9.2 (see p. 140). Answer E refers to the steepest slope on the graph.

32M In the circuit diagram *C* is a 1 μF capacitor holding a

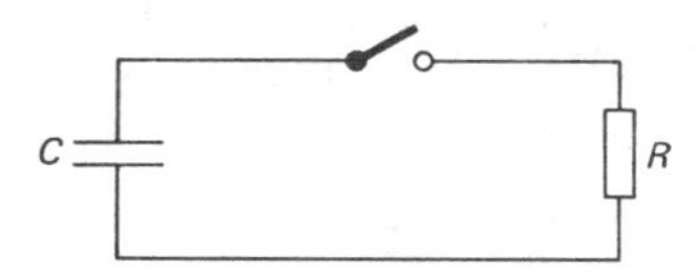

charge of 10^{-5} C and *R* is a 10 Ω resistor. If the switch is suddenly closed, the initial current flowing in the circuit will be:

A zero. B 10^{-5} A. C 10^{-1} A. D 1 A.
E 10 A.

(**London**: *Oxford*, O and C* and all other Boards*)

Calculate the pd across the charged capacitor of *C* from E7.13. Initially, this will also be the pd across *R*. See p. 105 and F7.19 if necessary.

33M When a 240 V rms ac supply is connected to the terminals

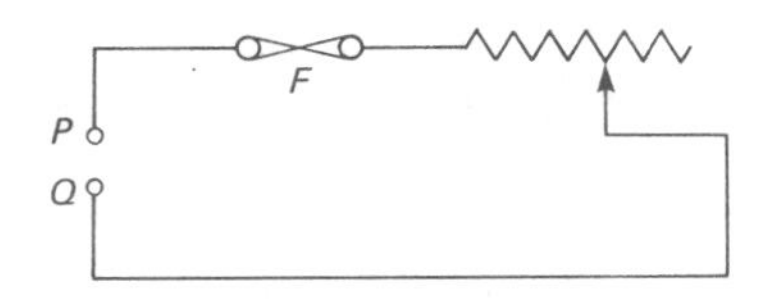

PQ in the circuit above, the fuse *F* breaks the circuit when the current just exceeds 13 A rms. If the supply were replaced with a 120 V dc source, an identical fuse would break the circuit when the current, in amperes, just exceeds:

A 13/2. B $13/\sqrt{2}$. C 13. D $13\sqrt{2}$. E 26.

(**NISEC**: *all other Boards*)

How is rms current defined? Disregard the voltage values, and think about the basic meaning of rms current. See p. 106 if necessary.

34M The temperature of a hot liquid in a container of negligible thermal capacity falls at a rate of 2° C per minute just before it begins to solidify. The temperature then remains steady for 20 min by which time the liquid has all solidified. The quantity $\frac{\text{specific heat capacity of liquid}}{\text{specific latent heat of fusion}}$ is equal to:

A 1/40 K^{-1}. B 1/10 K^{-1}. C 1 K^{-1}. D 10 K^{-1}.
E 40 K^{-1}.

(**Cambridge**: *and London, JMB, AEB Oxford and WJEC*)

Assume mass = *m*, specific heat capacity of liquid = *c* and specific latent heat of fusion = *l*.

Write down an expression for the heat loss per minute of mass *m* of liquid when cooling at 2° C per minute. See E6.3 if necessary. From your expression, write down the heat loss in 20 min of mass *m* of liquid cooling as above.

When the liquid solidifies, assume its rate of loss of heat is the same as just before solidifying started. Then the heat loss is the same as above, and you can equate your expression to *ml*, etc.

35M The diagram shows the line spectrum from a hot gas. Which of the following reasons can account for line *Y* appearing much brighter than lines *X* and *Z*?

Red end of spectrum	X	Y	Z	Violet end of spectrum

A Line *Y* has the biggest wavelength.
B Line *Y* has the biggest frequency.
C Line *Y* originates in the hottest part of the gas.
D Line *Y* is the result of electrons making a much larger energy jump than those responsible for lines *X* and *Z*.
E Line *Y* is the result of more electrons making that particular energy jump than in the other two lines.

(**SEB H**: *and all other Boards except SEB SYS*)

The brightness of an individual spectral line depends upon the number of photons of that particular wavelength emitted. Remember that each emitted photon is produced when an electron makes a particular energy jump.

36M The uranium series of radioactive decays starts with an isotope of uranium, of mass number (atomic mass) 238 and of atomic number 92. What is the mass number and atomic number of the isotope, in the series, reached after a chain of decays involving a total of 3 α-particles and a β^--particle starting from a U-238 nucleus?

	A	B	C	D	E
Mass number (atomic mass)	226	226	230	230	231
Atomic number	85	87	85	87	86

(—: *NISEC and all other Boards*)

See E8.11 and E8.12 for the effects upon the nucleus of α-emission and β-emission.

37M When a spacecraft enters the gravitational field around a planet, which of the following quantities changes with distance from the centre of the planet according to the inverse rule (ie $1/d$):
1 The gravitational field strength acting on the spacecraft.
2 The **total** energy of the spacecraft.
3 The gravitational pe of the spacecraft.

A 1, 2, 3 correct. B 1, 2 correct only.
C 2, 3 correct only. D 1 correct only.
E 3 correct only.

(—: *all Boards*)

For **1**, and **3**, refer to the list of equations on p. 34. For **2**, remember that ke gain is equal to pe loss as the craft approaches the planet; you must consider the total energy (ie, ke + pe).

38M Some astronomers think that there was once a planet in circular orbit about the Sun at a distance of 2.8 × the distance from Earth to Sun. What would have been its time period, T, (ie, time to complete 1 orbit around the Sun) in Earth-Years?
A 0.21. B 1.00. C 1.80. D 2.80. E 4.70.

(—: *all Boards*)

The centripetal force on a planet (mv^2/R) is provided by the gravitational attraction between the planet and the Sun (GMm/R^2). Since the speed of the planet v = circumference ($2\pi R$)/time period (T), then it follows that $T^2 = \frac{4\pi^2}{GM}R^3$. ($M$ = mass of Sun, R = orbit radius of planet).

Since GM is the same for the 'mystery' planet as for the Earth, use the equation in the form $T = \text{constant}R^{3/2}$.

39M In the Hall-effect experiment, a current-carrying sample is placed in a magnetic field, as shown in the diagram. The charge carriers, which cause the current flow, are deflected by the magnetic field and a pd (the Hall voltage) builds up across the sides of the sample. Which of the following statements is/are correct?
1 The magnetic force (Bqv) is balanced out by the electric field force (qV_h/d) once the Hall voltage (V_h) has built up.
2 The Hall voltage is proportional to the magnetic field strength B.

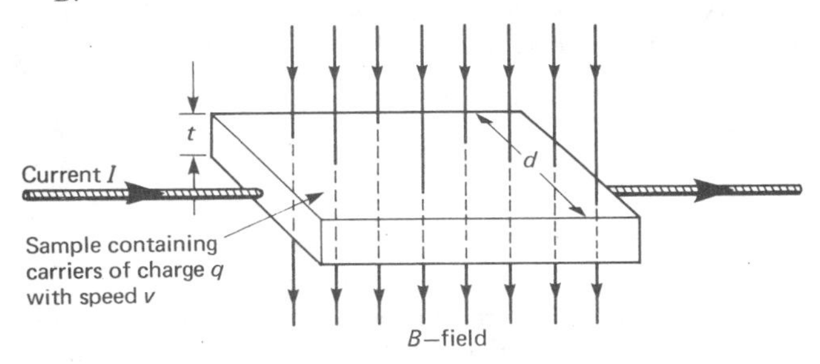

3 The Hall voltage is proportional to the current I.

A 1 only. B 1 and 2. C 1, 2 and 3. D 2 and 3.
E 2 only.

(—:*all Boards except Oxford, SEB H and NISEC*)

See p. 37 and E2.22.

40M Which of the graphs below best represents the current when a battery connected in series with a large self inductor is switched on (at p) and then off (at q)?

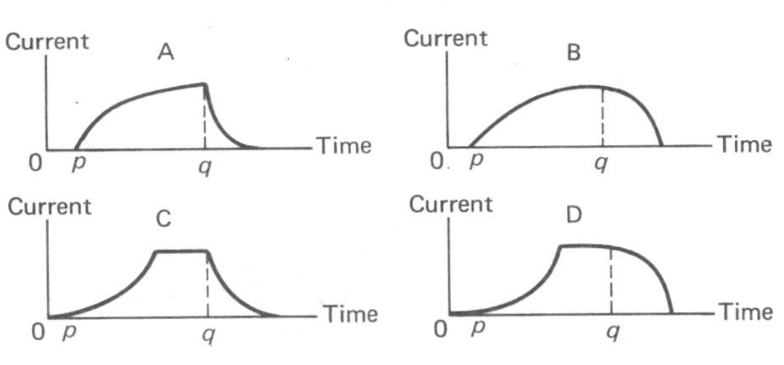

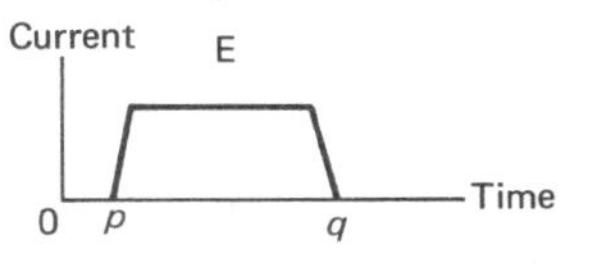

(**London**: *all other Boards*)

Referring to E2.29, the 'back emf' opposing the build-up of current must gradually decrease as more current flows, so from the equation it follows that dI/dt must similarly decrease as the current increases, ie, the slope of the graph must gradually get **less** with time. When the battery is switched off there will be an 'open-circuit' so the current will decrease rapidly, ie, a **large negative** dI/dt (or slope).

41M For the circuit shown in the diagram below, which graph A–E best represents the variation of supply current with time?

(—: *all Boards except O and C Nuffield and SEB SYS*)

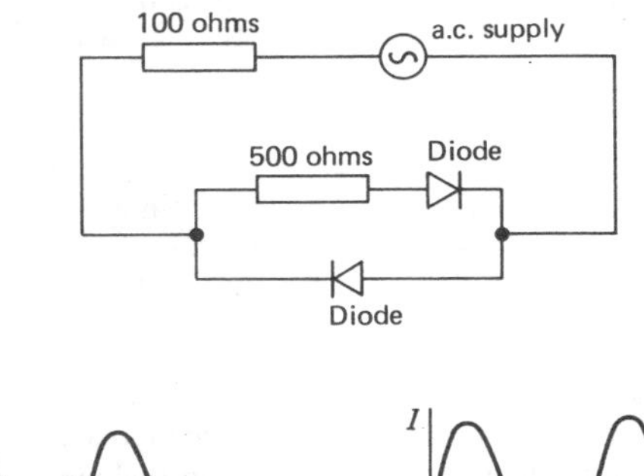

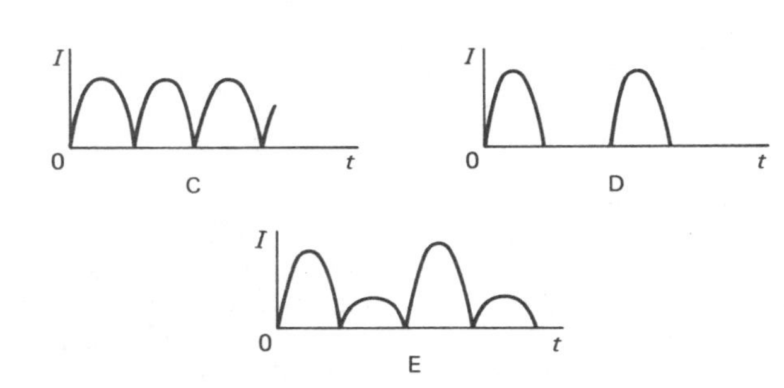

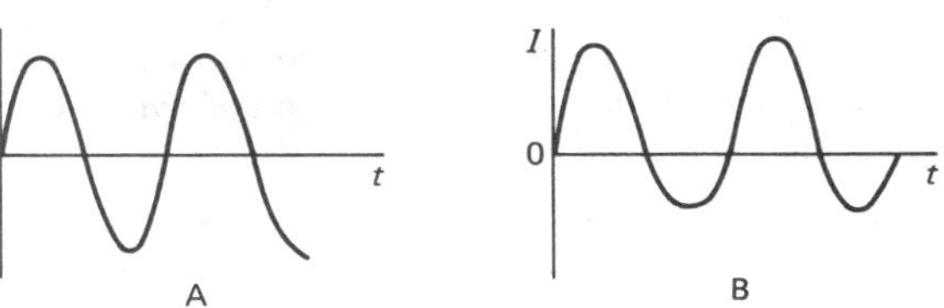

Consider each half-cycle in turn; the circuit has low resistance on one half-cycle and high resistance on the other half-cycle.

42M An ac source supplies an output of constant peak voltage and variable frequency f. Which of the following graphs shows the variation of current in the circuit as the frequency of the ac is increased?

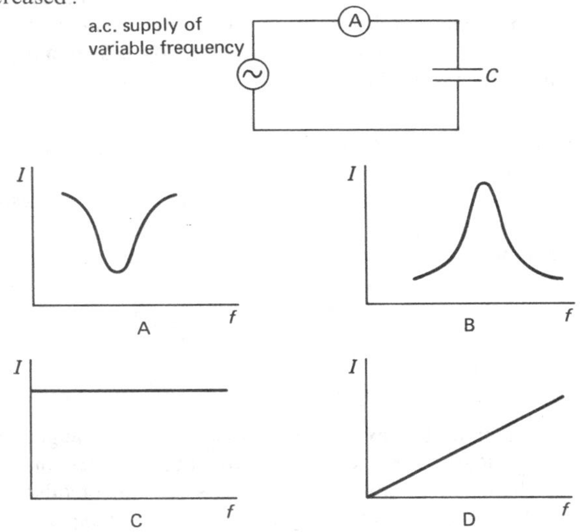

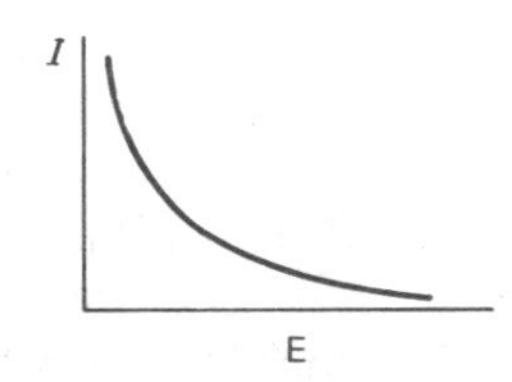

(**SEB H**: *and all other Boards*)

See E7.19 if necessary. Express the peak current I in terms of the peak pd V, the frequency f, and the capacitance C. Given that C and V are constants, choose the graph which best shows your expression for I in terms of f.

Alternatively, remember that $f = 0$ is dc. What is the current for $f = 0$?

Alternatively, remember that the reactance of the capacitor decreases with increased frequency, so the current must increase.

43M The electrons in a cathode ray tube are accelerated from cathode to anode by a pd of 2000 V. If this pd is increased to 8000 V the electrons will arrive at the screen with:
A twice the kinetic energy and four times the velocity.
B four times the kinetic energy and twice the velocity.
C four times the kinetic energy and four times the velocity.
D twice the kinetic energy and twice the velocity.
E twice the kinetic energy and the same velocity.

(**SEB H**: *and all other Boards*)

See p. 123 if necessary. If the accelerating voltage is increased by $\times 4$, then consider first the factor for the increase of ke. Then, remembering that the ke is proportional to (velocity)^2, consider the factor for the increase of velocity.

44M For a compound microscope in normal adjustment:
1 the image formed by the objective must lie at the focal point of the eyepiece.
2 the image formed by the objective is virtual.
3 the final image is 25 cm from a normal eye.
A If **1, 2, 3** correct. B If **1, 2** correct. C If **2, 3** correct.
D If **1** only. E If **3** only.

(—:*JMB, Oxford, Cambridge*)

See p. 68 and F4.20.

45M A radioactive isotope X initially contains 10^{20} atoms of X and none of its daughter product Y, which is stable. Each atom of X that decays releases 8×10^{-13} J of energy. Given that the half-life of X is 4 hours, the total energy released in the first 12 hours is:
A 8×10^7 J. B 4×10^7 J. C 6×10^7 J.
D 7×10^7 J. E 14×10^7 J.

(—: *NISEC and all other Boards except SEB SYS*)

How many atoms of X remain after 12 hours? Hence calculate the number that have decayed after that time has elapsed, and then the total energy released during that time. Watch out for simple arithmetical errors – division of 10^{20} by 2 does **not** give 10^{10} (answer is 5×10^{19} of course!).

46M In the Rutherford scattering experiment a collimated beam of alpha-particles was scattered by a thin gold foil, and the distribution of the scattered particles was studied. Which of the following statements is(are) a **direct deduction** from the results of this experiment?
1 The atoms in the foil are arranged in a regular lattice.
2 Alpha-particles have an associated wavelength.
3 Atoms have nuclei that are much smaller than the atom itself.
A **1** only. B **3** only. C **1** and **3** only.
D **2** and **3** only. E **1, 2** and **3**.

(**NISEC**: *and all other Boards except Oxford and O and C*)

This refers to the experiment carried out by Geiger and Marsden for Rutherford (see p. 126). Note that the question does not ask whether the numbered statements are true or false, but whether they follow directly from the results of the experiment.

47M The diagram shows a metal bar which has been heated and

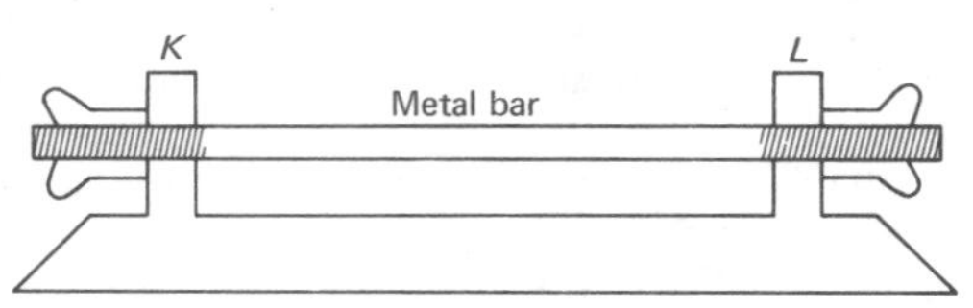

clamped rigidly between two supports to prevent it contracting as it cools. The stress in the bar will depend upon:
1 the linear expansivity of the metal.
2 the Young's modulus for the metal.
3 the distance KL between the supports.
A **1, 2, 3** all correct. B **1, 2** only correct.
C **2, 3** only correct. D **1** only correct.
E **3** only correct.

(**AEB** Nov 81: *AEB only*)

Stress is given by E5.1. Remember that the thermal expansion is equal to the extension when cool. Thus if L_0 is the original length of that portion of the bar that on heating expands to length KL, then the amount of expansion is given by $x = aL_0\Delta\theta$ where a is the coefficient of thermal expansion (see p. 90) and $\Delta\theta$ is the temperature rise. When cooled to the original temperature the strain is $x/L_0 = a\Delta\theta$.

48M The two insulated conductors shown have the same cross-sectional area but Y is twice as long as X, and has twice the thermal conductivity. The junction temperature, in °C, will be:
A 20. B $33\frac{1}{3}$. C $66\frac{2}{3}$. D 25. E 50.

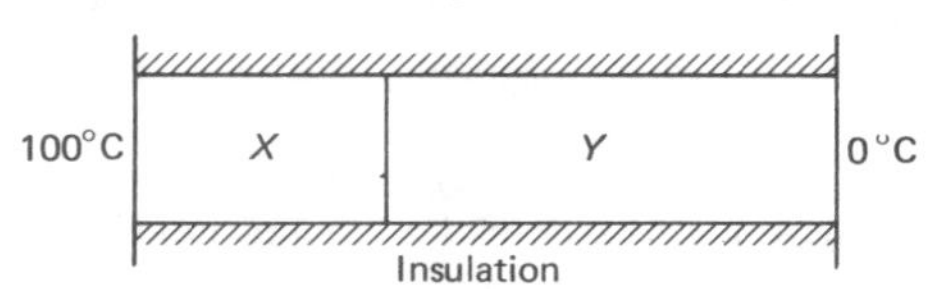

(—: *all Boards except SEB H, SEB SYS*)

Since the heat flow/second through X = heat flow/second through Y, equate $\frac{kA\Delta\theta}{L}$ (see E6.11 for meaning of symbols if necessary) for X to the corresponding expression for Y using an assumed interface temperature θ. Hence calculate θ.

49M A sphere of radius a moving with a velocity v under streamline conditions in a viscous fluid experiences a retarding force given by $F = kav$, where k is a constant. The dimensions of k are:
A $ML^{-2}T^{-1}$. B $ML^{-2}T^{-2}$. C $ML^{-1}T^{-1}$.
D MLT^{-1}. E MLT^{-2}.

(**Cambridge**: *and all other Boards except SEB H*)

Not all exam Boards write dimensions inside square brackets; indeed occasionally square brackets are used to analyse units as well. The method of balancing an equation to obtain dimensions is explained in 9.1 (see p. 138). Hints on obtaining the dimensions of force are given in 9.7M.

50M The temperature difference between a liquid and its surroundings ($\Delta\theta$) is measured as the liquid cools. A mercury-in-glass thermometer is used to measure the temperature of the liquid and its surroundings. A graph of $\ln(\Delta\theta)$ is plotted against time (t). It is expected that it will be a straight line; however two of the experimental points (K and L) fall above the line as shown.

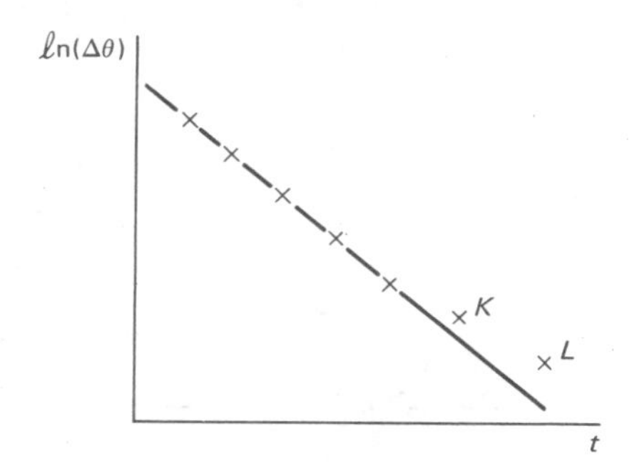

The following suggestions are given to explain the position of points K and L:

1 Some of the liquid was solidifying at both the points K and L.

2 The temperature of the surroundings was lower for the points K and L than for all the other points. (The temperature of the surroundings was measured at the start of the experiment).

3 The thermometer used is in error at 'high' temperatures.

Which of the above suggestions could be correct?

A **1** only. B **2** only. C **3** only. D **1** and **2** only.
E **1** and **3** only.

(—: *all other Boards except SEB H*)

Remember that if the liquid is solidifying, the temperature of the solid/liquid mixture will stay constant until there is no liquid left. Is this indicated by the points K and L? If the temperature of the surroundings dropped would this increase or decrease the temperature difference of a cooling substance? Is this what the graph shows? Do the points K and L represent high or low temperatures of the liquid? Correct answers to these questions should lead to the right choice.

It is important in questions of this nature to use the identification of a right or wrong statement; for example, if 3 is discovered to be true then 2 cannot be true as that combination is not allowed for in the answers. It is frequently possible to identify the correct answer even when you are still unsure as to whether one of the statements is true or false.

Short-answer questions

There are 16 questions arranged in two sets. For each set, you should be able to complete six questions in about 1 hour.

Set 1: 51S – 58S

51S In driving a pile into the ground, a hammer of mass 500 kg falls freely from rest through a height of 5.0 m onto a pile of mass

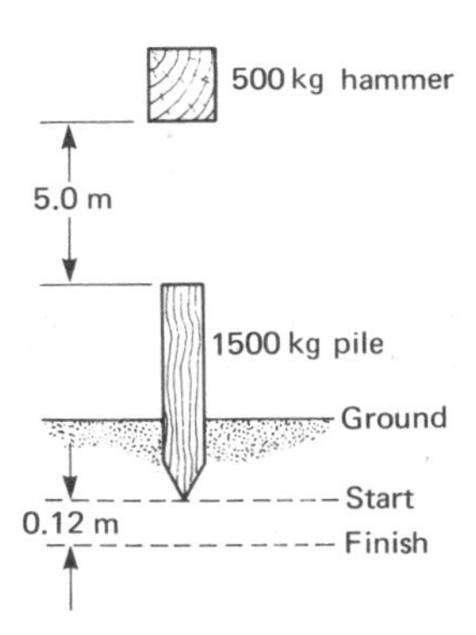

1500 kg. The pile and hammer then move together as the pile is driven 0.12 m into the ground.

(a) Determine the speed of the hammer just before it hits the pile.

(b) (i) Using the principle of conservation of momentum, calculate the common speed of pile and hammer immediately after the collision. (ii) State one assumption which you must make to justify your application of momentum conservation in part (b) (i).

(c) From the moment just after the collision until the system comes to rest, what is the change in: (i) the total kinetic energy of pile and hammer; (ii) the total potential energy of pile and hammer?

(d) By considering these energy changes, or otherwise, calculate a value for the average resistive force which the ground offers to the motion of the pile during its movement into the ground.

(**SEB H**: *and all other Boards*)

(a) Just before impact, all the initial pe is converted into ke. Hence v can be found.

(b) See p. 18 if necessary. Assume that the hammer and pile move together after impact and that there is no 'bounce'; thus, the momentum of the hammer just before impact = momentum of the hammer plus pile just after impact.

(c) Use your value for the common speed to calculate the ke of the system just after impact. Since the final ke is zero, then the change of ke is numerically equal to the ke of the system just after impact. For the change of pe from just after impact to the rest position, remember that the system falls 0.12 m after impact.

(d) Using the relationship Average force × distance moved = change of energy, the average resistive force can be found.

52S A preliminary stage of spacecraft Apollo 11's journey to the moon was to place it in an Earth parking orbit. This orbit was circular, maintaining an almost constant distance of 189 km from the Earth's surface. Assuming the gravitational field strength in this orbit is 9.4 N kg^{-1}, calculate:

(a) the speed of the spacecraft in this orbit; and

(b) the time to complete one orbit.

(Radius of the Earth = 6370 km.)

(**London**: *all other Boards except SEB H*)

(a) The centripetal force (mv^2/r) to keep the satellite in orbit is supplied by the gravitational attraction (mg), giving $g = v^2/r$. (Watch your units!).

(b) Simply use: time = distance round one orbit/speed.

53S

(a) Optical interference can be observed by the *superposition* of light waves from *coherent* sources. Explain the meaning of the words in *italics*.

(b) Describe, with the aid of a diagram, **one** experimental arrangement for producing two coherent light sources, explaining why the sources are coherent.

(**JMB**: *Oxford* and all other Boards except O and C Nuffield and SEB H*)

(a) You should firstly explain the meaning of 'coherent sources'; the key points here relate to the frequency of each source, and to the phase difference between light emitted from each source. See p. 64. Superposition is discussed on p. 53.

(b) See p. 63 for a description of Young's double-slits experiment. The arrangement of apparatus is shown in F4.2. In addition to describing the experimental arrangement, you must explain why the two slits of the double-slits arrangement act as coherent emitters of light waves. See p. 64.

54S A vertical steel wire is kept in tension by a piece of iron attached to one end. The wire is set in transverse vibration and emits a note of frequency 200 Hz. The iron is now completely immersed in water and the frequency of the note changes to 187 Hz. If the density of the water is 1000 kg m^{-3}, calculate the density of the iron.

(**AEB** June 82: *all other Boards except SEB H*)

The formula for the frequency of the fundamental note may be derived from E3.7 and E3.13, assuming the wire length L is equal to one-half of the wavelength; hence establish the link between the tension in the wire T and the frequency f, in the form $T = k f^n$ where k is constant in this situation, and n has a value which you can work out from E3.7 and E3.13. Use the given frequency values to determine the initial and final tensions in terms of k. The initial tension is equal to the weight of the iron, and the difference between the tensions is equal to the upthrust of the water on the iron. Since the upthrust is equal to the weight of water displaced, then the volume of the iron (= volume of water for total immersion) can be calcuated in terms of k. The final step is to calculate the density of the iron from its mass and volume.

55S Describe the path traversed by an electron when it is projected at right angles to **(a)** a uniform magnetic field, **(b)** a uniform electric field.

A uniform electric field is superimposed on a uniform magnetic field of 5.0×10^{-2}T so that the two fields allow an electron to travel through them in a straight line with velocity 4.0×10^{6} m s^{-1}. What is the strength of the electric field? State clearly the relation between the directions of the two fields and the direction of motion of the electron.

(**SUJB**: *all other Boards except SEB H*)

See F8.1 and F8.2 for the paths, if necessary.

An electron will travel straight through only if its velocity is such that the electric force balances the magnetic force. See p. 123 and E8.3 if necessary.

56S Fig. 1 shows an arrangement for investigating the characteristics of a transistor circuit. The input voltage V_i is varied using the potentiometer, P. The corresponding output voltage V_o is shown graphically in Fig. 2.

Fig. 1

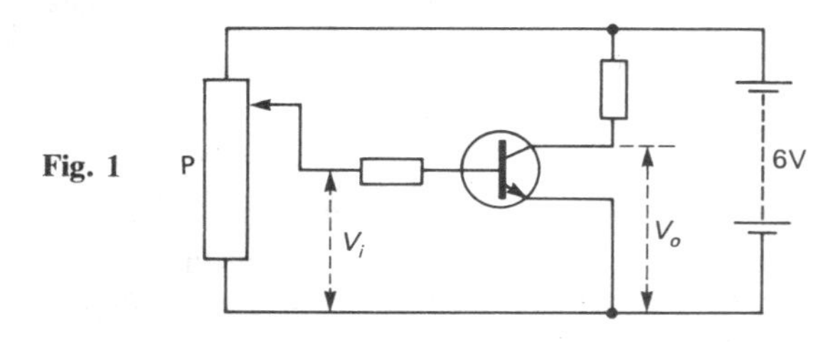

Fig. 2

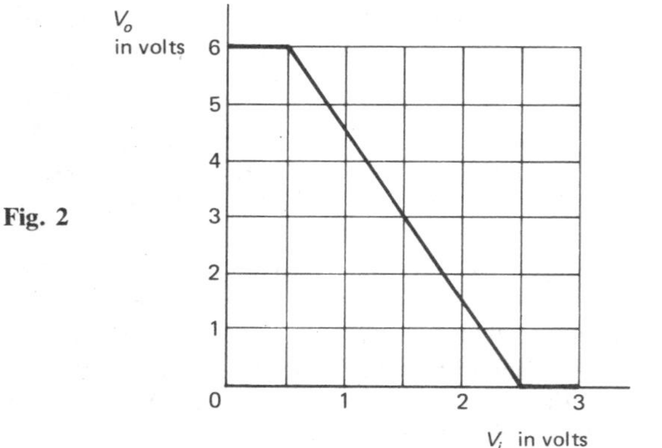

The circuit is to be used as an alternating voltage amplifier. The input voltage must first be fixed at a suitable value by adjusting P.
(a) Suggest the most suitable value for this fixed input voltage, explaining your answer.
(b) A sinusoidally alternating voltage of amplitude 0.5 V is superimposed on this fixed voltage. What will be the amplitude of the output voltage variations? Will the output variations be sinusoidal? Justify your answers.
(c) Sketch one complete cycle of the output voltage which would be obtained if the amplitude of the superimposed sinusoidal voltage were increased to 1.5 V

(**AEB** June 83: *all other Boards except SEB H and SYS*)

(a) Choose so that V_o is at the mid-point of its range. See p. 109 if necessary.

(b) The extra input signal causes the input voltage to range from the value in (a) by 0.5 V either way. Use Fig. 2 to locate the maximum and minimum input voltages and the corresponding output voltages. Hence determine the amplitude of the output voltage. Does the output voltage reach saturation? See p. 109.

(c) See pp. 109–110 if necessary.

57S A closed square coil consisting of a single turn of area A rotates at a constant angular speed, ω, about a horizontal axis through the mid-points of two opposite sides. The coil rotates in a uniform horizontal magnetic flux density, B, which is directed perpendicularly to the axis of rotation.
(a) Give an expression for the flux linking the coil when the normal to the plane of the coil is at an angle α to the direction of B.
(b) If at time $t = 0$ the normal to the plane of the coil is in the same direction as that of B, show that the e.m.f., E, induced in the coil is given by

$$E = BA\,\omega \sin \omega t$$

(c) With the aid of a diagram, describe the positions of the coil relative to B when E is (i) a maximum (ii) zero. Explain your answer.

(**JMB**: *all other Boards except SEB H*)

(a) Start by making a quick sketch of the coil and field lines, as seen from along the coil axis. Mark the normal to the coil and label angle α. The flux per turn is given by the coil area $\times$ B-component along the normal; hence, derive the required expression.

(b) Substitute $\alpha = \omega t$ into your expression for total flux from **(a)**. Then use E2.25 to obtain the expression for the induced e.m.f.

(c) Maximum e.m.f. is when the sides of the coil cut directly across the field lines, and this ought to be when $\sin \omega t = +1$ or -1. Which points of each cycle do these values correspond to? Likewise, zero e.m.f. is when the coil sides move along the field lines, so which points of each cycle correspond to zero e.m.f.? Give a diagram showing a view along the coil axis for (i), and a separate diagram for (ii), then explain the e.m.f. in terms of the sides cutting the field lines.

58S Find the shortest wavelength of X-radiation that might be produced by electron bombardment of the screen of a television set in which the accelerating potential is 2.0 kV.

(**Cambridge**: *all other Boards except SEB H*)

Assume all the energy of a single electron (eV) is given to a single photon. Hence calculate the wavelength of such a photon. See E8.8 if necessary.

Set 2: 59S–66S

59S In answering the questions below you will have to estimate various quantities and then use your estimates in order to obtain the required answers.

Show clearly your estimates and all the steps in your calculations. Always include units in estimates and in answers.

The men's world high jump record is about 2.3m.
(a) How much gravitational potential energy would an athlete have to supply in order to jump this height?

estimates *calculations*

(b) State two other factors that must be taken into account in order to obtain a more accurate value for the total energy required.
(c) Accepting the answer in **(a)**
(i) What would be his vertical take-off speed?
(ii) What would be the average vertical force he exerted on the ground during take-off?

estimates *calculations*

(**O and C Nuffield**: *all other Boards*)

(a) Estimate the athlete's mass. Then calculate the gain of pe.
(b) Does the athlete's centre of gravity also rise by about 2.3 m? Why does the athlete take a run up to the high jump bar?
(c) (i) Assume ke is converted to pe.
(ii) You need to estimate the distance moved by the force which gives the initial 'lift-off'. Assume work done by the force goes to ke of 'lift-off', and so calculate the force.

60S A uniform rod AB of length 50 cm and mass 10 kg is freely hinged at A to a fixed point, and is supported in a horizontal position by a string BC inclined at an angle of 60° to the horizontal. Calculate the tension in the string.

What is the work done when the string is pulled so as to raise the rod to the position AB', where it is inclined at an angle of 60° to the horizontal?

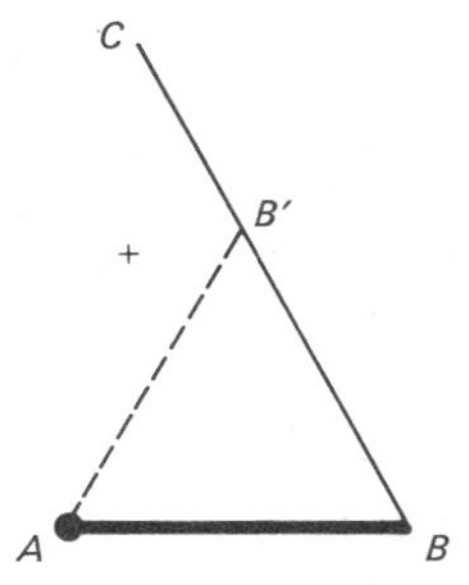

(**SUJB**: *all other Boards*)

Start by making a sketch diagram showing the three forces acting on the rod, which are its weight (acting at the C of G), the tension in the string, and the reaction at the hinge. To calculate the tension, take moments about A so eliminating the reaction from the equation. Remember that the moment of a force about a point is given by force × *perpendicular* distance from the point to the line of action.

Note that the three forces acting on the rod form a triangle, each side representing a force vector. It is a general rule, sometimes called the **closed polygon** rule, that the force vectors for an object in equilibrium form a closed polygon – in this example, the closed polygon is a triangle because there are only three forces acting on the rod.

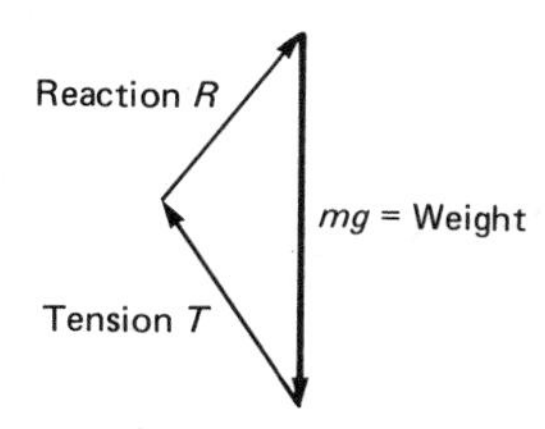

To calculate the work done, calculate the height gain of the Centre of Gravity of the rod, then calculate the gain of pe of the rod for the work done.

61S A horizontal wire AB is situated 5 cm vertically above a compass needle C; the wire lies in a north – south direction. In which direction does the compass needle deflect when a current of 5 A, flowing from south to north, is switched on in this wire? (There is no current in the other wire.) A second horizontal wire DE,

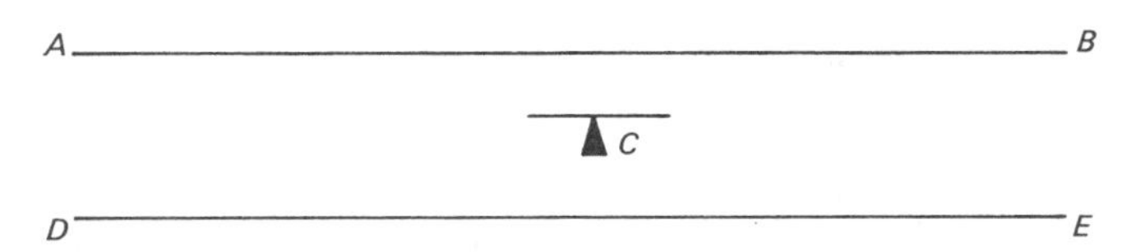

parallel to AB, is situated 7.5 cm vertically below the compass needle. What is the direction and magnitude of the current in DE which, together with the current of 5A in AB, produces no deflection of the compass needle?

What is the magnitude and direction of the force per unit length between the two wires carrying these currents?

(**SUJB**: *all other Boards except SEB H*)

The compass needle will turn so that it lies along the magnetic field line through C; see p. 37 for the field pattern, if necessary.

No deflection of the compass needle must mean that, at C, the magnetic field strength produced by DE is equal and opposite to that produced by AB. What is the formula for the magnetic field strength near a long straight wire? See E2.17. The formula shows that, in this example, the value of (I/d) for AB is equal to that for DE. (I is the current in the wire, d is the perpendicular distance from C to the wire). To work out the current direction, consider an 'end' view as in the diagram shown; the field at C due to DE must be opposite in direction to that due to AB, so must the two sets of field lines go round each wire in the same 'sense' (eg. both clockwise) or in opposite senses?

To determine the force per unit length, calculate the magnetic field strength at the position of one wire (eg. AB) due to the other wire using E2.17. Then use E2.15 to calculate the force per unit length on the first wire; note that the angle between current and field is 90°, so the equation simplifies. See E2.20 and subsequent comments for the force direction.

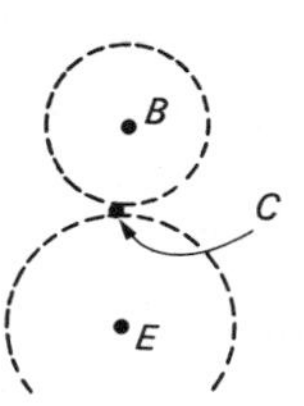

62S **(a)** Sketch a graph showing how the product pV varies with θ, where V is the volume occupied by one mole of an ideal gas at a pressure p and a Celsius temperature θ. Explain the significance of the gradient and of the intercept on the temperature axis.

How, if at all, would the graph change if (i) a second mole of the same gas were added to the first, (ii) the original gas were replaced by one mole of another ideal gas having half the relative molecular mass of the first?

(b) An ideal gas has a relative molecular mass of 4.00. The total translational kinetic energy of the molecules of a certain mass of this gas is 374 J at a temperature of 27°C. Calculate (i) the total translational kinetic energy of the molecules at a temperature of 127°C, (ii) the mass of gas present.

Molar gas constant = 8.32 J mol^{-1} K^{-1}.

(**JMB**: *all other Boards except O and C Nuffield and SEB SYS*)

(a) Use the Ideal Gas Equation modified by substituting $\theta + 273$ for T, the absolute temperature. Then consider the equation in the form $y = mx + c$ with $y = pV$ and $x = \theta$, and so give the required sketch graph etc.

(i) With two moles, use E6.13 to consider how the equation would change, and how the new gradient relates to the gradient for one mole. However, does the intercept on the temperature axis alter?

(ii) E6.13 does not include the relative molecular mass, so if the number of moles is the same, the equation is the same – therefore, is the graph the same?

(b) See p. 92. Remember to convert to Kelvins!

63S A certain solenoid has a resistance of 6.0 Ω and an inductance 2.5H.

(a) Draw sketch graphs to show how the resistance R and the reactance X of the solenoid depend on frequency f.

(b) At what frequency is the resistance of the solenoid 1% of its reactance?

(**Cambridge**: *all other Boards*)

(a) See p. 107 if necessary. The reactance graph is a straight line through the origin (y = Reactance, x = frequency) with a slope which you can calculate. Alternatively, calculate the co-ordinates of one point other than the origin, and so mark each axis with the origin and each co-ordinate of the chosen point.

(b) Use E7.20 if necessary.

64S In an experiment to determine the specific latent heat of vaporization of benzene, it was found that when the electrical power input to the heater was 82 W, 10.0 g of benzene was evaporated in 1 minute; when the power input was reduced to 30 W, the rate of evaporation was 2.0 g per minute. Calculate the specific latent heat of vaporization of benzene.

The saturation vapour pressure of benzene is 1.0×10^5 Pa at a temperature of 80°C; at the same temperature, the saturation vapour pressure of acetone (propanone) is 1.8×10^5 Pa. Which of these two compounds has the higher boiling point, and why? (atmospheric pressure = 1.0×10^5 Pa.)

(**SUJB**: *and London, JMB, AEB, Oxford, Cambridge and WJEC*)

The basis of the calculation is provided by E6.9 and related comments. Remember to convert time in minutes to seconds.

The temperature at which a liquid boils is when its svp is equal to the external pressure. Assume boiling point in the question is at atmospheric pressure. Remember that svp increases (though **not** linearly) with temperature. See p. 93 for further details if necessary.

65S In the circuit shown in the diagram, cell E_1 has an emf of 2.00 V and an internal resistance of 1.50 ohms. The battery E_2 has an emf of 3.00 V and an internal resistance of 0.50 ohms. R_1 and R_2 are resistors of resistance 5.00 and 13.00 ohms, respectively. V is a very high resistance voltmeter and can be assumed to take no current. Calculate the reading of the voltmeter V.

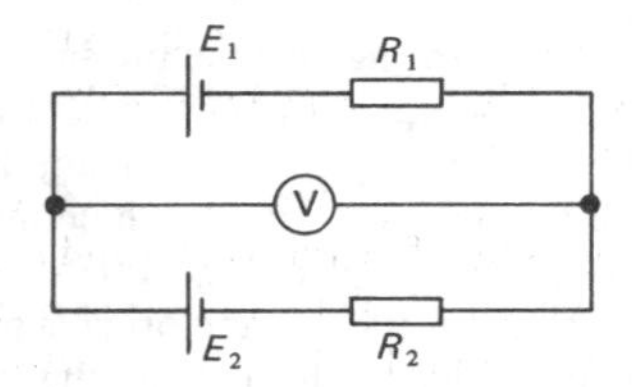

(**AEB** Nov 81: *all other Boards*)

V has a very high resistance and can therefore be assumed to take negligible current. Under these circumstances the current through each cell and resistor is the same and can be assumed to flow clockwise around the circuit since E_2 is greater than E_1. The circuit should be redrawn showing the internal resistances of the cells, the emfs and the pds. What is the resultant emf bearing in mind that E_1 opposes E_2? What is the total resistance of the loop? (Ignore the voltmeter for the total resistance). Hence calculate the current, and then determine the pd across the internal resistance of E_2 and also across R_2. Since the emf of E_2 less the pd across its internal resistance (ie, lost voltage; see p. 102) is equal to the pd across R_2 + the voltmeter reading, then calculate the reading.

It is worth considering E_1 before leaving the question; since current is forced **into** E_1, then its emf E_1 **plus** the pd across its internal resistance is equal to the voltmeter reading **minus** the pd across R_1. Check it, but remember that it is easier to calculate the reading from E_2!

66S

(a) (i) Describe briefly the Rutherford–Bohr model of a hydrogen atom. (ii) State one feature of this model that cannot be explained by the laws of classical physics.

(b) Give a simplified sketch of the energy levels of the electron in this atom. Use your diagram to explain why the spectrum of light from a hydrogen discharge tube contains lines.

(c) The ionization potential of hydrogen is about 14 eV. (i) Estimate the longest wavelength of electromagnetic radiation which could cause the ionization of hydrogen. (ii) A hydrogen atom may also be ionized by an inelastic collision with an electron of suitable energy. Estimate the speed of the slowest electron that could cause ionization.

(Planck's constant $h = 6.6 \times 10^{-34}$ J s; speed of light $c = 3.0 \times 10^8$ m s^{-1}; electron charge $e = 1.6 \times 10^{-19}$ C; electron mass $m_e = 9.1 \times 10^{-31}$ kg.)

(**NISEC**: *all other Boards except SEB H*)

(a) (i) Rutherford's main contribution was showing that almost all the mass of an atom is concentrated in a small positively charged nucleus with orbiting electrons accounting for most of the atom's volume. Bohr's main contribution was showing that electrons could move only in certain fixed orbits, and energy was absorbed or emitted when an electron jumped from one allowed orbit to another. See your textbook if necessary. (ii) This is concerned with the existence of discrete orbits. According to classical theory, electromagnetic radiation should be emitted continuously by an electron on a circular path.

(b) See p. 125.

(c) (i) Since the value given is in eV (electron volts), it is, strictly speaking, the ionization **energy**. See p. 124 for the definition of an electronvolt. This indicates that the ionization energy can be expressed in joules by multiplying by the value of electron charge e. The energy given to an electron by electromagnetic radiation can be estimated from hf_0 where h is Planck's constant and f_0 is the minimum frequency required. Since $v = f\lambda$, then v/λ (or in this example c/λ_0) can be substituted for f_0 making the energy hc/λ_0. Equating this to the ionization energy (in joules) gives a value for λ_0. (ii) The ionization energy must be supplied by the electron that collides with the atom. Equate the required ionization energy in joules to the kinetic energy of the electron, and hence calculate speed.

Long-answer questions

The questions are in three sets. For each set, choose three questions to answer in one session. Allow 30 mins per question except for Nuffield questions which are allocated 45 mins.

Set 1: 67L–72L

67L This question is about the behaviour of water waves. The diagram shows a wave in water deep enough for the wavelength to be much less than the depth. As the wave moves to the right, the water at P acquires in succession the velocities v_1, v_2, v_3, v_4 and v_5 ($v_5 = v_1$) and it can be shown that it moves in a circle with constant speed, where the radius of the circle is equal to the wave amplitude A.

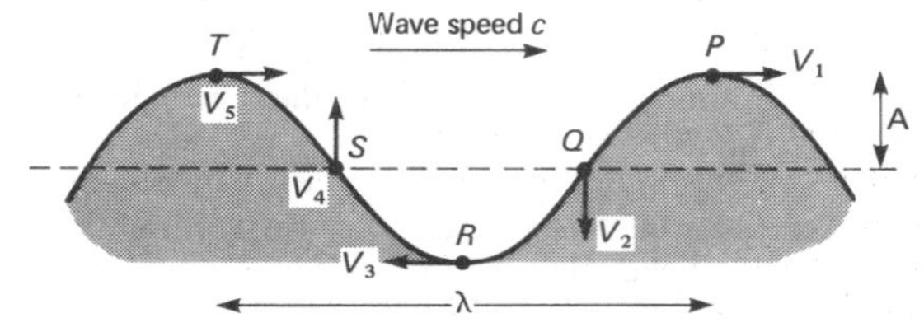

(a) (i) How would you find the time it takes for a water particle to go once round in a circle? (ii) If the water particles are moving in circles, how would you find the magnitudes, and what are the directions, of their accelerations at P, and at Q?

(b) The speed c of deep water waves is given by $c^2 = \lambda g/2\pi$. Sketch graphs of (i) speed c, and (ii) frequency, against wavelength λ, for deep water waves in the range $\lambda = 1$ m to $\lambda = 100$ m. Indicate on your sketch the orders of magnitude of speed and frequency for wavelengths 1 m and 100 m. What would you plot in order to obtain straight line graphs relating wavelengths, λ, to (i) speed, (ii) frequency?

(c) (i) Suppose a storm at sea generates waves of wavelengths in the range 1 m to 100 m. What wavelengths of waves from the storm will be felt by a ship 100 km from the storm during the 24 hours following the onset of the storm? (ii) A small boat at sea has to ride such waves. What will be the speed of the water around circles in waves of wavelengths 100 m and amplitude 10 m? Describe the motion of such a boat (short compared to the wavelength) which rides such waves and also travels forward in the direction of travel of the waves at a mean speed of about 2 m s^{-1}. (iii) What will be the maximum vertical acceleration of the water in such waves? How will the maximum force on the yacht causing this acceleration compare with the weight of the yacht?

(d) If the wavelength λ is larger than the depth d, the water no longer moves in circles and the wave speed for such shallow water waves is given by $c^2 = gd$.
Calculate the wave speed for waves of wavelengths 1 m and 1000 m in a sea of depth 100 m.

Combine these results and those from **(b)** above to draw a rough sketch-graph to show how wave speed will vary with wavelength in the range $\lambda = 1$ m to $\lambda = 1000$ m for a sea of depth 100 m so that the waves are 'deep' water waves for short λ and 'shallow' waves for long λ. Label and explain the main features of your sketch.

(**O and C Nuffield**: *all other Boards except SEB H*)

(a) (i) Once round the circle (P to T) is the period of the wave. (ii) The magnitude of circular acceleration is v^2/r; what are v and r in this case? The acceleration direction is at right angles to the velocity.

(b) The relationship between c and λ is given; use $c = f\lambda$ to get the relationship between f and λ ($(f\lambda)^2 = \lambda g/2\pi$). Calculate f and c at wavelengths of 1 m and 100 m; then use the relationships just described to deduce the shape of each graph between the two known points (F9.7 may help). Plot λ on the x-axis. Note that the question effectively warns you that neither graph is a straight line! Straight lines can be in the form:

$y = (\text{Constant}) . x$

eg. $c^2 = (g/2\pi) . \lambda$

Hence a graph of c^2 plotted against λ will be a straight line. The relationship between f and λ needs to be treated in the same way.

(c) (i) The waves must travel 100 000 m in 24 hours, from which the lowest speed of wave that can reach the ship in that time can be calculated. Find the wavelength of these waves; as longer wavelengths travel faster, all the larger wavelength waves can also reach the ship. (ii) Find the period of the waves ($1/f$). This gives the time taken to travel round a circle whose circumference can also be found (radius 10 m); hence the circular speed is found. The boat's motion is a combination of the circular motion of the water and the steady 2 m s^{-1} velocity in the direction of the waves; you must describe what motion results. (iii) Although the acceleration (v^2/r) has constant magnitude its changing direction means that it is acting vertically only at P, R and T. Using $F = ma$ it should become apparent that to compare the maximum force and the boat's weight (mg) it is only necessary to compare the maximum acceleration to the acceleration due to gravity.

(d) Calculate the wave speeds using the right formulae; $c^2 = \lambda g/2\pi$ for the 1 m waves, and $c^2 = gd$ for the 1000 m waves (notice that this value for shallow waves is independent of the wavelength). The graph of c against λ from part **(b)** must be extended up to wavelength of 1000 m where it will have a steady speed as already calculated. From about $\lambda = 50$ to 200 m the graph undergoes transition from the relationship previously plotted in **(b)** to the horizontal line (speed on vertical axis) of the constant speed. This needs to be explained; it might be a good idea to label some wave speeds with their wavelengths (using the values that you have already calculated).

68L A rocket is caused to ascend vertically from the ground with constant acceleration, a. At a time, t_s, after leaving the ground the rocket motor is shut off.

(a) Neglecting air resistance and assuming that acceleration due to gravity, g, is constant, sketch a graph showing how the velocity of the rocket varies with time from the moment it leaves the ground to the moment it returns to ground. In your sketch, represent the ascending velocity as positive and the descending velocity as negative. Indicate on your graph (i) t_s, (ii) the time to reach maximum height, t_h, (iii) the time of flight, t_f.

(b) Account for the form of each portion of the graph and explain the significance of the area between the graph and the time axis from zero time to (i) t_s, (ii) t_h, (iii) t_f.

(c) Either by using the graph or otherwise, derive expressions in terms of a, g and t_s for (i) t_h, (ii) the maximum height reached, (iii) t_f.

(**JMB**: *all other Boards*)

(a) Before sketching the graph, you should try to establish the situation clearly by breaking the motion into three parts:

1 From $t = 0$ to t_s : the rocket speed increases steadily from rest.

2 From $t = t_s$ to t_h: the rocket still moves upwards, but its speed falls steadily until it comes to momentary rest at its highest point.

3 From $t = t_h$ to t_f: the rocket falls to earth with steadily increasing speed.

Your sketch graph should show y = velocity; x = time. Remember that the slope of each part is equal to the acceleration in that part ($+a$ in 1, $-g$ in 2 and 3).

(b) In your account, two key points should be applied to each part of the motion:

1 Slope = acceleration.

2 Area under line = displacement. Since the rocket falls back to earth at $t = t_f$, then the area above the x-axis should equal the area beneath the x-axis (since area above represents + ve displacement, etc.) up to $t = t_f$. For the significance of the area up to $t = t_s$ and up to $t = t_h$, remember what t_s and t_h represent.

(c) GRAPHICALLY: write down an expression for the speed when the motors are shut off in terms of a and t_s; since that speed drops to zero in time ($t_h - t_s$) at a rate $-g$, then you should be able to write an expression for ($t_h - t_s$) in terms of a, g and t_s. For the maximum height, remember that the triangle area of parts 1 and 2 gives the displacement to maximum height; triangle area = $\frac{1}{2}$ base $\times$ height. For t_f use the fact that the area under part 3 = the area under the other two parts. Remember to give your answers in terms of a, g and t_s.

OTHERWISE: apply the dynamics equations E1.1 to 1.4.

69L This question is about explaining ideas in physics.

Choose **one** of the three subjects, (a) (b) or (c) below. For the one that you choose you should give a careful explanation of each of the three topics listed. Your explanation should be suitable for a friend studying A-level physics who missed the teaching of this particular subject.

Show also how your explanations could help your friend to understand **one** everyday application of the subject.

Subjects:

(a) Electronics	Topics: block diagrams, logic, feedback.
(b) Thermodynamics	Topics: number of ways, temperature, entropy.
(c) Radioactivity	Topics: decay, radiation, isotopes.

(**O and C Nuffield**: *(c) only for all other Boards except SEB SYS*)

There is a temptation for candidates answering this style of question to write long, rambling, unplanned answers that cover their entire knowledge of the required subject area. This question is intended to test knowledge, but perhaps of greater importance is the ability to explain information logically and also to identify terminology and explain its meaning: this is the idea behind the stipulation that the answer is 'for an A-level student who missed the teaching of the subject'. A planned answer is essential, so this commentary provides such a plan for the most popular subject – Radioactivity. Information on the other subjects, Electronics and Thermodynamics, can be found in Units 7 and 6, respectively, of this book and in Nuffield Student Books 6 and 9.

The method of planning used for this question is to list key points under each topic heading. Words that need explanation (terminology) have been printed in **bold** type (you could underline them in your plan). The application was chosen before the rest of the plan; this enables the quoted examples/formulae/equations, etc., to feature the application; this will save time and improve the sense of the answer. Finally each point is numbered in the order in which it will be written into the answer to provide a logical explanation.

APPLICATION: 7 Radioactive carbon dating with carbon-14.

DECAY: 3 One element turns into another element.

5 Example decay equation (features application) ${}^{14}_{6}C \rightarrow {}^{14}_{7}N + {}^{0}_{-1}\beta$.

4 **Mass number** and **Charge on nucleus**; use ${}^{14}_{6}C$ as an example.

2 **Stable** and **Unstable** nuclei.
1 Radioactivity involves nuclei of atoms.
6 **Half-life** (carbon-14's is 5600 y) and decay graph.
8 Mathematical model of decay $N_t = N_0 e^{-\lambda t}$.

RADIATION: 10 Radioactive radiations are **alpha, beta, gamma**.
9 Radiation means energy carried by streams of waves or particles.
11 Dangers of radiation.

ISOTOPES: 12 Same element different nucleus due to **neutron number** (eg, $^{14}_{6}C$ and $^{12}_{6}C$).

70L In the model of a crystalline solid the particles are assumed to exert both attractive and repulsive forces on each other. Sketch a graph of the potential energy between two particles as a function of the separation of the particles. Explain how the shape of the graph is related to the assumed properties of the particles.

The force F, in N, of attraction between two particles in a given solid varies with their separation, d, in m, according to the relation

$$F = \frac{7.8 \times 10^{-20}}{d^2} - \frac{3.0 \times 10^{-96}}{d^{10}}$$

State, giving a reason, the resultant force between the two particles at their equilibrium separation. Calculate a value for this equilibrium separation.

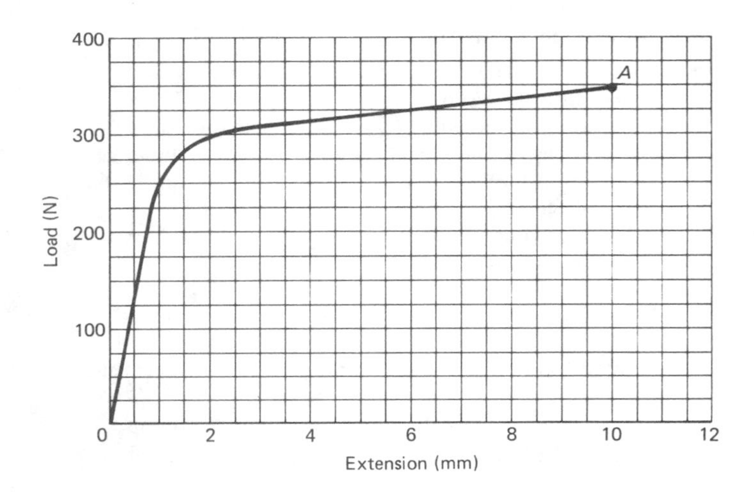

The graph displays a load against extension plot for a metal wire of diameter 1.5 mm and original length 1.0 m. When the load reached the value at A the wire broke. From the graph deduce values of (a) the stress in the wire when it broke,
(b) the work done in breaking the wire,
(c) the Young modulus for the metal of the wire.

Define *elastic* deformation. A wire of the same metal as the above is required to support a load of 1.0 kN without exceeding its elastic limit. Calculate the minimum diameter of such a wire.

(O and C: *all other Boards except SEB H and SEB SYS*)

See 2.4 and F2.5(b) in particular.

One term of the equation represents the repulsive interaction, the other the attractive interaction so at equilibrium the two terms balance one another out. Hence calculate d at equilibrium.

(a) See 5.1 if necessary, **(b)** The area under the curve gives the work done, so count squares – also, you need to work out how much work each square corresponds to.

(c) Use E5.1 if necessary but remember to use the section of the graph from 0 to the limit of proportionality.

Elasticity is discussed in 5.2. From the graph, read off the load corresponding to the elastic limit, then calculate the stress at that point. Hence calculate the area needed to support 1kN, and so determine the diameter of the second wire.

71L State *Kirchhoff's laws*. Explain how each is based on a fundamental physical principle.

Use the laws to deduce values of the currents I_a, I_b, I_c and I_d as shown in the circuits below (Figs 1 and 2).

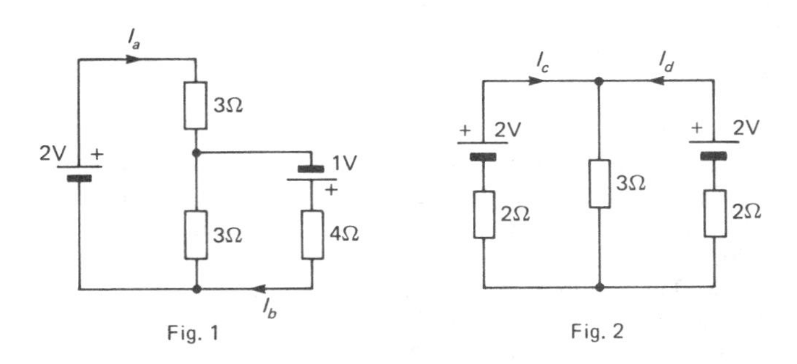

A battery, switch and uniform wire of variable length l are connected in series. A bar magnet is suspended vertically above the wire by a fibre fixed to a torsion head fitted with a scale to measure angular rotation. The magnet is initially aligned along the wire. When the switch is closed, the magnet twists but it may be restored to its original position by rotation of the torsion head through a known angle φ. Values of φ for total lengths l of the wire are given.

φ (°) . . .	33.2	20.1	14.3	11.1
l (m). . .	1.00	2.00	3.00	4.00

(a) Explain why the magnet twists when current flows.
(b) Explain why φ increases with increasing current.
(c) Plot a graph of $1/\varphi$ against l and find the intercept on the l-axis.
(d) Discuss the significance of this intercept with reference to the battery.

(Cambridge: *all other Boards except SEB H*)

Kirchhoff's 1st Law is that the sum of the currents into a junction is equal to the sum of the currents out of the junction. Use 'conservation of charge' to explain the law. The 2nd Law is that for a closed loop of a circuit, the sum of the emfs round the loop is equal to the sum of the pds across the resistances of the loop. Because emf and pd are both defined in terms of energy per unit charge, use 'conservation of energy' together with the meaning of emf and pd to explain the second law.

Fig. 1: Assume the current through the lower 3 ohm resistor is downwards, and write this current in terms of I_a and I_b. There are three possible loops but because there are only the two unknown quantities (ie. I_a and I_b), it is only necessary to consider two of the three loops. One such loop is made of the 2 V cell and the two 3 ohm resistors; its only emf is that of the cell, the pd. across the top 3 ohm resistor is $3I_a$ (from E7.7) and $3(I_a - I_b)$ across the lower 3 ohm resistor – hence use of Kirchhoff's 2nd Law gives $2 = 3I_a + 3(I_a - I_b)$. Now consider one of the other two loops to give another equation containing I_a and I_b. Take care to sum pds in accordance with the current directions. With two equations, solve for I_a and I_b.

Fig. 2: Start by writing an expression for the current through the 3 ohm resistor in terms of I_c and I_d. Then consider two loops, and for each loop write an equation using Kirchhoff's second law. Hence solve for I_c and I_d.

(a) In an external magnetic field, a bar magnet will always try to align itself along the field line direction – with its N-pole trying to move in the same direction as the field, and its S-pole in the opposite direction. What is the direction of the magnetic field around a straight wire? Hence give your explanation.

(b) Increased current will cause the magnetic field of the wire to alter. How? A bigger angle is necessary if the torque on the bar magnet due to the wire is bigger.

(c) 'Plot' means use graph paper – as opposed to 'sketch' when you can use axes drawn on ordinary paper.

(d) φ is proportional to the magnetic field strength which is proportional to the current I: l is proportional to the *wire* resistance R. For the given circuit, $E = I(R + r)$ where E is the battery emf and r is its internal resistance. The intercept on the l-axis therefore gives a length of wire equal to which resistance?

72L Give the theory of the action of a diffraction grating on a parallel monochromatic beam of light incident normally on the grating.

A pure spectrum is one in which there is no overlapping of light of different wavelengths. Describe how you would set up a diffraction grating to display a pure spectrum on a screen. Explain the purpose of each optical component used.

When a diffraction grating is used on a spectrometer in normal adjustment, the light from a sodium lamp is diffracted through 16° in the first order.

(a) What is the highest order of diffraction that can be seen, and what is the angle of diffraction in this order?

(b) How would the angle of diffraction in the highest order be changed if the air between the grating and the telescope were replaced by a gas of refractive index (relative to air) of 1.002?

(c) If the ruled lines of the diffraction grating are all 1 cm long, and the rulings extend over 2 cm, what is the minimum diameter of the object glass of the telescope if it is to be capable of receiving all the light diffracted in the highest order?

(**SUJB**: *all other Boards*)

Use a sketch like F4.8 to give a proof of the diffraction grating equation E4.8. To display a pure spectrum on a screen, consider how F4.23 may be adapted to focus each diffracted beam onto a screen. Essentially, only the objective of the telescope is required instead of the whole telescope, so where should the screen be placed.

(a) Use the given information and E4.8 to calculate λ/d (d = distance between adjacent slits). Then use E4.8 again with the condition that $\sin\theta \leqslant 1$ to give the highest order etc. See p. 66 if necessary.

(b) The wavelength in the gas λ' is related to the wavelength in air λ by the refractive index. See p. 67 if necessary. Hence determine the value of λ'/d from λ/d and the refractive index. Then use E4.8 to calculate the angle of diffraction of the highest order.

(c) What is the width of highest order parallel beam which emerges at the angle calculated in **(a)** to the 2 cm wide grating? Is the beam width greater or less than its height? The minimum diameter of the object glass is determined by whichever is the greater.

Set 2: 73L–78L

73L This question is about the storage of energy as part of the national electricity system.

The graph (Fig. 1) shows the way in which the demand for electrical power varies during a typical working day in the USA.

(a) One way of meeting the demand is to build enough generating stations so that their combined power can meet the peak demand. Use the graph to estimate the number of hours per day for which at least one quarter of the power generation system would **not** be required. Explain how you arrive at your estimate.

Give two reasons why such a system would be wasteful.

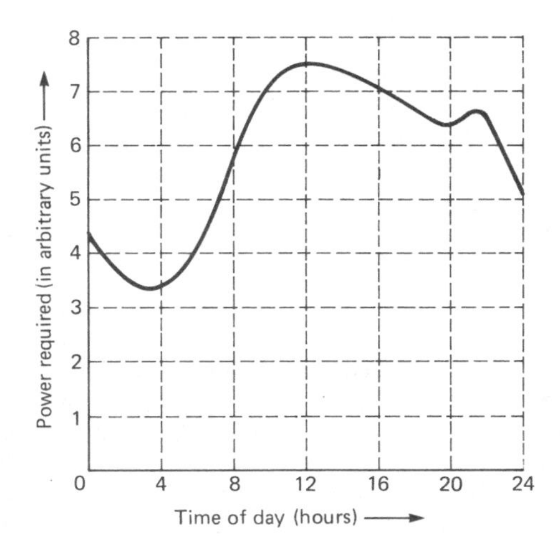

(b) An alternative way is to provide two types of power station. The first type is a conventional power generator which can be kept running at full power for 24 hours per day. The second type is one in which electrical power is *used* for part of the day *to store* energy so that for other parts of the day power can be *generated* using the energy stored. Explain the advantages of this second idea. What can you say from the graph in Fig. 1 about the maximum power rating required for the storage systems (second type) compared with the conventional systems (first type)?

(c) It has been shown that a useful power for storage systems in the U.S.A. would be about 100 000 MW. Estimate the energy in kW hours that the storage systems should be able to store, explaining how you make the estimate using Fig. 1 (this graph can be assumed to apply for the U.S.A. as a whole – ignore any effects of time zones).

(d) One idea for a storage power station is to use off peak power to produce a large volume of compressed air and then to reverse the system so that this air drives a generator. One such system compresses air to a pressure of 70 atmospheres (1 atmosphere ~ 10^5 N m^{-2}) into an underground cavern of volume 3×10^5 m^3. Estimate the volume of this air before compression, and sketch a graph of P against V assuming its temperature to be constant. If the stored energy is given by the area under the P–V curve, estimate its magnitude given the formula [area = $P_1V_1\ln(P_2/P_1)$] where P_1 and P_2 are the low and high pressures respectively and V_1 the volume before compression.

(e) Batteries could also be used as storage systems. Lead acid batteries can store about 40 watt hours per kilogramme, and generate a peak power of 70 watts per kilogramme. Discuss briefly the design of a battery system to meet the target in **(c)**.

(f) The graph (Fig. 2) shows how cost per kW of the battery and air systems varies with daily discharge hours at their power rating. Explain why the cost versus hours relationships for the battery systems might be different from that for air.

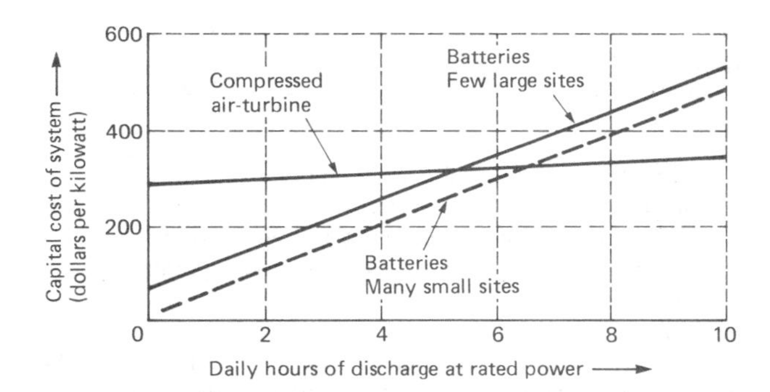

(**O & C Nuffield**: *all other Boards*)

(a) You need to estimate the number of hours each day when the power curve of Fig. 1 falls below 75% of its peak value. The system is wasteful if the generators produce peak power all the time – what happens to this power at 'slack' periods? What about the cost of building and manning generators which waste power?

(b) The first type supplies the mean power continuously, so when the power required drops below the mean power, should the second type store or generate power? How is the second type used when the power required exceeds the mean power? To compare power ratings of each type, sketch a line on the graph where you think the mean power level lies, and then consider the maximum variation from that level.

(c) How many hours per day is the power required less than the mean power? Hence calculate the energy stored in kW hours for useful power of 10^5 MW.

(d) See E6.13 and F6.11 if necessary.

(e) The target in **(c)** is the energy to be stored then released each day. Each kilogramme of battery will store 40 W hours (= 0.040 kW hours) so how many kilogrammes are needed? If the batteries generate at a peak of 70 W per kilogramme, can the no. of kilogrammes above generate useful power of 10^5 MW?

(f) If each battery stores 40 W hours per kg, and then generates power at 70 W per kg, how long will a battery last when it generates? Hence explain why the longer the daily hours of discharge are, the more batteries will be needed – and how will that affect the capital cost of the battery system? However, the air system can supply energy up to the total stored by letting the pressure down to a certain level – you don't need more caverns, so how does this point affect the capital cost?

74L Two bodies are in thermal equilibrium. What does this statement mean? State the Zeroth law of thermodynamics and explain its importance in relation to the use of a thermometer to measure temperature.

(a) The temperature of a hot liquid, measured on the empirical centigrade scale of a certain resistance thermometer, is 68.4°.

(i) What is meant by describing the resistance thermometer scale as *empirical*?

(ii) Write down an equation which defines the centigrade scale of the thermometer in terms of resistance readings.

(iii) Draw a diagram of a simple circuit for use in resistance thermometry. How may the resistance be determined accurately?

(b) The element of the resistance thermometer in **(a)** is of mass 0.013 kg and has a specific heat capacity of 4.5×10^2 J K^{-1} kg^{-1}. Initially, it was at room temperature, for which a reading of 17.1° was obtained. It was then completely immersed in 0.30 kg of liquid of specific heat capacity 2.5×10^3 J K^{-1} kg^{-1}, giving an equilibrium reading of 68.4°

(i) What was the temperature of the liquid just before the thermometer was immersed? (For the range of temperature of the experiment, assume that the specific heat capacities of the thermometer and the liquid are independent of thermodynamic temperature and that the empirical scale of the resistance thermometer is linear with respect to thermodynamic temperature. Neglect the heat capacity of the container.)

(ii) How could the cooling effect of the thermometer be made less significant?

(**Cambridge**: *all other Boards except O and C Nuffield and SEB SYS –* (**b**) *only for SEB H*)

Thermal equilibrium is discussed on p. 87. See the commentary of 6.24L for the Zeroth law, if necessary.

(a) (i) The resistance thermometer scale is given by E6.1, so explain why it is empirical (ie. based on experiment).

(ii) See E6.1

(iii) See p. 105 if necessary.

(b) (i) When the thermometer is put into the liquid, the thermometer gains heat energy from the liquid until their temperatures become equal. Let the initial temperature of the liquid be θ, and write down expressions for the heat loss of the liquid in terms of θ, and the heat gain by the thermometer. Assume heat loss from the liquid equals heat gain by the thermometer, and hence determine θ.

(ii) Consider how your calculation in (i) would differ if the initial temperature of the thermometer was greater.

75L

(a) (i) Explain how stationary transverse waves form on a stretched string when it is plucked.

(ii) State the factors that determine the frequency of the fundamental vibration of such a string and give the formula for the frequency in terms of these factors.

(b) When two notes of equal amplitude but with slightly different frequencies f_1 and f_2 are sounded together, the combined sound rises and falls regularly.

(i) Explain this, and draw a diagram of the resulting wave form.

(ii) show that the frequency of these variations of the combined sound is $f_1 - f_2$

(c) A car engine has four cylinders, each producing one firing stroke in two revolutions of the engine. The exhaust gases are led to the atmosphere by a pipe of length 3.0 m.

(i) Assuming that vibration anti-nodes occur near each end of the pipe, calculate the lowest engine speed (in revolutions per minute) at which resonance of the gas column will occur.

(ii) What may happen at higher speeds?

(iii) Where, in the gases in the pipe, will the greatest fluctuations of pressure take place at resonance?

(Take the speed of sound in the hot exhaust gases to be 400 ms^{-1})

(**Oxford**: *all other Boards except SEB H*)

(a) (i) Explain how the arrangement produces two travelling waves passing through one another, give the conditions for the formation of stationary waves, and explain how these conditions are met here. See p. 52 if necessary.

(ii) See 3.26L if necessary.

(b) (i) See p. 54 for an explanation of 'beats'.

(ii) Suppose the two notes are in phase at $t = 0$, so giving a 'beat'. For each complete cycle of the higher note, the lower note falls behind in phase by a set amount per cycle of the higher note. Eventually, the lower note is one complete cycle behind the higher note so bringing the two notes back into 'zero phase difference' and the next beat. Let T = time between successive beats, so write down the number of cycles completed by each note in that time, and use the fact that one completes one cycle less than the other in that time. Hence derive an expression for the beat frequency $1/T$.

(c) (i) the situation is like F3.13 (a)

(ii) See p. 55

(iii) Pressure is proportional to density so where does the density vary most?

The diagram below shows the variation of position of each 'layer' at intervals over one full cycle, so locate the points of maximum compression and rarefaction in relation to the displacement nodes and antinodes.

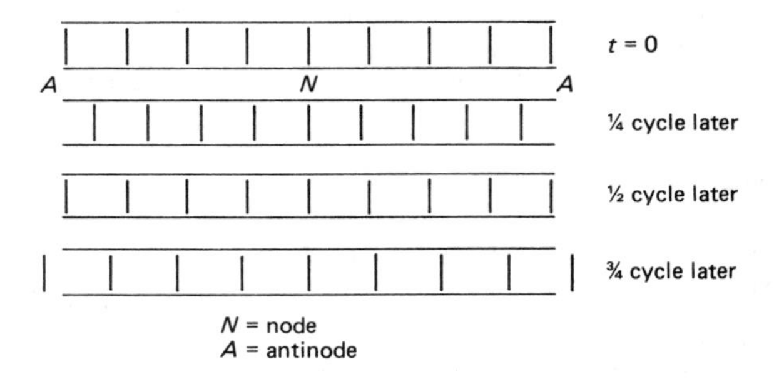

76L

(a) Explain why, when a magnet is plunged into a coil, the direction in which the induced current flows is consistent with the principle of the conservation of energy. A compass needle is pivoted above a horizontal copper disc which can be made to rotate about a vertical axis through its centre. The axis about which the needle rotates coincides with that of the disc. Explain what happens to the needle as the copper disc rotates.

(b) Describe an experiment by which the law of electromagnetic induction relating to the magnitude of the induced emf may be verified.

(c) A copper disc of radius 10.0 cm rotates in a uniform magnetic field of flux density 0.030 T, the direction of the field being normal to the plane of the disc. If the emf induced between the centre and rim of the disc is 2.0 mV what is the rate of rotation of the disc?

(**AEB** June 81: *all other Boards except SEB H*)

(a) Remember that the induced current sets up a magnetic field in and around the coil. The coil magnetic field must try to stop the magnet moving towards the coil – otherwise the coil field would pull the magnet in, so making the magnet move faster, so making the induced current and field stronger, so pulling the coil in even more, so . . ., etc.; this would mean more and more electric energy and ke out of nowhere! – so the point is that work must be done to push the magnet in against the coil field. You should relate the current direction to the coil field direction with the aid of a simple diagram, and then comment as indicated above. In the second part, when the copper disc rotates, the disc cuts the magnetic flux due to the magnet. Hence induced currents flow in the disc. Will the magnetic field produced by these induced currents act against the compass or will the compass be attracted and hence 'pulled round' by the disc? Remember that the induced field direction must be such as to 'oppose the change that causes it' – in other words, to lessen the relative motion between disc and compass.

(b) See your textbook for a suitable experiment. You should give a labelled diagram to show the means of producing and varying a changing magnetic flux linkage and of measuring the induced emf. Give an account of the procedure explaining how the rate of change of flux linkage is measured and varied, and indicate that the emf is measured for various rates. Then explain how the measurements are used to verify Faraday's law.

(c) Use E2.25. Note that the flux swept out per second is the product of the flux density and the area swept out per second by an imaginary radius of the disc.

77L In this question you are asked to consider whether facts or theories come first in explaining pieces of physics.
Here are two opposing views about facts and theories.

1 'Facts come first. You can't make up any theories until you have a lot of facts to go on.'
2 'Theories come first. In practice, theories tell you which facts to collect, and without theories you get a heap of meaningless information.'

(a) Imagine you had to teach a new set of sixth formers about any **two** of the topics listed in (i) to (iv) below. Discuss the two examples you choose in detail, explaining the 'facts' and 'theories' involved at the various stages in each.

(b) For your two topics, explain briefly whether you think view 1 or view 2 is more appropriate, or whether you think it more sensible to adopt neither view, or a mixture of both views.

Topics
(i) Energy levels in atoms.
(ii) The wave/particle models of light.
(iii) Electromagnetic waves.
(iv) The nuclear model of the atom.

(**O and C Nuffield**: *all other Boards*)

Some comments on the four topics may help you in planning your answer.

(i) The first ideas about energy levels arose from experiments (by Franck and Hertz) involving excitation potentials in gases. The theory derived from this work was confirmed by further experiments involving emission and absorption atomic spectra (see 8.3 and 8.4). Further evidence arises from studying behaviour of electron streams in thyratrons (anomalous drops in tube current can be due to excitation); see Nuffield Unit 2.

(ii) The original theory for light was a particle model which had to be modified to a wave theory as a result of experiments (such as Young's 'slits') that proved wave properties, (see p. 63). However the photoelectric effect (see p. 124) showed that light had also to be explained using a particle model; hence the development of a joint theory (De Broglie's equation E1.18 and Planck's equation E1.16) that treats particles and waves as equivalent.

(iii) It was some time before it was realized that light, ultraviolet, infrared, X-rays, etc. were all members of the same family of radiation. It was experiments that showed them all to have the same speed (and also to be transverse waves) that provided the link. The method of production of electromagnetic waves was understood theoretically long after the practical discovery of the 'family' of waves. There were some early theories of particle streams that had to be discounted by showing that the radiation had no 'mass'. See 3.7 for a brief discussion of the nature of the waves.

(iv) The original model of the atom was the 'currant bun model' proved wrong by Rutherford's alpha-particle scattering (see your textbook). The theory had to be modified to a 'nuclear' theory as a result of this experiment. Later studies of radioactivity suggest that the nucleus must consist of discrete particles experimentally supported by the discovery of the neutron and the proton. Electrons had been identified earlier but initially were not understood to be 'bits' of atoms.

78L Explain what is meant by capacitance.
The plates of a parallel-plate capacitor each have an area of 25 cm^2 and are separated by an air gap of 5 mm. The electric field intensity between the plates is $7 \times 10^4\ V\ m^{-1}$. If one plate is at zero potential relative to earth, find the potential of the other plate and indicate on a sketch some equipotential surfaces in the gap. What is the potential half way between the plates? (Ignore end effects.)
What is the capacitance of the capacitor, and how much electrical energy is stored in it?
A slab of dielectric of relative permittivity 15 is introduced into the isolated capacitor so as to exactly fill the gap. What are the new values of **(a)** the potential difference, **(b)** the capacitance and **(c)** the energy?
Give a labelled sketch of a practical form of variable capacitor and indicate **one** use.
($\varepsilon_0 = 8.85 \times 10^{-12}\ F\ m^{-1}$)

(**WJEC**: *and all other Boards except SEB H*)

Capacitance is explained and defined on p. 105. See E2.11.

The electric field between charged parallel plates is uniform. See 2.8 for a full discussion of uniform electric fields. Use E2.9 to calculate V (the plate pd). F2.1 shows the field pattern for parallel plates. Note that electric field intensity is the same as electric field strength.

To calculate the capacitance of the parallel plates, use E2.12. The equation for energy can be calculated from E2.14, using the values for capacitance (C) and pd (V) between the plates which you have already calculated.

With the dielectric and the isolated capacitor, you should first realize the significance of the word 'isolated' here; it means that the charge on the plates is fixed at the value it has in the first part (calculated from your values for C and V with equation E2.11). The problem is most easily tackled by calculating the new capacitance first; remember that filling the spacing with dielectric of relative permittivity k increases the capacitance to C_0k (C_0 = original capacitance). Then, calculate the new pd across the plates, using the value of charge as above and the new value of capacitance; see E2.11 if necessary. Finally, use the new values of pd and capacitance to calculate the new energy stored; see E2.14 if necessary.

For a practical form of a variable capacitor and its use, consult your textbook. **Set 3; 79L–83L**

79L Describe, suggesting suitable values for the separation of the components, how you would set up a double-slit apparatus to demonstrate interference of light. Explain the reasons for your choice of method of (i) illumination of the slits and (ii) observation of the fringes.

How accurate a value for the wavelength of the light (approximately 600 nm) would you expect to obtain using the apparatus you describe?

(a) Coherent light of wavelength 590 nm illuminates double slits of separation 10^{-4} m. A graticule is placed 0.2 m from the plane of the slits, and an eyepiece, focused on the graticule, is used to observe the interference fringes. Calculate the spacing of the fringes as measured by using the graticule.

(b) The eyepiece and graticule are now attached to one end of a tube having a converging lens at the opposite end, forming a telescope focused at infinity. The interference fringes are viewed through the telescope. Explain why the separation of the fringes seen by the eyepiece is now independent of the distance of the telescope from the slits. What focal length of the objective lens will give the same fringe separation as in (a)?

(**O and C**: *all other Boards*)

See F4.2 for a suitable arrangement. Explain why a *narrow* single slit is necessary to illuminate the double slits – displacing the single slit to one side slightly would displace the fringes slightly to the opposite side, so using a wide single slit equivalent to several narrow slits would 'blur' the fringe pattern. Observation of the fringes may be directly from the screen, or using an eyepiece.

The wavelength is calculated using E4.3; estimate the % accuracy with which each of the three measurements can be determined. The % accuracy for wavelength is the sum of the individual % estimates.

(a) Use E4.3

(b) Because the telescope is in normal adjustment, the fringes are formed in the focal plane of the objective so that the viewer sees a magnified image of the fringes by looking through the eyepiece. Light to a bright fringe from the double slits must therefore follow parallel paths at an angle to the axis equal to an integer times λ/d where d is the slit spacing. Thus the first bright fringe off the axis, assuming a central bright fringe on the axis, is formed at angle λ/d to the objective lens, as illustrated below. Hence calculate the focal length required to give the required fringe spacing, using λ and d as in **(a)**.

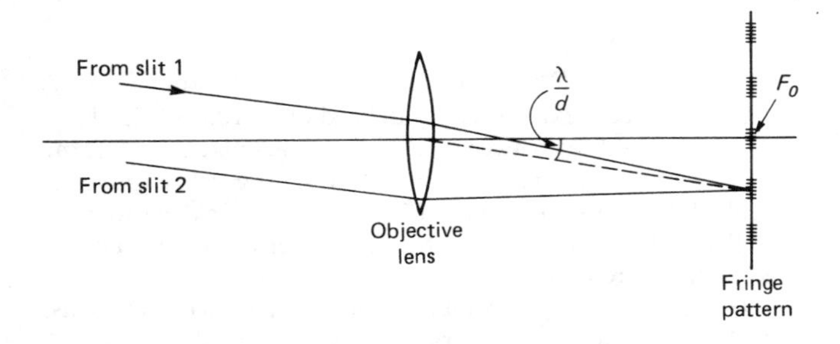

80L

(a) Draw a labelled diagram of a moving-coil galvanometer. Explain those features of its design that (i) determine its current sensitivity, (ii) prevent the coil from oscillating when the current changes.

(b) Two otherwise identical moving-coil galvanometers P and Q are fitted with different coils; the coil of P has 1000 turns of fine wire with a resistance of 200 Ω, while the coil of Q has 60 turns of thicker wire with a resistance of 1.5 Ω. Calculate the ratio of the deflections they will give if they are connected in turn to a thermocouple of negligible resistance and e.m.f. 5 mV.

(c) A bicycle wheel of effective radius 0.30 m is rotated in an east-west vertical plane at 5 revolutions per second. Electrical contacts are made at the hub and the rim. (Take the horizontal component of the Earth's magnetic field to have a flux density of 1.8×10^{-5} *T*)

(i) Calculate the e.m.f. produced between the contacts.

The wheel has 40 spokes, each of length 0.30 m and cross-sectional area 2.5 mm^2, and made of material of resistivity 2.0×10^{-7} Ωm.

(ii) Calculate the resistance between the rim and the hub of the wheel.

(iii) What is the maximum power available to an external load at the above speed?

(Oxford: *and JMB, AEB Cambridge for (a) and (b)*)

(a) Beware of spending too much time on your diagram – about 5 minutes is all that should be needed to sketch and label the essential features. Consult your textbook if necessary before making a sketch. The equation for current sensitivity is given in the commentary of 2.14M so the features are represented by the factors in that equation. Proof of the equation is not asked for here. Coil oscillations are prevented by choice of material for the coil frame to allow eddy currents to be induced to oppose the motion when the coil frame cuts the field lines. So what sort of material should the coil frame be made from?

(b) Field strength B, coil area A and spring strength k are the same for the two meters. Calculate the current taken by each meter, and then use the expression for current sensitivity to obtain an expression for the deflection of each coil in terms of B, A and k; hence determine the ratio of deflections.

(c) (i) Calculate the area swept out by a spoke in 1 second, and hence determine the magnetic flux swept out per second so giving the induced e.m.f. See E2.25 if necessary.

(ii) Calculate the resistance of a spoke. See E7.8 if necessary. The total resistance is that of 40 spokes in parallel – n equal resistances R in parallel have combined resistance R/n.

(iii) The wheel is a generator of e.m.f. and internal resistance as calculated in (i) and (ii) respectively. The Maximum Power Theorem which is required by only a few Boards must be used here; it states that maximum power is delivered by a generator when the load resistance is equal to the internal resistance of the generator. Hence calculate the maximum power – see E7.6 if necessary.

81L In this question you are asked to discuss the different ways in which you might use graphs in thinking about a piece of physics.

In **(a)**, **(b)**, **(c)** below you are shown three graphs and told about the data from which they are derived. For each of these three graphs provide as many of the items (i) to (iv) below as you think relevant. You should also comment on similarities or differences in the way the three different graphs may be used.

(i) A concise description in words of the facts of the situation as they can be seen simply by looking at the graph.

(ii) A discussion of the aspects of the graph which you ought to be able to understand from the basic physics of the system.

(iii) An explanation of the useful data which can be derived from various features of the graph: slopes, curvatures, intercepts, turning points and so on.

(iv) A consideration of new graphs that might be plotted, using the data from the original graph, to reveal something more about the situation.

Graphs

(a) Recordings of distance (s) against time (t) for a new rocket fired vertically into the sky.

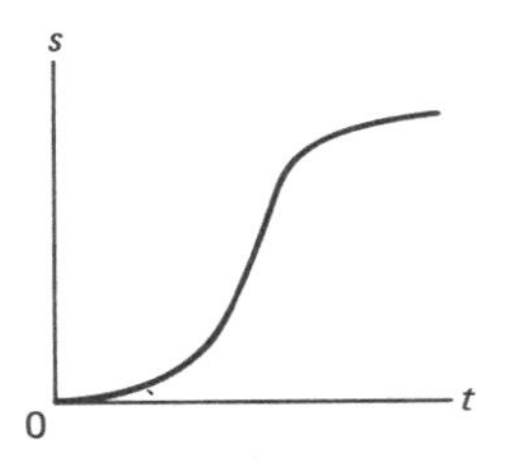

(b) Readings of extension (x) against stretching (+) or compressing (−) force (F) for a coiled wire spring.

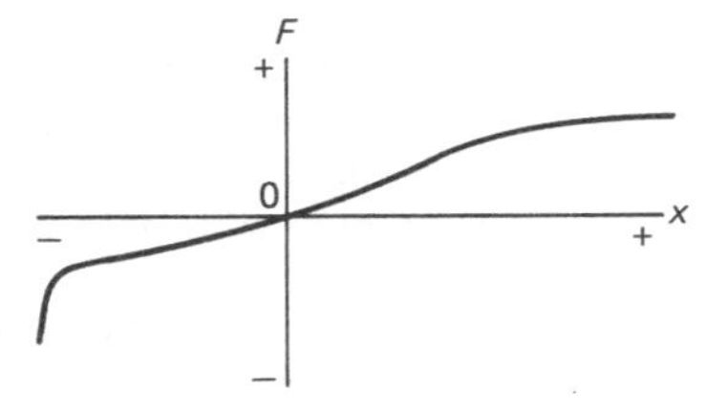

(c) Readings of the number of ions (I) per unit length of path versus path length (x) for an alpha-particle in air as it travels away from its parent nucleus.

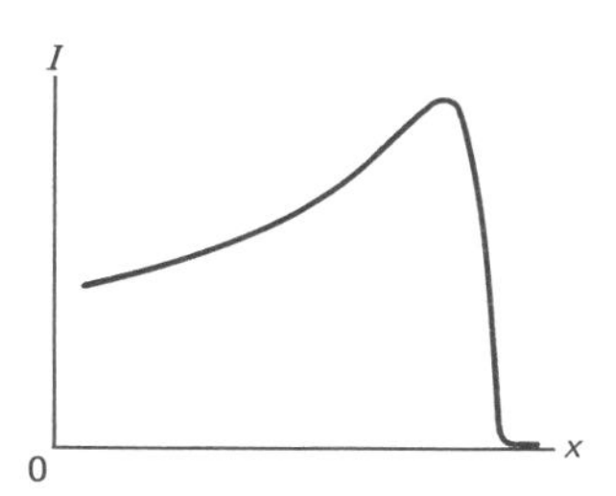

(**O and C Nuffield**: *N Ireland* and all other Boards*)

Topic 9.2 will be helpful to those who have difficulty analysing graphs. However the following observations about the three graphs and the physical situations that they describe may be of further assistance.

(a) This is a distance–time graph, not a displacement–time graph; hence it cannot have a negative slope at any point as the total distance travelled can never decrease. Note that the speed of the rocket (slope of the graph) gradually slows down at large t; if the rocket were falling back to earth it would be speeding up. Where does the rocket stop accelerating, and why? It is possible to get a speed–time graph and a graph of magnitude of acceleration against time, how?

(b) Does Hooke's law apply? At large compressions and extensions the graph shows very different behaviour. Describe this behaviour and explain it; remember that it is a 'coiled wire spring'. From the graph it is possible to deduce both the spring constant k and the energy stored in the spring for a given extension or compression (see F9.4). Could some useful graphs be plotted involving these quantities? Would a graph of k against x be sensible (for example)?

(c) Here the physics of the situation is complicated and needs some explaining. Remember that the bigger the distance x travelled by an alpha-particle the slower it will be travelling; eventually it stops, this range is defined by the graph. Why do the alpha-particles slow down? From the graph it can be shown that slower moving alpha-particles can create more ions than higher energy alpha-particles; to explain this, the analogy that two magnets will exert greater impulses on each other if they travel slowly past one another may help. How can you deduce the total number of ions produced from the graph, could this lead to a further graph, would this tell you anything about the total energy of the alpha-particle?

82L (a) What do you understand by 'half-life $T_{\frac{1}{2}}$', and 'decay constant λ', for a radioactive substance. Deduce the relationship between them.

(b) (i) The first part of the decay series of the artificially produced neptunium isotope $^{237}_{93}$Np involves the following sequence of emissions: $\alpha\beta\alpha\alpha\beta\alpha$. Illustrate these changes on a plot of N, the number of neutrons in the nucleus, against Z the atomic number. Using the table of elements below, identify (by its symbol) the last element in this portion of the decay series, and make clear which isotope of this element is produced.

Element	Bi	Po	At	Rn	Fr	Ra	Ac	Th	Pa	U
Atomic number	83	84	85	86	87	88	89	90	91	92

(ii) Three other nuclides also initiate decay series. The half-lives and abundances are as shown below.

Element	Half-life/years	Natural abundance
^{232}Th	1.4×10^{10}	abundant
^{235}U	7.1×10^{8}	rather rare
^{237}Np	2.2×10^{6}	not found
^{238}U	4.5×10^{9}	abundant

Comment on the age of the Earth in the light of these data.

(**Cambridge**: *NISEC and all other Boards except SEB SYS–SEB H (b) only*)

(a) See p. 129 and your textbook if necessary. The relationship is that the product $T_{\frac{1}{2}} \times \lambda$ equals $\log_e 2$; to deduce this, start from the equation $N = N_0 e^{-\lambda t}$ where N = the number of atoms of the radioactive isotope left after time t, and N_0 is the initial number of radioactive atoms.

(b) (i) Remember that the atomic number (Z) is the number of protons in the nucleus, and in the isotope symbol it is the lower number. The upper number (237 here) is the number of neutrons and protons in the nucleus. Also, each α-emission means that the nucleus loses 2 neutrons and 2 protons, and each β-emission involves a neutron changing into a proton. Your plot should cover the range of neutron number and proton number involved in the sequence. Mark clearly the position of each nucleus (on the plot) in the sequence, and indicate the changes by an arrow from one nucleus to the next nucleus in the sequence. Indicate on the arrow if the change is an α- or a β-emission. The final element is identified by the atomic number of the last isotope in the sequence. (ii) The key is the abundance; assuming that all these nuclides were present at the time of the Earth's formation, those with half-lives much shorter than the present age of the Earth ought to have decayed by now to a negligible fraction of the initial amount.

83L In this question you are asked to make quantitative estimates about proposals to use solar radiation as a source of power. The data given at the end of the question may be useful for your calculations.

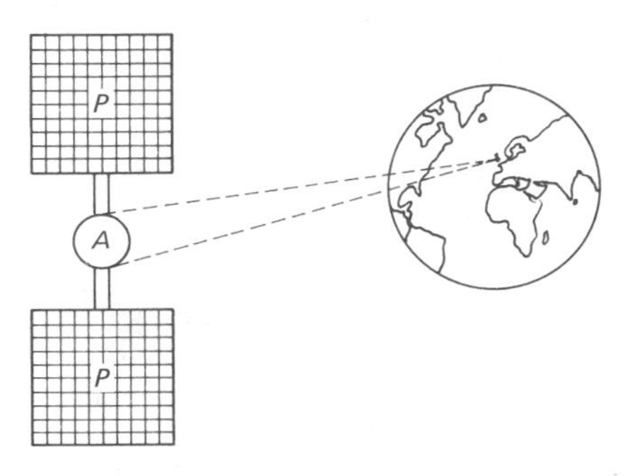

Energy may be obtained from the Sun's radiation by use of solar cells, which produce electrical power when illuminated by the Sun. Such cells can convert 15% of the power of the radiation into electrical power and they cost about £5000 per square metre. One way of using these cells to make a power station is to construct a space station out of two panels of cells P–P as shown, with a dish-shaped aerial A between them which will transmit the electrical energy as a beam of radiation of wavelength 10 cm. The station will be kept in orbit at a distance chosen so that it stays permanently above a fixed receiving station on the Earth. It is proposed that the panels P–P be each of dimensions 4 km by 4 km and that the aerial A be of diameter 1 km; the orbit will have to be at 3.6×10^{7} m above the Earth's surface.

You are asked to think about the implications of this idea, considering the points listed below and making as many quantitative estimates as you can to illustrate or explore your ideas.

(a) The number of such stations needed to supply the needs of the United Kingdom (about 5×10^{10} W).

(b) The fraction of time that such a station would be in the Earth's shadow.

(c) The area on the ground needed for a receiving aerial.

(d) The area needed to generate the same mean power with solar cells on the Earth's surface.

(e) The costs and other factors that you would need to consider in deciding between a solar-cell power station in space and a solar-cell power station on the Earth's surface.

(f) The possible advantages and disadvantages of having the space station in an orbit much farther from or much nearer to the surface of the Earth.

Data

Radiation incident on the Earth's outer atmosphere has power 1400 W m^{-2}.

Radiation incident on the Earth's surface in the U.K. in full daylight has power 600 W m^{-2}.

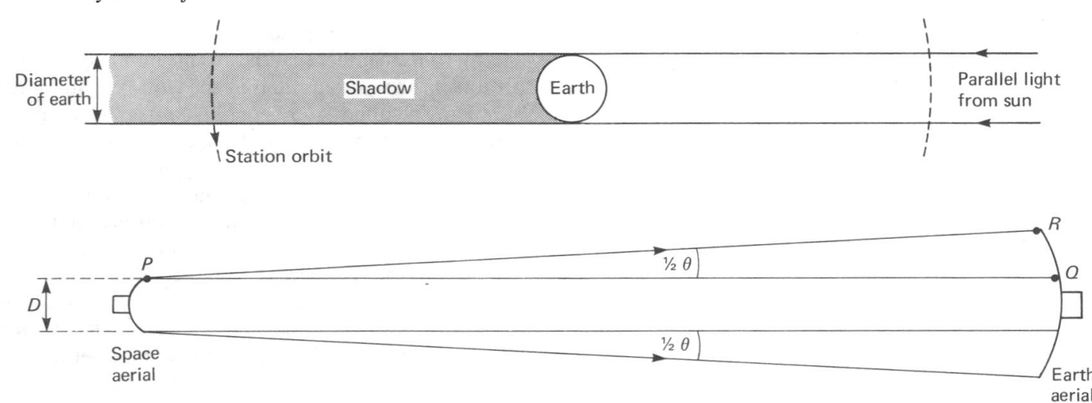

The efficiency of aerial transmission is such that about 75% of the electrical power produced in the space station can be produced as electrical power in the receiving station on earth.
The diameter of the Earth is 1.3×10^7 m.

(**O and C Nuffield**: *all other Boards*)

(a) From the area of the solar panels (2*P*) the power of radiation collected in the Earth's outer atmosphere can be found. 15% of this becomes electricity and 75% of this electricity reaches the Earth. Divide the power reaching the Earth by the UK requirement and remember that you cannot have a fraction of a space station!

(b) From the diagram it can be seen that for a distance equal to the diameter of the Earth the station is in shadow; the fraction of a complete orbit that this represents is the answer needed for this part.

(c) This requires an application of E4.13. Resolving power is an angle (θ) and the same formula can be applied to a parabolic aerial. The aerial should give out a parallel beam but will not because of diffraction, in fact the angle of spread (θ) is roughly given by $\theta/2 = \lambda/D$ if D is the aerial diameter and λ is the wavelength of transmission. The diagram should make it clear how to calculate the diameter of the Earth aerial needed, as $\theta = 2\,(QR/PQ)$.

(d) Remember that the power received on Earth is only 600 W m^{-2} and that the 15% efficiency term is needed; but not the 75% term!

(e) The relative cost of constructing the panels can be compared, but allow for the cost of getting the space station into orbit. The space station will be weightless and surrounded by the 'vacuum' of space. For how long will the Earth station be in shadow? Consider problems over transmission of power and security of the stations, and what about maintenance needs? Will the size of the stations cause a problem? What about the weather? Remember that a space station also requires an Earth receiving station.

(f) What will the effect of a more distant orbit be on the following: the time for which the station is in shadow, the radiation received by the station, the size of the aerial on the Earth, the cost of building the station, the energy lost in transmission (absorbed by the atmosphere), etc.

Practical problems questions

Although these questions are all taken from Nuffield papers they will be useful to all students. In the examination 11 minutes per question is allowed, but this does include the time for taking the measurements. The measurements are shown in black.

84P A piece of wood and a piece of metal are provided. Put a sheet of graph paper on the wood and then put a sheet of carbon paper sensitive side downwards, on the graph paper, as shown in the diagram.

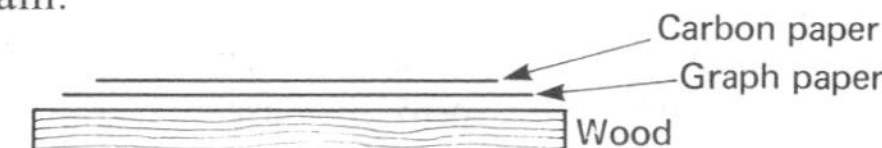

(a) Release the glass marble from a height of 0.5 m so that it lands on the carbon paper and makes a mark on the graph paper. What is the area of the mark?
Area of mark = 20×10^{-6} m^2

(b) Replace the wood by the metal and repeat **(a)** so that a mark is made, measure the area again.
Area of mark = 3×10^{-6} m^2

(c) Assume that the area is proportional to the depth of the dent produced and compare as quantitatively as you can the force exerted by the marble on the wood with the force exerted by the marble on the metal.

(d) State any assumptions you have made, in addition to that given, in doing part **(c)**. (**O and C Nuffield**: *all other Boards*)

(c) Because the glass marble is dropped from the same height in each experiment it will have the same ke to lose each time it hits the carbon paper. This ke loss can be measured as work done (force × distance) where the distance involved is the depth of the dent. It should be easy now to compare the forces involved in the two cases; note that the deeper dent does not represent the bigger force!

(d) The following points may help you: The 'work done' formula applies to constant forces only; will the force be constant in these experiments? Does the bouncing of the ball matter? (E1.5 may be applied to help solve this.) Is all the ke of the ball converted into helping create the dent?

85P The circuit shown in the diagram is set up on the bench. *X* and *Y* are two resistors, and *S*1 and *S*2 are two switches.

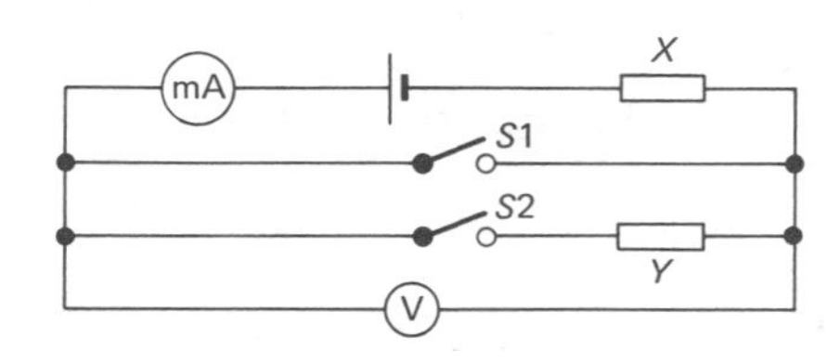

(a) Take readings to complete the following table.

	Voltmeter reading	Ammeter reading
(i) *S*1 open, *S*2 open	2.0 V	0 mA
(ii) *S*1 closed, *S*2 open	0 V	62 mA
(iii) *S*1 open, *S*2 closed	0.5 V	46.5 mA

(b) Deduce a value for *X*, stating any assumptions you make.
(c) Deduce a value for *Y*, stating any assumptions you make.

(**O and C Nuffield**: *all other Boards*)

(b) The key is to understand the significance of the readings (i) (ii) and (iii) in part **(a)**. It may help to redraw the circuit diagrams (in rough) simplifying them to allow for the effects of opening and closing the switches; an example for (iii) is below. One of the sets of readings of **(a)** gives a value for the emf of the battery (assuming the voltmeter draws a negligible current) and another gives a value for the current through X when the full emf of the battery (assuming that the internal resistance of the battery and the resistance of the ammeter are negligible) is dropped across X; hence the value of X can be determined. Note that closing *S*1 connects the voltmeter across a wire which, having effectively no resistance, will have zero pd across it so giving a zero reading on the voltmeter; as long as the ammeter's internal resistance and the internal resistance of the battery are negligible, then the only resistance in the circuit is X which must therefore have the full battery emf across it.

(c) The third experiment of part **(a)** gives the current through X and Y when they are connected in series (same current) and the pd across Y. Here the internal resistance of the cell and the resistance of the ammeter will not affect the result, however if the voltmeter's resistance is not negligible then the ammeter will read incorrectly.

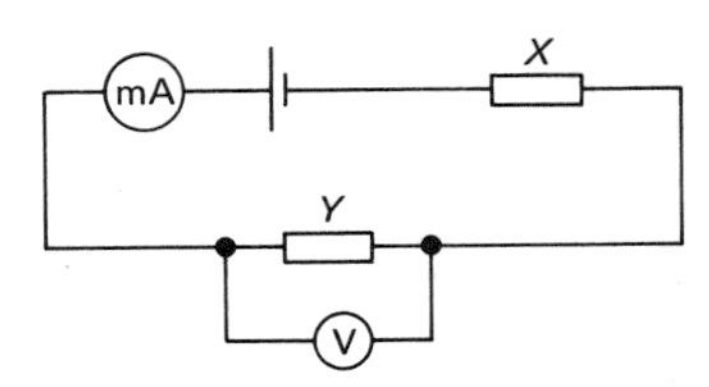

86P Masses totalling 200 g are secured to the end of a metre rule which is clamped to the bench. A stopwatch is provided.

(a) Make as accurate a measurement as possible of the period of oscillations of the masses on the end of the rule. Show your readings and working.

Three timings are made for 50 oscillations: 19.6 s, 19.5 s, 19.3 s.

(b) The period T of the oscillations is given by the formula

$$T = 2\pi\sqrt{\frac{m}{k}}$$

where m is the moving mass and k is the restoring force per unit displacement.

(i) Using a spring balance provided measure a value of k for this system as accurately as possible. Say what measurements you take and how you use them to obtain your value of k.

The amplitude of the ocillating masses is about 2 cm so the following results are taken:

Deflection/cm	+ 2.0	+ 1.0	− 1.0	− 2.0
Force needed/N	+ 1.8	+ 1.0	− 0.9	− 1.7

(ii) Substitute for m and k into the formula given to determine a value for T.

(iii) Comment on whether the measured value for T and the calculated value compare as you would expect.

(**O and C Nuffield**: *all other Boards*)

(a) When calculating T avoid the temptation to quote your answer to too many significant figures, use the variations in timings of 50 oscillations to guide you (see also 9.3).

(b) (i) Note that k is the force needed to displace the masses by unit distance (ie, 1 metre), but the displacements are measured in centimetres! (ii) Convert the masses into kg! Quote the answer to a sensible number of significant figures. (iii) Do the results agree? If not, why not? Do they agree after allowing for the likely experimental errors? Can the values of m and k be wrong? What about the mass of the ruler – would this affect the results in the way you have found in your calculations?

87P This question is concerned with the absorption of light by coloured filters. The circuit shown is set up on the bench, and a 'white light' lamp and two identical red filters are supplied. When light falls on the photoresistor, its resistance changes, causing the reading of the meter to change.

(a) With the lamp switched off, note the reading of the meter.0 A.

(b) Switch on the lamp and, without any filters in position, adjust its distance from the photoresistor until full-scale deflection of the meter is obtained. Write down the value of this current.10 mA.

(c) (i) Support one red filter in front of the lamp and note the reading.4.5 mA.

(ii) Say why the reading of the meter falls when the filter is put in position.

(d) (i) Add a second red filter and note the reading.3.3 mA.

(ii) Why is the change in the reading produced by adding the second filter less than the change produced by adding the first filter?

(e) Use the results obtained above to deduce what you think would be the effect of adding a third red filter.

(**O and C Nuffield**: *all other Boards*)

Obviously the resistance of the photoresistor drops when light shines on it, hence the current in the circuit will increase. In part **(b)** white light (ROYGBIV) is allowed to hit the photocell. However the red filter absorbs most of the colours apart from red (OYGBIV). So with only red light hitting the photocell a current of 4.5 mA is recorded. A second filter reduces the brightness of this red light so the photocell reads 3.3 mA. If the current reading were proportional to the brightness of the light then introducing a filter cuts the red light brightness by about a quarter (from 4.5 to 3.3 mA). So a third filter should reduce the brightness by a quarter again. Beware the trap of wrongly suggesting that as the current dropped by 1.2 mA (4.5 – 3.3) it will do so again.

88P This experiment requires you to charge an electroscope and to do some calculations about capacitances. The apparatus shown is already set up on the bench. Plug the lead into the battery so that the bulb lights and casts a shadow of the gold leaf on to the screen. Make sure the electroscope is discharged by touching the cap with your finger.

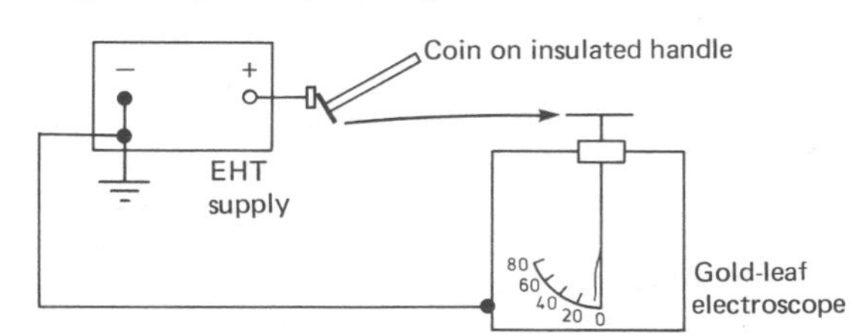

(a) Touch the 2p coin on to the positive terminal of the eht supply and transfer the charge to the cap of the electroscope. Repeat this three more times and note the total deflection of the leaf caused by the four transfers of charge. 65°.

(b) Discharge the electroscope again and transfer the charge in the same way using the ½p coin this time, until the same deflection, or as nearly as possible the same deflection, is obtained. Write down the number of transfers of charge needed and the deflection caused by the total charge. 6 transfers produce a deflection of 65°.

(c) Use the information you have obtained in (a) and (b) to estimate the ratio: (capacitance of a 2p coin)/(capacitance of a ½p coin).

Set out your reasoning clearly and state any assumptions you make.

(**O and C Nuffield**: *all other Boards*)

The scale on the electroscope could be calibrated in volts; ie, the electroscope is charged up to the same voltage in parts **(a)** and **(b)**. Using $Q = CV$ we can see that the same charge must be transferred to the electroscope leaf on each occasion. So four transfers of charge from the 2p coin are equivalent to six from the ½p coin. But coins behave as capacitors and they are also charged up to the same voltage each time (the eht supply voltage) before their charge is transferred to the gold leaf. This makes it easy to compare the capacitances of the two coins. However, the comparison can only be made if all the charge from the coins is transferred to the electroscope; under what conditions will this happen? Also what about the effect of charge leaking from the coins before the transfer is made – could this affect the results?

89P The circuit shown is already set up. The total resistance of the microammeter and resistor R connected in series is 20×10^3 ohms. The cell has an emf of 1.5 V.

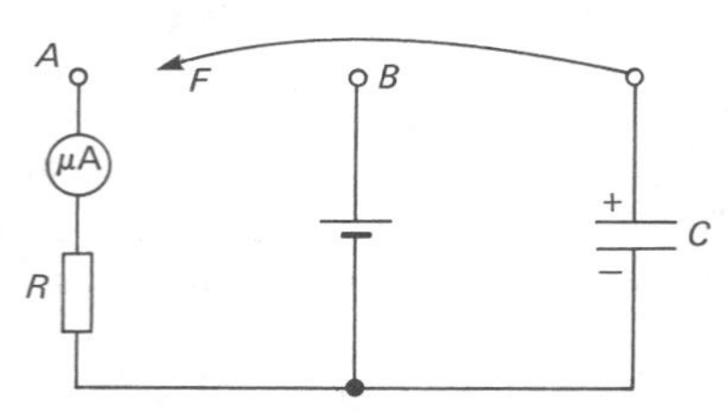

(a) Connect the flying lead F to the battery terminal B. Now connect F to A and record what happens to the deflection of the meter.

On connecting F to A the needle flicks across to about 45 μA, and the reading gradually drops taking about 10 seconds to reach zero.

(b) Compare the deflection of the meter with what you would expect. Give your reasoning clearly.

(c) Estimate the value of the capacitance C. Explain your reasoning and mention any assumptions you make.

(**O and C Nuffield**: *Oxford*, O and C and all other Boards*)

(b) The current – time characteristic for a discharging capacitor is shown in F7.19. The starting value of the current on discharge (I_0) can be calculated from the resistance in the circuit (20×10^3 ohms) and the voltage to which the capacitor is charged (1.5 V). This value does not agree with the reading in part **(a)**. Why not? (Inertia of the meter is the key).

(c) The capacitance can be estimated from the length of time it takes for the capacitor to 'empty'. This should be about 5 time constants ($5 \times CR$). This enables a value of C to be obtained. Note that a polarized capacitor has been drawn in the circuit diagram, so expect a reasonably large capacitor value (greater than about 10 μF). Is it necessary to assume in addition that the capacitor was fully charged to 1.5 V for this estimation to be valid? (No! – the time taken to fill or to empty capacitors is independent of the voltage used).

In this experiment you are asked to compare the masses of two trolleys by a dynamic method. Most of the marks are awarded for part **(b)**.

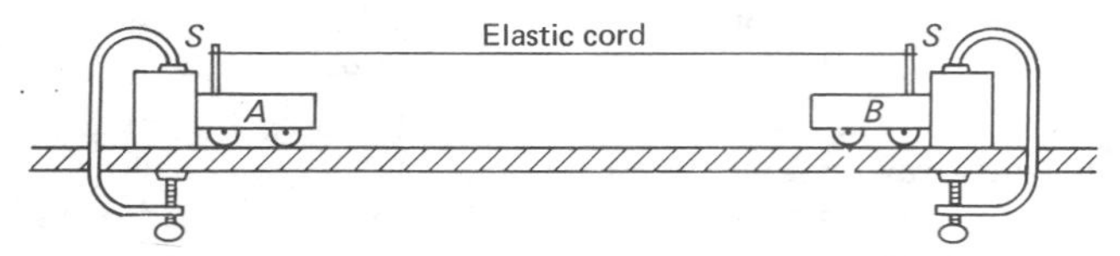

(a) Pull the trolleys apart until they reach the stops S, S as shown in the diagram. The positions of S must not be changed throughout the experiment. Release the trolleys simultaneously and note how far the trolleys move before colliding. Make three sets of observations.

Distance moved by trolley A/cm	36.5	35.2	35.5
Distance moved by trolley B/cm	24.0	25.3	25.0

(b) Work out what you can about the relative masses of A and B.

(**O and C Nuffield**: *all other Boards*)

The trolleys both travel for the same length of time (t) before they collide. Furthermore the trolleys experience the same strength forces but acting in opposite directions (action and reaction are equal and opposite); this is due to the elastic cord that connects the trolleys. Remember that the starting velocity (u) of each trolley is zero and use $F = ma$ and $s = ut + \cdots$ (E1.3) for each of the trolleys to get a value for the ratio of their masses. Although the force and acceleration are NOT constant as the second formula requires, it is still reasonable to assume that the distance travelled is proportional to the acceleration.

91P This experiment is about forces caused by electromagnetic induction. Most of the marks are awarded for part (c). The apparatus shown in the diagram is set up on the bench.

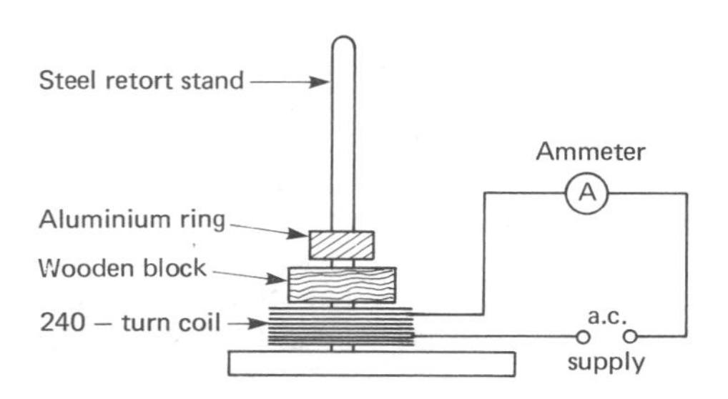

(a) Switch on the power supply and adjust the current until the solid aluminium ring is just on the point of lifting off the wooden block. Slide a split aluminium ring on top of the first ring and adjust the current until they are on the point of lifting off the wooden block. Repeat with the third ring (also split) on top. Record the current in each case.

Number of rings	1	2	3
Current I in A	2.6	4.1	5.5

(b) Say, with reasons, whether or not you think your results show that the current varies linearly with the number of rings.

(c) What would you have to think about in trying to explain your results? Marks will be awarded for sensible and relevant comments, but a complete theoretical explanation is not expected.

(**O and C Nuffield**: *all other Boards*)

(b) Be careful; the results do not form a straight-line graph passing through the origin ($2.6/1 \neq 4.1/2 \neq 5.5/3$). However, plot the results on a sketch graph and the answer ought to be obvious! Alternatively, note that each time the number of rings is increased by 1, the current rises by roughly the same amount. See 9.2 if necessary.

(c) The system behaves as a transformer with one turn (the unsplit or solid ring) on the secondary in each experiment. The current readings in **(a)** give the primary current but the secondary current must be proportional to this, ie, the current flowing around the solid ring. This secondary current in the aluminium ring creates a magnetic field which interacts with the magnetic field due to the current in the primary (240 turn coil). Thus the ring is repelled by the coil so lifting the ring. No current is induced in the split rings (as there is no electrical 'circuit') so the bottom ring does all the lifting (the rings weigh the same). As the lifting force must be proportional to the magnetic field strength which is proportional to the current, the linear increase in current is predictable. However, we would expect these results of **(a)** to form a straight line through the origin – why does this not happen in this experiment? In a 'perfect' transformer, all the magnetic flux linking the primary passes through the secondary – is the flux through the aluminium ring the same in each case?

92P In this experiment you are required to take two quick measurements and then to criticise and suggest improvements. Marks will be awarded for the criticisms and suggested improvements.

The human body has a density which is close to that of water. A large fraction of the body is water so perhaps its electrical conductivity is the same as that of water. The circuit shown is set up on the bench to start an investigation of this.

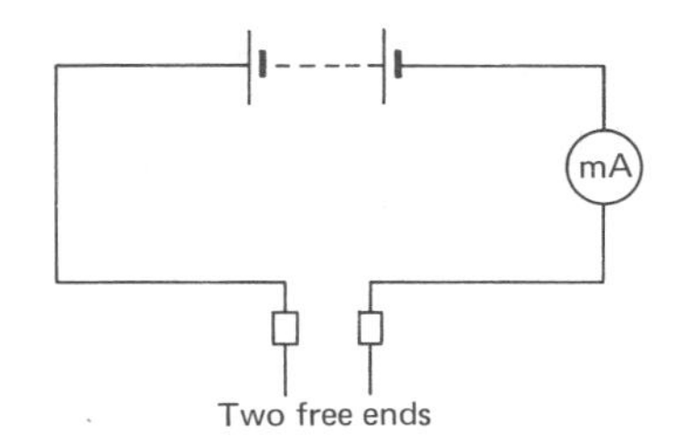

Dip the two free ends of the wires into the beaker of tap water about 5 cm apart and note the current which flows. Current is 3.4 mA.
Now dry the ends of the wires and put them about 5 cm apart on one of your fingers and note the current that flows. You may need to change the range of the meter to do this. Current is 0.09 mA.

Criticise this experiment as a means of comparing electrical conductivities and say what you would do to improve it.

(**O and C Nuffield**: *all other Boards*)

See E7.8 for conductivity (conductivity = 1/resistivity).

In the experiment the same voltage is used, but the currents obtained can only be used to compare conductivities if specimens (finger and water) of the same size were used; was this the case? Furthermore a laboratory wire has a very low resistance but its plastic insulating sleeve has a very high resistance; apply the same idea to the human body and the apparent conflict of results and theory in this experiment can be better explained. Ideas for improving the experiment should follow naturally when the faults have been identified.

93P This experiment is about the effect of a baffle on the sound from a loudspeaker. The baffle consists of a large square piece of card with a circular hole cut in its centre; the hole is the same diameter as the cone of the circular loudspeaker. The audio oscillator is set at a frequency of 340 Hz; do not alter this setting.

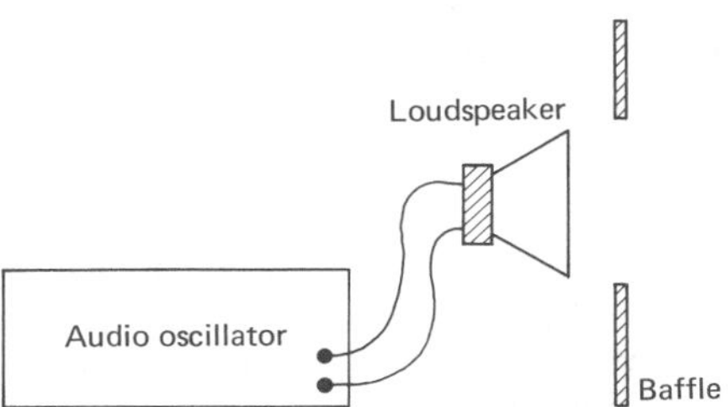

(a) Calculate the wavelength of sound emitted by the loudspeaker at this frequency. (Assume that the velocity of sound is 340 m s^{-1}.)
(b) Turn up the volume control until the sound is JUST audible. Then put the baffle up against the loudspeaker touching it with the hole in front of the cone. Try different positions of the baffle. What effect does the baffle have on the loudness, and where is it effective and ineffective?
The sound from the loudspeaker is much louder when the baffle touches the loudspeaker but the effect wears off very quickly as the baffle is moved away about 4 to 5 cm, producing no audible effect.
(c) The sound from the back of the loudspeaker is out of phase with that from the front (the cone moves in at the back when it moves out at the front). Use this fact to give as full an explanation as you can of the observations made in **(b)**.
(d) What effect would you expect if the frequency were ten times greater?

(**O and C Nuffield**: *all other Boards*)

(c) Without the baffle, sound can travel to the ear from both the back and front of the loudspeaker; using the baffle cuts the sound from the back. Superposition of waves and interference are the keys to this part.

(d) The change of wavelength (from 1 m at 340 Hz to 0.1 m at 3400 Hz) is the important point. The loudspeaker has a diameter of about 0.1 m. Wherever the listener is placed, at 340 Hz waves from the back are going to destructively interfere with those from the front. However in the second experiment the baffle will be found to have little effect. Considering a wave from the front of the loudspeaker compared with one from the back should help explain why.

Comprehension papers

There are two comprehension papers. The time allowance for each is 1 hour.

94C (1 hr) This article is about those aspects of noise implied in the well known definition, 'sound undesired by the recipient'. There is in principle no technical difficulty in expressing, in detail, the physical events that constitute a noise. This may be done to any desired degree of accuracy. For example, oscillograph records of the wave forms of the pressure variations at each ear of the hearer will contain all the information of an acoustic nature about the noise.

There are only three basic dimensions of acoustic sensation: pitch, loudness and quality. Thus two sounds may differ from each other with respect to one of these dimensions while being the same with respect to the other two. Loudness is a subjective quantity that corresponds rather closely with the physical intensity of sound. Loudness represents the size of the sensation and is not concerned with the pleasantness or unpleasantness of the sound.

Measurements of subjective values must be kept as simple as is consistent with the facts. One example is the simplified rule relating loudness, S, in sones to loudness-level, P, in phons:
$P - 40 = 33 \log_{10} S$
On the sone scale the number is proportional to the average person's estimate of the magnitude of the loudness sensation. On the phon scale the number depends on the physical intensity of a 1000 Hz reference tone which the average person judges to be of the same loudness as the noise.

The scale of noticeable differences is much coarser in the subjective world than in the physical. This has important consesequences; for example, it takes a rather large physical reduction of a noise to achieve a modest degree of subjective improvement. At the same time, if difficulties in enforcing noise limits are to be avoided, these limits must be clearly defined in terms of meter readings. There must not be too great a tolerance on the maximum allowed reading.

The test of subjective measurability depends on checks of consistency. For example, if in the opinion of a listener, A and B are equally noisy, and B and C are equally noisy, then one can verify experimentally that he also finds A and C equally noisy. In a similar way, the possibility of measurement on the sone scale depends on the experimental verification of propositions such as this: if A is three times as loud as B, and B is twice as loud as C, A must be six times as loud as C. In practice we exact a further condition of measurability, namely a reasonable consensus of opinion between different 'normal' listeners. because we are ordinarily concerned with decisions taken on behalf of communities of people rather than individuals. Loudness passes these tests of measurability.

Questions

1 Are the following quantities physical or subjective? Give a brief explanation of each of your answers.
(a) The variation of sound pressure at the ear.
(b) The energy transmitted by the sound wave.
(c) The loudness-level of a sound.
(d) The 'annoyingness' of a sound. **(10 marks)**

2 Explain in your own words what you have learned from the article about the following:
(a) loudness.
(b) the sone scale. **(6 marks)**
3 Give an example of a physical and a subjective quantity from a branch of physics not included in the article. **(4 marks)**
4 Explain in your own words the meaning of the phrase 'the scale of noticeable difference is much coarser in the subjective world than in the physical'. **(3 marks)**
5 Sketch the kind of oscillograph record you would expect to obtain from:
(a) a short burst of a pure tone.
(b) starting the engine of a lorry. **(4 marks)**
6 The loudness-level of a sound x is 73 phons.
(a) What is the loudness of x?
(b) If the loudness of x is four times the loudness of a sound y, what is the loudness of y? **(3 marks)**

(**London**: *all other Boards*)

1 Measurement of a physical quantity is independent of whoever makes the measurement; however, the measure of a subjective quantity would vary from one person to another. With this point in mind, answer (a) to (d) in turn. Your background knowledge of pressure and energy should enable you to explain your answers to (a) and (b). For (c), the last sentence of the third paragraph is the key to your explanation – judgement is involved! For (d), everyday experience ought to enable you to give a brief explanation.
2 (a) Look closely at the second and final paragraphs before you answer. (b) See the third paragraph and the first part of the fourth paragraph. Remember you must explain in your own words.
3 Medical physics is one branch which has many examples of a physical stimulus giving a subjective response. Think about one of the human senses other than 'sound'.
4 Give a simple example to illustrate your answer; if the intensity of a sound is doubled, will its subjective measurement also double?... or will it be more or less than double? A much coarser scale means greater difficulty in its measurement.
5 (a) Remember that a pure tone is a single frequency note. (b) The sound would contain a range of frequencies, and as the lorry 'revs up' (engine speed increases), the higher frequencies would predominate.
6 (a) Use the equation given in the passage.
(b) Read the final paragraph before answering the question.

95C First read carefully the passage below and afterwards answer the questions that follow.

Thermoelectric cooling

A thermocouple develops an emf if the junctions of the two wires of dissimilar metals composing it are maintained at different temperatures. This is the Seebeck effect and an important term concerning it is thermoelectric power which may be defined as the emf developed per degree difference in temperature between the junctions. It is possible to relate graphically the thermoelectric power to the temperature difference between the junctions. This has been done for a copper – iron thermocouple in Fig. 1 and the expected emf for a given temperature difference can be calculated from the area under the graph.

The converse of the Seebeck effect is the Peltier effect in which heat is given to, or absorbed from, the surroundings if an electric current is passed through a junction between two dissimilar materials. The Peltier effect can be made the basis of a heat pump, ie, a device that transfers heat from a cold junction to a warmer one. This is best done by using specially prepared substances known as n-type and p-type semiconductors as the effect is then so large as to make thermoelectric cooling of practical interest where small size and absence of mechanical movement are desirable. Fig. 2 shows a single cooling unit consisting of such semiconductor elements, n and p, joined with suitable contacts. T_0 is the ambient temperature, ie, the temperature of the surroundings. When a current I flows as shown, heat is pumped from the common junction which reaches a temperature T which is

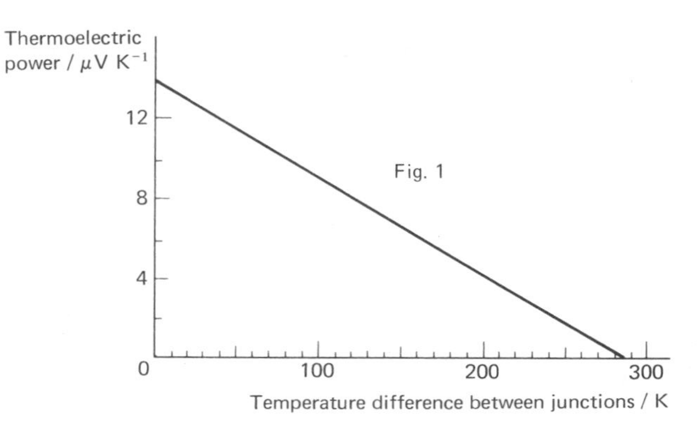

Fig. 1

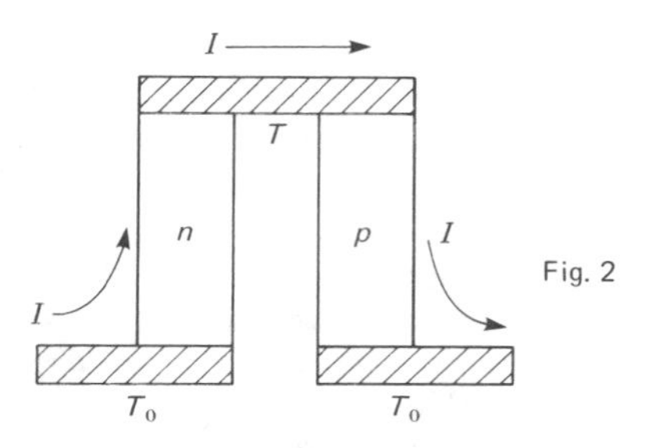

Fig. 2

less than the ambient temperature by an amount which we shall denote as ΔT.

The magnitude of the Peltier cooling effect is reduced by heat conducted down the elements and by the normal heating effect which accompanies an electric current (Joule heating). In the absence of external heat sources, the cooling effect is balanced by these heat losses and it can be proved that the relevant equation is:

$$QTI = \lambda \Delta T + \tfrac{1}{2}I^2R \qquad \text{(i)}$$

where Q is the thermoelectric power, λ the thermal conductance and R the total electrical resistance of the semiconductor elements. From equation (i) it can also be proved that the maximum value of ΔT is given by:

$$\Delta T_{max} = \frac{Q^2 T^2_{max}}{2\lambda R} \qquad \text{(ii)}$$

where T_{max} is the temperature of the common junction when ΔT is a maximum. In an experiment with a thermoelectric cooling unit using n-type lead telluride and p-type lead telluride the following results were obtained:

I/A	1.0	2.0	2.5	3.0	3.5	4.0	5.0
$\Delta T/K$	4.9	9.3	11.0	11.3	11.0	10.1	7.9

In such a cooling arrangement, a figure of merit, F, based on the expression for the maximum temperature difference, ΔT_{max}, may be defined as:

$$F = \frac{Q^2 \sigma}{k} \qquad \text{(iii)}$$

where σ is the electrical and k is the thermal conductivity. For the semiconductor material lead telluride typical values are $\sigma = 10^5\,\Omega^{-1}\,m^{-1}$ and $k = 4\,W\,K^{-1}\,m^{-1}$ and the predicted maximum temperature difference, ΔT_{max}, is 11.5 K. An even more promising thermoelectric material is bismuth telluride for which a figure of merit of $3 \times 10^{-3}\,K^{-1}$ has been obtained, leading to a maximum temperature difference of 80 K. High values of ΔT_{max} are necessary since additional heat loads, such as would be experienced in actual devices, act to reduce the temperature difference.

Semiconducting materials having high thermoelectric powers can also be made into effective generators of electricity by applying the Seebeck effect. These are particularly attractive with high F materials since the same figure of merit applies. In fact, however, it is usual for different materials to be used for power generation since they are required to withstand higher temperatures.

Marks

1 (a) Explain in your own words what is meant by the statement that the Peltier effect is the converse of the Seebeck effect **3**

(b) What does the passage describe as the practical advantages of thermoelectric cooling? 2

2 (a) What would be the effect on the thermoelectric cooling unit shown in Fig. 2 if the n-type semiconductor were replaced by a p-type element equivalent to that already present? 2

(b) What would be the effect of reversing the current through the unit in Fig. 2? 2

(c) What is meant by Joule heating? 2

3 By consideration of dimensions or otherwise, derive the units of λ, thermal conductance (equation (i)), and explain the relationship between thermal conductance and thermal conductivity, k. 4

4 Use Fig. 1 to estimate the emf of a copper–iron thermocouple when the cold junction is at 0°C and the hot junction is at (*a*) 285°C, (b) 200°C. 4

5 (a) For lead telluride, plot a graph of the quoted experimental values of ΔT against the corresponding values of current. Include on your graph the predicted value of ΔT_{max} and comment briefly on whether your graph fits in with the accepted theory. 4

(b) Rewrite ΔT in terms of the other quantities given in equation (i) and hence show that the maximum value of ΔT occurs when $I = QT/R$. 2

6 If the figure of merit for a unit constructed of lead telluride is only one-twelfth of that for bismuth telluride, use the numerical information given in the passage to calculate the thermoelectric power of lead telluride. 3 = 28

(**JMB**: *all other Boards except SEB H*)

1 (a) The Seebeck effect is described at the beginning of the passage, the Peltier effect in the second paragraph. Do not forget to put the explanation into your **own** words

(b) See paragraph 2.

2 When current passes through n-type material, the end at which the (conventional) current leaves the material becomes cooler; for p-type material, the end which the current enters becomes cooler.

(a) Would the new piece of p-type material try to make the junction cooler or warmer?

(b) See initial comments.

(c) See the text before equation (i); you should explain what 'normal' electrical heating is.

3 Use equation (i); each term of the equation is a power term (eg, $\frac{1}{2}I^2R$ is in watts).

4 The y-axis is in μV K^{-1}; the x-axis is in K. What does the area under the line represent?

5 (a) Use equation (ii) to calculate a value for ΔT_{max}.

(b) Differentiate the expression for ΔT with respect to I. What is the value of $\frac{d(\Delta T)}{dI}$ when ΔT has its maximum value? Hence prove $I = QT/R$.

6 Use equation (iii).

Data analysis papers

96D (1½ hours) Read the following account of an experimental investigation and then answer the questions at the end.

Corresponding measured values of potential difference, V, and current, I, for a semiconductor diode are given in the Table.

Potential difference, V/V	Current, I/μA
0.255	0.40
0.315	1.60
0.345	3.6
0.385	8.9
0.410	18.2
0.455	52.2
0.475	90.3
0.495	140
0.505	182
0.515	223
0.530	310

Questions

1 Using the values in the Table, plot a graph with I as ordinate against V as abscissa, determine, for the point on the graph corresponding to $V = 0.500$ V.

(a) the rate of change of I with V, and hence

(b) the percentage change in I corresponding to a 1% change in V. **(12 marks)**

2 The following theoretical equation (the 'rectifier equation') applies for certian types of semi-conductor diode:

$$I = I_0(e^{aV} - 1) \quad (1)$$

where I_0 and a are constants.

If V is sufficiently large, $I \simeq I_0 e^{aV}$ (2)

so that $\ln I \simeq \ln I_0 + aV$

or $\log_{10} I \simeq \log_{10} I_0 + 0.434\,aV$.

From an appropriate table of values plot

either a graph of ln (I/μA) as ordinate against V as abscissa.

or a graph of $\log_{10}$ (I/μA) as ordinate against V as abscissa.

From your graph derive values for I_0 and a. **(11 marks)**

3 Explain for which value of potential difference in the Table the approximation made in equation (2) will be most serious. Using the values of I_0 and a derived in Question 2, calculate the current at this pd using the exact equation (1). (e = 2.72.) Plot the corresponding point on your second graph.

State, giving your reasons, whether you consider that use of the approximate equation (2) was justified in analysing the results in Table 1. **(7 marks)**

(**London**: *all other Boards*)

NB Before starting, read through **all** the questions so that you know exactly what you have got to do!

1 Make sure you draw a **smooth curve** through the points.

(a) Draw a tangent at 0.500 V and extend it to give you a **large** triangle from which to measure the gradient (which has units!).

(b) Find 1% of V (ie, 1% of 0.500 V), then from your value of the gradient at this point, find the corresponding **change** in I. Express this change in I as a % of the value of I at 0.500 V.

2 The question says **from an appropriate table of values**, so you must show a table of V, I and ln($I/\mu A$) or $\log_{10}$ (I/μA). Before drawing the graph, read question 3 very carefully!

$\ln I = \ln I_0 + aV$ can be re-arranged as

$\ln I = aV + \ln I_0$ ie, of the form

$y = mx + c$

The equation will be more accurate for the larger values of V, so the slope in this region should be found to give a. The scale of your graph would be too small if you were to include the origin and the intercept c, so c must be calculated by substitution of suitable values (ie, large V) of y and x into $y = mx + c$. Remember that c will then be $\ln I_0$, not I_0.

3 The approximation is **best** for **large** values of V. Some indication of the % error incurred by using the approximation should be included in your discussion.

97D (45 mins)

ISOTOPE	R in cm	E in MeV	T in s	lg R	lg T
$^{228}_{90}$Th	4.02	5.38	5.98×10^{7}	0.604	7.78
$^{224}_{88}$Ra	4.35	5.68	3.15×10^{5}	0.638	5.50
$^{220}_{86}$Em	5.06	6.28	5.46×10^{1}	0.704	1.74
$^{216}_{84}$Po	5.68	6.77	1.60×10^{-1}	0.754	−0.80
$^{212}_{83}$Bi	4.79	6.05	3.61×10^{3}	0.680	3.56

The table contains data referring to a number of α-emitting isotopes in the thorium series. R is the range in air of the alpha particles emitted by each isotope, E is their energy and T is the half-life of the isotope.

Geiger and Nuttall suggested the following relationship:

$$\lg T = \mathrm{m} \lg R + \mathrm{B} \qquad \text{(i)}$$

where m and B are constants.

(a) Plot a graph of lg T (y-axis) against lg R (x-axis) and determine the gradient of the line.

(b) Determine values for m and B in equation (i). Show how you arrived at your answers.

(c) Geiger suggested another possible relationship:

$$R = \mathrm{a}\, E^{3/2} \qquad \text{(ii)}$$

State, with reasons, the quantities which you would plot on a graph to test this relationship.

(d) Draw up an appropriate table of values for the graph you have chosen.

(e) Plot this graph.

(f) Explain whether this graph has confirmed equation (ii).

(**AEB** June 83: *all other Boards*)

(a) and **(b)** Since equation (i) is of the form $y = m\,x + \mathrm{c}$, your knowledge of graphs and their equations should enable you to predict the shape of the graph of $y = \lg R$ against $\lg T$ if the suggested relationship is correct. The y-value is from 0.604 to 0.754 so do you need to include $y = 0$? Use the slope (ie. gradient) and y-intercept to obtain values for m and B – see p. 140 if necessary.

(c) See E9.10 if necessary. By plotting a graph of $y = \lg R$ against $x = \lg E$, explain how you could then test the '3/2' relationship.

(d) and **(e)** Your y-scale must cover the values 0.604 to 0.754, and the x-scale is only over a limited range as well. Do you need to include $x = 0$, $y = 0$, and if you did, how would it affect the accuracy of your test?

(f) What value ought the slope have if the '3/2' relationship is correct? Hence measure the slope as accurately as possible to test the relationship.

Part V For reference

Answers to numerical problems and multiple-choice questions

UNIT 1

1.1M B; **1.2M** D; **1.3M** B; **1.4M** C; **1.5M** A; **1.6M** A; **1.7M** D; **1.8M** D; **1.9M** E; **1.10M** E; **1.11M** D; **1.12M** E; **1.13M** C; **1.14M** B; **1.15M** E; **1.16M** A;

1.17S **(a)** 45 m, **(b)** 400 N; **1.18S** **(a)** 1 N, **(b)** (i) 2 N (ii) 2.5 ms^{-2}, **(c)** (i) 0.2 J (ii) 0.25 J, **(d)** 0.05 J; **1.19S** **(b)** (i) 0.035 to 0.040 ms^{-1}; **1.20S** (ii) 1.8×10^{6} ms^{-2}, (iii) 1.3×10^{3} ms^{-1}; **1.21S** **(a)** (i) 0.38 m (ii) 3 ms^{-2}, **(b)** (i) 5750 N (ii) 5000 N; **1.22S** **(a)** 2.3 Hz **(b)** Highest; **1.24S** **(a)** 7.3×10^{-5} rad s^{-1}, **(b)** 470 ms^{-1}, **(c)** 0.034 ms^{-2}; **1.25S** 13 kg m^{2}; **1.26S** 480 N

1.27L **(g)** 0.02 m, **(h)** 2.00 m; **1.28L** 122; **1.29L** **(c)** 2.7 s (i) 0.25 J (ii) 0.47 ms^{-1}; **1.30L** **(c)** (i) 1800 N towards the centre of rotation; **1.31L** **(b)** (i) 2.4 Nm (ii) 120 J (iii) 240 J, **1.32L** **(b)** (i) 7.9 Nm^{-1} (ii) 0.63 ms^{-1} (iii) 0.039 J (iv) 4.0 m s^{-2} (v) 2.8 N, 0.12 m; **1.33L** **(a)** 300 N, **(b)** 50 000 J, 10 000 kg m s^{-1}, 50 000 N; **1.34L** **(b)** 14.5^{0}, 0^{0}, 30^{0}, **(c)** 62.5J; **1.36L** **(b)** (i) 10.5 J (ii) 2.2 ms^{-1} (iii) 18 rad s^{-1}, **(c)** (i) 3.3 ms^{-2} (ii) 5 ms^{-2}

UNIT 2

2.1M E; **2.2M** C; **2.3M** D; **2.4M** D; **2.5M** D; **2.6M** C; **2.7M** A; **2.8M** B; **2.9M** B; **2.10M** C; **2.11M** A; **2.12M** E; **2.13M** E; **2.14M** B; **2.15M** A; **2.16M** C; **2.17M** E; **2.18M** E; **2.19M** D; **2.20M** D; **2.21M** E; **2.22M** C; **2.23M** A; **2.24M** D;

2.25S **(i)** 1.3×10^{-11}N, **(ii)** 7.5×10^{8} N C^{-1}; **2.26S** **(d)** (i) 220 N (ii) 13 J; **2.28S** **(a)** (i) 0, **(b)** 0.61 N kg^{-1}, **(c)** 2.24×10^{9} J; **2.29S** **(a)** 0.70 A, **(b)** 7.8×10^{-4} J;

2.31L **(c)** (i) 5 μC (ii) 4.5×10^{5} V; **2.32L** **(c)** 155 C; **2.33L** **(a)** 8.9×10^{-9} C, **(b)** 4.4×10^{-9} C, 3.3×10^{-6} J; 562 V, 3.7×10^{-6} J; **2.34L** **(b)** (iii) 1.17×10^{7} m, **(c)** 3.83×10^{8} m; **2.35L** **(a)** (i) 1.00 (ii) 1.10, **(b)** (i) 15 (ii) 20, 1.07; **2.36L** **(b)** (ii) 4×10^{-4}T (iii) 3.2×10^{-19} N; **2.37L** **(c)** 61 μT at 71° to the vertical; **2.38L** **(b)** (ii) 0.5 A, 1.5 A s^{-1}

UNIT 3

3.1M A; **3.2M** C; **3.3M** E; **3.4M** E; **3.5M** D; **3.6M** C; **3.7M** D; **3.8M** A; **3.9M** D; **3.10M** E; **3.11M** A; **3.12M** A; **3.13M** D; **3.14M** B; **3.15M** B; **3.16M** C; **3.17M** D;

3.19S 321 m s^{-1}; **3.21S** 100 Hz;

3.24L **(b)** (i) 2.86 cm (ii) 2.86×10^{-6} m; **3.26L** 3/2; **3.27L** 2 ms, 44 mm; **3.28L** **(b)** 3×10^{7} m s^{-1} away, **(c)** (i) 550 N – 0.005 nm (ii) 500 N – 0.005 nm; **3.30L** **(b)** (ii) 7.2 m s^{-1} or any multiple of 7.2

UNIT 4

4.1M E; **4.2M** B; **4.3M** A; **4.4M** E; **4.5M** B; **4.6M** A; **4.7M** D; **4.8M** C; **4.9M** E; **4.10M** E; **4.11M** C; **4.12M** A; **4.13M** C; **4.14M** B; **4.15M** C; **4.16M** D;

4.17S **(b)** (iii) 664 nm; **4.18S** 0.01 rad; **4.19S** **(a)** 56; **4.21S** 30 cm above lens, 6 mm; **4.22S** **(b)** $f = 20$ cm; **4.24L** 0.15 mm

4.25L (b) (ii) 486 nm, 580 nm **4.27L** 460 nm, 690 nm, 67° (violet); **4.28L (b)**(i) 625 nm (ii) 470 nm (iii) $\alpha = 22.1°$ (iv) $\beta = 48.8°$; **4.29L (b)**(i) 912 mm, × 75 (ii) Increase separation by 1.04 mm (iii) 93.2 mm

UNIT 5

5.1M E; **5.2M** B; **5.3M** B; **5.4M** C; **5.5M** A; **5.6M** E; **5.7M** A; **5.8M** E; **5.9M** C; **5.10M** D; **5.11M** D; **5.12M** B;

5.13S (a)(i) 7×10^5 N m^{-2}, (ii) 0.07, **(b)**(i) 0.63; **5.14S** 6.02×10^{28} m^{-3}, 2.55×10^{-10} m; **5.15S (b)**(i) 0.8, (ii) 4.8×10^{23}, (iii) 1.7×10^{-29} m^3, (iv) 3×10^{-10} m, **(c)**(i) 4.03×10^{-10} m, (ii) 22.6°; **5.16S (b)** bigger, **(c)**(i) 26°, (ii) 0.18 nm; **5.17S (a)** 0.045 eV

5.19L (a) about 300 J, **(c)** 42 J for wood, 12 J for concrete, **(f)** about 7 kN and 4 kW; **5.20L** 5.83×10^4 J m^{-3}; **5.21L** 6.45×10^{10} N m^{-2}; **5.22L** 0.8 nm; **5.23L (b)** (i) 8.46×10^{28} m^{-3}, (ii) 2.03×10^{-10} m, (iii) 5.63×10^{-19} J atom^{-1}; **5.24L (a)**(i) 0.3 nm; (ii) 2.0×10^{-20} J **5.25L (b)** from **(a)** (ii) $v = 0.9$ m s^{-1} and from **(a)**(iii) $v = 1.3$ m s^{-1}, **(d)** measurements suggest diameter should be 0.8 cm and not 1.0 cm; **5.26L (c)**(i) 2 m s^{-1} (ii) 0.039 kg s^{-1}; **5.27L** -1.44×10^{10} Pa s^{-1}

UNIT 6

6.1M E; **6.2M** D; **6.3M** A; **6.4M** C; **6.5M** B; **6.6M** B; **6.7M** D; **6.8M** B; **6.9M** A; **6.10M** A; **6.11M** B; **6.12M** D; **6.13M** C; **6.14M** D; **6.15M** A

6.16S 7.8 s lost; **6.17S (a)**(i) 657 W; **6.18S** About 20 K; **6.19S (b)** 49°C (internal heating assumed); **6.20S** 480 m s^{-1}; **6.21S (a)** 2.26 MJ, **(b)** 0.17 MJ, **(c)** 2.09 MJ; **6.22S** 101 kPa

6.23L 4.00; **6.24L (d) (ii)** 21° **6.25L** 200 s; **6.26L (b)** (i) 2.51 kW, (ii) 72%; **6.27L** (i) 530 W, (ii) 250 W, Interface temp. = 9.5°C; **6.28L (b)** (i) 13.3 km, 11.6 km, (ii) 13.3% **6.29L** 44 300 kPa, 890 K; **6.30L** 35/36

UNIT 7

7.1M B; **7.2M** B; **7.3M** B; **7.4M** A; **7.5M** E; **7.6M** C; **7.7M** B; **7.8M** A; **7.9M** E; **7.10M** A; **7.11M** A; **7.12M** B; **7.13M** D; **7.14M** C; **7.15M** A; **7.16M** C; **7.17M** E; **7.18M** E; **7.19M** C; **7.20M** E; **7.21M** E; **7.22M** C;

7.23S (ii) 25.7 kW; **7.24S (b)** (i) 1Ω, (ii) 3.75 J; **7.25S** 50.5mΩ, 1.36 A; **7.26S (a)** (i) 100k Ω
7.27S (a) 11 000 ohms, **(b)** 0.27 mA, **(c)** $\frac{3}{4}$ of the way down; **7.28S (b)** 8×10^{-6} Ωm; **7.29S (a)** (v) 0.01 J, **(b)** (i) 5 μF (ii) 40 V (iii) 1.2 μF, **(c)**(i) S; **7.30S** $V_1 = 4/7$ V, $V_2 = 10/7$ V, $E_1 = E_2 =$ 290 V m^{-1}; **7.32S (a)** 47.2 A, **(b)** 8.91 kW

7.34L 0.5 V in both cases; **7.35L (b)** (i) 0.5 V (ii) 190 ohms in series (iii) 10 mA; **7.36L (b)** (i) 22.5 W (ii) 2400 K (iii) 4.3 ohms, **(c)** 3.9 A if in series with 4.27 ohm resistor at 20 V; **7.37L (a)** 1.0×10^4 s, **(b)** (i) 1 A in $ST{:}SX{:}SV{:}YZ{:}WZ{:}UZ$, 0.5 A in remainder, (ii) 2.5×10^{-6} ohms; **7.38L (b)** n = 12; **7.40L** 50W, 50 W, 1.1 H; **7.41L** 39.8 Hz, 1.0 A; I is 56.3° ahead of V; 79.6 Hz; **7.42L (a)** 10 V, 429 ohms, **(b)** (i) 1732 ohms (ii) 1.0 mH, 159 kHz, **(d)** 20 mA; **7.43L (a)** (i) 4.7kΩ, 150kΩ (ii) 33; **7.45L** Voltage gain = −8

UNIT 8

8.1M C; **8.2M** E; **8.3M** E; **8.4M** B; **8.5M** A; **8.6M** B; **8.7M** B; **8.8M** B; **8.9M** D; **8.10M** D; **8.11M** D; **8.12M** B; **8.13M** C; **8.14M** A; **8.15M** D; **8.16M** E; **8.17M** C; **8.18M** B; **8.19M** D; **8.20M** C; **8.21M** D; **8.22M** E;

8.23S 12; **8.24S** 1.76×10^{11} C kg^{-1}; **8.25S** 6.7×10^{-34} J s; **8.27S** 8.0×10^{18} Hz; **8.28S (a)** 2.67×10^{16}, **(b)** 6.6×10^5 s; **8.29S (a)** 1.28 W, **(b)** 0.32 W, **(c)** 4×10^7 J; **8.30S (b)** 1.08×10^{17} s; **8.31S** 1.82×10^{-12} J

8.32L 6.4×10^{-18} A; **8.33L (a)** (ii) 4.8×10^{-19} C (iii) 20 m s^{-2}, **(b)** (i) 2.0×10^7 m s^{-1} (ii) 2.2×10^{-3} m; **8.34L (c)** (i) 1.1×10^{-3} T (ii) 0.517 m; **8.35L (c)**(i) 1.0×10^4 eV (ii) 1.23×10^{-11} m (iii) 1.6×10^{-15} J, 1.24×10^{-10} m; **8.36L (d)** 3.3×10^{-19} J, 2.7×10^{-19} J; **8.38L** 6.55×10^{-30} kg, Photon energy = 1.35×10^{-13} J, Max. ke = 4.55×10^{-13} J; **8.39L** 6.8 MeV (^{4}He): 2.4 MeV (^{3}He); **8.40L** 0.58 MeV

UNIT 9

9.1M E; **9.2M** E; **9.3M** C; **9.4M** B; **9.5M** E; **9.6M** E; **9.7M** C; **9.8M** E; **9.9M** B; **9.10M** B; **9.11S** $A = 6.67$ m kg^{-1}, $m = 0.9$ kg; **9.12L (a)** (i) 21.14 cm, 20.97 cm (ii) ± 0.01 cm (iii) 24%; **9.13D (d)** $b = 0.12$ mV°C^{-1}, $\theta_{p=0} =$ 300°C

TEST YOURSELF

Multiple choice

1M C	**2M** D	**3M** A	**4M** C	**5M** B	**6M** E
7M C	**8M** D	**9M** C	**10M** B	**11M** E	**12M** C
13M A	**14M** C	**15M** C	**16M** D	**17M** D	**18M** C
19M B	**20M** A	**21M** B	**22M** E	**23M** B	**24M** E
25M B	**26M** E	**27M** B	**28M** C	**29M** C	**30M** D
31M C	**32M** D	**33M** C	**34M** A	**35M** E	**36M** B
37M E	**38M** E	**39M** C	**40M** A	**41M** B	**42M** D
43M B	**44M** E	**45M** D	**46M** B	**47M** B	**48M** E
49M C	**50M** C				

Short answers
51S (a) 10 ms^{-1}, **(b)**(i) 2.5 ms^{-1}, **(c)** (i) 6250 J (iii) 2400 J, **(d)** 7.2×10^4N;
52S (a) 7.85 km s^{-1}, **(b)** 5250 s; **54S** 7950 kg m^{-3};
55S 2×10^5 V m^{-1};
56S (a) 1.5 V **(b)** 1.5 V, Yes; **58S** 6.19×10^{-10} m; **59S (a)** 1380 J for 60 kg mass, **(c)**(i) 6.8 ms^{-1} (ii) about 5000 N including his weight; **60S** 100 N, 21.7 J; **61S** 7.5 A in DE in the same direction, 6×10^{-5} N (attraction);
62S (b) (i) 499 J, (ii) 4×10^{-4} kg; **63S (b)** 38.2 Hz; **64S** 390 J g^{-1};
65S 2.325 V; **66S (c)** (i) 88 nm (ii) 2.2×10^6 ms^{-1}

Long answers
67L (b) for 1 m wavelength $c = 1.25$ ms^{-1} and $f = 1.25$ Hz, for 100 m wavelength $c = 12.5$ ms^{-1} and $f = 0.125$ Hz, **(c)** (i) All (ii) 7.85 ms^{-1} (iii) 6.2 ms^{-1}, **(d)** for 1 m wavelength c = 1.25 ms^{-1} and for 1000 m wavelength $c = 31$ ms^{-1}
70L At equilibrium $F = 0$, d = 0.281 nm, **(a)** 217 MN m^{-2}, **(b)** 3.0 J, **(c)** 1.40×10^{11} N m^{-2}, 3.0 mm; **71L** $I_b = 4/11$ A, $I_a = 17/33$ A, $I_c = I_d = 1/4$ A, **(c)** − 1.51 m
72L (a) $m = 3$, $\theta_3 = 55° 47'$ **(b)** − 10′ **(c)** 1.14 cm; **73L (a)** 9 hrs, **(b)** about 40%, **(c)** 9×10^8 kW hours, **(d)** 2.1×10^7 m^3, 8.92×10^6 MJ
74L (b) (i) 68.8°
75L (c) (i) 33.3 Hz; **76L (c)** 2.12 Hz; **78L** 350 V, 175 V; 4.43 pF, 2.71×10^{-7} J; **(a)** 23.3 V, **(b)** 66.5 ρF, **(c)** 1.82×10^{-8} J;
79L (a) 1.18 mm **(b)** 0.2 m ;
80L (b) $\theta_q = 8\theta_p$ **(c)** 25.5 μV, 0.6 mΩ, 0.27 μW; **82L (b)** $^{221}_{86}$Rn;
83L (a) 10 stations, **(b)** about 1/20th of the time, **(c)** about 0.9×10^8 m^2, **(d)** about 5.6×10^8 m^2

Practical problems
84P Force on metal is about 7 × that on wood;
85P (b) 32.2 ohms **(c)** 10.8 ohms; **86P (a)** 0.39 s, **(b)** (i) 90 N m^{-1} (ii) 0.30 s;
87P (e) 2.4 mA; **88P** 6:4; **89P** 100 μF approx; **90P** $A{:}B = 2{:}3$;
93P (a) 1 m

Comprehension papers
94C $Q6$, **(a)** 10 sones, **(b)** 2.5 sones;
95C $Q3$ W K^{-1} $Q4$, **(a)** 1.995 mV, **(b)** 1.800 mV $Q6$ 1.0×10^{-4} W A^{-1}K^{-1}

Data analysis papers
96D 1(a) 3.7×10^{-3} Ω^{-1}, **(b)** 12%; **2 24** V^{-1}, 1 nA;
97D (a) − 55.0 **(b)** − 55.0, 40.8

Index

References in colour refer to questions (including commentaries). The type of question is indicated by a letter on the end according to the following code:

M–Multiple choice **S**–Short answer **L**–Long answer
P–Practical problem **C**–Comprehension **D**–Data analysis.

The questions are also coded by unit; eg, 1.3M means the third question in Unit 1, which happens to be of the multiple-choice type

Questions without a unit code can be found in Part IV (Test yourself); eg, 53S is the 53rd question in 'Test yourself', a short-answer question.

If the letter E is at the front of a reference it indicates an equation; eg, E1.3 means the third equation in Unit 1.

If the letter F is at the front of a reference it refers to a Figure (diagram, graph, or table) in the text; eg, F1.3 means the third figure in Unit 1.

Sometimes reference is given to an entire topic; eg, 1.3 means the third topic of Unit 1.

Page numbers may be given, in which case the number will appear on its own; eg, 13 means page 13. Page numbers have no letter code.

All particularly important references have been featured by printing them in bold type.

THE TABLE OF CONSTANTS AND NUMERICAL VALUES

The table of constants and numerical values appears on the following pages. Entries define a quantity and its units; commonly used symbols for the quantities are given together with a numerical value (if available) that is more accurate than that shown by the position in the table. Values quoted to the nearest order of magnitude are called 'typical' if they can vary by more than an order of magnitude. All important constants have been printed in blue.

Two examples will illustrate the use of the table:
Example 1 (From the 10^8 row in the table):

The numerical value of the speed of light is roughly 3×10^8
A more accurate value is 3.00×10^8 (3 significant figures instead of one)
The units are m s^{-1}
The commonly used symbol for the velocity of light is c.
It is an important constant (printed in blue)

Example 2 (From the 10^{-12} row in the table)

A typical gamma-ray wavelength has a numerical value of 10^{-12}
The units are m, a commonly used symbol is λ
The numerical value might vary by at least ten times, it is a 'typical' value

It is suggested that users of this book should add to this table if the opportunity arises. They can fill some blank spaces with any quantities that are important to the Course that they are studying.

Multiplying factor	To the nearest order of magnitude	1	2	3	4
$\geqslant 10^{24}$	n charge carriers (m^{-3}) in metals ($\sim 10^{28}$)		M_s mass of the Sun (2×10^{30} kg)		W_s Sun's radiated power (3.9×10^{26} W)
$\times 10^{23}$					
$\times 10^{22}$					
$\times 10^{21}$		M mass of Earth's oceans (kg) (1.42)			
$\times 10^{20}$	f_γ typical gamma-ray frequency (Hz)				
$\times 10^{19}$			V volume of the Moon (m^3) (2.2)		
$\times 10^{18}$					
$\times 10^{17}$		age of the Earth (s) (1.4)			
$\times 10^{16}$					
$\times 10^{15}$	f frequency of light (Hz)				
$\times 10^{14}$					
$\times 10^{13}$	f atomic vibration frequency (Hz)				
T $\times 10^{12}$	R typical input impedance of a dc electrometer (Ω)				
$\times 10^{11}$	E Young's modulus of metals (Nm^{-2})	distance from Earth to Sun (m) (1.496)	$T_{1/2}$ half life of carbon − 14 (s) (1.8)		
$\times 10^{10}$	age of the universe in years (y)				one Curie (s^{-1}) (3.7)
G $\times 10^{9}$					V pd of lightning flash (V)
$\times 10^{8}$	v_β typical speed of β-particles (ms^{-1})			c speed of light in a vacuum (ms^{-1}) (3.00)	
$\times 10^{7}$		T Temperature at Sun's centre (K) (1.36)	P_c critical pressure of water (Pa) (2.2)	seconds in a year (s) (3.15)	h height of a parking orbit (m) (3.6)
M $\times 10^{6}$	v_α typical speed of α-particles (ms^{-1})		l_v sp. lat. ht. vap. of water ($J\ kg^{-1}$) (2.3)	E breakdown field of air ($V\ m^{-1}$)	
$\times 10^{5}$	ρ resistivity of pure silicon (Ωm)	P_0 atmospheric pressure (Pa) (1.013)		l_f sp. lat. ht. fus. of water ($J\ kg^{-1}$) (3.3)	
$\times 10^{4}$		ρ density of lead ($kg\ m^{-3}$) (1.13)	f highest audible sound frequency (Hz)		
k $\times 10^{3}$	k typical ionic bond constant ($N\ m^{-1}$)	ρ_w density of water ($kg\ m^{-3}$) (1.0)	T_m melting point of steel (K) (1.63)	P svp of water at 25°C (Pa) (3.17)	c sp. ht. cap. of water ($J\ kg^{-1}\ K^{-1}$) (4.2)
$\times 10^{2}$	c sp heat capacity of metals ($J\ kg^{-1}\ K^{-1}$)	field in Earth's atmosphere ($V\ m^{-1}$) (1.3)		v velocity of sound in air (ms^{-1}) (3.3)	
$\times 10^{1}$	k thermal conductivity of alloys ($W\ m^{-1}\ K^{-1}$)	g aceeleration of free fall (ms^{-2}) (0.981)	f lowest audible sound frequency (Hz)	monatomic molar ht. cap. ($J\ K^{-1}\ mol^{-1}$) (2.5)	
(×1) $\times 10^{0}$	λ typical VHF radio wavelength (m)	ρ_a density of air ($kg\ m^{-3}$) (1.3)	γ c_p/c_v for helium (1.66)	π (3.142)	

5	6	7	8	9	**Multi-plying factor**
	M_E mass of the Earth (6×10^{24} kg)				$\geqslant 10^{24}$
	L N_A Avogadro constant (mol^{-1}) (6.02)				$\times 10^{23}$
		M_m mass of the Moon (kg) (7.3)			$\times 10^{22}$
					$\times 10^{21}$
					$\times 10^{20}$
					$\times 10^{19}$
M mass of Earth's atmosphere (kg)	number of eV in one Joule (eV) (6.25)				$\times 10^{18}$
					$\times 10^{17}$
					$\times 10^{16}$
				metres in 1 light year (m) (9.46)	$\times 10^{15}$
					$\times 10^{14}$
		E energy released in fission ($J\ kg^{-1}$)			$\times 10^{13}$
					tera $\times 10^{12}$
					$\times 10^{11}$
					$\times 10^{10}$
				f atomic clock frequency (Hz) (9.192 631 770)	**giga** $\times 10^{9}$
		R_S radius of the Sun (m) (6.96)			$\times 10^{8}$
E energy released by burning coal ($J\ kg^{-1}$)	f UHF TV channel frequency (Hz)				$\times 10^{7}$
	R_E radius of the Earth (m) (6.5)				**mega** $\times 10^{6}$
					$\times 10^{5}$
		P svp of ether at 25°C (Pa)			$\times 10^{4}$
v velocity of sound in steel ($m\ s^{-1}$)	ρ_E mean density of Earth ($kg\ m^{-3}$) (5.51)				**kilo** $\times 10^{3}$
	T_c critical temperature of water (K) (6.47)	μ_r relative permeability or iron (7.0)	c_p for carbon dioxide ($J\ kg^{-1}\ K^{-1}$) (8.2)	number of MeV in 1 amu (MeV) (9.31)	$\times 10^{2}$
$T_{1/2}$ half life of radon −220 (s) (5.4)	k bond constant for steel atoms ($N\ m^{-1}$)				$\times 10^{1}$
	ε_r relative permittivity of porcelain		R molar gas constant ($J\ K^{-1}\ mol^{-1}$) (8.31)	breaking strain of rubber	($\times 1$) $\times 10^{0}$

Multiplying factor	**To the nearest order of magnitude**	**1**	**2**	**3**	**4**
$\times 10^{-1}$	λ typical microwave or radar wavelength (m)	B field at end of bar magnet (T)			1 mph in metres per second (ms^{-1}) (4.5)
$\times 10^{-2}$	ρ typical semiconductor resistivity (Ω m)		volume of a mole at stp (m^3) (2.24)		
m $\times 10^{-3}$		η viscosity of water (Pa s) (1.0)		k_W Wien displacement constant (mK) (2.9)	temp. coeff. of resistance of Cu (K^{-1}) (+ 4.3)
$\times 10^{-4}$	λ typical infrared wavelength (m)				
$\times 10^{-5}$	B strength of Earth's magnetic field (T)		η viscosity of air (Pa s) (1.7)		
μ $\times 10^{-6}$	C max. capacitance of non-electrolytics (F)	μ_0 magnetic constant ($H\ m^{-1}$) (1.26)			
$\times 10^{-7}$	ρ typical resistivity of metals (Ω m)		F force used to define the ampere (N) (2.0)	λ_B wavelength of blue light (m)	
$\times 10^{-8}$	λ typical ultraviolet wavelength (m)				
n $\times 10^{-9}$	F typical force exerted by electron beam (N)		D diameter of an oil molecule (m)		
$\times 10^{-10}$	λ typical X-ray wavelength (m)	D diameter of a carbon atom (m) (1.3)			
$\times 10^{-11}$	C typical parallel plate cap. in air (F)				
p $\times 10^{-12}$	λ typical gamma-ray wavelength (m)				
$\times 10^{-13}$					
$\times 10^{-14}$	D diameter of medium-size nucleus (m)		E min. ke of 'fast' neutron (J)		
f $\times 10^{-15}$	E typical X-ray (1 keV) photon energy (J)				
$\times 10^{-16}$					
$\times 10^{-17}$					
a $\times 10^{-18}$					
$\times 10^{-19}$			e charge of an electron (C) (− 1.6)		
$\leqslant 10^{-20}$	ρ density of outer space ($\sim 10^{-21}$ kg m^{-3})	k Boltzmann constant (1.38×10^{-23} J K^{-1})	m_n m_p neutron or proton mass (1.67×10^{-27} kg)		E typical ke of slow neutron ($\times 10^{-21}$ J)

5	6	7	8	9	Multi-plying factor
γ surface tension of mercury ($N\ m^{-1}$) (5.0)	k thermal conductivity of water ($Wm^{-1}\ K^{1}$) (6.0)	tyre/dry road kinetic friction coefficient		therm. cond. of glass ($Wm^{-1}\ K^{-1}$) (0.93)	$\times 10^{-1}$
		γ surface tension of water ($N\ m^{-1}$) (7.3)		ρ density of hydrogen ($kg\ m^{-3}$) (9.0)	$\times 10^{-2}$
					milli $\times 10^{-3}$
					$\times 10^{-4}$
					$\times 10^{-5}$
					micro $\times 10^{-6}$
			λ_R wavelength of red light (m)		$\times 10^{-7}$
	σ Stefan's constant ($W\ m^{-2}\ K^{-4}$) (5.67)				$\times 10^{-8}$
					nano $\times 10^{-9}$
					$\times 10^{-10}$
		G gravitational constant ($N\ m^{2}\ kg^{-2}$) (6.67)			$\times 10^{-11}$
				ε_0 electric constant ($F\ m^{-1}$) (8.85)	**pico** $\times 10^{-12}$
					$\times 10^{-13}$
					$\times 10^{-14}$
					femto $\times 10^{-15}$
					$\times 10^{-16}$
					$\times 10^{-17}$
					atto $\times 10^{-18}$
					$\times 10^{-19}$
		h Planck constant (6.63×10^{-34} J s)		m_e electron mass (9.11×10^{-31} kg)	$\leqslant 10^{-20}$